Fundamentals of Electroceramics

Fundamentals of Electroceramics

Fundamentals of Electroceramics

Materials, Devices, and Applications

R. K. Pandey
Ingram Professor Emeritus, Texas State University
Cudworth Professor Emeritus, The University of Alabama
Professor Emeritus, Texas A&M University

WILEY

The American Ceramic Society

This edition first published 2019.

Registered Office
John Wiley & Sons, Inc., 111 River Street, Hoboken, NJ 07030, USA

Editorial Office
111 River Street, Hoboken, NJ 07030, USA

For details of our global editorial offices, customer services, and more information about Wiley products visit us at www.wiley.com.

Wiley also publishes its books in a variety of electronic formats and by print-on-demand. Some content that appears in standard print versions of this book may not be available in other formats.

Library of Congress Cataloging-in-Publication Data

Names: Pandey, R. K., 1937- author.
Title: Fundamentals of electroceramics : materials, devices, and applications / R. K. Pandey, Ingram Professor Emeritus, Texas State University, Cudworth Professor Emeritus, The University of Alabama, Professor Emeritus, Texas A&M University, Wiley-American Ceramic Society, The American Ceramic Society.
Description: Hoboken, NJ, USA : John Wiley & Sons, Inc., 2019. | Includes bibliographical references and index. |
Identifiers: LCCN 2018029650 (print) | LCCN 2018031928 (ebook) | ISBN 9781119057352 (Adobe PDF) | ISBN 9781119057284 (ePub) | ISBN 9781119057345 (hardcover)
Subjects: LCSH: Electronic ceramics.
Classification: LCC TK7871.15.C4 (ebook) | LCC TK7871.15.C4 P36 2019 (print) | DDC 621.381–dc23
LC record available at https://lccn.loc.gov/2018029650

Cover Image: Courtesy of R. K. Pandey
Cover Design by Wiley

Set in 10/12pt WarnockPro by SPi Global, Chennai, India

Printed in the United States of America

V10006425_112918

Dedicated to my wife, Dr. Christa Pandey, whose love, support and inspiration for the last 50 plus years have been instrumental to my professional achievements and personal happiness.

Contents

Preface

Let us remember: One book, one pen, one child, and one teacher can change the world.

Malala Yousafzai

The word *ceramic* may be the most misunderstood scientific concept so far as its public image is concerned. Most people, when they hear the word ceramic, are likely to think of such things as coffee mugs, glazed pottery, floor tile, or bathroom toilets. It is largely unknown to the public, or even to many scientific communities, that the use of ceramic materials goes far beyond these products.

Ceramic materials by definition are based on inorganic raw materials. Oxides form the leading group of electroceramic materials. Aluminum oxide (Al_2O_3), silicon carbide (SiC), silicon nitride (Si_3N_4), titanium oxide (TiO_2), iron oxides (FeO, Fe_2O_3, Fe_3O_4), zinc oxide (ZnO) and tin oxide (SnO_2) are typical examples. As for non-traditional applications ceramics are used in aerospace and other extreme temperature applications due to their excellent thermal properties; in the medical field, ceramics are used because of their compatible with the human bone; and, in the military ceramics are used for applications such as body armor due to their extreme hardness.

The subject of this book is *Electroceramic* which is a special category of electronic materials. Electroceramic materials, as the name suggests, conduct electric currents obeying various physical mechanisms of current transport. These materials can exhibit a host of physical properties including high temperature superconductivity, magnetism, semiconductor, electro-optic, acousto-optic and nonlinear dielectrics. Because of their multi-faceted physical and mechanical properties these materials are poised to impact the advancement of ultrafast computer memory technology, green energy technology, sensors and detector technology as well as many other emerging areas of applied sciences and engineering. The field of electroceramic devices and applications is vast and diverse. It already impacts many areas of engineering and basic sciences such as microelectronics, solid-state sciences, microwave engineering, communication engineering, signal processing, actuators and sensors and micro-electro mechanical systems (MEMS) technology.

Electroceramic materials possess many interesting physical phenomena and that is what makes them fascinating and attractive for scientific discoveries leading to novel innovations. The physical principles involved in the origin of electroceramic phenomena are intriguing and so diverse that its intellectual challenges can be felt in a wide range of engineering and basic science disciplines. Yet electroceramic is not the household word even among electrical engineers, materials scientists, and applied physicists. Hardly anywhere a course devoted to electroceramics is offered in the US universities for electrical engineers, materials scientists, and physicists. As a result students are deprived of the knowledge in topics specific to electroceramic and its applications. Some examples may include noncentrosymmetric crystals, symmetry elements, piezoelectricity, ferroelectricity, pyroelectricity, multiferroic materials and phenomena, magnetoelectronics, spintronics, coupled hybrid devices, colossal magnetoresistive effect, giant photovoltaic effect, and energy harvesting.

The ignorance about electroceramics is pervasive. For example, many of the electrical engineering graduates and practitioners have never heard of varistor devices though almost all electrical engineering curriculums include a course or two on solid-state materials and devices. It is disappointing and yet amusing. Bipolar varistors diodes are very useful devices that are present in practically all electrical and electronic circuits as circuit protectors against abrupt surges of current or voltage. Not only that varistors are forerunners of transistors, which are named so because of certain attributes they share with varistors.

Those of us working in the area of electroceramic materials and devices believe that it deserves its own separate presence in the curriculum of electrical engineering, materials science and applied physics; as well as of ceramic engineering, and perhaps also of mechanical and chemical engineering. The nature of elecroceramics is interdisciplinary and therefore, such a course could be cross listed to be taught in many technical disciplines.

The motivation to write this book came to me in the Fall of 2010 when for the first time I offered a special topic course on electroceramics for electrical engineering and physics students at Texas State University, San Marcos, TX. After teaching for 30 plus years, courses on electronic materials and solid-state sciences at graduate and undergraduate levels at Texas A&M University and at the University of Alabama, I was confident that I would have no problems in handling this course. I had more than enough of my own lecture notes and homework problems on topics related to electroceramics. But finding a good textbook to recommend to students became a formidable enterprise. There are many books on electroceramics, but just one or two that could qualify for a text book. But they are simply too old by now and therefore inadequate for a text book. Many of the new advancements and discoveries made during the last 10–15 years are conspicuously absent in these books. As a result my resolve became stronger to undertake the task of writing a text book on electroceramics. I discussed this informally with many of my friends and colleagues at different universities, and all of us agreed that we need a new text book on the subject. Mr. Mark Mecklenborg and Mr. Greg Geiger of the American Ceramic Society also encouraged me to write such a book.

The book is divided in 11 chapters beginning with the essential elements of solid-state science and ending in electro-optics and acousto-optics. I have done my best to develop each chapter in such a way that a good student can follow the materials easily and enjoy learning about them. Separate chapters are devoted to materials processing, characterization, and crystallography including noncentrosymmetric crystals and symmetry elements. Coupled dielectric phenomena such as piezoelectricity, ferroelectricity, and pyroelectricity have been covered in two chapters; a separate chapter is devoted to electroceramic semiconductors, and so are individual chapters on green energy, magnetism, electro-optics and acousto-optics. Each chapter includes some practical examples that should be helpful in understanding the theoretical concepts discussed. I have purposely tried to keep the use of advanced mathematics just adequate enough to make the theoretical concepts understandable without intimidating the students.

Each chapter also includes suggestions for advanced reading for the benefit of interested readers. At the end of each chapter, a glossary of technical terms has been added. Students are advised to look at them first before beginning to read the materials covered in the chapter. Also a set of homework problems have been added to each Chapter. I encourage students to work out all of them thoroughly and completely because it will solidify and amplify their insight into the subject matter and give them confidence and pleasure of learning.

I owe gratitude to a long list of individuals and institutions. First and foremost, I must thank my dear wife, Dr. Christa Pandey, who has been an inspiring influence throughout my long journey in life and supportive of all my professional endeavors. Her unwavering confidence in my abilities to excel professionally has been my strength and has given me the confidence to undertake the challenges that comes naturally in executing a project such as writing a text book.

I also must thank all my former students at Texas A&M University at College Station, and at the University of Alabama at Tuscaloosa, AL, whom I had the honor of teaching one or more courses during the span of 30 years. That experience and the challenges I had to face made me a better class room teacher and a good researcher. But the driving force behind this book were my students at Texas State University at San Marcos, TX, whom I had the pleasure of teaching a special topic course on electroceramics three times. Their enthusiasm and dedication to learning and doing research in labs even as undergraduate seniors were contagious. To all these students, I owe gratitude and give my very sincere thanks for the doors they opened. Thanks also to two of my former graduate students, Dr. Jian Zhong and Dr. Hui Han, who critically reviewed one chapter each and pointed out to some of the lapses which I subsequently corrected.

My gratitude and thanks also to my friends and colleagues, Professor Rick Wilkins of Prairie View A&M University and Professors Ravi Droopad, Harold Stern and William A. Stapleton of Ingram School of Engineering at Texas State University. They were gracious and patient enough to go over two or more chapters and provide me with their suggestions and comments. All of them were valuable suggestions and I have revised these chapters incorporating all of their suggestions,

My thanks go also to my gifted grand-daughter, Dr. Alysha Kishan, for proposing multiple designs for the cover page and for teaching me the fine points of making good graphics. This has added to the looks and quality of the book. I am thankful to all the publishers who have been gracious enough to grant us permission for reproducing their copyrighted materials. I also owe thanks to institutions such as Sterling C. Evans Library at Texas A&M University and Rodgers Library for Science and Engineering of the University of Alabama for the privilege of accessing journals and books remotely online that expedited the progress of the book. I have also used resources available in Wikipedia and Google Search and I thank them too for this privilege.

My sincere thanks and gratitude to Ingram School of Engineering at Texas State University for allowing me to hold the rank of Ingram Professor for six years which of course facilitated the success of this project.

Finally I must thank my Production Editor of Wiley Publishers, Ms. Jayashree Saishankar in Chennai, India for her patience and efficient execution of this project that has culminated into the publication of this book. Also I extend my thanks to other editors, Ms. Beryl Mesiadhas, and Mr. Michael Leventhal, both of Wiley, for their assistance and help.

It is my sincere hope that this book will serve the students for whom it has been written and will be an important part of their university education and scholarly experience.

I realize that inspite of all the precautions and thorough review of each chapter many mistakes and errors might have crept in the book. By no means is it a reflection on the competence of any of the reviewers. I alone and no one else is to be blamed for these lapses. I offer my apologies to the readers in advance. I encourage them to point out the shortcomings of the book to me so that I can correct them in future.

November 04, 2018 *R. K. Pandey*
Austin, Texas

About the Companion Website

This book has a companion website hosted by Wiley:

www.wiley.com/go/Pandey/Fundamentals_Electroceramics

The website contains the following periodically updated information:

1) Solution Manual
2) PPT slides

1

Nature and Types of Solid Materials

CHAPTER MENU

Do not worry about your difficulties in Mathematics, I can assure you mine are still greater.
Albert Einstein

1.1 Introduction

In this chapter, we will learn about the fundamental nature of solids and how their defining properties are associated with quantum mechanical concepts of electrons and their energy. The exposure to the most essential concepts of solid-state physics will greatly help us in understanding the nature of electroceramics and the multiple physical phenomena they can exhibit that form the basis for a large number of novel device applications that impact electronic and sensor technology. We have purposely tried to avoid the intricacies of mathematical models in describing these concepts because the goal here is not to produce another book on solid-state physics but rather to make use of the essential features of various theoretical models in understanding the transport properties of electrons, uniqueness of semiconductors, and the scientific basis behind the dielectric properties of materials.

1.2 Defining Properties of Solids

Solids can be broadly classified as conductors, semiconductors, and insulators of which dielectrics are a subset. Another important group of solids are classified as *high temperature superconductors*. Because of the unique physical mechanisms involved in the origin of superconductivity, these materials are of a special category and will be treated as an independent class of materials. We will devote a section on superconductivity later in this chapter. So far as the other three groups are concerned, we can differentiate between them on the basis of their defining properties. For example, a conductor is defined by its capacity to facilitate the transport of an electrical current associated with the inherent material property that we call *resistance*. Similarly a semiconductor is defined by its energy gap (also, called bandgap) and a dielectric by its dielectric property. We discuss in this chapter, the origin of these properties and how they add uniqueness to materials.

1.2.1 Electrical Conductance (*G*)

All materials tend to resist the flow of an electric current by virtue of its built-in resistance. The magnitude

Fundamentals of Electroceramics: Materials, Devices, and Applications, First Edition. R. K. Pandey.

Companion website: www.wiley.com/go/Pandey/Fundamentals_Electroceramics

of current, I, is dictated by the resistance, R (or, conductance, G) when a voltage, V, is applied between the two ends of a solid sample. This relationship is given by the famous law of physics universally known as the Ohm's law that was conceived in 1825–1826 by Gerog Ohm of Germany. It states that the current (I) generated between the two fixed points of a conductor (such as a metal) is directly proportional to the potential applied and inversely proportional to its resistance. Mathematically, it is expressed as Eq. (1.1).

$$I = \frac{V}{R} = GV \tag{1.1}$$

Here G being the conductance that is simply the inverse of resistance. From the above equation, we can conclude that I increases as R decreases or it increases with the increase in conductivity G. The resistance (R) changes as two reference points between which it is measured is changed. For example, it increases with the increase in the distance between the reference points and decreases if the distance between these points is reduced. That means that the resistance (or, conductance) is dependent upon the geometry of the sample. In other words, neither resistance nor conductance is the intrinsic property of the sample under consideration. Unless we can develop the concept of intrinsic resistance of a material, we would not be able develop theoretical models that are independent of sample geometry. To accomplish this goal, let us introduce now a parameter which we shall call *resistivity*. It is defined as follows:

$$\rho = R\left(\frac{A}{L}\right) \tag{1.2}$$

Here ρ is the resistivity, L the sample length, and A the cross-sectional area. The unit of the resistivity is Ω m. We can see from the above equation that the resistivity becomes an intrinsic property of materials. No two materials would have the same value of resistivity.

While defining the resistivity, we assumed the sample to be uniform in which the current flows uniformly. However, in reality that may not always be the case. We therefore need to develop a more basic definition of resistivity. We can imagine that an electric field prevails inside the sample when it experiences a potential difference between any two fixed points. It is actually the electric field (E) that enables the current flow within the sample, and therefore, the resistivity must be associated with the current density (J) that exists within the sample. We can then redefine the resistivity with respect to E and J as in Eq. (1.3).

$$\rho = \left(\frac{V}{L}\right)\cdot\left(\frac{A}{I}\right) = \frac{E}{J} \tag{1.3}$$

The inverse of the resistivity is called *conductivity* (σ) and its unit is S m^{-1} or (Ω m)$^{-1}$. Replacing the resistivity with conductivity, we can rewrite Eq. (1.3) in its alternative formulation as follows:

$$J = \sigma E \tag{1.4}$$

Metals have the highest conductivity among all solids, and it is greater than 10^5 (S m^{-1}). In comparison, in semiconductors, it varies from $10^{-6} < \sigma < 10^5$ (S m^{-1}). The dielectrics have very small conductivity that is smaller than 10^{-6} (S m^{-1}). Based on this information, we can now distinguish between the three types of solids as in Eq. (1.5).

$$\sigma_{\text{metal}} \gg \sigma_{\text{semiconductor}} \gg \sigma_{\text{dielectric}} \tag{1.5}$$

In Table 1.1, a list of materials with their electrical conductivity is presented.

1.2.2 Bandgap, E_g

The defining property of a semiconductor is its energy bandgap that exists between the valence band and the conduction band. The width of the bandgap is expressed in electron volt with the symbol of E_g. The unit of electron volts for energy is defined as the work done in accelerating an electron through 1 V of potential difference. For converting 1 J of energy to electron volts, we need to divide it by the charge of an electron that is 1.602×10^{-19} C.

The concept of energy being in bands of solids instead of just being discrete is based on the band theory of solids to which we will introduce our readers later in this chapter. For the time being, let us be satisfied with the assumption that electrons and other charge carriers (e.g. holes) can reside only in the valence band or the conduction band. It is forbidden for any charge carrier to be found in the bandgap at absolute zero. The Fermi–Dirac distribution function (also known as F-D

Table 1.1 Room temperature electrical conductivity of selected solids.

Materials	Electrical conductivity, σ (S m^{-1})
Aluminum (Al)	3.5×10^7
Carbon (graphene)	1.00×10^8
Carbon (diamond)	$\approx 10^{-13}$
Copper (Cu)	5.96×10^7
Gold (Au)	4.10×10^7
Silver (Ag)	6.30×10^7
Platinum (Pt)	9.43×10^6
Germanium (Ge)	2.17 (depends on doping)
Silicon (Si)	1.56×10^{-3} (depends on doping)
Gallium arsenide (GaAs)	1.00×10^{-8} to 10^3

statistics) with its enormous importance to the quantum nature of solids completely excludes the possibility that any electron can be found in the bandgap. Not only that, this theory also predicts that at absolute zero (0 K), all electrons are frozen in valence band, and the conduction band is completely empty. We will deal also with this magnificent theory later in this chapter. However, it is also probable that some electrons might get sufficient kinetic energy to escape the valence band and migrate to the conduction band. But this probability is allowed only at temperature ≫0 K according to the F-D statistics.

In general, metals have almost no bandgap, whereas insulators have large bandgaps. The bandgaps of semiconductors lie between these two extremes. If the bandgap is greater than 2 eV, the material is thought to be an insulator, though this notion is not always supported by facts. For example, there are many semiconductors with $E_g > 2$ eV, and they are classified as wide bandgap semiconductors and not insulators.

The semiconductors are normally classified as narrow bandgap, midlevel bandgap, and wide bandgap. In Figure 1.1, a qualitative picture of bandgap is given, which can serve for distinguishing among metals, semiconductors, and dielectrics.

We see in Figure 1.1 that dielectrics have much larger bandgaps than semiconductors, whereas metals have no bandgap at all. In fact, the two bands merge in metals causing an overlapped region, where electrons are shared by the two bands. We also find in this figure that besides the bandgap, there is another parameter labeled as *Fermi level,* which lies between the upper (conduction band) and lower (valence band) energy bands. It is defined as the sum of the potential energy and kinetic energy. For convenience, for example, in discussing the semiconductor properties, the potential energy is set at zero corresponding to the bottom of the valence band.

It is important to know that all solids have Fermi energy, and its location with respect to the bandgap is commonly referred to as Fermi level. We can now summarize that

$$E_{g,\text{dielectric}} \gg E_{g,\text{semiconductor}} \gg E_{g,\text{metal}} \tag{1.6}$$

In Table 1.2, values for the bandgap for some common semiconductor materials is given at 300 K.

1.2.3 Permeability, ϵ

From Figure 1.1, we can also conclude based on the arguments advanced in the previous section that the large bandgap of a dielectric material would inhibit the electrical conduction since it would be difficult for electrons to gain sufficient energy to overcome the bandgap at room temperature. This is certainly consistent with our everyday experience that dielectrics are very poor carriers of electricity. However, one need to remember that theoretically even the best of dielectric can conduct electricity when subjected to a large potential difference, but the magnitude of the resulting current would be so small as to be of any practical interest.

The defining property of a dielectric material is the *permittivity,* which is also known by its other name of *dielectric constant* with the universal symbol of ϵ. All materials will get polarized when subjected to an electric

Table 1.2 Some semiconductor materials and their bandgap.

Materials	Bandgap (eV)
Ge	0.661
Si	1.12
InSb	0.17
InP	1.344
GaAs	1.424

Source: From http://hyperphysics.phy-astr.gsu.edu/hbase/hph.html.

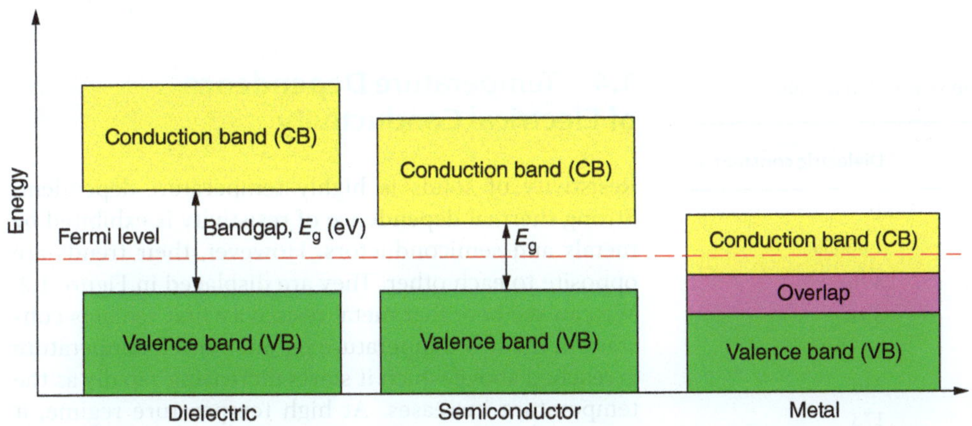

Figure 1.1 Comparative representation of insulators, semiconductors, and metals on the basis of their energy bandgaps.

field. We know that the relationship between the electric displacement (D), and the electric field (E) is given by the fundamental equation of electromagnetics which states that

$$D = \epsilon_0 E + P \tag{1.7}$$

where ϵ_0 is the permittivity of vacuum with the value of 8.85×10^{-12} F m^{-1} and P the electric field-induced polarization. At low electric field, the product $\epsilon_0 E$ is a very small number, and therefore, we can approximate $D \approx P$. Therefore, for low electric fields, Eq. (1.7) takes the form of Eq. (1.8).

$$P \approx \epsilon_r \epsilon_0 E \tag{1.8}$$

The parameter ϵ_r is the relative dielectric constant that is a unitless quantity and is equal to $\epsilon \cdot \epsilon_0^{-1}$, where ϵ is the permittivity of the material. The permittivity is specific to a material similar to the electrical conductivity. Therefore, we can also use this parameter to distinguish between the three types of solids as shown in the relationship in Eq. (1.9).

$$\epsilon_{r,\text{dielectric}} \gg \epsilon_{r,\text{semiconductor}} \gg \epsilon_{r,\text{metal}} \tag{1.9}$$

In Table 1.3, a list of relative dielectric constant (ϵ_r) for selected materials is presented.

1.3 Fundamental Nature of Electrical Conductivity

We defined in Eq. (1.4) the electric current, I. This derivation was based on geometrical considerations of a sample of finite size and length. The question now arises what causes the onset of current and how do we understand its true nature. To accomplish this goal, we need to consider that the current is generated when electrons move from one point to another under the influence of an applied electric field. Such a movement will obviously involve a velocity and mobility.

Table 1.3 Dielectric constant of some selected materials.

Materials	Dielectric constant, ϵ_r
Vacuum	1
Air	1.00059
Mica	3–6
Polyvinyl chloride	3.18
Germanium (Ge)	16
Strontium titanate ($SrTiO_3$)	310
Titanium dioxide (TiO_2)	173

Source: From http://hyperphysics.phy-astr.gsu.edu/hbase/hph.html.

We can easily visualize a picture in which a traveling electron will encounter thermally generated *phonons* in a crystal lattice and then will acquire an average velocity that is also called the drift velocity, v_d. But what are *phonons* and where do they come from? It is quantum mechanical concept and refers to the unit of vibrational energy originating from the oscillations of atoms within a crystal lattice. The atomic oscillations increase with increasing temperature resulting in larger number of thermally generated phonons. Phonons are the counterpart of photons and both being quantum mechanical concepts. They are the two main types of elementary particles associated with solids.

The magnitude of the drift velocity will be proportional to the applied electric field. The coefficient of proportionality is called the electron mobility (μ_e). Alternatively, it can also be defined with the help of the following equation:

$$\mu_e = \left(\frac{\Delta v_d}{\Delta E}\right) \tag{1.10}$$

The electron mobility is a very important property and plays a vital role in designing a transistor. Materials with larger values of mobility are desired because that translates to faster transistors. We will discuss this parameter again in Chapter 7. Its unit is m^2 V^{-1} s^{-1}.

We can easily visualize that electrical conductivity (σ_e) and electron mobility (μ_e) to be related somehow. We can in fact find this relationship simply by assuming that there are n number of electrons involved and their transport from one point to another is facilitated by the onset of mobility (μ_e) and the applied electric field (E) such that

$$\sigma_e = ne\mu_e \tag{1.11}$$

where e is obviously the electronic charge. Equation (1.11) is the standard expression and gains a special importance while dealing with semiconductor materials where the conductivity is the sum of the contributions made by electrons and holes. This is discussed also in Chapter 7.

1.4 Temperature Dependence of Electrical Conductivity

Resistivity of solids is highly temperature-dependent. Strong thermal dependence of resistivity is exhibited by metals and semiconductors. However, their trends are opposite to each other. They are displayed in Figure 1.2. We can see here that metal resistivity first remains constant in the low temperature regime until a temperature is reached above which it starts increasing rapidly as the temperature increases. At high temperature regime, it follows approximately a linear relationship with temperature yielding a positive temperature coefficient of resistivity $\left(\frac{\Delta\rho}{\Delta T} = \eta\right)$. The semiconductor resistivity, on the other hand, increases rapidly with decreasing temperature following an exponential thermal dependence. At sufficiently low temperatures, all semiconductors become good insulators. At higher temperatures, its resistivity decreases at a vastly reduced rate such that the change is almost monotonous. Resistivity of a typical insulator follows qualitatively the same temperature dependence as semiconductors. Obviously, the resistivity of an insulator is much greater than that of semiconductors as can be concluded from Figure 1.1.

In Figure 1.2, we have included the temperature dependence of resistivity also for a superconductor simply to demonstrate the distinction one can make between metals, semiconductors, and superconductors based on the behavior of their electrical resistivity with temperature. In superconductors, the resistivity goes through a phase change at a critical temperature, called the superconducting *transition point* below which a normal metal becomes superconducting. Its resistance vanishes and the material acquires infinite conductivity and remains in the superconducting state so long as temperature remains below the transition point. Above the critical temperature, it loses its superconducting nature and behaves like a normal metal. The thermal behavior of solids, as shown in Figure 1.2, can be easily explained on the basis of physics as describe below.

1.4.1 Case of Metals

The thermal behavior of electrical resistivity of metals can be expressed empirically by Matthiessen's rule that is given by Eq. (1.12).

$$\rho_{\text{net}} = \rho_0 + \rho(T) \tag{1.12}$$

where ρ_0 the temperature-independent part and $\rho(T)$ the temperature-dependent part. The origin of temperature-independent part of the resistivity lies in the presence of impurities and imperfections in the sample. It dominates at low temperatures following the $\rho_0 \propto T^5$ law. Below a certain temperature called, the Debye temperature, it remains constant. Above the Debye temperature, the resistivity increases linearly with temperature obeying the $\rho \approx \eta T$ relationship. The temperature-dependent part is due to the thermal vibrations of the lattice. At high temperatures, more and more phonons are excited impacting the thermal behavior of resistivity. The knowledge of the thermal dependence of metal resistivity above room temperature gives us the value of the temperature coefficient, η, which has important practical applications in temperature measuring devices such as thermocouples and thermistors. We can easily determine its value by measuring the resistance at some well-defined temperatures. Let us say that at temperature T_0, the resistance is R_0, and it is R at temperature T, which is greater than temperature T_0. Then η can be expressed as in Eq. (1.13) (Table 1.4).

$$\eta = \frac{(R - R_0)}{R(T - T_0)} = \left(\frac{\Delta R}{\Delta T}\right) \cdot \frac{1}{R} \tag{1.13}$$

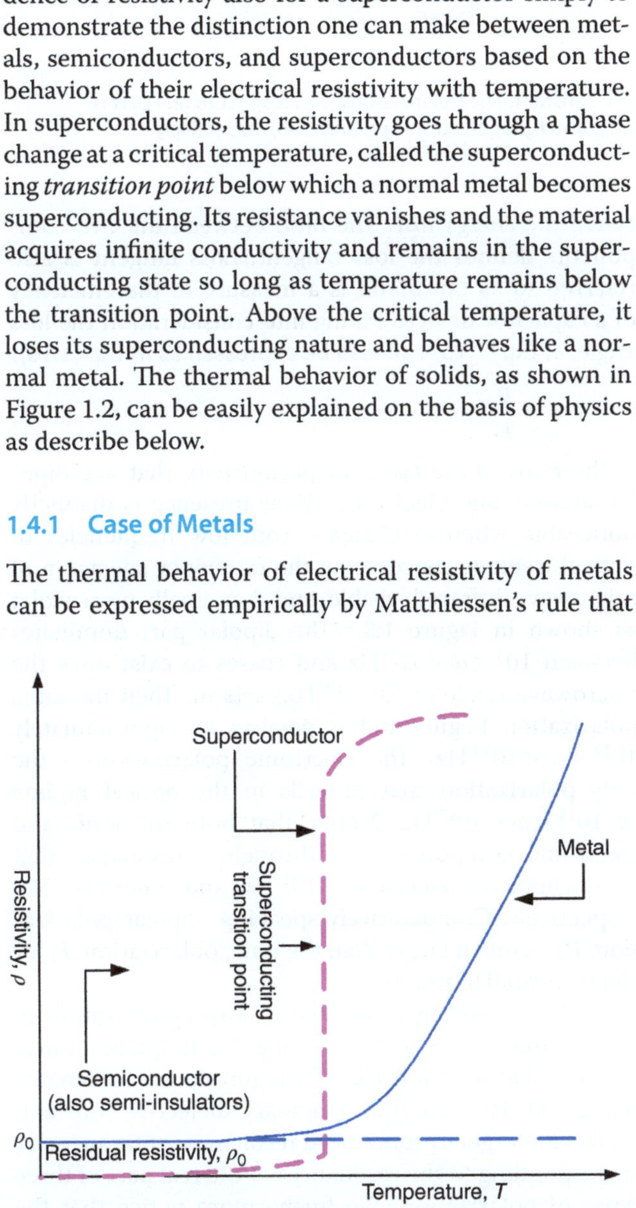

Figure 1.2 Temperature dependence of resistivity of metals, semiconductors, and superconductors.

1.4.2 Case of Semiconductors

For intrinsic semiconductor, the conduction can only take place when electrons closest to the surface of the bandgap acquire sufficient energy to escape the bandgap and reach the conduction band. The temperature dependence of the resistivity (ρ) is given by Eq.(1.14).

$$\rho = \rho_0 \exp\left(-\frac{E_g}{2k_B T}\right) \tag{1.14}$$

Table 1.4 Temperature coefficient of resistivity (η) of some common metals.

Metals	$\eta \times 10^{-3}$ (per °C)
Silver, Ag	3.8
Copper, Cu	3.9
Gold, Au	3.4
Aluminum, Al	4.3
Iron, Fe	6.5
Tungsten, W	4.5
Platinum, Pt	3.92

Nichrome is an alloy of Ni and Cr.
Source: From http://hyperphysics.phy-astr.gsu.edu/hbase/hph.html.

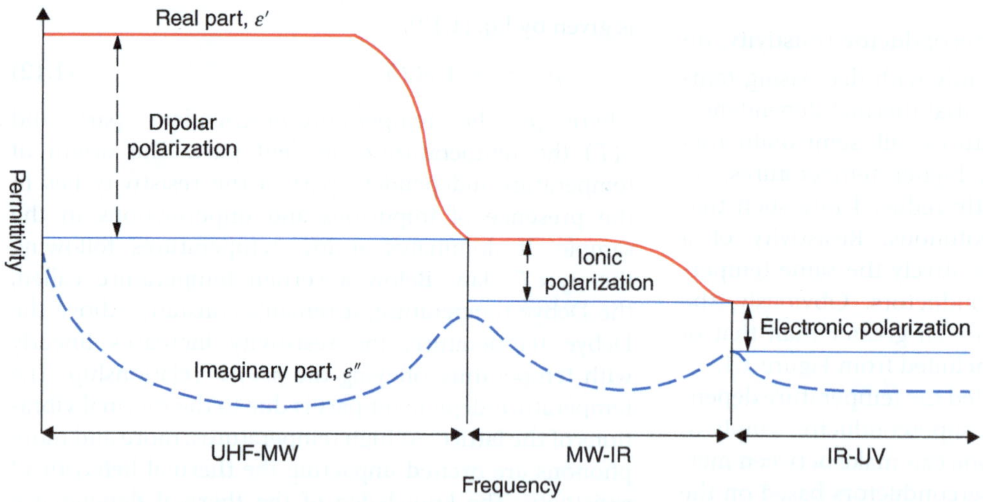

Figure 1.3 Frequency dependence of real and imaginary parts of dielectric constant. The polarizations with respect to real part of permittivity are shown as P_d for dipolar polarization, P_i for ionic polarization, and P_e for electronic polarization, respectively.

In Eq (1.14), ρ_0 is the temperature-independent part of the resistivity, E_g the bandgap, and k_B the Boltzmann constant. Equation (1.14) tells us that the resistivity of a semiconductor material increases exponentially as the temperature decreases. This can be seen from Figure 1.2 as well.

1.4.3 Frequency Spectrum of Permittivity (or Dielectric Constant)

So far we have paid more attention to metals and semiconductor, while discussing the nature of electrical conductivity. Let us now consider the case of an insulator. We may recall that even a standard semiconductor material can become a good insulator when cooled to very low temperatures. The electrical conductivity is of no special interest while discussing the nature of insulators. It is the dielectric constant, or polarizability, that is of greater interest for understanding the dielectric nature of electroceramics. Comparatively speaking, electroceramics show much higher permittivity than semiconductors. Equation (1.7) gives us an expression for the displacement (D) when an insulator is subjected to an external electric field (E). Permittivity is strongly dependent upon the frequency of the applied electric field. Permittivity measured at any frequency (ω) consists of real and imaginary components as shown in Eq. (1.15).

$$\epsilon(\omega) = \epsilon'(\omega) + j\epsilon''(\omega) \tag{1.15}$$

Here $\epsilon(\omega)$ is the measured permittivity at frequency (ω), $\epsilon'(\omega)$, the real part and $\epsilon''(\omega)$ the imaginary part. The real part is related to the stored electrical energy of the medium such as a capacitor, and imaginary part is related to the dissipation of the energy which is also called the *energy lost*. The ratio between the two components defines the loss tangent. Loss tangent is also referred to as tan δ and is a measure of the efficiency of a capacitor device. Taking into consideration the loss angle, δ, Eq. (1.15) can also be expressed as in Eq. (1.16).

$$\epsilon = \frac{P}{E}(\cos\delta + i\sin\delta) \tag{1.16}$$

There are three types of permittivity that are dipolar, atomic, and electronic. Their presence is distinctly noticeable when ω changes from low frequencies to optical frequencies covering the frequency spectrum of microwave, infrared, visible, and then finally ultra-violet as shown in Figure 1.3.[1] The dipolar part dominates between $10^3 < \omega < 10^9$ Hz and ceases to exist once the microwave range ($\approx 10^{11-13}$ Hz) sets in. Then the ionic polarization begins and it persists for approximately $10^{12} < \omega < 10^{13}$ Hz. The electronic polarization is the only polarization that prevails in the optical regime of $10^{14} < \omega < 10^{17}$ Hz. Notice that both the ionic and electronic components go through a resonance that occurs approximately at $\omega \approx 10^{12}$ Hz and at $\omega \approx 10^{15}$ Hz, respectively. Comparatively speaking, dipolar polarization, P_d, is much larger than the ionic polarization, P_i, or electronic polarization, P_e.

We find a strong resonance of ionic polarization in the infrared (IR) regime covering the frequency range between 300 GHz and 430 THz (equivalent wave lengths being 700–10^6 nm). The imaginary dielectric constant, ϵ'' also undergoes pronounced resonances at frequencies corresponding to the resonances of the real part of three types of polarization. We furthermore notice that the

1 https://en.wikipedia.org/wiki/Permittivity

imaginary part decreases with increasing frequency and the largest change occurs in the dipolar region. What actually causes the onset of these polarizations? We shall discuss this now.

The *dipolar polarization* that is also known as orientation polarization is randomly oriented dipoles in the absence of an electric field. However, when an external electric field is applied, these dipoles orient themselves in the direction of the field. At frequencies (ω) below 10^{11} Hz the applied field can induce periodic rotation. At moderately high fields, the rotations become rapid and energy is absorbed as heat. This is the basis for microwave heating. As can be seen from Figure 1.3, this polarization dominates in the frequency domain covering $0 < \omega < 10^{11}$ Hz.

At moderately high frequencies, molecules may not rotate because the high energy that is absorbed results in the emergence of the resonance peaks. On the application of electric field with frequencies in the microwave domain ($10^{11} < \omega < 10^{13}$ Hz), small displacements of ions from their equilibrium positions can result. As a consequence, a net dipole moment can be induced. These dipoles will absorb energy that like in the previous case will give rise to resonance. Such a response is mostly present in the infrared region. The *electronic polarization*, on the other hand, sets in at the optical frequencies with $\omega > 10^{14}$. The dipolar and ionic polarizations are negligible at optical frequencies because of the inertia of molecules to respond to optical frequencies. At optical frequencies, the square of the refractive index, n, equals the value of the relative dielectric with good accuracy. Some examples are given in Table 1.5.

Then there is another type of polarization that occurs only in heterogeneous materials such as in a ceramic where grain boundaries are present or at the interface such as material–metallic electrode interface. Its origin lies in the limited movements of charges under the influence of an applied electric field at very low frequencies. Charges accumulate at the grain boundaries or at the interfaces giving rise to *interfacial polarization*. This effect has no fundamental value but is of considerable importance to electronic technology.

Table 1.5 Comparison of refraction index and permittivity for some materials.

Materials	ϵ_r	n^2	Comments
Diamond, C	5.7	5.85	Electronic
Germanium, Ge	16	16.73	Electronic
Sodium chloride, NaCl	5.9	2.37	Electronic and ionic
Water, H_2O	80	1.77	Electronic, ionic and dipolar

1.5 Essential Elements of Quantum Mechanics

We are well familiar with the fact that quantum mechanics is a powerful branch of physics that provides us with the requisite tools for understanding the physical phenomenon that cannot be adequately described by classical physics. Since its emergence in the early twentieth century, it has dominated the field of solid-state sciences of which electroceramics is a part. In this section, we will try to appreciate the essential concepts that form the pillars of quantum mechanics leading to a better understanding of the physical properties of electroceramic materials. Of particular interest are the concepts of Planck's radiation law, Einstein's photoelectric effect, Bohr's theory of the hydrogen atom, de Broglie principle of duality of matter and waves, Schrödinger's equations, Heisenberg's uncertainty principle, and the quantum mechanical interpretation of the periodic table of elements.

1.5.1 Planck' Radiation Law

Until 1900, classical physics could explain satisfactorily most of the physical phenomena observed. However, a time came when it was not possible to explain some of experimental results using the concepts of classical physics. One of them was the true nature of emitted radiation from a black body. Ideally, a black body is a perfect radiator and an absorber of energy at all electromagnetic wavelengths. Energy is considered to be continuous according to classical physics. However, physicists at the time failed to explain black body radiation using the concepts of classical physics. In 1901, Max Planck of Germany took a bold step and postulated that light energy is not continuous, but rather it exists in discrete packets which he called quanta. The emitted energy (E) is proportional to the frequency of emitted radiation (ν).

$$E = nh\nu \tag{1.17}$$

where $n = 1, 2, 3, \ldots$ and h Planck's constant which is equal to 6.625×10^{-34} J s.

This simple equation tells us that the radiated energy from a black body can only assume values in integral steps of $h\nu$ with $n = 1, 2, 3, \ldots$ In the vocabulary of quantum mechanics, such a situation is described as quantized. That makes radiative energy a quantized parameter that is one of the cornerstones of quantum mechanics. The concept of quantization plays a vital role in quantum mechanics. With this assumption, Planck was finally able to explain successfully the nature of radiated energy from a black body. Planck received Nobel Prize in Physics in 1918 for this very fundamental contribution. Equation (1.17) can be written in other forms as well; one of them being as in Eq. (1.18).

$$E = nh\nu = n\frac{h}{2\pi}(2\pi\nu) = n\hbar\omega \tag{1.18}$$

The symbols $\hbar$ and ω are reduced Planck's constant and angular frequency, respectively. From Eq. (1.18), it follows that the photon energy, E_{ph}, between any two successive quantum number is given by

$$E_{ph} = nh\nu - (n-1)h\nu = \hbar\omega \tag{1.19}$$

It is interesting that neither Planck nor Einstein later, in explaining the photoelectric effect, used the word *photon* in place of *light quanta*. It was Gilbert N. Lewis, an American Physical Chemist, coined the word *photon* in 1926 to describe light quanta. Ever since, this word has been in use universally to mean light quanta.

1.5.2 Photoelectric Effect

The photoelectric effect was discovered by Heinrich Hertz of Germany in 1887 while experimenting with electromagnetic waves whose existence he conclusively proved. Electromagnetic waves were theoretically predicted in 1864 by James Clark Maxwell of England in his celebrated "electromagnetic theory of light." It was Heinrich Hertz of Germany who had discovered the photoelectric effect in 1887 while illuminating metallic surfaces with ultraviolet light. He noticed during his experiments, the emission of bursts of sparks. It is the same Hertz who had also discovered radio waves and experimentally showed the existence of electromagnetic waves predicted by Maxwell. Today, in his honor, Hz (Hertz) is used as the unit for frequency.

The photoelectric effect phenomenon could not be explained on the basis of classical physics. It offered a dilemma to the physicist of the time and remained unexplained until 1905 when Albert Einstein successfully explained the effect for which he received the Nobel Prize in Physics in 1921. It is interesting to note that though he had earlier developed the "special theory of relativity" that gained him international stature and respect, it was his work on the photoelectric effect that was recognized by the Nobel Committee and not the celebrated "special theory of relativity." The photoelectric effect is defined as the emission of electrons or other charged particles from a material when irradiated by light of suitable frequency. This effect can be observed by doing a simple experiment with the setup similar to the one shown in Figure 1.4.

When a cathode made of a metal is irradiated by photons (light quanta of Planck) of suitable energy, electrons are emitted. These electrons are collected at the positively charged anode resulting in the onset of a photocurrent,

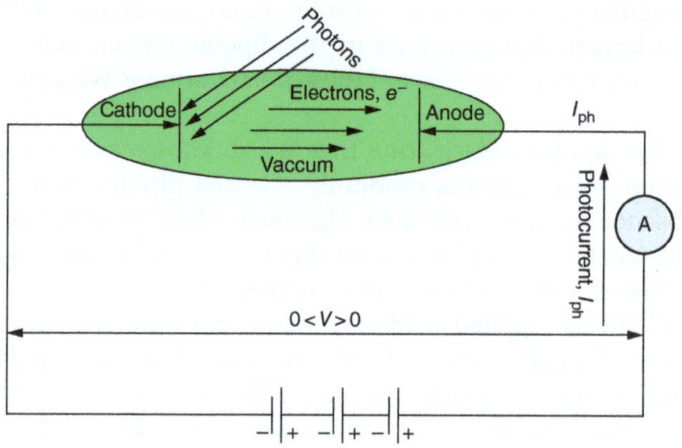

Figure 1.4 Sketch of experimental set up for photoelectric effect.

I_{ph}. However, the emission can take place only when the Einstein's equation of electron emissivity is obeyed which states that

$$h\nu = \Omega_{max} + W \tag{1.20}$$

Here Ω_{max} is the maximum kinetic energy of the emitted particles and W the work function which is a material constant. From this equation, we can infer that for photoemission to set in the threshold energy equivalent to W must be overcome. That is W must be equal to the photon energy of $h\nu_0$, where ν_0 is the frequency corresponding to the threshold energy. Then Eq. (1.20) takes the form of Eq. (1.21).

$$\Omega_{max} = h(\nu - \nu_0) \tag{1.21}$$

Equation (1.21) tells us that the maximum kinetic energy of emitted electrons is directly proportional to frequency with the slope of the straight-line giving us the experimental determination of the value of Planck's constant, h. This is another important implication of Einstein's equation of photoemission. In Figure 1.5,

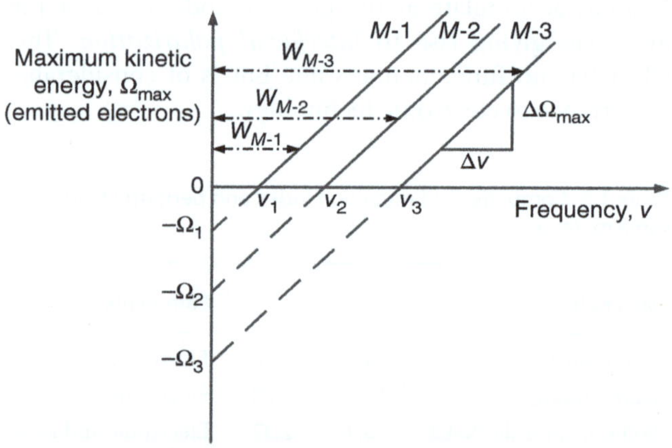

Figure 1.5 Kinetic energy of emitted electron vs. frequency for different metals M-1, M-2, and M-3.

the maximum kinetic energy as a function of radiation frequency for three arbitrary metals (M-1, M-2, and M-3) is plotted. We can easily find that the slope of the plots gives us the value of the Planck's constant. The intercepts on the x-axis gives the values of the threshold frequencies for the three metals, respectively, which are labeled as ν_1, ν_2, and ν_3. The intercepts on the negative side of the y-axis and identified as Ω_1, Ω_2, and Ω_3 are the potentials that must be applied to stop the photoelectric effect entirely. It is important to remember that photoemission is a frequency-dependent function and is independent of the photo-current, I_{ph},

When a voltage, $V > 0$ is applied in the circuit of Figure 1.4 the photocurrent, I_{ph}, will be amplified and similarly a negative potential will make it smaller. This is shown in Figure 1.6. From this figure, we also find that the photocurrent increases with the increase in the intensity of light. However, the process of photoemission itself remains unaffected by the intensity of light.

As the positive potential increases, the photocurrent is first amplified and keeps on increasing until it begins to saturate. However, exactly the opposite happens when the sample is biased with a negative potential. The photocurrent, as expected, becomes smaller and finally disappears completely when the photoemission stops. This characteristic negative potential, $-V_s$, is called the "stopping potential." The work done by an electron in transporting against the "stopping potential" must be equal to its maximum kinetic energy, Ω_{max}. Substituting it in Eq. (1.21), we get Eq. (1.22).

$$h\nu = eV_s + W \tag{1.22}$$

When V approaches the stopping potential, the photoemission stops so that for $\nu = 0$, $V_s = -\frac{W}{e}$. In Figure 1.5, the intercepts along the y-axis at $\nu = 0$ correspond to kinetic energies at the stopping potentials which are $-\Omega_1 \equiv \frac{W_{M\text{-}1}}{e}$, $-\Omega_2 \equiv \frac{W_{M\text{-}2}}{e}$, and $-\Omega_3 \equiv \frac{W_{M\text{-}3}}{e}$. This enables us to determine the work function of a metal accurately because V can be measured more accurately than the kinetic energy.

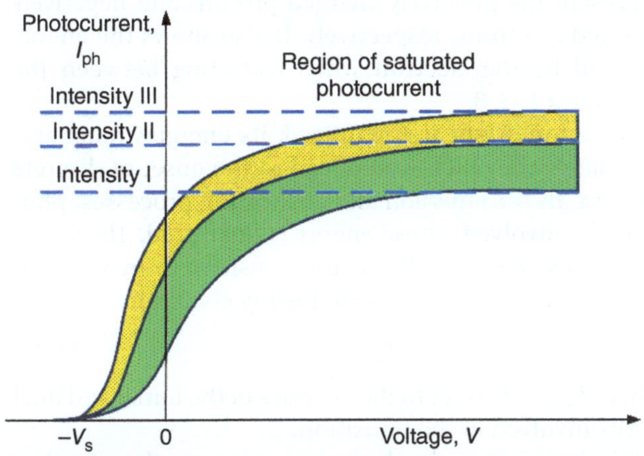

Figure 1.6 Photoelectric current vs. voltage for three different intensities of light at constant wavelength.

Work function is an important physical parameter that plays crucial roles in solid-state electronics, field emission, thermodynamics, and chemical processes. It is defined as the minimum energy required for an electron to escape from the surface of a solid to reach the vacuum level. By convention the energy of the vacuum level is assigned the value of infinity. Its experimentally determined values vary from one technique to another depending upon the method used. We present its value for some selected group of metals which are commonly used in electronics. A list is presented in Table 1.6. There are many good applications based on the photoelectric effect. Some of them are night vision devices, image sensors, and photomultipliers.

Exercise 1.1

In a photoelectric effect experiment, a polished surface of Ca with work function of 2.9 eV is radiated with the ultraviolet (UV) radiation having the wavelength of 250 nm. What is the velocity of the emitted electrons?

Solution

We have from Eq. (1.20) $\Omega_{max} = h\nu - W = \frac{ch}{\lambda} - W$. Here, Ω_{max} is the maximum kinetic energy of the emitted electron, c = velocity of light = $3 \times 10^8\ \mathrm{m\,s^{-1}}$, h = Planck's constant = $6.63 \times 10^{-34}\ \mathrm{J\,s}$, W the work function of Ca = 2.9 eV. Substituting these values in Eq. (1.20) we get

$$\Omega_{max} = \left(\frac{3 \times 10^8 \times 6.63 \times 10^{-34}}{1.60 \times 10^{-19} \times 250 \times 10^{-9}}\right) - 2.9 = 2.1\ \mathrm{eV}$$

Now, $\Omega_{max} = \frac{1}{2} m_e (v_m)^2$ where $m_e = 9.1 \times 10^{-31}$ kg.

Table 1.6 Work function of some commonly used metals.

Metal	Work function, W (eV)	Average value (eV)
Silver, Ag	4.26–4.74	4.50
Aluminum, Al	4.06–4.26	4.16
Gold, Au	5.1–5.47	5.29
Copper, Cu	4.53–5.10	4.82
Platinum, Pt	5.12–5.93	5.53
Palladium, Pd	5.22–5.6	5.41
Iron, Fe	4.67–4.81	4.74

Source: https://en.wikipedia.org/wiki/Work_function. Licensed under CC BY 3.0.

Substituting this for Ω_{max}, we get the maximum velocity, v_m, for emitted electron to be

$$v_m = \sqrt{\frac{2\Omega_m}{m_e}} = 8.6 \times 10^5 \text{ m s}^{-1}$$

1.5.3 Bohr's Theory of Hydrogen Atom

In 1911, Lord (Ernest) Rutherford of England (originally, from New Zealand; Nobel Prize in Chemistry in 1909) proposed a model for an atom in which he compared an atom to an ultra-miniaturized prototype of our solar system. According to this model, an atom consists of a nucleus that is surrounded by a number of orbits. The entire mass of the atom is densely packed at the core of the nucleus that consists of many subatomic particles of which neutrons and protons are just two examples. Proton is positively charged, whereas neutron is electrically neutral. Both of them are of approximately equal mass, and each is roughly 1840 times heavier than an electron with the mass of 9.1×10^{-34} kg. The atomic number, Z, of an element is equal to the number of protons residing at the nucleus. A very strong Colombian force between the proton and the electron holds the atom together and gives stability to the structure.

Niels Henrik David Bohr, a Danish physicist, used Rutherford's model of atomic structure to develop his celebrated theory of the hydrogen atom for which he received the Nobel Prize in Physics in 1922. This theory is also considered to be one of the pillars of quantum mechanics. In the field of optical spectroscopy, it was well known that the wavelengths of hydrogen spectrum obeyed an empirical relationship as given in the following equation.

$$\frac{1}{\lambda} = R\left(\frac{1}{n_i^2} - \frac{1}{n_f^2}\right) \tag{1.23}$$

where λ is the wavelength of light, R the Rydberg constant that is equal to 1.097×10^7 m^{-1}, n_i and n_f are integers associated with specific spectral series. For example, when $n_f = 2$, then $n_i = 3, 4, 5, \ldots$, then the spectral series is called the Balmer series. The next series is called the Paschen series with $n_f = 3$ followed by the Lyman series with $n_f = 4$. There are many more spectral series for hydrogen atom ($Z = 1$), and we need not account for all of them. It is possible that the integer n_i can assume the value of infinity. We would agree that this type of empirical explanation does not offer a sound scientific reasoning. Obviously, it was beyond the capacity of classical physics to come forward with a sound scientific theory to explain the experimental results found by spectroscopists of the time. This must have inspired Bohr to look at this problem from a completely different angle, and for this, he made use of the concept of quantized photon energy proposed earlier by Planck. Bohr made three assumptions:

Assumption 1: The electrons can traverse around the orbits but without emitting or absorbing any radiation. The order of orbits in an atom, beginning with the first orbit nearest the nucleus, follow the ascending order of the *principal quantum number*, *n*, which can only have only the integral values of 1, 2, 3, …

Assumption 2: The electrons can transit from one orbit to another. Because the energy of each orbit is different, during the process of transition, the electrons can either absorb or emit radiation in order to satisfy the law of conservation of energy. In either case, Planck's radiation law must prevail, and as such the photon energy must be equal to $h\nu$.

Assumption 3: The angular momentum, L, is quantized and can have only the values equal to integral multiples of $\hbar$. This was his boldest assumption and has the same importance as Planck's quantized energy. Quantized L is called the *orbital quantum number*.

Mathematically, we can express the third assumption in the form of Eq. (1.24).

$$L_n = m_e r_n^2 \omega_n = \frac{nh}{2\pi} = n\hbar \tag{1.24}$$

where m_e is the electron mass, r_n the radius of the nth circular orbit, ω_n, its angular velocity and $\hbar = \left(\frac{h}{2\pi}\right)$. It follows from Eq. (1.24) that $\omega_n = \frac{n\hbar}{m_e r_n^2}$. That gives us $r_n = \left(\frac{n\hbar}{m_e \omega_n}\right)^{\frac{1}{2}}$. Using this relationship, Bohr accurately calculated the radii of the orbits and their respective angular momenta for different spectral series, and these calculations were found to be in agreement with experimentally determined values. In Figure 1.7, the Bohr's model of hydrogen atom is shown. Here p^+ and e^- represent the positively charged protons and negatively charged electrons, respectively. It also shows the energy emitted by the electron when transiting between the orbits $n = 1, 2, 3$.

Since the orbits are quantized, its energies must also be quantized, which would lead to the onset of discrete spectra. In the emission and absorption processes, photons are involved whose energy is quantized. Therefore, the change in energy during the transition from one orbit to another must satisfy the following condition.

$$\Delta E = h\nu = E_f - E_i \tag{1.25}$$

where E_i and E_f refer to the energies of the initial and final orbits involved in the transition.

We know that the hydrogen atom is the simplest element of the periodic table having the atomic number, $Z = 1$. Bohr's elaborate calculation resulted in the

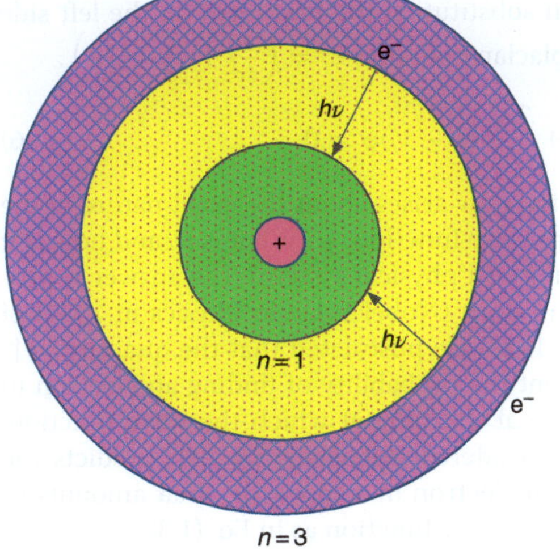

Figure 1.7 Bohr's model of hydrogen atom.

ground-state energy, E_0 (when $n = 1$) for the hydrogen atom to be

$$E_0 = \frac{1}{h^2}\left(\frac{m_e e^4}{8\epsilon_0^2}\right) = 13.6\ \text{eV} \tag{1.26}$$

Similarly, the energy of the nth orbit is given by Eq. (1.27).

$$E_n = -\left(\frac{13.6}{n^2}\right)\text{eV} \tag{1.27}$$

Substituting these values in Eq. (1.27) we arrive at Eq. (1.28).

$$h\nu = 13.6\left(\frac{1}{n_i^2} - \frac{1}{n_f^2}\right) \tag{1.28}$$

Since $c = \nu\lambda$ where c is the velocity of light and λ its wavelength, we can rewrite Eq. (1.28) as Eq. (1.29).

$$\frac{1}{\lambda} = \frac{13.6}{ch}\left(\frac{1}{n_i^2} - \frac{1}{n_f^2}\right) \approx R\left(\frac{1}{n_i^2} - \frac{1}{n_f^2}\right) \tag{1.29}$$

By substituting the values of the universal constants c and h in ($\frac{13.6}{ch}$) one gets $1.097 \times 10^7\ \text{m}^{-1}$, which is the value of the Rydberg constant used in spectroscopy.

In summary, we can conclude that the Bohr's theory satisfactorily explains the experimental results of the atomic spectra of the hydrogen atom and thereby could solve the longstanding problem of classical physics. Therefore, his assumptions were intuitively correct that the orbits are quantized and that the Rydberg constant is not an arbitrary number to fit an experiment but rather it is the combination of fundamental constants like the charge of the electron, its mass, Planck's constant and the permittivity of vacuum.

It should also be recognized that in spite of the success of Bohr's theory in explaining the spectrum of the hydrogen atom and giving us the concept of the quantization of angular momentum it has some fundamental flaws. For example, it is in violation of the Heisenberg Uncertainty Principle, and it cannot explain the Zeeman Effect when the spectral lines split up in several components in the presence of a magnetic field.

1.5.4 Matter–Wave Duality: de Broglie Hypothesis

Planck gave the concept of energy being quantized, Einstein gave the concept of photons that can behave like a particle yet has no mass, and Bohr advanced quantum mechanics that was in its infancy by proposing the angular momentum to be quantized. All these groundbreaking concepts point to the particle-like behavior of light that was well established to be wave-like by electromagnetic theory. This leads us to the question of how can a particle (such as an electron) also acquire wave-like characteristics?

The answer was provided by Louis de Broglie of France in 1923 in his famous hypothesis of matter–wave duality. This led to the development of the famous Schrödinger's equation that firmly anchored quantum mechanics as the new physics. de Broglie based his hypothesis on two well-established results. He considered Einstein's matter–energy equation and Planck's theory of light quanta.

Einstein's matter–energy equation states that

$$E = m_e c^2 \tag{1.30}$$

And Planck's law of radiation says that

$$E = h\nu \tag{1.31}$$

de Broglie argued that since particles and waves have the same traits, the two energies must be equal. Combining these two equations Eq. (1.32) results.

$$m_e c^2 = h\nu \tag{1.32}$$

Considering that the real particles, such as an electron, cannot acquire the velocity equal to the velocity of light, we need to modify the above equation slightly to use it for real particles. Let us assume that we are dealing with electrons with the mass, m_e and velocity, v_e. Furthermore, we can substitute for the frequency $\left(\frac{v_e}{\lambda_e}\right)$, where λ_e is the wavelength associated with an electron. These manipulations result in Eq. (1.33).

$$\lambda_e = \frac{h}{m_e v_e} = \frac{h}{2\pi}\left(\frac{2\pi}{p_e}\right) = \frac{2\pi\hbar}{p_e} \tag{1.33}$$

where p_e is the momentum of the electron. We can also write Eq. (1.33) as Eq. (1.34) which is the standard form of de Broglie's relationship.

$$p_e = \frac{h}{\lambda_e} = \hbar k_e \tag{1.34}$$

where k_e is the wave number which by definition is $\left(\frac{2\pi}{\lambda_e}\right)$.

This simple equation derived from another two very simple equations may look humble, but it has far-reaching consequences in solid-state physics and electronics. de Broglie was awarded the Nobel Prize in Physics in 1929 for this contribution. Its experimental proof was given by Clinton Davisson and Lester Germer, both American Physicists, in 1925 confirming the wave nature of electron. For this contribution, they too shared the Nobel Prize in Physics in 1937.

1.5.5 Schrödinger's Wave Equation

What Newton's laws of motion and his concept of *conservation of energy* are to classical physics so is the Schrödinger's equation to quantum mechanics. He is one of the giants of physics of the twentieth century and belongs to the class of Sir Isaac Newton. The matter–wave duality hypothesis of de Broglie is the nucleating factor for Schrödinger's equation. Schrödinger argued that since particles can have a wavelength associated with them, they must be represented by a wave equation.

Schrödinger's equation predicts the future behavior of electrons in a dynamic frame work. It is the probability of finding an electron in events to come. A partial differential equation describes how the quantum state of a quantum system changes with time. This is the cornerstone of quantum mechanics that opened up multiple avenues to evolve and advance. It was formulated in 1926 by Erwin Schrödinger, a brilliant theoretical physicist of Austria. It earned him, of course, the Nobel Prize in physics in 1933. It should be remembered that there is no formal derivation of Schrödinger's equation. It is intuitive and Schrödinger simply wrote it. It was immediately accepted by other geniuses of his time and has never been challenged. One of the greatest theoretical physicists of our time, Richard Feynman, is quoted to have said, "Where did we get that from? It is not possible to derive it from anything you know. It came out of the mind of Schrödinger?"

Let us now write the one-dimensional form of Schrödinger's equation.

$$\frac{d^2\psi}{dx^2} + \frac{2m_e}{\hbar^2}(E - V)\psi = 0 \tag{1.35}$$

Here ψ is the wave function, E the total energy, and V the potential energy. The kinetic energy of the electron then is equal to $(E - V)$.

In its three-dimensional form, Eq. (1.35) becomes Eq. (1.36) on substituting the first term on the left side with the Laplacian operator $\nabla^2 = \left(\frac{\delta^2}{\delta x^2} + \frac{\delta^2}{\delta y^2} + \frac{\delta^2}{\delta z^2}\right)$.

$$\nabla^2\psi + \frac{2m_e}{\hbar^2}(E - V)\psi = 0 \tag{1.36}$$

The question now arises about the exact nature of the Schrödinger's wave function, ψ. What is it, and how is it significant in a real physical system? The answer is provided by Max Born (Nobel Prize in Physics in 1954) of Germany in 1926. He postulated that the quantity $|\psi|^2$ must represent the probability of finding an electron in a unit volume at the time at which the wave function, ψ, is being considered. Alternatively, $|\psi|^2$ predicts the presence of an electron in a space, dv. That amounts to normalizing the wave function as in Eq. (1.37).

$$\int_{-\infty}^{+\infty} |\psi|^2 dv = 1 \tag{1.37}$$

Equation (1.37) sets the boundary conditions that the solutions for wave function, ψ, must obey. The other boundary conditions imposed on the wave functions are (i) they must be continuous and (ii) mathematically well behaved. This amounts to telling us that $\psi(x)$ must be a continuously varying function of x and its first derivative with respect to x, $d\psi/dx$, must also be a continuous function of x.

Another form of the Schrödinger's equation can be represented using the Hamiltonian operator that is the sum of the kinetic and potential energy in quantum mechanics. The operator is named after Sir William Hamilton, a reputed physicist of Ireland who lived in the nineteenth century. He is best known for the development of Hamiltonian mechanics that is essentially the reformulation of Newton's mechanics. If T is the kinetic energy and V the potential energy, then the corresponding Hamiltonian takes the form of Eq. (1.38).

$$\hat{H} = \hat{T} + \hat{V} \tag{1.38}$$

Here the potential energy operation $\hat{V}$ is equivalent to the space and time variants of the potential energy, V. The momentum, p, in operator form is written as:

$$\hat{p} = -i\hbar\nabla \tag{1.39}$$

Similarly, the kinetic energy operator form is as in Eq. (1.40).

$$\hat{T} = \frac{\hat{p}^2}{2m_e} = -\frac{\hbar^2}{2m_e}\nabla^2 \tag{1.40}$$

Substituting these two equations in Eq. (1.38), we get the Hamiltonian operator, $\hat{H}$, as in Eq. (1.41).

$$\hat{H} = -\frac{\hbar^2}{2m_e}\nabla^2 + V(r, t) \tag{1.41}$$

We can now rewrite the time-independent Schrödinger's equation in terms of the Hamiltonian $\widehat{H}$, as

$$\widehat{H}\psi_i = E_i\psi_i \tag{1.42}$$

Here ψ_i is called the eigenfunctions and E_i the eigenvalues of energy.

The Hamiltonian operator also lead us to the time-dependent Schrödinger's equation which is given by Eq. (1.43).

$$\widehat{H}\psi = i\hbar\left(\frac{\delta\psi}{\delta t}\right) \tag{1.43}$$

The probability of finding an electron in the volume element (dx dy dz) at a time t is then given by

$$|\psi(x, y, z, t)|^2 \mathrm{d}x\,\mathrm{d}y\,\mathrm{d}z$$

Exercise 1.2

Express the time-independent Schrödinger's equation in terms of the momentum.

Solution

We have the standard form of the time-independent Schrödinger's equation containing the energy term in Eq. (1.36).

$(E - V)$ is the kinetic energy T of the electron; then it follows from Eq. (1.36) hat

$$\nabla^2\psi = -\left[\frac{\hbar^2}{2m_\mathrm{e}}T\right]\psi \tag{a}$$

Equation (b) gives us the kinetic energy in term of the momentum, p_e of the electron.

$$T = \text{k.e.} = \frac{1}{2}m_\mathrm{e}v^2 = \frac{1}{2}\left(\frac{m_\mathrm{e}^2v^2}{m_\mathrm{e}}\right) = \frac{p_\mathrm{e}^2}{2m_\mathrm{e}} \tag{b}$$

Substituting this in Eq. (a) and after a little rearrangement, we get Eqs. (c) and (d).

$$\nabla^2\psi = -\frac{\hbar^2}{2m_\mathrm{e}}\left(\frac{p^2}{2m_\mathrm{e}}\right)\psi = -\left[\left(\frac{\hbar}{2m_\mathrm{e}}\right)^2 p_\mathrm{e}^2\right]\psi \tag{c}$$

$$\nabla^2\psi = -\left(\frac{\hbar p_\mathrm{e}}{2m_\mathrm{e}}\right)^2\psi \tag{d}$$

Equation (d) is the momentum form of Eq. (1.36).

1.5.6 Heisneberg's Uncertainty Principle

We learned in the previous section that the Schrödinger's equation is statistical in nature and can predict the probability of an event happening but cannot predict accurately either the position of an electron or its velocity. Similarly, it is not possible to predict either the momentum in a particular space in which the electron finds itself nor the energy it might acquire in a particular instant of time. The reason being that the *uncertainty principle* forbids the measurements of two complimentary parameters concurrently with arbitrary accuracy. The theory was developed by Werner Heisenberg (Nobel Prize in Physics in 1932) of Germany in 1927.

The essence of this theory is that the product of two complimentary variables cannot be less than a constant value. For example, if position x and momentum p are considered, then the uncertainty in position Δx and momentum Δp is given by the following inequality.

$$\Delta x \times \Delta p \geq \hbar \approx 10^{-34}\ \mathrm{J\,s} \tag{1.44}$$

Similarly, the uncertainty in energy E and time t can be expressed as follows:

$$\Delta E \times \Delta t \geq \hbar \approx 10^{-34}\ \mathrm{J\,s} \tag{1.45}$$

One can draw the conclusion that if one tries to measure one physical parameter with arbitrarily high precision, the uncertainty in measuring the other conjugate parameter becomes larger. The more the particle becomes smaller such as atomic and subatomic particles, the accuracy in determining their two complimentary variables cannot exceed the limits of $\approx 10^{-34}$ set by the uncertainty principle. One should remember that it is not a reflection on the inaccuracy of measurement instruments or the methods used for experimentation. It is simply inherent in the quantum mechanical interpretation of nature. As the particles approach macroscopic scales, the uncertainty decreases drastically. To illustrate this point, let us consider a mass m, which is 10^3 times greater than the mass of an electron. If the velocity of the particle is v, its momentum $p = mv$ and $\Delta p = m\Delta v$. Substituting this in Eq. (1.44) for ΔP, we get Eq. (1.45).

$$\Delta x \times \Delta v \geq \frac{\hbar}{10^3 m_\mathrm{e}} \approx 10^{-4}\ \mathrm{J\,s\,kg^{-1}} \tag{1.46}$$

The uncertainty has decreased by 1000-fold for a macroscopic system whose characteristics can be determined individually with greater accuracy. Nevertheless, it is to be learned from Eq. (1.46) that both the velocity of a particle and its position cannot be measured with arbitrary accuracy at the same instant.

1.6 Quantum Numbers

The wave function, ψ, describes the probability of finding an electron at certain energy levels within an atom. Since it is associated with an electron in an atom it is also called the *atomic orbital*. It defines a region in space in which the probability of finding an electron is high. To every such electron, there are four quantum numbers associated with it which are its defining characteristics. We have already discussed two of them; the principal quantum number, n, and the orbital quantum number, l.

The other two are magnetic quantum number, m_l, and the spin quantum number, s. We now describe all four in some detail.

I. *Principal quantum number, n*: Allowed values are only integers ranging from 1 to ∞. It determines the total energy of the electron; and the number of orbitals ($=n^2$) having different energy levels.

II. *Orbital quantum number, l*: Allowed values are from 0 to $(n-1)$.

The second quantum number is the orbital quantum number and is directly associated with the principal quantum number, n. It is also referred to as angular momentum quantum number and azimuthal quantum number. We already discussed previously that it also is allowed to have only integral values. It divides the shells into smaller group of subshells identified by letters such as s, p, d, f, g, etc. The origin of such a nomenclature lies in optical spectroscopy where the emission or absorption processes were identified as s (sharp), p (principal), d (diffused), f (fundamental), g (ground), etc. After the discovery of quantum mechanics, it was realized that these spectral series correspond to specific values of the orbital quantum numbers as shown in Table 1.7.

The term "subshells" are preferred by chemists, whereas physicists prefer the term "orbitals." The other designation assigned to the subshells or orbitals with certain values of l are called the Bohr designation of atomic subshells with the letter of K, L, M, N, etc. This designation is followed by experts of X-ray diffraction. The total number of orbitals (or subshells) is given by $2n^2$. That is, there are two orbitals for $n = 1$; 8 for $n = 2$, 18 for $n = 3$, 32 for $n = 4$, and so forth. Table 1.7 lists them all.

If $n = 1$, then $l = 0$, the orbital is called 1s; if $n = 2$ and $l = 0$, the orbital is called 2s; and if $n = 3$ and $l = 0$, the orbital is 3s. Other identifiers follow the same logic. So far as the orbital energy (E) is concerned, it increases with increasing orbital quantum number, l. It follows the sequence: $E_s < E_p < E_d < E_f < E_g$. Their relative energy levels follow the sequence of 1s < 2s < 2p < 3s < 3p < 3d < 4s < 4p < 4d < 4f < 5s – and so forth.

III. *Magnetic quantum number, m_l*: Its allowed values are $m_l = 0$ to $\pm l$ with total number of m_l being $(2l+1)$.

Table 1.7 Correspondence of spectral series with orbital quantum number, *l* from 0 to 4.

Orbital quantum number, *l*	**0**	**1**	**2**	**3**	**4**
Spectral series	s	p	d	f	g

From Amperé's law, we know that a moving charge generates an electric current which in turn can induce a magnetic field when enclosed in a loop (such as an orbit). That is the reason that this quantum number is called the magnetic quantum number and as such it is supposed to be directional. It can assume any of the $(2l+1)$ different directions. This was indeed shown to be the case experimentally by Otto Stern and Walther Gerlach, both German physicists, in 1922. They confirmed that the magnetic moments are quantized and can orient only in certain directions. For this ground breaking work Stern was recognized with the Nobel Prize in Physics in 1943, but Gerlach was excluded apparently because of his association with the Nazi Regime.

IV. *Spin quantum number, s*: Allowed values $+\frac{1}{2}$ or $-\frac{1}{2}$.

In an atomic system, electrons can reside in different orbits. They are allowed to move around the orbit while at the same time spinning around its own axis. A spinning electron generates a magnetic field with two well defined orientations. These orientations are designated as either "up (↑)" or "down (↓)." Alternatively, it can only have the values of $=\pm\frac{1}{2}$.

In Table 1.8, we present a list of values for orbital and magnetic quantum numbers with respect to the values of the principal quantum number. Their resulting spectrographic and Bohr designations are also given there.

1.7 Pauli Exclusion Principle

The four quantum numbers define a wave function of an electron fully and completely. They define its quantum state, its energy, and almost any other characteristics associated with it. The orbital quantum number, l, and magnetic quantum number, m_l, can each have multiple values for any fixed value of the principal quantum numbering as outlined in Table 1.8. What happens when there are a large number of electrons present in a system? This can cause an enormous challenge to sort out their quantum states leading to utter confusion.

This is where the selection rule conceived by Wolfgang Pauli of Austria in 1925 comes to our rescue. This rule is universally known as the Pauli's Exclusion Principle for which he received the Nobel Prize in Physics in 1945. It states that: *No two electrons in an atom can have exactly the same set of four quantum number; the spins must be antiparallel.* This simply means that there can be two electrons for each combination of n, l, and m_l, but their spin orientations must be antiparallel. Following this rule, we can assign 2 electrons to each s-state, 6 to each p-state, 10 to each d-state, 14 to each f-state, and

Table 1.8 Relationship between n, l, and m_l, and their spectrographic and Bohr designations.

Principal quantum number, n	Orbital quantum number, l	Magnetic quantum number, m_l	Number of values for magnetic quantum number, m_l	Electrons per orbital = $2n^2$	Spectrographic designation	Bohr designation of shells
1	0	0	1	2	1s	K
2	0	0	1	2	2s	L
	1	0, ±1	3	6	2p	
3	0	0	1	2	3s	M
	1	0, ±1	3	6	3p	
	2	0, ±1, ±2	5	10	3d	
4	0	0	1	2	4s	N
	1	0, ±1	3	6	4p	
	2	0, ±1, ±2	5	10	4d	
	3	0, ±1, ±2, ±3	7	14	4f	

Source: From Leonid 1963 [1]. Azaroff and Brophy (1963).

so forth. They vary in arithmetic progression with four being the common difference. It is important to note that this selection rule is not arbitrary, rather it is based on sound mathematical principles of quantum relativistic physics. A full mathematical treatment of Pauli exclusion principle is beyond the scope of this book.

1.8 Periodic Table of Elements

The periodic table of elements was originally developed by the Russian chemist with the name of Dmitri Mendeleev in 1869. He arranged all the elements known until that time (about 60) in rows and columns according to their atomic weight and chemical properties. Many more elements have since been discovered since, and they all can be arranged in the periodic table on the basis of their atomic numbers, chemical properties, and electronic configurations. The periodic table is an indispensable tool available to scientists and engineers engaged in the study of chemical systems and materials. The modern periodic table consists of eight columns and seven rows.

To take the full benefit of the subject matter covered in this section, it is advisable that readers should have a modern copy of the periodic table readily available. There are many sources from which one can get a good copy of the Periodic Table. The NIST (National Institute of Standards and Technology) in the United States may be a reliable source.

The discovery of quantum numbers greatly shaped the periodic table resulting in advancement to the fields of chemistry, physics, and materials science. Elements found in the same column are referred to as belonging to the same *group* such as Groups I, II, III, IV, V, etc., as they are similar in their chemical properties. There are a total of eight groups, many of which are subdivided in A and B subgroups in many of today's periodic tables. The rows in the periodic table are called *periods*. There are seven periods in which elements are arranged with increasing values of atomic numbers. For example, hydrogen with its atomic number (Z) of 1 is the first element of the periodic table, then comes He with $Z = 2$ followed by lithium with $Z = 3$, and so on. Currently, the highest atomic number of $Z = 118$ belonging to the artificially synthesized *ununoctium* (also known as *eka-radon*) with the chemical symbol of *Uuo*. It is radioactive and very unstable. With its discovery in 2002, the seventh period of the periodic table is complete and a new period begins. The heaviest naturally found element is uranium-238 (U-238) with $Z = 92$. It is a well-established radioactive material with the half-life time of ≈4.5 billion years. The heaviest stable element is bismuth (Bi) with $Z = 83$ and density = $11.34\,g\,cm^{-3}$.

Each element has its unique atomic electronic configuration based on the number of the principal quantum number, n, its atomic number, and the number of electrons in each orbit as dictated by the Pauli exclusion principle. For example, our lightest element is hydrogen with $Z = 1$ only and its electronic configuration is written as $1s^1$. The rule for writing the electronic configuration of an element can be described as follows: "The integer on the left refers to the value on the principal quantum number, n; followed by the orbital (s, p, d, f, etc.) and then a superscript giving the number of electrons found in each orbital." The filling sequence follows in the order of s, p, d, f, etc.

The electrons in the outermost orbital are technically called the *valence electrons*. The valence electrons play a decisive role in initiating a chemical reaction and in forming chemical bonds between atoms which makes a structure stable. They can be shared with other atoms giving rise to chemical bonds known as ionic, metallic, and covalent. The concept of valence electrons is also very important to solid-state sciences, materials science, and electroceramics because they help us in developing models and theories for understanding electronic properties and physical phenomena displayed by materials.

Following the rule stated above, electrons in an atom can be divided between different orbitals. Let us now write electronic configurations for the second and third elements of the periodic table. The second element is He ($Z = 2$), and its electronic configuration is $1s^2$, and the third element Li ($Z = 3$) has the configuration of $1s^2 2s^1$ which is equivalent to $[He]2s^1$. This short cut simply tells us that the first two s-electrons of Li have the same configuration as He and the third electron moves to the higher orbital. This makes it easier to assign electronic configurations of elements with higher values of atomic numbers.

Let us now consider the elements of the Group VIII that is the home of the seven noble gases. Each of them represents the completion of the period in which they reside and the beginning of the next period. We present their electronic configuration in Table 1.9.

These elements are called the noble gases because they are to a great extent chemically inert. They represent the configurations with maximum allowable electrons in each subshell leaving no vacancy at all.

We stated already that many of the physical properties and phenomena exhibited by materials can be best explained based on the value of the valence electrons present. The tables that follow include some elements are of great interest (Table 1.10).

Notice that each of these elements have just one s-valence electrons and represents the beginning of a new group. Chemically, the alkali metals are highly reactive (Tables 1.11–1.13).

Ga and In also form very important semiconductor materials when alloyed with certain members of Group V. Al, of course, is a heavily used metal for transmission of electrical power and makes good contacts with semiconductors and dielectrics (Table 1.14).

There is large number of elements classified as transition metals, and they are found in Groups III through VIII. We include here in our table only those found in the fourth period with $Z = 22–29$. They are characterized by the occupancy of their 3-d subshell. They exhibit interesting magnetic properties. Chemically, they have multiple oxidation states (Table 1.15).

Fe, Co, and Ni are the only 3-d elements that are also strongly ferromagnetic (FM). Ti and Mn are paramagnetic (PM), Cr is antiferromagnetic (AFM), whereas the magnetic nature of V is unknown but as V_2O_5 it is

Table 1.9 Elements of Group VIII.

Elements	Helium, He	Neon, Ne	Argon, Ar	Krypton, Kr	Xenon, Xe	Radon, Rn
Atomic number, Z	2	10	18	36	54	86
Atomic electronic configuration	$1s^2$	$[He]2s^2 2p^6$	$[Ne]3s^2 3p^6$	$[Ar]3d^{10} 4s^2 4p^6$	$[Kr]4d^{10} 5s^2 5p^6$	$[Xe]4f^{14} 5d^{10} 6s^2\ 6p^6$

Table 1.10 Group I – alkali metals.

Elements	Lithium, Li	Sodium, Na	Potassium, K	Rubidium, Rb	Cesium, Cs	Francium, Fr
Atomic number, Z	3	11	19	37	55	87
Electronic configuration	$[He]2s^1$	$[Ne]3s^1$	$[Ar]4s^1$	$[Kr]5s^1$	$[Xe]6s^1$	$[Rn]7s^1$

Table 1.11 Group IIA – alkali earth metals.

Elements	Beryllium, Be	Magnesium, Mg	Calcium, Ca	Strontium, Sr	Barium, Ba	Radium, Ra
Atomic number, Z	4	12	20	38	56	88
Electronic configuration	$[He]2s^2$	$[Ne]3s^2$	$[Ar]4s^2$	$[Kr]5s^2$	$[Xe]6s^2$	$[Rn]7s^2$

Here the valence electrons are 2s electrons.

Table 1.12 Group IIB – important industrial materials.

Elements	Zinc, Zn	Cadmium, Cd	Mercury, Hg
Atomic number, *Z*	30	48	80
Electronic configuration	$[Ar]3d^{10}4s^2$	$[Kr]4d^{10}5s^2$	$[Xe]4f^{14}5d^{10}6s^2$
Significant physical properties	Industrial material when alloyed with Group VI elements, it becomes a good semiconductor. Examples: ZnS, ZnSe, and ZnTe	Industrial material when alloyed with Group VI elements, it becomes a good semiconductor. Examples: CdS, CdSe, and CdTe	First material in which superconductivity was discovered, when alloyed with Te, it becomes a semiconductor

They form alloys with Group VI elements to become good semiconductors.

Table 1.13 Group III – an important group of elements used for doping Group IV semiconductors and for making contacts.

Elements	Boron, B	Aluminum, Al	Gallium, Ga	Indium, In	Thallium, Tl
Atomic number, *Z*	5	13	31	49	81
Electronic configuration	$[He]2s^2 2p^1$	$[Ne]3s^2 3p^1$	$[Ar]3d^{10}4s^2 4p^1$	$[Kr]4d^{10}5s^2 5p^1$	$[Xe]4f^{14}5d^{10}6s^2 6p^1$

Table 1.14 Group IV – a very interesting group that begins with highly conductive element (C) and ends with one of the heaviest elements (Pb) which also becomes a superconductor at low temperatures.

Elements	Carbon, C	Silicon, Si	Germanium, Ge	Tin, Sn	Lead, Pb
Atomic number, *Z*	6	14	32	50	82
Electronic configuration	$[He]2s^2 2p^2$	$[Ne]3s^2 3p^2$	$[Ar]3d^{10}4s^2 4p^2$	$[Kr]4d^{10}5s^2 5p^2$	$[Xe]4f^{14}5d^{10}6s^2 6p^2$
Significant physical property	Highly conductive: in diamond form, it is an excellent semiconductor material	Leading elemental semiconductor: king of microelectronics	Very good elemental semiconductor material: acquires excellent properties when alloyed with silicon	An important industrial material; exhibits superconductivity at low temperatures	Heavily used in industrial products, becomes a superconductor at low temperatures: when alloyed with Titania, it becomes ferroelectric/nonlinear dielectric material

Table 1.15 Some important transition metals (found in the fourth period).

Element	Titanium, Ti	Vanadium, V	Chromium, Cr	Manganese, Mn	Iron, Fe	Cobalt, Co	Nickel, Ni	Copper, Cu
Atomic number, *Z*	22	23	24	25	26	27	28	29
Electronic configuration	$[Ar]3d^2 4s^2$	$[Ar]3d^3 4s^2$	$[Ar]3d^4 4s^2$	$[Ar]3d^5 4s^2$	$[Ar]3d^6 4s^2$	$[Ar]3d^7 4s^2$	$[Ar]3d^8 4s^2$	$[Ar]3d^{10}4s^2$
Magnetic characteristics	PM	V_2O_5 is DM	AFM	PM	FM	FM	FM	DM

diamagnetic (DM). Copper is also diamagnetic and has completely filled 3d-subshell that excludes the presence of ferromagnetism. All other members of the transition metal period have partially filled 3d-subshell.

Rare earth elements: These are also transition elements and are found in the sixth period. They are also called the lanthanides. Some examples of rare earths are cerium, Ce ($Z = 58$); neodymium, Nd ($Z = 60$);

samarium, Sm ($Z = 62$); europium, Eu ($Z = 63$); and gadolinium, Gd ($Z = 64$).

Radioactive elements: Elements of the seventh period are called the actinides or radioactive elements. They are also classified as transition elements. Some of these include the following thorium, Th ($Z = 90$); uranium, U ($Z = 92$); plutonium, Pu ($Z = 94$), and americium, Am ($Z = 95$).

1.9 Some Important Concepts of Solid-State Physics

1.9.1 Ceramic Superconductivity

The superconductivity was discovered in 1911 at the University of Leiden in the Netherlands by Heike Kamerlingh Onnes. He was awarded the Nobel Prize in Physics in 1913 for the production of very low temperatures. It was in mercury that he found the resistance became nonexistent when cooled to 4.2 K. He also discovered superconductivity properties in lead and tin. The table that follows lists the ground-breaking advancement in the field of superconductivity since its discovery. Since 1911 until 1987 superconductivity was found mostly in metallic systems at very low temperatures. It was in 1987 that superconductivity was observed for the first time in ceramic compounds at relatively high temperatures. This landmark discovery dramatically changed the field of superconductivity from being a curiosity of fundamental science to be of great importance to technology. Some of these issues we discuss in this section. The nature of this book does not allow us to discuss this topic in detail. However, interested readers may wish to consult a good book on superconductivity for advanced studies; we recommend the book by Orlando and Delin [2]. This field has produced a number of Nobel laureates. We list them in Table 1.16.

As already stated until 1985, superconductivity was observed only in metals and their alloys with the uppermost critical temperature of 20 K. Then it increased to 35 K with the discovery of superconductivity in a ceramic sample of Ba–La–Cu-oxide. The discovery was made in January of 1986 at IBM Zurich Laboratories by Georg Bednorz and K. Alex Müller who were awarded Nobel Prize in Physics in 1987. This landmark discovery was a paradigm shift in solid-state physics for two reasons: first, the critical point for superconductivity crossed the boiling point of Ne (27 K ≈ −246 °C) and second, the superconductivity was found in a ceramic system against all prevailing concepts of physics at the time. By now many more oxide superconductors have been discovered, and we list some of them in Table 1.17.

It is interesting that the crystal structure of these oxides happen to be perovskite (ABO_3), which is the leading group in which prominent nonlinear dielectrics such as ferroelectrics are found.

So far as an explanation of this interesting physical phenomenon is concerned, there is only one unified theory that can explain superconductivity and even that is not

Table 1.16 List of Nobel Prize in Physics awarded for superconductivity.

Name	Year	Contribution
Heike Kamerlingh Onnes	1913	Discovery of superconductivity
John Bardeen, Leon N. Cooper, and J. Robert Schrieffer	1972	BCS theory of superconductivity
Leo Esaki, Ivar Giaever, and Brian D. Josephson	1973	Josephson tunneling effect
Georg Bednorz and K. Alex Müller	1987	High temperature superconductivity (ceramic superconductivity)
Alexi A. Abrikosov, Vitaly L. Ginzburg, and Anthony J. Leggett	2003	Theory of superconductivity and superfluids

Table 1.17 List of some ceramic superconductor materials.

Ceramic superconductors	Critical point, T_C (K)	Number of Cu–O planes/unit cell	Crystal structure	Crystal unit cell
123 YBCO ($YaBa_2Cu_3O_7$)	93	2	Perovskite	Orthorhombic
2122 BCSCO ($Bi_2CaSr_2Cu_2O_8$)	85	2	Perovskite	Tetragonal
2223 BCSCO ($Bi_2Ca_2Sr_2Cu_3O_{10}$)	110	3	Perovskite	Tetragonal
2223 TCBCO ($Tl_2Ca_2Ba_2Cu_3O_{10}$)	125	3	Perovskite	Tetragonal
1223 HBCCO ($HgBa_2Ca_2Cu_3O_8$)	134	3	Perovskite	Tetragonal

Note all critical points are above the liquid nitrogen temperature of 77 K (−196 °C).
Source: https://en.m.wikipedia.org/wiki/Hightemperature_Superconductivity.

adequate to handle the superconductivity found in oxide systems. In 1972, almost 60 years after the discovery of superconductivity, a macroscopic theory was developed by three American Physicists named John Bardeen, Leon Copper, and J. Robert Schrieffer that has been successful in explaining the superconductivity found in metallic systems at low temperatures. The theory is also known as BCS theory, and the three physicists were awarded Nobel Prize in Physics in 1972 for developing this elegant theory. This was the second Nobel Prize for John Bardeen, the first one was for the discovery of transistors in 1956.

The BCS theory requires a sound knowledge of advanced physics and therefore is beyond the scope of this textbook. The central point of this theory is the concept of so-called Copper pairs. In the superconducting phase Cooper pairs can form when two electrons couple with antiparallel spins. Cooper pairs can behave very differently than single electrons that must obey the *Pauli Exclusion Principle*, whereas the Cooper pairs behave more like *bosons* that can condense in the same energy levels. The Copper pairs are also called superconducting electrons. Though the BCS theory has been very successful in explaining conventional superconductivity satisfactorily, it appears not to be applicable to ceramic superconductivity. Various groups of theoreticians are currently working on this problem, and we hope one day soon we might have a good theory of superconductivity found in electroceramics.

1.9.2 Superconductivity and Technology

Superconductivity is a unique physical phenomenon poised to play a vital role in the evolution of new technology. A large number of devices and applications have been proposed based on magnetic and electronic properties of superconductivity that we will discuss subsequently in this chapter. Production of very high magnetic fields, Josephson junctions, and superconducting quantum interference device (SQUID) magnetometers are so far the most established technologies based on low temperature superconducting materials. They operate only at cryogenic temperatures. However, a large number of applications, from highly sophisticated to straightforward and simple, have been proposed based on ceramic superconductors with critical temperatures far above cryogenic temperatures. For these proposals to be more useful, and commercially viable, room-temperature superconducting materials have to be discovered. The hope is pinned on new ceramic materials because it will be possible to produce them in high volumes and in high quality at reasonable costs.

Superconducting magnets can produce fields far greater than those generated by the most powerful electromagnets. Currently, the highest sustained magnetic fields achieved are about 8.3 T ($=8.3\times10^5$ G) by niobium–titanium (Nb–Ti) superconducting magnets that operate at the extremely low temperature of 1.9 K. The magnetic fields are measured in the units of Tesla and Gauss, and they are abbreviated as T and G, respectively. Superconducting magnets are universally used in magnetic resonance imaging (MRI) machines that is a powerful diagnostic tool indispensable to health professionals as well as of great significance to scientists for new discoveries. One of the most intriguing aspects of these magnets is the onset of *persistent currents*. Once the magnet is energized the windings of Nb–Ti become superconducting closed loops at about 1.9 K giving rise to a persistent current following the Faraday's law of induction. This law states that a magnetic field can induce a current in a conducting loop according to $L\left(\frac{\mathrm{d}I}{\mathrm{d}t}\right) = -a\left(\frac{\mathrm{d}B}{\mathrm{d}t}\right)$ where L is the inductance, I the current, and B the magnetic flux. The current generated in a superconducting loop can flow for months even in the absence of an external magnetic field. At this point, the external power supply can be turned off, and the magnetic field is sustained by the persistent current. We will learn more about it later while studying the magnetic properties of superconducting materials.

Another very powerful superconducting device is the SQUID magnetometer based on the Josephson junction effect that establishes the tunneling of the Copper pairs. There are many other unique applications based on this effect, and we will discuss some of them when we study the Josephson effect.

1.10 Signature Properties of Superconductors

Temperature, magnetic field, and pressure are three external agents that can greatly alter the fundamental nature of a superconducting material by switching them from normal phase to superconducting phase, and vice versa. Infinite electrical conductivity and the onset of diamagnetism below a critical temperature are the two most important properties of a superconductor material. We discuss both these properties in the next sections here. It is known that a large number of elements of the periodic table become superconducting at a critical temperature and a critical pressure. Recently, it has been reported that H_3S becomes superconducting with a critical temperature (T_C) of 203 K at 150 GPa of pressure.[2] This is the highest transition point reported so far for

2 Reported in *Physics Today* of July 2016 about this discovery made by Mikhail Eremetes and his team at the Max Planck Institute of Chemistry in Germany.

any superconducting material. H_3S belongs to the same family of chemicals as hydrogen sulfide (H_2S) which is present in almost any chemistry laboratory and has a noxious smell.

1.10.1 Thermal Behavior of Resistivity of a Superconductor

Superconductivity can be defined most simply by stating that below a critical temperature a superconductor completely loses its resistivity resulting in the conductivity to be infinite. The *critical temperature* is also known as *critical point* or *superconducting transition temperature* or even *superconducting transition point*. Its universal symbol is T_C. The temperature dependence of resistance of a superconductor material is presented in Figure 1.8.

Here we find two distinct phases to exist: one above the critical point, T_C, and the other below this point. As we can see from the figure, a superconducting material goes from its normal state at $T > T_C$ to its superconducting state at $T < T_C$ undergoing a phase change at $T = T_C$. The transition from the normal state to the superconducting state is governed by the relationship described by Eq. (1.47).

$$\sigma \approx \frac{C}{T - T_C} \tag{1.47}$$

When $T = T_C$ the conductivity, σ, is infinite. Thermodynamically, it is a phase change of the second order. Equation (1.47) is the standard form of the Curie–Weiss law that is obeyed by ferromagnetic and ferroelectric materials where we also encounter similar phase transitions at their respective critical temperatures called the Curie temperatures, T_C. We will learn about ferroelectricity and ferromagnetism in the other chapters in this book.

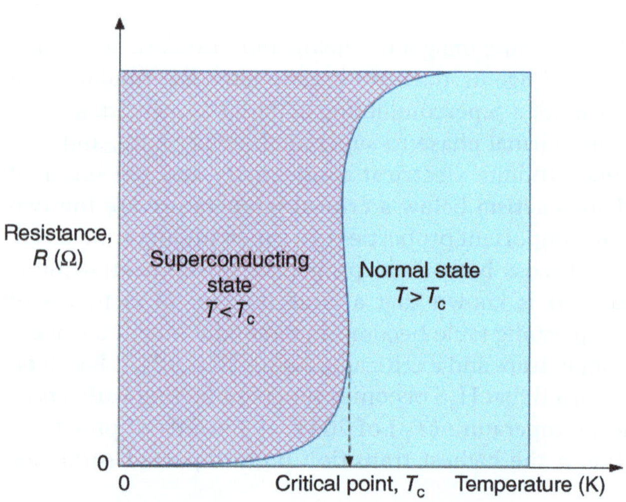

Figure 1.8 Resistivity as a function of temperature for a superconducting material.

The state of infinite conductivity persists so long as the temperature is below the critical point of T_C. It is a significant result and obviously of great importance to power transmission technology. All metals lose part of the original electric power by Joule heating ($\approx I^2R$) due to a nonzero resistance. Over the course of time, this loss can be significant. The hope is that one day we will be able to use superconducting wires instead of metallic wires to transmit electric power from one point to another and thereby completely eliminate the loss of power.

We can easily imagine the benefit of such a system to consumers and industry. This dream can only be fulfilled if a room temperature superconducting material is discovered. Today, we are not there, but science is working diligently to reach that goal post. The discovery of ceramic superconductor has given us a roadmap to reach this destination and the race is intense.

1.10.2 Magnetic Nature of Superconductivity: Meissner–Ochsenfeld Effect

We concluded in the previous section that all superconductors become ideal conductors ($\sigma \to \infty$) below their critical temperatures. They also become ideal diamagnetic materials (magnetic susceptibility, $\chi = -1$) in the superconducting phase. This remarkable property was confirmed by the Meissner–Ochsenfeld effect, which is referred erroneously in many circles only as the Meissner effect. The Meissner–Ochsenfeld effect was discovered in Germany in 1933 by Walther Meissner and Robert Ochsenfeld. It states that in the superconducting phase, the magnetic flux lines are expelled after it reaches a certain depth below the surface. This is called the London penetration depth and is associated with the superconducting electron density. It decays exponentially within the interior of the superconducting phase. Once the London penetration depth is reached, the magnetic field stops penetrating the sample, which then becomes an ideal diamagnetic material. Diamagnetism is characterized by the capacity of a material to oppose the penetration of magnetic flux lines below its surface. This is in contrast to the nature of a ferromagnetic or a paramagnetic material in which magnetic flux lines can penetrate unopposed. The Meissner–Ochsenfeld effect cannot be explained by classical physics, and the explanation is provided by quantum mechanics establishing the fact that the superconductivity is a quantum mechanical phenomenon. Mathematically, the London-penetrating depth is given by

$$\lambda_L = \left(\frac{\epsilon_0 m_e c^2}{ne^2} \right)^{\frac{1}{2}} \tag{1.48}$$

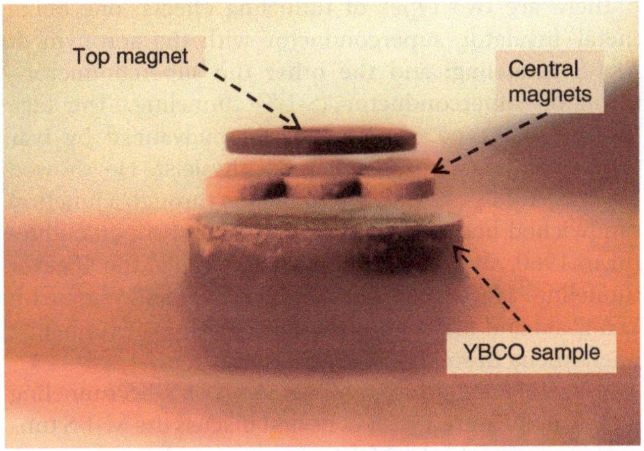

Figure 1.9 Magnetic double levitation by 123 YBCO in superconducting state.

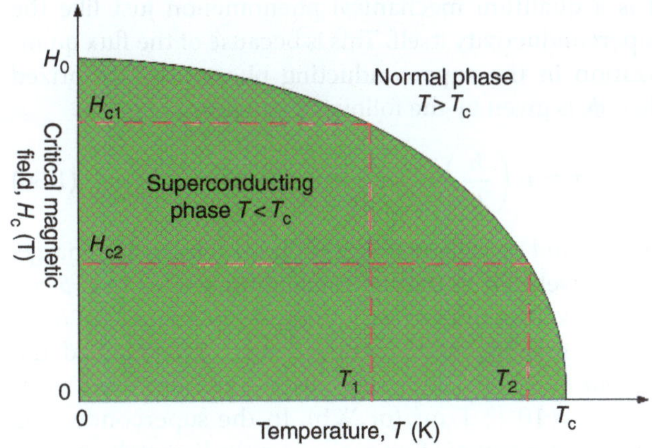

Figure 1.10 Temperature dependence of critical magnetic field in a superconductor material.

where n is the superconducting electron density. The parameters ϵ_0, c, m_e, and e are physical constants as defined already above.

Another critical property associated with superconductivity is called the coherence length. It is related to the energy gap and the Fermi velocity of the superconducting phase. It bears no relationship with the London penetration depth. Both the coherence length and the London penetration depth have been experimentally determined for a number of superconducting materials. Its magnitude usually is in the nanometers range.

The Meissner–Ochsenfeld effect can be demonstrated easily by cooling a 123 YBCO ceramic sample below its critical point using liquid nitrogen (≈77 K).

Once superconductivity sets in, the sample becomes diamagnetic facilitating the levitation of a magnet in air above the sample surface, which can be seen in Figure 1.9. As expected, the magnetic flux lines are expelled resulting in the magnetic levitation. In Figure 1.9, the double levitation was achieved by floating a ceramic magnet on top of another magnet.[3]

It is not only the temperature that defines the superconducting phase. Also under the influence of an external magnetic field, H, the superconducting phase can switch to the normal phase. The magnitude of the field that can induce this phase change in called the critical magnetic field (H_C), and it is found to be strongly temperature-dependent as we can see from Figure 1.10. Here, each combination of H_{C1}-T_1 and H_{C2}-T_2 is capable of inducing switching of the superconducting phase to the normal phase.

At absolute zero, the critical magnetic field, H_C, reaches its maximum value of (H_0). Experimentally, it has been established by studying a large array of superconducting materials that the temperature-dependence of the critical magnetic field is described by Eq. (1.49).

$$H_C = H_0\left(1 - \left(\frac{T}{T_C}\right)^2\right) \tag{1.49}$$

We can conclude from Figure 1.10 that at any temperature below the superconducting critical temperature, T_C, the superconducting phase can be destroyed by the application of a magnetic field greater than the critical field, H_C. This observation plays an important role in designing a superconducting magnet that can produce very large fields without being plagued by any loss of power as is the case for normal electromagnets using metallic wires.

Exercise 1.3
Find the temperature at which the superconducting phase will switch to the normal phase if the needed critical field, H_C, is 80% of the maximum field, H_0 and the critical temperature of the superconducting material is 90 K.

Solution
From Figure 1.10, we can infer that $H_C < H_0$ and $T < T_C$ in Eq. (1.49). By substituting the values of the parameter given in the problem in this equation, we find that: $\frac{T}{90} = \sqrt{0.2}$. That is, $T = 40.5$ K.

Another very interesting result of the Meissner–Ochsenfeld effect is the fact that a *persistent current* can be produced in the superconducting phase that can last for months even in the absence of an applied magnetic field. It is estimated that it can last for approximately 10^5 years. This phenomenon cannot be understood with the help of the electromagnetic theory. It is found that

3 R.K. Pandey, unpublished results. This experiment was done in the Electronic Materials Labs of Electrical and Computer engineering department at Texas A&M University circa 1989.

it is a quantum mechanical phenomenon just like the superconductivity itself. This is because of the flux quantization in the superconducting phase. The quantized flux, ϕ, is given by the following equation:

$$\phi \approx n\left(\frac{h}{2e}\right) \quad \text{with } n = 1, 2, 3, \ldots \tag{1.50}$$

Here, h and e are Planck's constant and electronic charge, respectively. From the above equation, we see that ϕ can assume values only as $\phi_1 = h/2e$, $\phi_2 = h/e$, $\phi_3 = 3h/2e$, and so on. The first quantized flux, ϕ_1, is called the *fluxoid* and normally written as ϕ_0. Its numerical value is 2.068×10^{-15} T m^2 (or Wb). In the superconducting phase, the magnetic flux can exist only discretely in units of the fluxoid. This is a remarkable result originating from the Meissner–Ochsenfeld effect. Quantization of magnetic flux is another unique property of a superconducting material and is an important consideration in designing a superconducting magnet. It also plays an important role in finding applications of the Josephson junction in high-speed data transfer. We will describe the Josephson effect in the next section.

1.10.3 Josephson Effect

It was in 1962 that Brain David Josephson of Great Britain predicted theoretically the tunneling of Copper pairs in a junction sandwiched between two superconducting arms. This is a macroscopic quantum phenomenon universally associated with the onset of a current because of tunneling through a junction by the Copper pairs. The uniqueness of the effect lies in the fact that the tunneling current flows even when no potential is applied to the junction. One should be careful in not confusing the tunneling current associated with Copper pairs with the *persistent current* we discussed with respect to superconducting magnets.

The Josephson junction is also known as a *weak link* in engineering. The Josephson effect is the basis for such practical and very important devices as SQUID magnetometers, superconducting qubits, rapid single flux quantum (RSFQ) digital electronics that can operate at very high speeds with minimal of power consumption, oscillators, and voltage calibrators. The standard for 1 V is based on the Josephson effect. Apparently NIST produces this standard by connecting thousands of Josephson junctions in series. Josephson was awarded the Nobel Prize in Physics in 1973 for his landmark prediction of the tunneling by Copper pairs. Electron tunneling is one of the landmark contributions of quantum mechanics. It is based on the fundamental idea of the matter–wave nature of electrons as predicted by de Broglie.

There are two types of tunneling effects: one being metal–insulator–superconductor with the acronym of M-I-S tunneling; and the other the superconductor–insulator–superconductor (S-I-S) tunneling. The idea for the first type of tunneling was advanced by Ivar Giaever, a Norwegian-American physicist. He showed experimentally the electron tunneling through a junction sandwiched between a normal metal and superconductor in 1960, and so it is appropriately called the Giaever tunneling. He shared the Nobel Prize in Physics with Leo Esaki and Brian Josephson in 1973 for the tunneling phenomena in solids.

A typical configuration for an M-I-S or S-I-S tunneling is shown in Figure 1.11. Let us first discuss the M-I-S tunneling process. In the M-I-S experiment, the upper sample is a normal metal. When a current, I, is injected in the junction, its transport first follows the path from a metal to an insulator. If the insulation layer is thick, no voltage drop will occur in the configuration. If it is, however, between 1 and 2 nm, then the electrons can acquire sufficient energy to overcome the barrier and reach the superconducting part of the structure and then finally exit it to complete the circuit. If the superconductor is in the normal phase ($T > T_C$), then the I–V relationship will be governed by the Ohm's law as shown in Figure 1.12 by the dashed line.

But when the superconducting arm is kept at $T < T_C$, then the electron tunneling will happen manifesting itself in the appearance of a tunneling current at the apparent potential equivalent to $\left(\frac{\Delta}{e}\right)$ where Δ is the bandgap of the superconductor and e the electron charge. The structure will remain in this state as long as $T < T_C$. As soon as the superconductor reverts back to its normal phase, the tunneling current will disappear and we will be left with the ohmic I–V characteristics of the device.

Now let us discuss the S-I-S tunneling that is the backbone of the Josephson effect. If we replace the metal in the upper arm of the structure of Figure 1.11 with a superconducting material that is identical to the superconducting material in the lower arm of the configuration, then

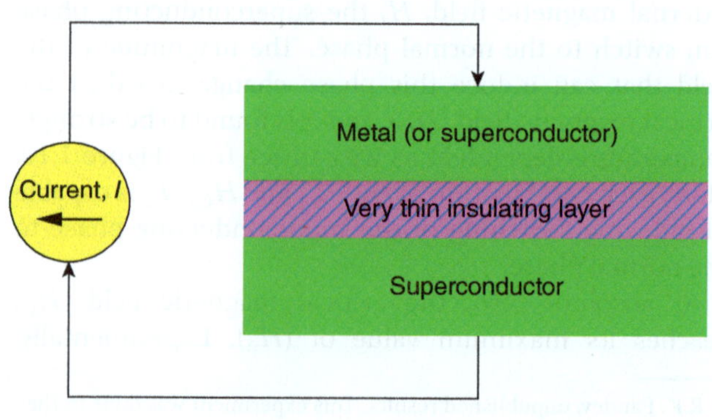

Figure 1.11 Configuration of M-I-S or S-I-S tunneling junctions.

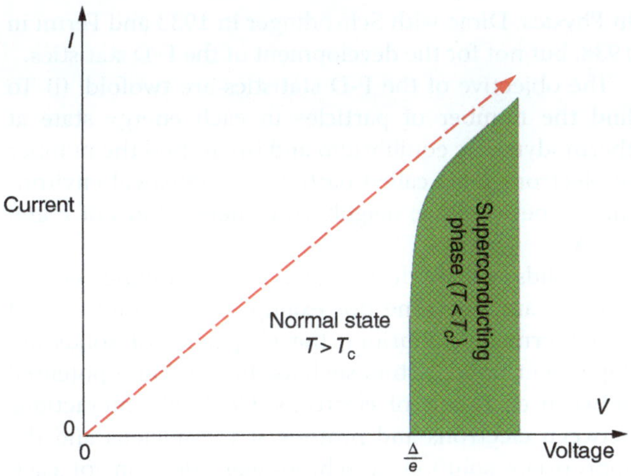

Figure 1.12 *I–V* characteristics of Giaever junction consisting of S-I-M configuration.

we will have a Josephson junction. Once again, we apply a current to the structure, but this time we do not have normal electrons available to give us as the characteristic ohmic plot. Noting happens no matter how large a current we inject so long as the structure is at $T > T_C$. As soon as the samples are cooled to $T < T_C$ we notice the appearance of a current, $\pm I_c$, even when $V \approx 0$ which is a situation unique to superconductivity because there is no resistance present in the superconducting state and therefore ohmic current does not exist. This is shown in Figure 1.13.

The current that develops at $V \approx 0$ is the result of the Cooper pair tunneling through the very thin insulating barrier. This current is called the Josephson current; it

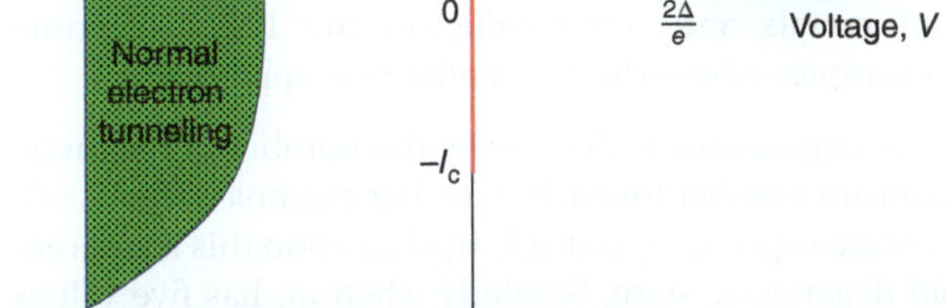

Figure 1.13 *I–V* characteristics of Josephson junction consisting of S-I-S configuration.

was he who on the basis of the famous BCS (Bardeen, Copper, and Schrieffer) theory of superconductivity predicted that the probability of Copper pairs tunneling through a barrier is the same as that of an electron. It is an ordered coherent process in which a macroscopic wave function travels from one superconductor to the other. The experimental proof of the Josephson junction was provided by Phillip Anderson and John Rowell, both of the USA, in 1963.

But what are the Cooper pairs? In the BCS theory, Leon Cooper argued that in the superconducting phase, two electrons with anti-parallel spin bind loosely to form a pair and move with the same speed, but in opposite directions. This assumption is the core of the BCS theory which successfully explained the quantum nature of superconductivity.

Now back to the Josephson junction story. After the development of the Josephson current, the Cooper pairs wander until they gain sufficient energy to overcome the bandgap, Δ. At that point, the binding energy of the Cooper pairs breaks down releasing the normal electrons to travel through the barrier. As soon as $V \approx \frac{2\Delta}{e}$, the current appears, which is the result of the normal electron tunneling while the structure is maintained in the superconducting phase by keeping the temperature $T < T_C$. The tunneling occurs at $V = \pm\frac{2\Delta}{e}$ positions as shown in Figure 1.13.

We now need to understand the very nature of the Josephson current, $\pm I_c$. We know that this is the current that develops even when there is no potential applied to the circuit. But it has one limitation; it cannot keep on increasing in magnitude uncontrolled. As soon as it reaches a critical value, it stops increasing and a voltage appears at the junction. The structure switches from a state of zero voltage to a state of a finite voltage. This is the basis for high-speed digital electronics with superior attributes than those found in the best of semiconductor devices that dominate the present-day field of microelectronics. A Josephson junction can be manipulated by the applications of a potential, both DC and AC, as well as by a magnetic field resulting in some interesting applications. Let us briefly examine some of these cases.

Case I: Application of a DC potential

If the potential is of such a magnitude that it exceeds the value of $\left(\frac{2\Delta}{e}\right)$, the Cooper pairs will disintegrate into two normal electrons giving rise to the tunneling effect as shown in Figure 1.13. However, if the equivalent energy, $E \leq \frac{2\Delta}{e}$, the Cooper pairs will be retained intact but will begin to oscillate back and forth emitting electromagnetic waves with a specified frequency. This is the basis for fabricating oscillators that are extremely accurate and can be integrated in many applications.

Case II: Application of an AC potential
If instead of a DC potential, we now subject the Josephson junction to an electromagnetic field, then we will naturally induce an AC potential across the junction. Since frequency can be measured more accurately than a voltage, AC modulated Josephson junctions are used for producing very accurate voltmeters, some of which are used for calibrating other voltmeters.

Case III: Application of an external magnetic field
The Josephson junctions are extremely sensitive to even the smallest magnetic field applied to it. This unique response to the magnetic field is exploited in the design and construction of the world's most sensitive magnetometer universally known as the SQUID. They are so sensitive that they can even detect the feeble and very weak magnetic fields present in organs such as the human heart (10^{-10} T ≈ 0.1 μG) and brain (10^{-13} T ≈ 10 nG). These highly sensitive magnetometers are indispensable tools for scientific studies requiring determination of extremely small magnetic fields. The magnetic response of the Josephson junction resulted in the discovery of quantized magnetic flux that is defined in Eq. (1.50). Some caution should be exercised when referring to the Bohr magneton, μ_B, and quantized magnetic fluxoid, ϕ_0. Bohr magneton refers to the magnetic moment of an electron having a value of 9.274×10^{-21} J T^{-1}, whereas the fluxoid is the magnetic flux (B) at the quantized level corresponding to the first principal quantum number (n = 1). Both are physical constants.

1.11 Fermi–Dirac Distribution Function

The F-D distribution function or F-D-statistics is a beautiful piece of theoretical work and impacts solid-state physics and solid-state electronics in a very significant way. It helps us in understanding how the population and depopulation of quantum energy states vary with temperature which becomes instrumental in understanding the properties of conductors and semiconductors at a thermal equilibrium. Before this powerful theory was published in 1926 first by Enrico Fermi of Italy and soon after that by Paul Dirac of England, it was not possible to fully understand the contributions made by electrons in the specific heat of solid and magnetic susceptibility. Classical physics was not at all in the position to solve these physical properties of solids. The picture changed drastically once the Pauli exclusion principle became the accepted fact of life from 1925 onward. Fermi and Dirac applied the restrictions imposed on electrons by the Pauli exclusion principle and were successful in developing the F-D statistics. Both received the Nobel Prize in Physics, Dirac with Schrödinger in 1933 and Fermi in 1938, but not for the development of the F-D statistics.

The objective of the F-D statistics are twofold: (i) To find the number of particles in each energy state at thermodynamic equilibrium and (ii) To find the number of electrons (also called particles in statistical environments) between the neighboring energy states of E and $E + \Delta E$.

In solids, we are dealing with a large number of particles at any one time (for example, $\approx 10^{26}$ mol^{-1}), and at a thermal equilibrium, the properties of solids are dependent upon factors such as the Coulomb potential between each pair of electrons, Coulomb interactions between electrons and protons at the nucleus and the appropriate solution of Schrödinger equation for each set under consideration. We can easily visualize what a daunting task it would be without resorting to statistical methods.

There are two statistical models that are useful when dealing with solids: (i) The classical model of Maxwell–Boltzmann statistics and (ii) the F-D distribution of quantum statistics. Now let us examine the fundamental difference between the particles dealt with by each of these two types of statistical models.

The classical model of Maxwell–Boltzmann: Here, the particles are assumed to be atoms, ions, and molecules. They are distinguishable from each other because they do not interact mutually, and their energy is continuous. All phenomena dealing with the classical particles at thermal equilibrium can be explained satisfactorily by the classical model.

The quantum mechanical model of F-D: This model is applicable to electrons that obey the Pauli exclusion principle with spin being $\pm\frac{1}{2}$. They are indistinguishable from each other because they can electrostatically interact mutually. Since they are subjected to restrictions imposed by the principles of quantum mechanics, their energy states must be considered degenerate which means that they are not continuous but discrete. The electrons that obey the F-D-distribution function are called *fermions*. It was Dirac himself who coined the name *fermions*; apparently in honor of Enrico Fermi. Thermal properties of metals and semiconductors are the two prime examples where the F-D statistics is applicable.

The degeneracy is defined by the number of magnetic quantum number for each state. For example, when $l = 1$, the state is p and m_l is 0, ±1. By definition this is a threefold degenerate state. Similarly when m_l has five values (for $l = 2$ and state being d) the degeneracy is fivefold. By inference, then when $l = 0$, the state is s, and there are no corresponding values for m_l; then the energy is said to be single-fold degenerate.

Besides the F-D statistics, there is another quantum statistics called the Bose–Einstein distribution function. Here the particles are also indistinguishable, but with integral values of spin. These particles are called *bosons*. This model is used for those cases where bosons are the particles. Atomic and nuclear physics as well as chemistry take full advantage of this model.

Now let us try to make some mathematical distinction between these three statistical models. For this, we consider that there are N_i particles that must be distributed in E_i energy states at the thermal equilibrium with temperature T.

For classical particles that obey the M-B statistics, the formulation of Eq. (1.51) is used.

$$N_i = A \exp\left(-\frac{E_i}{k_B T}\right) \tag{1.51}$$

where A is a constant and k_B the Boltzmann constant.

Generalization of the above expression can be done by using the concept of probability, $f(E)$, which simply tells us the statistical probability of finding a particle with energy, E. Then we can convert the above equation into Eq. (1.52).

$$f(E) = A \exp\left(-\frac{E}{k_B T}\right) \approx \frac{A}{\exp\left(\frac{E}{k_B T}\right)} \tag{1.52}$$

When, however, degeneracy is present as is the case with quantum statistics, we can express the probability simply by defining the probability as: $f(E) \approx \left(\frac{N_i}{g_i}\right)$. The probability functions for the two quantum statistics can then be expressed as in Eqs. (1.53) and (1.54).

$$f(E) = \frac{1}{A \exp\left(\frac{E}{k_B T}\right) + 1} \quad \text{for F-D statistics} \tag{1.53}$$

and

$$f(E) = \frac{1}{A \exp\left(\frac{E}{k_B T}\right) - 1} \quad \text{for B-E statistics} \tag{1.54}$$

Having made the distinctions between the three types of statistical models, we need to concentrate now on the F-D statistics that is of paramount importance to our needs. Let us consider once again that there are N_i electrons to be distributed in E_i available energy states having degeneracy of g_i when the system is in thermodynamic equilibrium at temperature, T. Then the general formulation of the F-D statistics is described by the expression of Eq. (1.55).

$$N_i = \frac{A g_i \exp\left(\frac{-E_i}{k_B T}\right)}{1 + A \exp\left(\frac{E_i}{k_B T}\right)} \tag{1.55}$$

Let us at this point introduce two new terms, the Fermi function, $F(E)$ which is nothing else than a substitute for the general term probability, $f(E) = (N_i/g_i)$ and the Fermi energy, E_F. We are using the term Fermi function, $F(E)$, and not the general term of probability, $f(E)$, just to be consistent with the convention. By setting $A = \exp\left(\frac{E_F}{k_B T}\right)$, and after some rearrangement, we can rewrite Eq. (1.55) as Eq. (1.56).

$$F(E) = \frac{1}{\left(\exp\frac{(E_i - E_F)}{k_B T}\right) + 1} \tag{1.56}$$

What is the Fermi energy (E_F) but what exactly does it mean? While dealing with semiconductor materials we constantly encounter the terms Fermi energy and Fermi level, and it is not uncommon to confuse one with the other. It is important to distinguish between them for the sake of accuracy.

Fermi energy is defined as the difference in kinetic energy that exists between the highest and the lowest occupied quantum states at absolute zero temperature ($T = 0$ K). In general, the lowest occupied state is defined as the state with zero kinetic energy. In semiconductors, the lowest energy state corresponds to the top of the valence band, whereas for metals it is the bottom of the conduction band. The Fermi energy is the kinetic energy, whereas the Fermi level is the sum of both the kinetic energy and the potential energy. Furthermore, Fermi energy is always defined with respect to absolute zero temperature, whereas the Fermi level can be referred to at any temperature. In contrast to the Fermi energy, the Fermi level for a metal at absolute zero corresponds to the highest occupied state.

Let us now go back to Eq. (1.56) and interpret its importance. We discuss three specific conditions which are the following:

Case I: When $T = 0$ K and $E_i < E_F$, the exponential term in the denominator becomes ≈ 0 making $F(E) = 1$. That means that all energy states are fully occupied below the Fermi energy.

Case II: When $T = 0$ K and $E_i > E_F$. The exponential term now approaches ∞ rendering $F(E) = 0$. This means that all energy states above E_F are totally empty.

Case III: If $E_i = E_F$, then $F(E) = \frac{1}{2}$. This means that the probability of occupancy at $E_i = E_F$ is always 50% no matter what the temperature at equilibrium might be.

Case IV: At high temperatures, $E_i - E_F \ll k_B T$. Then Eq. (1.56) reduces to

$$F(E) = \exp\left\{-\left(\frac{E_i - E_F}{k_B T}\right)\right\} \approx \exp\left(-\frac{\Delta E}{k_B T}\right) \quad (1.57)$$

We can easily recognize it as another way for writing the M-B distribution function (see Eq. (1.52)). We infer from this that at sufficiently large thermal energies, the F-D statistics yields the same result as the M-B statistics. This then means that at elevated thermal equilibrium, electrons can be statistically described by the M-B distribution just as we would do for the atoms, ions, and molecules.

Case V: For very low temperatures but above $T = 0\,\text{K}$, if $E_F - E_i \gg k_B T$, then the Fermi function is given by

$$F(E) \approx 1 - \exp\left(\frac{E_i - E_F}{k_B T}\right) \approx 1 \quad (1.58)$$

This simply means that even at very low temperatures, the probability of occupation is 100%, which is the same as at $T = 0\,\text{K}$ (see Case I above).

Figure 1.14 is the typical plot for the F-D distribution. This, in fact, gives us the graphical picture of all the five cases we just discussed. Notice that the transfer of fermions from the upper half of the curve to the lower half occurs when $T > 0\,\text{K}$ and passes through the 50% point of the $F(E)$ axis. The population-depopulation of energy states must go through the $\frac{1}{2}$-point. This situation will reproduce itself time, and again, as the temperature rises and more and more electrons migrate from the upper half of the plot to its lower half. The probability of 100% occupancy of all available energy states is guaranteed at absolute zero. The Fermi energy, E_F, can now be defined also as the cut-off point between the populated and depopulated energy states.

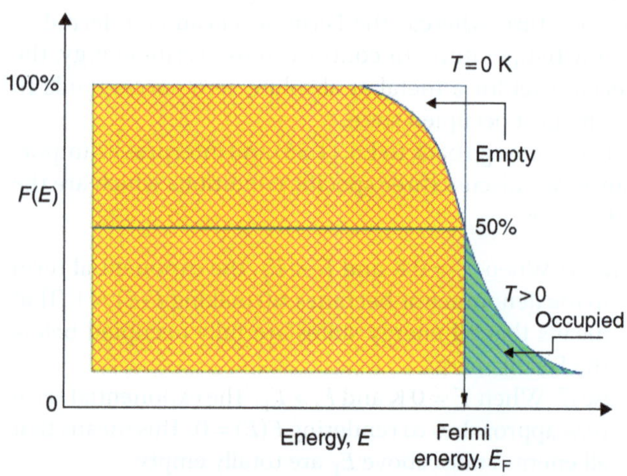

Figure 1.14 Fermi–Dirac distribution plot: Fermi function vs. energy at $T = 0$ and $T > 0$.

We have so far discussed the conditions for population and depopulation of energy states according to the F-D statistics. Now we need to examine: (i) How many energy states might be there between the energy levels E and $E + \text{d}E$ and (ii) What is the quantitative nature of the Fermi energy, E_F.

The technical term for the number of energy states found between the E and $E+\text{d}E$ levels is *density of states*, $Z(E)\text{d}E$. The calculation to find a mathematical expression for $Z(E)$ is quite involved. We shall leave it for Chapter 7, where we will study the essential elements of semiconductors. The concept of density of states plays an important role in understanding the physical principles involved on semiconductor devices and therefore it will be more beneficial to deal with this topic there. For the time being, let us just give its mathematical formulation which is shown in Eq. (1.59).

$$Z(E)\text{d}E = \left[\frac{4\pi V(2m_e)^{\frac{3}{2}}}{h^3}\right] \cdot E^{\frac{1}{2}} \cdot \text{d}E = A \cdot E^{\frac{1}{2}} \cdot \text{d}E \quad (1.59)$$

where V = the volume with N number of electrons contained therein, m_e and h being the electron mass and the Planck constant, respectively. A is a constant containing all the parameters under the square brackets. The numerical expression for the Fermi energy can be derived from the density of state, $Z(E)\text{d}E$ and is given by Eq. (1.60).

$$E_F \approx \left(\frac{h^2}{8m_e}\right) \cdot \left(\frac{3N}{\pi V}\right)^{\frac{2}{3}} \approx \frac{h^2}{8m_e}\left(\frac{3n}{\pi}\right)^{\frac{2}{3}} \quad (1.60)$$

Here n = number of electrons per unit volume.

The temperature corresponding to the Fermi energy is called the Fermi temperature, T_F; and the velocity with which the electrons travel at the Fermi energy is called the Fermi velocity, V_F. The Fermi temperature is defined as $T_F \approx \left(\frac{E_F}{k_B}\right)$ and the Fermi velocity as $V_F \approx \left(\frac{p_F}{m_e}\right)$, where p_F is the Fermi momentum which is equal to $\sqrt{2m_e E_F}$. The values of these parameters for some common metals is given in Table 1.18 [3].

Exercise 1.4

Consider a system with fivefold degeneracy and number of electrons to distribute in these energy states are only 3. Find the permissible distribution function, Ω.

Solution

For the first electrons, there are five choices available; for the second there are four choices, and for the third there are three choices. The distribution function is then:

$$\Omega = \frac{5.4.3}{3!} = \frac{5!}{3!2!} = 10$$

Table 1.18 Fermi energy, Fermi temperature, and Fermi velocity for some common metals [3].

Metal	E_F, Fermi energy (eV)	T_F, Fermi temperature $\times 10^4$ (K)	V_F, Fermi velocity $\times 10^6$ (m S^{-1})
Na	3.24	3.77	1.07
Cu	7.00	8.16	1.57
Ag	5.49	6.38	1.39
Au	5.53	6.42	1.40
Al	11.7	13.6	2.03

Exercise 1.5

Consider the case of sodium with atomic weight is 23, density equal to 0.968 g cm^{-3}, and the electronic configuration [Ne]3s. Calculate its Fermi energy, Fermi temperature, and Fermi velocity.

Solution

The Fermi energy can be found using Eq. (1.60) and from its value then we can calculate the other two parameters. The electronic configuration tells us that Na is monovalent. Therefore, the number of atoms per unit volume is simply given by $n = \frac{dN_A}{A}$, where d is the density, N_A = Avogadro's number $= 6.0 \times 10^{23}$ per atomic weight, A (here A is grams per mole). Substituting for these parameters in Eq. (1.60) we get

$$n = \frac{dN_A}{A} = \frac{(0.968 \times 6.02 \times 10^{23})}{23}$$
$$\approx 2.54 \times 10^{21}\ \text{cm}^{-3}$$

Substituting it then in Eq. (1.60) and dividing the value of Fermi energy obtained in joules by electron charge ($=1.60 \times 10^{-19}$ C), we get $E_F = 3.38$ eV for Na. This is in agreement by 96% of the experimentally obtained value of 3.24 eV.

The Fermi velocity, v_F is obtained by using the relationship $v_F \approx \frac{\sqrt{2m_e E_F}}{m_e}$. Substituting for m_e and E_F, we get $v_F = 1.11 \times 10^6$ m s^{-1}. Note E_F needs to be multiplied by electron charge to convert it from electron volts to joules.

Finally, the Fermi temperature $T_F \approx \frac{E_F}{k_B} \approx \frac{(3.24 \times 1.6 \times 10^{-19})}{1.38 \times 10^{-23}}$ $\approx 3.75 \times 10^4$ K.

In conclusion, the calculated values for Na are $E_F = 3.24$ eV, $V_F = 1.11 \times 10^6$ m S^{-1} and $T_F = 3.75 \times 10^4$ K. These values are in good agreement with the values reported in literature.

1.12 Band Structure of Solids

The free electron theory is capable of explaining almost all physical phenomena associated with metals. But it fails when it comes to insulators and semiconductors. In metals electrons are supposed to be free so that they can cause electrical conduction to take place. This is not the case for insulators where the electrons are bound and not free to roam around to produce electrical conduction. A similar picture we can visualize for semiconductors which by definition are bad insulators. The failure of the free electron theory made it essential to find a suitable theoretical model that could explain the basic nature of solids other than metals. So far we have learned that discrete energy states exist in single atoms, but the picture is very different when many atoms are involved as is the case of solids. A solid can be visualized in which positive and negative ions are present simultaneously and the electrons are bound. When separation between two atoms becomes infinitesimally small, the available energy states lose their respective discrete states and form bands. This is the consequence of the Pauli Exclusion Principle.

Three theoretical models are often used in solid-state sciences to understand how these energy bands originate in solids. They are Kronig–Penney model, Ziman model, and Feynman model. Each of them explains the physical mechanisms for the formation of allowed and forbidden bands. But they all require solid skills in mathematical manipulations. We will not go into those details here and will limit ourselves to some simpler approach that would satisfy our needs. That would be the approach by considering a simplified version of the Kronig–Penney model. This model was developed in 1931 by Ralph Kronig of Germany and William Penney of England.

The solids in this model are treated as a highly ordered periodic three-dimensional structure built on the basis of unit cells that repeat themselves in space. Each cell is identical in every respect to the other. We will recognize such a well-defined solid to be nothing else than a highly symmetric crystal lattice. To simplify the calculation Kronig and Penney considered the case of a one dimensional lattice of atoms. In such an arrangement, the atoms will experience the effects of periodic potentials varying with the lattice period. Figure 1.15 gives a graphical

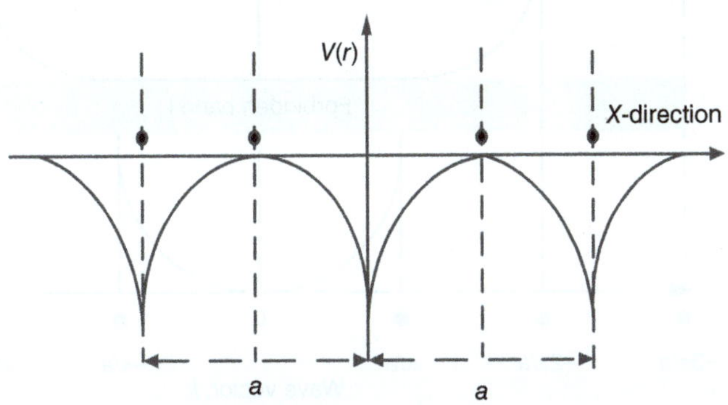

Figure 1.15 Linear array to atoms with a period of a.

representation of this concept. Here the atoms are found along the x-axis with the period of a.

By careful analysis of the time-independent Schrödinger equation, we can find the eigenfunctions and eigenvalues that would satisfy the mathematical conditions for the electrons in the potential field with the period of a. The wave function that would satisfy this condition was proposed by Felix Bloch of Switzerland, and it is expressed mathematically as follows:

$$\psi_{nk}(r) = \exp(ik \cdot r)u_{nk}(r) \tag{1.61}$$

where k is the wave vector and $u_k(r)$ is the periodicity of the lattice such that $u_k(R+r) \approx u_k(r)$, where R is defined as the position vector capable of generating an infinite number of lattice points. The concept of a position vector is related to the reciprocal lattice in crystallography. We will visit this topic in Chapter 4, where we well cover the essential elements of crystallography. The parameter n is as always equal to an integer, 1, 2, 3, … For each value of k, there could be multiple solutions for the Schrödinger equation corresponding to the values of n. If $u_{nk}(r) = 0$ and $n = 1$ then Eq. (1.61) reduces to $\psi_k(r) \approx \exp(ik \cdot r)$, which represents a plane wave of a free electron.

The most important results that follow from the Kronig–Penney model are (i) Presence of forbidden and allowed energy bands in a solid; (ii) Switch from one energy state to another is discontinuous; and (iii) The discontinuity occurs at $k = \pm n\left(\frac{\pi}{a}\right)$. Notice that it also tells us that in solids, the wave vectors are also quantized. With respect to these values of k, the Schrödinger's wave function can be given by Eqs. (1.62) and (1.63).

$$\psi_1 = \exp\left\{i\left(\frac{n\pi x}{a}\right)\right\} + \exp\left\{-i\left(\frac{n\pi x}{a}\right)\right\} = 2\cos\left(\frac{n\pi x}{a}\right) \tag{1.62}$$

$$\psi_2 = \exp\left\{i\left(\frac{n\pi x}{a}\right)\right\} - \exp\left\{-i\left(\frac{n\pi x}{a}\right)\right\} = 2\sin\left(\frac{n\pi x}{a}\right) \tag{1.63}$$

For ψ_1 the maxima occurs when $x = 0, a, 2a, \ldots, na$. Exactly, at these values of x, the minima for the wave-function ψ_2 occur. From this, we can infer that there are two values of wave function for the same value of k, indicating that for the same values of k, there could be two values of energy, E, as well. We present graphically these results in the E–k diagram as depicted in Figure 1.16.

We find in this figure that for each set of k, there is a corresponding allowed energy band. Sandwiched between the successive allowed energy bands, there are the forbidden energy bands where no electrons are allowed to reside. The first allowed band is called the first Brillouin zone; similarly the second and third bands are called the second and third Brillouin zones, respectively. They are named after the French physicist with the name of Léon Nicolas Brillouin who made valuable fundamental contributions of in many fields of physics.

Theoretically, it is possible to have infinite number of bands and as such an infinite number of energy states. But because of the limited number of electrons available in all types of solids, in reality there are only a few allowed and forbidden energy bands. The most important bands relevant to solid-state electronics and optics are the

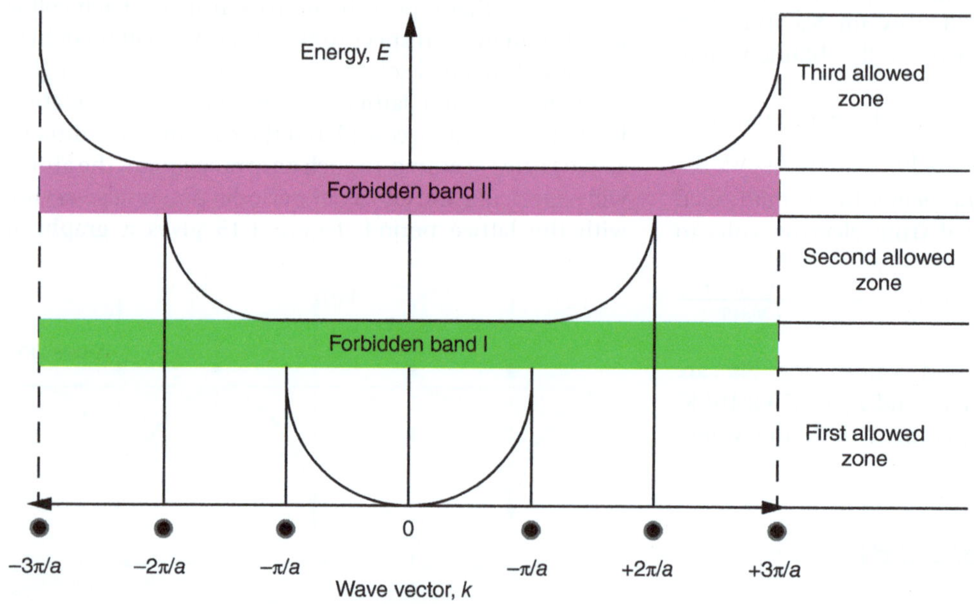

Figure 1.16 E–k diagram showing allowed and forbidden energy bands in a solid.

valence band, conduction band, and the bandgap. We all are familiar with these bands as they are the fundamental nature of semiconducting materials. Of course, the same picture prevails also in insulators. In metals, however, the conduction band overlaps the valence band eliminating the possibility of the presence of a bandgap. The Fermi level is found in the forbidden gap of semiconductors and insulators, but it is buried inside one of the allowed bands for metals and semimetals. This is what we find in Figure 1.1.

Glossary

Allowed band In solids according to quantum mechanics, energy states exist in bands and are not discrete as is the case with electrons in an isolated atom.

Bandgap It is the energy band that is sandwiched, for example, in a semiconductor material, between the valence band and conduction band. No electrons are allowed to find themselves in this band. Therefore, it is also called the forbidden gap. It width is measured in electron volt.

Brillouin zones The allowed energy bands in a solid are grouped as the first allowed band, the second allowed band, the third allowed band, and so forth. They are also referred to as the first Brillouin zone, the second Brillouin zone, the third Brillouin zone, respectively. Léon Nicolas Brillouin (1869–1969) was a brilliant French physicist who gave the concept of Brillouin zones in a crystal lattice. He made also many more ground-breaking contributions.

Cooper pairs In superconducting materials, electrons form pairs with antiparallel spins and do not obey Pauli Exclusion Principle like the normal electrons. Leon N. Cooper, one of the recipients of Nobel Prize in Physics in 1972 with John Bardeen and Robert Schrieffer, showed that the Copper pairs are responsible for the superconductivity phenomenon.

Critical field It refers to the critical magnetic field that can induce switching of the superconducting state into normal state. This is a signature property of a superconducting material.

Critical point The critical temperature at which a superconducting material becomes a normal conductor. This is the other signature property of a superconducting material.

Fermi energy This is the difference in kinetic energy between the highest and lowest occupied states of solid at absolute zero (0 K). This is the characteristic property of all solids and plays a vital role in describing the fundamental nature of metals and semiconductors at a thermal equilibrium.

Fermi level By definition is the sum of the kinetic energy and potential energy of electrons in a solid. Unlike Fermi energy, it can be referred to at any temperature. In semiconductor, it is common to refer to Fermi level instead of Fermi energy. Here the bottom of the valence band corresponds to the potential energy equal to zero. Then the Fermi level for an intrinsic semiconductor becomes equal to the width between the bottom of the valence band and the position of the Fermi level in the bandgap. Fermi level changes with temperature and also with the doping level of the extrinsic semiconductor.

Forbidden band The energy band in which no electrons are supposed to be present. Same as the bandgap of a semiconductor material.

Frequency spectrum Dielectric constant vs. frequency plot of an insulator in which the dipolar, ionic, and electronic components of the permittivity are identified.

Josephson effect It is the physical phenomenon specific to superconductors which was discovered by Brian David Josephson of England in 1962. The effect describes the tunneling of the Copper pairs through a very thin insulating layer sandwiched between two superconducting layers. It has many practical applications including detecting extremely small magnetic fields.

Magnetic levitation When a normal material becomes superconducting below the critical temperature, its acquired diamagnetism prevents the penetration of magnetic fluxes within the superconducting material. As a result fluxes are expelled back causing a magnet to be levitated above the superconducting surface and being suspended in air (as in Figure 1.9). This also has many applications and represents another signature property of a superconductor.

Matter–wave-duality This is the true nature of an electron as proposed by Louis de Broglie of France in 1923. According to his hypothesis, an electron can behave simultaneously as a particle and a wave. This concept is the corner stone of quantum mechanics, and it has led to the discovery of the celebrated Schrödinger equation. de Broglie received the Nobel Prize in Physics in 1929 for this contribution.

Persistent current A current generated by a magnetic field in a superconducting ring that can persist for a very long time even after the originating magnetic field is withdrawn.

Phonon In a crystal lattice, temperature can set up oscillations of atoms. The unit to measure the resulting vibrational energy is referred to as phonons and is a quantum mechanical concept. It is considered to be an elementary particle associated with a solid and is used to describe a mechanical wave.

Photon What a phonon is to a mechanical wave so is a photon to an electromagnetic wave, or more precisely, to an optical wave. It is a quantum of energy associated with light and originally introduced by Planck.

Potential field In a solid with perfectly ordered lattice and periodicity in space, the potential energy can vary from a minimum to a maximum value. This facilitated the discovery of allowed and forbidden energy bands in a solid using the model proposed by Kronig and Penney.

Quantization In quantum mechanics, when a parameter can assume values in steps of 1, 2, 3, … (equivalent to the value of the principal quantum number, n), then the parameters are called quantized. Some examples are orbital quantum number, l; Planck's photon energy, $h\nu$; and magnetic flux, ϕ, in superconducting state.

Quantum numbers A set of four numbers identified as principal quantum number (n), orbital quantum number (l), magnetic quantum number (m_l), and spin quantum number $\left(\pm\frac{1}{2}\right)$ are jointly referred to as quantum numbers. They are used to describe the quantum mechanical properties of an electron.

SQUID It is the abbreviation for "superconducting quantum interference device" that is based on the Josephson junction. These are very precise magnetometers capable of detecting extremely small magnetic fields.

Superconductivity A special class of physical phenomenon occurring in solids below a certain temperature is called superconductivity. As the name suggests a superconducting material has zero resistance and as such infinite conduction. Such materials can deliver lossless electrical power ($P = I^2R$) at a long distance and as such they possess enormous technical value for mankind.

Work function It is a material parameter of enormous scientific and technical importance. Its commonly used symbol is W, and it can be defined in many different ways. For example, in a photoelectric experiment, it is the threshold energy that must be exceeded by the photon energy radiating a metallic surface before electrons can be emitted. It is also measured in electron volts. Its general definition is the following: this is the energy that an electron must acquire before it can escape a surface in order to reach its ultimate destination that is the vacuum level with infinite energy.

Problems

1.1 Consider a sample of GaAs with the following dimensions: length = 5 mm, width = 1 mm, and thickness = 1 mm. The electrical conductivity of GaAs is 1×10^{-8} S m^{-1}. Assume that the two parallel faces (5 mm long) are fully metallized. Find the electric field that must be applied to generate a current density of 1 μA m^{-2}.

1.2 Using the values of the dielectric constant given in Table 1.3 calculate the polarization that develops when 10 V is applied to a cube sample of 1 mm dimension of Ge, $SrTiO_3$, and TiO_2. Comment on the result.

1.3 Find the maximum kinetic energy with which the electrons will emit when the samples of metals listed in Table 1.6 are irradiated by a UV light of 0.2 μm wavelength. Comment on your result.

1.4 Find the stopping potential for photoemission for all the metals listed in Table 1.6 when the samples are irradiated by a UV light of 0.2 μm wavelength.

1.5 Describe the mechanism for population and depopulation of states involved in Fermi–Dirac statistics. Find the temperature at which there is 1% probability that a state with the energy of 0.1 eV above the Fermi energy will be occupied by an electron. Comment on your result.

1.6 Find the velocity and the momentum of the electron with de Broglie wavelength of 1 nm. Comment on your result.

1.7 Describe the Pauli exclusion principle and show how it leads to the quantum mechanical interpretation of the Periodic table of elements.

References

1 Azaroff, L.V. and Brophy, J.J. (1963). *Electronic Processes in Materials*. McGraw Hill.
2 Orlando, T.P. and Delin, K.A. (1991). *Foundations of Applied Superconductivity*. Addison Wesley.
3 Ashcroft, N.W. and Mermin, N.D. (1976). *Solid State Physics*. Sauders http://hyperphysics.phy-astr.gsu.edu/hbase/hph.html.

Further Reading

Solymar, L. and Walsh, D. (2010). *Electrical Properties of Materials*, 8e. Oxford University Press.

Kasap, S.O. (2006). *Principles of Electronic Materials and Devices*, 3e. McGraw Hill.

2

Processing of Electroceramics

CHAPTER MENU

Anyone who has never made a mistake has never tried anything new.

Albert Einstein

"Übung macht den Meister" (a German Proverb meaning "Practice makes it perfect").

2.1 Introduction

Like any other material of scientific and technical importance, electroceramics must also be processed with great care to exploit their reproducible physical and structural properties so that they can be considered for applications ranging from simple and straightforward to highly sophisticated high-density and high-speed computer memories. This naturally implies that the processing methods must be so designed that they can yield the best-quality materials.

Electroceramics, especially oxides, are produced in three principal categories. They are bulk ceramic, thin, and thick film and single crystal. We can add a fourth category of *nano-structured ceramics* that is a recent development and holds great promise for a host of new scientific breakthroughs as well as for a number of novel applications. There are large numbers of well-proven methods available to us to process oxides in any of these three categories.

It is a universal truth that a direct relationship exists between the structure of a material and its physical properties that in turn can be manipulated to meet the requirements for various technologies. We shall revisit this issue once again in Chapter 4 where we will learn about the essentials of crystallography. On the basis of this relationship, we will make a distinction between polar and nonpolar materials, and how they manifest themselves once subjected to external agents such as an electric field, a magnetic field, and stress or some combination of these three principal agents.

In the following processing, it is imperative that the materials be characterized for their crystal and surface structures followed by determination of their physical properties. In Figure 2.1, a roadmap is presented beginning with materials processing and ending in structural and surface characterization steps.

2.2 Basic Concepts of Equilibrium Phase Diagram

Thermodynamically, materials can exist in multiphases or single phase. When a piece of solid consists of more than one chemical composition it is said to be multiphased. That leads us to the definition of a single-phase material. It is of one unique composition. There are three phases of matter: they are solid, liquid, and vapor. A classic example of this is water (H_2O). It is our everyday experience that at atmospheric pressure it can exists in the physical states of solid below 0 °C, in liquid until 100 °C, and as vapor above this temperature. Also a solid is said to be in single phase if its crystal structure remains the same and changes only with the change in temperature provided the pressure is held constant at atmospheric pressure. An example of this is barium titanate crystal whose crystal structure changes at certain fixed temperatures. We will discuss this in detail in Chapter 6. A phase change can also occur in certain solids with a change of temperature without being

Fundamentals of Electroceramics: Materials, Devices, and Applications, First Edition. R. K. Pandey.

Companion website: www.wiley.com/go/Pandey/Fundamentals_Electroceramics

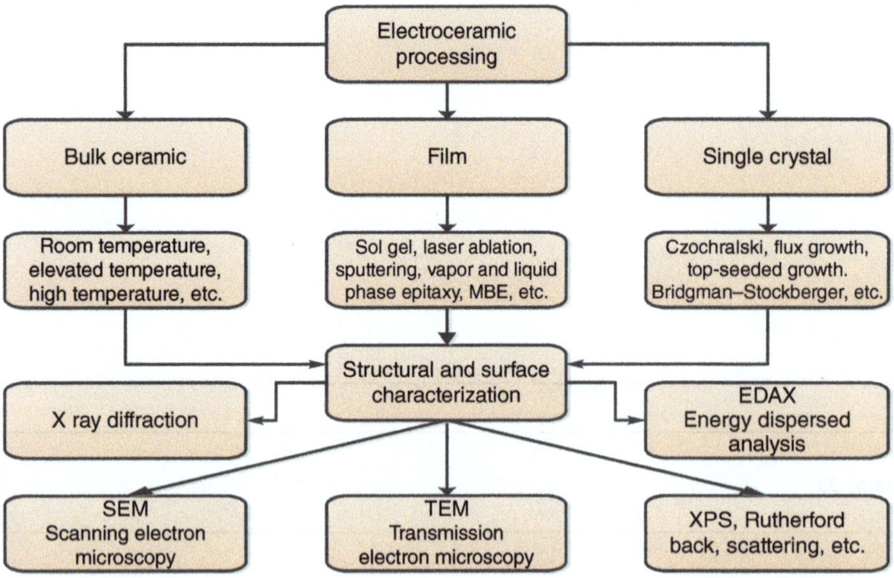

Figure 2.1 A roadmap outlining the processing and characterization steps for electroceramics.

accompanied by the change in its crystal structure. For example, a magnetic material becomes nonmagnetic at a certain temperature (called the Curie point), a superconducting material becomes a normal conductor (like a metal) at its critical temperature or it loses its superconductivity when subjected to a critical magnetic field even though the temperature remains below the critical point. We have already encountered these situations in Chapter 1 while discussing superconductivity (see Figures 1.8 and 1.10). It is worth remembering that in materials science and for all other practical considerations, we assume the pressure to remain constant, and the temperature is the only noncompositional variable that changes while discussing the phase equilibrium in materials. It is of utmost importance that a material processed for fundamental studies and technology must be produced in single phase.

2.2.1 Gibbs' Phase Rule

The golden rule of phase equilibria was given by Josiah W. Gibbs (between 1875 and 1878) purely based on thermodynamic considerations. It is a brilliant piece of theoretical work and sets the rule that must be obeyed when dealing with phase changes. It has found applications in fields such as chemistry, physics, materials science, metallurgy, and ceramic engineering. Gibbs introduced the concepts of four parameters that uniquely define a phase at any given thermodynamic equilibrium. They are (i) P, which stands for number of phases; (ii) F, is the degree of freedom; (iii) C, the number of compositional independent variables, and (iv) N, number of noncompositional variables such as temperature and pressure. The parameters P, C, and N are self-explanatory, but F needs more explanation. It is defined as the number of independent variables that can be varied simultaneously and arbitrarily without changing the number of phases in equilibrium. For example, we can vary pressure or temperature without inducing any changes in the number of phases. Equation (2.1) gives the mathematical formulation for the Gibbs phase rule.

$$P + F = C + N \tag{2.1}$$

Since pressure and temperature are the only two noncompositional variables, we can also rewrite the phase rule as Eq. (2.2).

$$F = (C + 2) - P \tag{2.2}$$

Let us examine this rule by applying it to water that is a single component system with $C = 1$ in its solid phase at $T \leq 273$ K. For this phase, $P = 1$, $C = 1$, and $N = 2$. For this phase, then $F = 2$. The same situation would prevail for the temperature range of $273 \leq T \leq 373$ K in which water is liquid, and for $T \geq 373$ K, when water is in its vapor phase. In case the system has only one chemical compound ($C = 1$), then the phase rule becomes $F = 3 - P$. If two phases are in equilibrium, then $F = 1$. A system for which there is only one degree of freedom, it is called *univariant*.

2.2.2 Triple Point and Interfaces

Let us now discuss a general case of a material. This we describe qualitatively with the help of a temperature–pressure (T–P) diagram as shown in Figure 2.2. The three phases of solid, liquid, and vapor are clearly marked in this diagram. There are three interfaces named solid–liquid, solid–vapor, and vapor–liquid. Each of them exists at its respective thermal equilibrium covering a part of the whole T–P diagram. They intersect

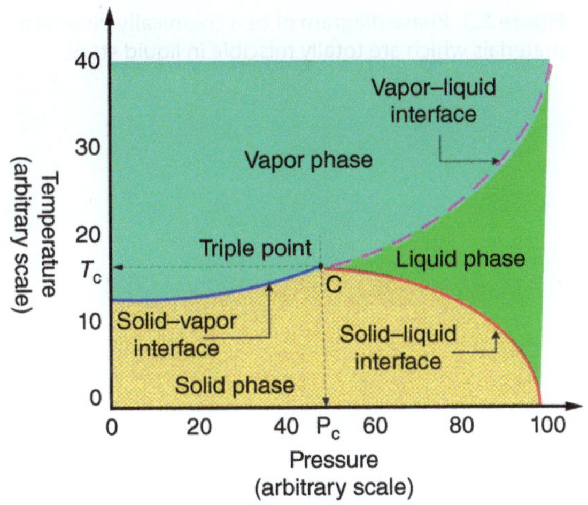

Figure 2.2 Temperature–pressure diagram of an arbitrary material.

Table 2.1 Triple points of selected materials.

Material	Chemical symbol	Temperature, T (K)	Pressure, P (kPa)
Water	H_2O	273.16	0.612
Argon	Ar	83.80	68.9
Carbon dioxide	CO_2	216.6	517
Hydrogen	H	13.80	7.04
Oxygen	O_2	54.30	0.152
Palladium	Pd	1 830	3.5×10^{-3}
Platinum	Pt	2 045	2×10^{-4}
Graphite	C	4 765	10 132

Source: https://en.wikipedia.org/wiki/Triple_point.

at point C at the critical pressure, P_C, and the critical temperature, T_C. This point is called the "triple point." Here the three phases coexist in thermodynamic equilibrium. At this point, the degree of freedom, F, becomes zero because $C = 3$, $P = 3$, and $N = 0$. The triple point is thermodynamically *invariant* because it has zero degrees of freedom.

The triple point of water forms the basis for defining the thermodynamic temperature with the unit of Kelvin (e.g. $0\,°\text{C}\approx +273$ K). Each material has its own triple point at a set combination of temperature and pressure. We present a list of triple points for some selected materials in Table 2.1. Note the unit of pressure used is Pascal (Pa) which is the standard international unit of pressure. One atmosphere (atm) of pressure is equal to 98.066 kPa.

Exercise 2.1
Find the number of degree of freedom (F) at the interfaces as shown in the above figure. Discuss the significance of the result.

Solution
Applying the Gibbs' phase rule as given by Eq. (2.2), it is easy to determine that at each of three interfaces, we have $P = 2$ (two single phases coexist), $C = 2$ (two compositions coexist), and $N = 1$. That makes $F = 1$. This means that for the interfaces if the temperature is fixed that automatically fixes the pressure as well. Similarly, choosing pressure will fix the temperature.

2.2.3 Binary Phase Diagrams

In the previous section, we discussed the phase relationship of a single compositional system and applied the Gibbs phase rule to understand how to interpret the pressure–temperature diagram for such a case. We will in the next two sections learn about systems that consist of two solids of different chemical compositions. Since electroceramics are always the result of two or more solids of dissimilar chemical compositions the knowledge of the phase diagram of binary systems will be very helpful.

We will discuss two phase diagrams that represent two different binary systems. The first in which two dissimilar solids are totally miscible in each other, and the second when they are only partially miscible. In the processing of electroceramics, these phase diagram help us in deciding about the sintering and annealing temperatures that are the crucial parameters for producing high-quality electroceramics. Phase diagrams are indispensable for crystal growth of oxides and other ceramics. Without the knowledge of good and reliable phase diagrams, a crystal grower is totally lost. There are many sources for finding good and reliable phase diagrams; the best that I would recommend are the multiple volumes produced by the American Ceramic Society.[1]

2.2.3.1 Totally Miscible Systems

Let us first consider the phase equilibrium diagram shown in Figure 2.3. A and B are two pure solids that are distinguishable from each other in their chemical formulations and differing in their respective melting points. Yet if A when in molten state can dissolve B such that it can form continuous series of mixture, then by definition it is said that A and B are *totally miscible*. A large number of intermediate compositions can be recrystallized, which can be represented by the general formula of $A_{1-x}B_x$. Here, the concentration x varying within the limits of $0 < x < 1$.

The compositions $A_{1-x}B_x$ are called *solid solutions*. It is defined as a uniform mixture of substances in a solid form such that two or more dissimilar atoms or molecules share a common crystal structure. Simply

1 http://ceramics.org/publications-and-resources/phase-equilibria-diagrams/phase-equilibria-diagrams-online/.

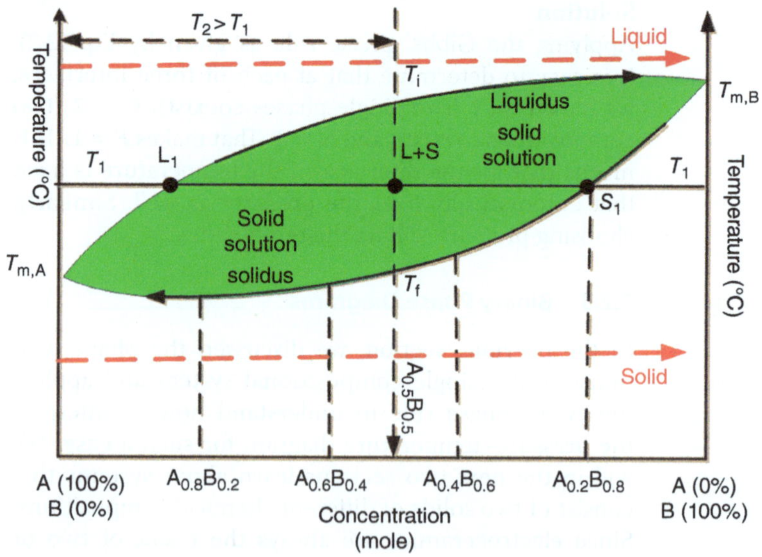

Figure 2.3 Phase diagram of two chemically dissimilar materials which are totally miscible in liquid state.

stated a solid solution is a mixture of two or more solids in which minor components are uniformly distributed within the crystal lattice of the major component. Many metals and oxides form solid solutions. It is a very important concept in electroceramics and plays a vital role in the discovery of new members of the ceramic family. It also forms the basis for processing electroceramics with desired properties. Some examples of prominent electroceramics that come as solid solutions are barium titanate, $BaTiO_3$; lead titanate, $PbTiO_3$; potassium niobate, $KNbO_3$; and strontium titanate, $SrTiO_3$. They all crystallize as perovskites.

The phase diagram depicted in Figure 2.3 resembles a convex lens and is typical of a system of two totally miscible solids. The phase diagram is called the temperature–concentration phase equilibrium diagram. It is constructed keeping the pressure constant, which is usually at one atmosphere. The concentration, x, is usually expressed as mole fraction, though at times, it can also be expressed as weight percentage. But for our discussion, we will assume that x is always in mole fraction unless it is clearly stated otherwise. The concentration axis is calibrated in units of $A_{1-x}B_x$. The member identified as $A_{0.8}B_{0.2}$ consists of 80 mol% A and 20 mol% B. Similarly, the solid solution $A_{0.5}B_{0.5}$, which is at the midpoint of the diagram, represents a composition in which there is 50 mol% of A and 50 mol% of B. Such a composition is also called an equimolar composition. The two end members are, of course, A ($x = 0$) and B ($x = 1$). That is, each being 100% pure with no presence of the other component. It is assumed that A melts at $T_{m,A}$ and B at $T_{m,B}$ such that $T_{m,A} < T_{m,B}$.

The upper solid curve labeled as *liquidus* is the interface between liquid (above this line) and the solid solution within the lens. Similarly, the bottom solid line of the lens is called *solidus*, which forms the interface between solid (right below it) and the solid solutions series within the lens. The mixture consists of liquid A in which solid B is dissolved; the dissolving power of A increases as we approach the end member B.

Now let us consider the dashed line corresponding to the mixture $A_{0.5}B_{0.5}$. While traveling along this line, say from temperature T_2 to T_1, we first encounter the liquids interface at temperature T_i, then enter the lens and finally stop at the L + S point without going all the way to T_f point at the solidus line. At the L + S point, liquid of composition L_1 is in equilibrium with the solid of composition S_1. Clearly, it is a two-phase system because each composition consists of liquid and solid components and therefore, it classifies as a univariant system. Here each composition has a definite concentration of two phases at each temperature.

We have now succeeded in crystallizing a solid with the composition of $A_{0.2}B_{0.8}$. If the cooling between T_i and L + S is very slow, and if we could harvest solid S_1 from the lens, then we would have succeeded in growing single crystals of $A_{0.2}B_{0.8}$. If we resume our travel to point T_f, we could then precipitate many more solids along the way. In principle, then we would be able to crystallize a solid of any composition we wish. For each point of equilibrium, we would generate a line similar in characteristics as the line L_1–(L + S)–S_1. Such a line is called the *tie line*. This line provides us a means to estimate the amounts of solid and liquid for each temperature. This is accomplished by applying the *lever rule* which is given by Eq. (2.3).

$$\frac{\text{Weight of liquid phase}}{\text{Weight of solid phase}} = \frac{[S_1 - (L + S)]}{[L_1 - (L + S)]} \tag{2.3}$$

If we now consider that W_{sp} = the weight of the solid phase; W_{lp} = the weight of the liquid phase; W_L = the weight of liquid at point L_1; W_S = the weight of solid at point S_1; and W_0 = the weight of the solid solution at point (L + S) which is the composition $A_{0.5}B_{0.5}$.

Then we can rewrite Eq. (2.3) as Eq. (2.4).

$$\frac{W_{lp}}{W_{sp}} = \frac{(W_S - W_0)}{(W_L - W_0)} \tag{2.4}$$

Exercise 2.2
Prove the validity of the lever rule using the parameters in Eq. (2.4).

Solution
This exercise is for students to work out. The instructor may wish to monitor.

Exercise 2.3
Find the molecular weight of the solid solution $A_{0.5}B_{0.5}$ considering that the molecular weight of A is 20 and of B is 10.

Solution
In solid solution AB, we have 10 units of A and 5 units of B. Therefore, its molecular weight is 15.

Exercise 2.4
Consider that in a phase diagram as shown in Figure 2.3, there is 15 g of $A_{0.5}B_{0.5}$. At equilibrium, the tie line at temperature T_1 is formed between L_1, L + S and S_1. Consider the ratio between the weight of the liquid phase, W_{lp} and the solid phase, W_{sp} to be equal to 0.5. Find the weights for the liquid at L_1 and solid at S_1.

Solution
To solve this, we can use Eq. (2.4). From the parameters given, we get: $\frac{W_{lp}}{W_{sp}} = \frac{1}{2} = \frac{W_S - W_0}{W_L - W_0}$ (1). At equilibrium, W_0 breaks into two components of W_L and W_S. That is, $W_L + W_S = W_0 = 15\,g$ (2). After some substitutions and rearranging (1) and (2) we get $W_S = 5\,g$ and $W_L = 10\,g$. On the tie line of T_1–L_1–(L + S)–S_1–T_1, the equilibrium is maintained by a liquid of 5 g and solid of 10 g.

2.2.3.2 Systems with Limited Solubility in Solid Phase

The phase diagram of Figure 2.4 deals with two solids, A and B, that are 100% miscible in liquid phase but have limited solubility in the solid phase. There are two different parts of the diagram separated from each other at the vertical dashed line labeled as the *eutectic concentration*, E_c. The point at which the two components solidify at a constant temperature is called the eutectic point, E_c. The eutectic composition represents an alloy between A and B. The corresponding temperature is called the

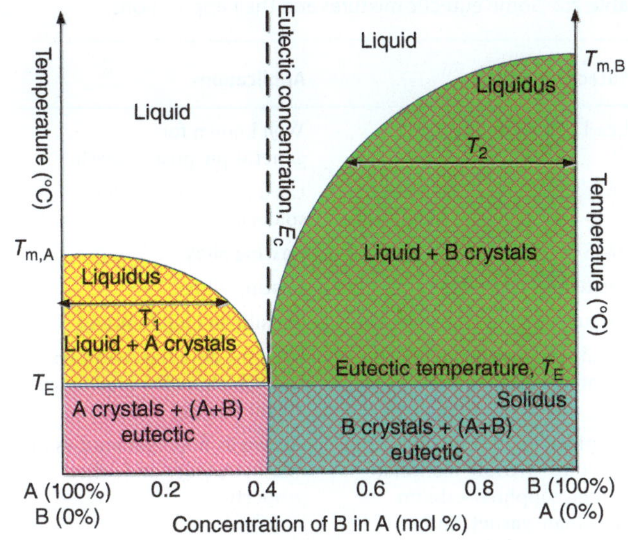

Figure 2.4 Phase equilibrium diagram of two solids with limited solubility in sold phase but unlimited solubility in liquid phase.

eutectic temperature, T_E. As can be seen from the diagram, it is the lowest temperature at which the eutectic concentration melts. The classic examples of eutectic alloys include a lead–tin alloy that is universally used for soldering and a salt–water mixture used for deicing in winter for roadways, automobiles, and airplanes in winter.

In the phase diagram, the solidus line is identical to the eutectic temperature line. Above it, we find two regions identified as liquid + A crystals on the left of the eutectic concentration and the other as liquid + B crystals on the right side. On the left side for the entire temperature range of the eutectic temperature, T_E and $T_{m,A}$ which is the melting point of A, only pure A can crystallize. If we drew a vertical line normal to the line identified as T_1, we would generate a tie line. Using the lever rule, we could then predict the amount of crystallized A. Similarly, on the right side, we could crystallize pure B for any temperature between its melting point of $T_{m,B}$ and the eutectic temperature T_E. At the eutectic point, solids of A and B would precipitate concurrently, as well as crystals of eutectic alloy A + B. Below the eutectic temperature, as we can see in the diagram, there is no liquid phase. In this region, only solids of A, B, and A + B eutectic alloy can be present. They will be so heavily intermixed that it will not be easy to harvest pure A or B. Therefore, the best approach would be to precipice them using either of the two parts of phase diagrams above the eutectic line. From Gibbs phase rule, it follows that in a eutectic system, $C = 1$, $N = 1$, $P = 2$, and therefore, $F = 0$. This implies that for such a system, the process of solidification can occur only at the melting point. In Table 2.2, we present some examples of eutectic mixtures and their applications.

Table 2.2 Some eutectic mixtures and their applications.

Eutectic mixtures	Applications
Lead (Pb)–Tin (Sn)	Well known for general-purpose soldering
$Sn_{0.63}Pb_{0.37}$	Used in electronics for soldering
Al–Si	Casting alloy
Austenite–cementite	Casting alloy
Si–Au	For silicon chip bonding
NaCl–H_2O with eutectic point at −21.2 °C	Used for de-icing roadways and vehicles; also used in ice cream making
Sapphire–oxide eutectics: sapphire–yttrium aluminate garnet, sapphire–erbium aluminate garnet, and sapphire–gadolinium aluminate garnet	Grown as fibers for studies of microstructure and other properties

2.3 Methods of Ceramic Processing

There are a large number of methods for ceramic processing. Some of the commonly used methods in industry are room temperature uniaxial and isostatic pressing, hot pressing, extrusion, slip casting, pressing, tape casting, and injection molding. The common denominator in all these different processing techniques with varying degree of complexity and effectiveness is to produce ceramics in predetermined shapes and sizes on a large scale. All require careful preparation of raw materials using high-purity grade chemicals, followed by shape forming, if necessary using binders. It is very important to ball mill the powder mixture to break the materials down into as small particles as possible. This facilitates the efficient diffusion process to set in while subjecting the powder mixture to different heat treatments. The mixture is heat treated to make it dense and rocklike, to induce the highest possible density, and finally to produce samples with well-defined microstructure and crystal structure. There are three heat treatment methods that differ in their subtleties. If the raw materials are in the form of carbonates, nitrates, or any other chemicals that undergo dissociation once heat treated, then the heat treating method is called *calcination*. An additional benefit of calcination is that moisture is driven off and the raw mixture becomes dry. Calcination usually takes place at relatively low temperatures chosen much below the melting points of individual species.

This produces the first stage of *green* ceramics. After this step is completed, the raw material is called a *charge*. The charge then goes through another heat treatment cycle, which is more extensive and of longer duration than the previous one. This step is called *sintering*. Sintering commonly requires heat treatment at relatively high temperatures but again, below the melting point of the charge. The common practice is to keep the maximum sintering temperature at around 80–85% of the melting point of the charge. One may allow some room for exception to take advantage of the solubility power of a liquid phase. In this case, called *liquid phase sintering*, the sintering temperature may be high enough to melt one or more constituents of the charge so that the other components may be dissolved in the liquid phase of the mixture. This results in producing homogeneous and single-phase ceramics. As attractive as this proposition may be, it is not always possible to find raw materials that can be subjected to liquid phase sintering.

The third and final heat treatment step is called *annealing* that is more stringent and time-consuming than sintering. It occurs at relatively high temperatures, but again obviously below the melting point of the sintered charge. The temperature is raised to the maximum desired level, held there constant for a number of hours, and then cooled slowly to room temperature. Annealing may be conducted in air, or in some special gases, such as argon, nitrogen, or oxygen, to provide neutral, oxidizing or reducing environments. This heat treatment first of all prevents cracking or even development of micro cracks. But the real advantages of annealing are many. It invariably produces samples of high density, and it is not uncommon to achieve a density of up to 95–98% of the theoretical density. But the real benefit of careful annealing results in ceramics with well-developed large grains with well-defined grain boundaries and crystallinity.

The range of ceramic products is large. We encounter ceramics daily in our modern life style. Much of our kitchenware is ceramic, and so is dinnerware, coffee, and tea cups, and so forth. In fact, if we look around, we would find ceramic all around us. They are universally used as insulators in electric and electronic circuits and transmission systems. They find applications in semiconductor technology as gate oxides for field-effect transistors and as capacitors for energy storage. Their scope of applications is so vast that it is impossible to count all of them. They range from mundane ceramic magnets to sophisticated applications in medicine, automobiles, avionics, and shipbuilding just to name a few. Ceramics are commonly machined and polished to meet the requirements of specific applications.

2.3.1 Room Temperature Uniaxial Pressing (RTUP)

Room temperature uniaxial pressing (RTUP) is one of the oldest methods to compress powder and granules.

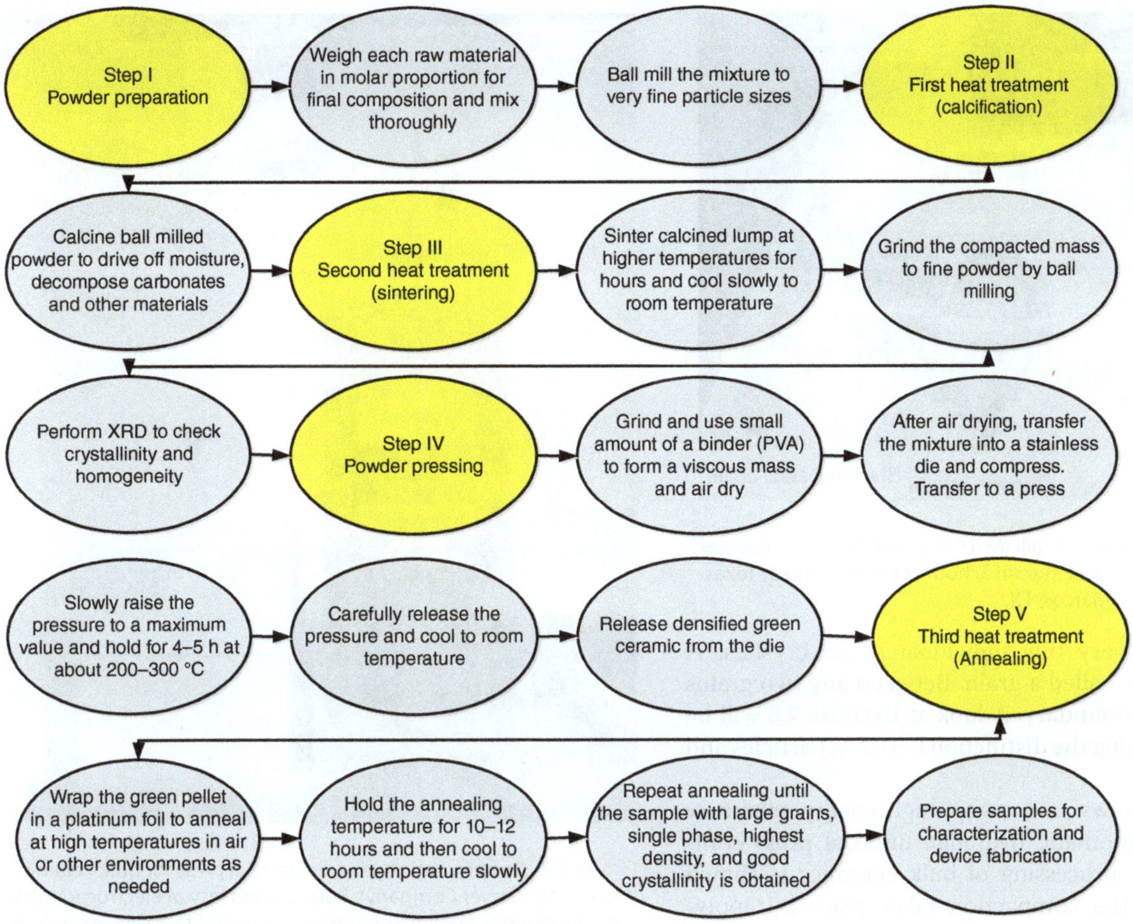

Figure 2.5 A block diagram outlining essential steps for processing of bulk ceramic for fundamental studies and device fabrication.

It is so widely used that it can be called the standard method for processing of bulk ceramic on large scales. It is also the simplest and an efficient method for producing bulk ceramics with reasonably high quality for fundamental studies and device fabrication. Because of these attributes, it is ideally suited for ceramic research in an academic setting. The multiple steps that must be taken to produce good samples of bulk ceramic are outlined in the block diagram of Figure 2.5. Each step must be carried out in a clean and contamination-free environment in order to produce ceramics with excellent structural and physical properties.

Figure 2.5 is self-explanatory except perhaps for emphasizing the importance of ball milling. From beginning to end, the whole process takes place in four distinct steps some of which have already been described earlier in this chapter. In the introductory portion, we have mentioned the role the particle size plays in producing ceramics with superior properties. The purpose of ball milling is to reduce the initial mixture of chemicals that come in different particle sizes to possibly one uniform size for the heat treatments of sintering and annealing.

These treatments are necessary in producing homogenous, dense, and rock hard bulk ceramics with large grain sizes and well-formed microstructures. This warrants the initial powder to be of as small particle size as possible. Ordinary ball milling cannot fulfill this requirement and also takes very long time to break down the particles to micrometer dimensions. High-energy and high-speed vibratory ball mills are capable of producing powders in nanometer scale in a relatively short time. Small charges may be reduced to acceptable particle size in less than 30 minutes instead of days required by slow speed ball mills. Even a bench model vibratory mill can break down powders into particles of 100–125 nm. Diffusion is accelerated in particles of such small dimensions, leading to the nucleation of tiny crystallites that when annealed could grow into larger grains.

At this point, we need to clarify the difference between the terms *particle* and *crystallite*. The difference is subtle. A particle may contain a number of tiny crystallites, but as agglomerate, it is polycrystalline in nature. But

Figure 2.6 A vibratory ball mill (MTI Company). Source: Courtesy of Electroceramics Lab. at Ingram School of Engineering at Texas State University, San Marcos, TX.

crystallites are very tiny individual single crystals. A crystallite is also called a grain. Between any two grains we find a grain boundary. A look at Exercise 2.5 will be helpful in clarifying the distinction between particles and crystallites.

Figure 2.6 shows a desktop high-energy vibratory ball mill. An automatic hydraulic uniaxial press commonly used for processing of bulk ceramics is shown in Figure 2.7. The compaction takes place by applying uniaxial pressure to the ceramic powder placed in a stainless steel die between two rigid metallic punches.

Figure 2.8 is a picture of a typical 1-in. diameter bulk ceramic sample that was produced following the protocol described in Figure 2.5 (R.K. Pandey, unpublished result, sample produced around 1988–1989 in Electronic Materials Labs of Texas A &M University, College Station, TX).

Exercise 2.5

After ball milling, the particle size for the charge for pressing strontium titanate ($SrTiO_3$) ceramic is found to be uniformly 100 nm in diameter. The crystal structure of $SrTiO_3$ is cubic and its lattice constant is equal to 0.39 nm. Find the number of unit cells in each particle.

Solution

Let us assume that the particles are spherical in shape of diameter = 100 nm. Therefore, its volume = $V_{\text{particle}} = \frac{4\pi}{3}\left(\frac{d}{2}\right)^3 \approx \frac{500\pi}{3} \times 10^3\ \text{nm}^3$. The lattice constant a = 0.39 nm of cubic $SrTiO_3$. Therefore, its unit cell volume = V_{cell} = $(0.39)^3 \approx 164 \times 10^{-3}\ \text{nm}^3$. The number of $SrTiO_3$ unit cells in each particle is = $n = \frac{\frac{500\pi}{3}\times 10^3}{164\times 10^{-3}} \approx 3 \times 10^6$. We see what a big number this is. We conclude that particle $\gg$ crystallite.

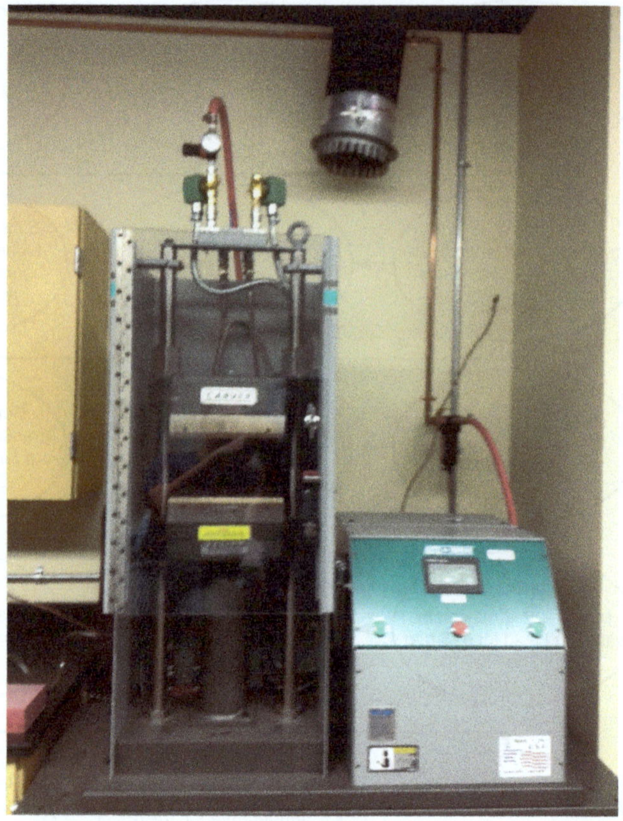

Figure 2.7 An automatic hydraulic uniaxial press for bulk ceramic processing (Carver Company). Source: Courtesy of Electroceramics Lab. at Ingram School of Engineering at Texas State University, San Marcos, TX.

Exercise 2.6

Calculate the weight of each raw material that you would need to prepare a charge of 100 g for the pressing of $SrTiO_3$ ceramic. Provide as much comment as you can about the chemical reactions and the end product of $SrTiO_3$.

Solution

We chose the raw materials to be strontium oxide (SrO) and titanium oxide (TiO_2). To get SrO, we need to first calcine strontium carbonate ($SrCO_3$). The chemical reactions are

$$SrCO_3 + TiO_2 = SrO + TiO_2 + \uparrow CO_2 \tag{2.5}$$

$$SrO + TiO_2 = SrTiO_3 \tag{2.6}$$

Considering the molecular weights, W, of the species that are 147.61 for $SrCO_3$, 79.87 for TiO_2, and 103.62 for SrO, we get the weight ratios between the species using Eqs. (2.5) and (2.6). They are

Figure 2.8 A sample of 123 YBCO ceramic superconductor. Source: Courtesy of Electronic Materials Laboratory, ECE Department, Texas A&M University, TX, ca. 1988.

From Eq. (2.5) we get $\frac{SrCO_3}{TiO_2} = \frac{W \text{ of } SrCO_3}{W \text{ of } SrO} \approx \frac{147.61}{79.87} \approx \frac{1.85}{1}$

And from Eq. (2.6) we have $\frac{SrO}{TiO_2} \approx \frac{W \text{ of } SrO}{W \text{ of } TiO_2} \approx \frac{103.62}{79.87} \approx \frac{1.3}{1}$.

Furthermore, $\frac{SrCO_3}{SrO} \approx \frac{147.61}{103.62} \approx \frac{1.42}{1}$.

We can use either of the two equations to calculate the charge of 100 g. From the weight ratio between $SrCO_3$ and TiO_2 of 1.85/1, the charge of 100 g will contain $SrCO_3 \approx 64.9$ g and $TiO_2 \approx 35.1$ g. Similarly, we will use $SrO \approx 56.52$ g and $TiO_2 \approx 43.08$ g to make a charge of 100 g.

Comments: (i) Both charges will give us 100 g of $SrTiO_3$ ($W = 183.49$) or an equivalent of approximately 0.55 mol of $SrTiO_3$ powder. (ii) Each charge should be calcined; from $SrCO_3$, it will drive off moisture and CO_2 gas ($W = 44$); (iii) In SrO, calcination will drive off moisture and make it dry. (iv) $SrTiO_3$ melts at approximately 2080 °C. That would require the sintering and annealing temperatures to be very high; at least around 1500 °C. (v) $SrTiO_3$ is cubic at room temperature and above and is centro-symmetric. It switches to tetragonal structure below −168 °C (≈105 K). It was the first oxide in which ceramic superconductor effect was discovered with the critical temperature of approximately 0.35 K. It is an excellent substrate material and is widely used as such. In crystal form, it is transparent.

2.3.2 Other Methods for Powder Compaction and Densification

Hot isostatic pressing (HIP), cold isostatic pressing (CIP), and low temperature sintering (LTS) are complimentary to the method described above. Each of them have some common features with the RTUP method and overlap each other in some subtle ways. Let us get familiarized with each of them in some detail.

2.3.2.1 Hot Isostatic Pressing (HIP)

The HIP method was developed about 65 years ago in the 1950s. Instrumentation for this method is obviously quite elaborate and far more expensive than for room temperature pressing. High-temperature heaters, thermal insulation, and choice of temperature resistance materials for dies must be carefully incorporated in the pressing system. The advantages of hot pressing are many, and for many applications it is the method of choice. High densification and compaction are the two most desirable attributes of this method. These can be achieved because the powder compaction takes place by applying pressure in multiple directions while the charge is concurrently maintained at high temperatures by placing it in the hot zone of a small furnace with precise temperature control. Simultaneously, the charge is also kept surrounded by a liquid or gaseous medium that facilitates isostatic pressing. As a result, the finished product is extremely hard and dense with practically no voids. Some of these applications of isostatic pressing are densification of castings, production of high density, and machine able ceramics for dental, medical, and other applications. HIP is also the method of choice to produce transparent bulk ceramic for electrooptical applications. The oxide lanthanum modified lead-zirconate-titanate (PLZT) is a classic example of a transparent bulk ceramic produced by hot pressing. It is widely used in a number of electro-optic applications including night vision goggles. Some other examples of hot pressed ceramics are aluminum-oxide/magnesium oxide (Al_2O_3–Mg), calcium fluoride (CaF_2), calcium oxide (CaO), and spinel ($MgAl_2O_4$).

2.3.2.2 Cold Isostatic Pressing (CIP)

This is a variation of HIP, and the process takes place at room temperature. Here the pressure is very high, between 100 and 600 MPa. The method is applicable to both metals and ceramic powders. The finished product can achieve a density as high as 100% of the theoretical density with practically zero voids. The green parts can be handled and machined even before sintering. Densification and compaction are also the objectives here (as with HIP), as well as, in uniaxial pressing. Obviously, the uniaxial pressing cannot produce ceramics with high compaction and densification as the HIP and CIP methods can. Its advantage, however, is rapid turn out of products of acceptable quality for many applications at a vastly reduced cost compared to products produced either by the HIP or CIP methods. The CIP method is used for compression of graphite, silicon nitride, silicon carbide, and other such materials. It can produce samples even larger dimensions for refractories and sputtering targets.

2.3.2.3 Low Temperature Sintering (LTP)

Around 2016 a paradigm-shifting paper authored by Clive Randall of Penn State University appeared in Angewandte Chemise (Applied Chemistry) describing this unique *process*. This pioneering invention is bound to impact the future of ceramic processing and technology. The process is carried out in two subsequent steps. First the ceramic powder is wetted with a few drops of water or acid solution resulting in the decomposition of particles and partial dissolution of powder to form a liquid phase at the particle–particle interfaces. The second step involves the application of the right combination of pressure and temperature to the slurry. The heat causes the dissolved particles to diffuse through the liquid and precipitate preferentially between the particles. This results in filling of the pores and allowing them to pack tightly together with well-organized atoms. The final product is a ceramic with high degree of crystallization and very high density. The essence of the method is to find the exact combination of moisture, pressure, and temperature for the process to be viable and successful. Obviously, it will vary from material to material. The method is applicable to metals, ceramics, and polymers. Compared to other conventional methods, it is energy efficient, cost-effective, and can be completed in much shorter time. The physical principle behind this process is somewhat related to the principle of hydrothermal growth of crystal which we will study later in the chapter.

2.3.3 Nanoceramics

Polycrystalline materials with grain size varying from 3 to 100 nm are loosely defined as nanostructured materials or simply NSMs. They were discovered in the 1980s, and ever since NSMs have become a topic of intense investigations all over the world. This is because of the fact that such materials exhibit both physical and mechanical properties that are far superior to those displayed by its conventional counterparts. Naturally therefore, they are the most sought-after materials for many applications and devices. It is not possible for us to go into any depth to discuss this topic within the framework of this book. However, we recommend for the benefit of advanced readers a chapter with the title of, "Nanostructured Materials" in the *Handbook of Nanoscience, Engineering and Technology* [1].

NSMs are also classified as nanocrystals and nanophased materials. They exist as metals, polymers, semiconductors, and, of course, as electroceramics. By now many oxides have been synthesized as NSMs with superior attributes as ferroelectrics, piezoelectrics, pyroelectrics, semiconductors, superconductors, and supercapacitors. This is where our interest lies in nanophased oxides.

Many different methods have been developed to synthesize nanophased materials. For processing of nanoceramics, the commonly used methods are solgel, sintering and two-photon-lithography. Vapor phase methods are also suited for those oxides which can sublime in vapor phase at reasonably low temperatures and pressures. Examples of such oxides include zinc oxide (ZnO), lithium oxide (LiO), tin oxide (SnO_2), and other binary oxides. As a good candidate for NSMs, ferroelectric/piezoelectric antimony-sulfo-iodide (SbSI) could be an excellent choice for vapor phase growth as nanoceramic. SbSI as single crystal has excellent ferroelectric, piezoelectric, and electrooptic properties.

Historically, the first nanophased ceramic was produced by the solgel method. This method is a chemical process and is also called the chemical solution process. Today, it is a widely used method for processing a large number of oxides and other electroceramics in film forms. We will discuss this method in some detail in the next section.

Sintering of nanosized ceramic powder is accomplished by microwave heating. The radiated energy is transferred almost instantaneously across the ceramic powder contained in an insulating chamber that does not impede the passage of the microwave radiation. In the process, heat is generated uniformly across the whole mass causing the sintering to occur. The sintering temperature is approximately 600 °C for most of the materials. It is also an energy-efficient process, and the entire process is completed in a short time.

The two-photon method is a laser-assisted lithographic technique to produce three-dimensional structures. This unique method can etch out lithographically three-dimensional nanophased structures. As the field is progressing and maturing, new and intriguing processing methods are being developed. Some of them even require simple and inexpensive instrumentation.

2.3.4 Thin Film Ceramics

Ceramic materials synthesized as thin or thick film forms are needed for basic studies and device development. The distinction between thin and thick films is not precise. As a crude estimation, we refer to films with a thickness in the nanometer range and up to a fraction of a micron as thin films. Films with a thickness from a few microns to submillimeter range may be classified as thick films. When we stack thin films layers of different materials over one another, we build integrated structures. These structures are unique in the sense that materials with dissimilar physical properties can be brought together in one integrated structure that becomes the vehicle for exploiting coupled behaviors. For example when one layer of a ferroic material with the spontaneous magnetization as its order parameter is stacked over another layer of a different ferroic material with the order parameter of spontaneous polarization, then we create an integrated structure consisting of two ferroic materials. Such a structure can be used to demonstrate coupling between the two-order parameters either by applying an external magnetic field or an electric field. Such a structure would be a model for the magnetoelectric phenomena. Such models are of enormous value to physicists and device engineers. We will discuss in-depth coupled phenomena in Chapter 5.

There are four types of films that are classified as amorphous, polycrystalline, textured, and epitaxial. The distinctions between them are subtle and are based on crystallographic orientations. The amorphous films have no degree of crystallinity and therefore, they have no long range order which is the characteristic of a crystalline solid. However, they have short range order at the atomic scale because of the nature of their bonding, Sometimes the amorphous state is also referred to as the glassy state. Amorphous films are also used in some applications.

The polycrystalline films exhibit crystallographic orientations that are totally random. In such films, the physical properties are isotropic. These films are totally devoid of any texturing. When a polycrystalline film is not totally random but exhibits some preferred orientations, then such a film is called textured. Texturing can be weak, moderate, or strong, depending on the degree of mismatch between the lattice constants of the film to those of the substrate on which it is deposited. If the substrate itself is polycrystalline or even amorphous the resulting thin films tend to be textured. The situation, on the other hand, is entirely different for *epitaxial* films.

Epitaxy is a Greek word. Epi means "above" and "taxis" meaning "an ordered manner." In our context, then we can define an epitaxial film as one whose crystalline orientation follows the orientation of the substrate. Here the crystallographic orientation of the film corresponds to that of the solid single crystal substrate. If the match between them is 0%, then the degree of epitaxy in the film is outstanding and approaches that of a bulk single crystal. If the mismatch is minor, then the degree of epitaxy is less than 100%. This mismatch can be accommodated through strain in which case the crystalline quality would be similar to the case when there is 0% lattice mismatch. However, for small lattice mismatch if the film thickness exceeds the critical thickness, the strain energy would be so large that the layer relaxes creating defects. The film is still of excellent crystallographic quality but is not exactly a single crystal. This suggests that the quality of epitaxy is itself a measure of lattice constant mismatch between the film and the substrate. The substrates for the growth of epitaxial films are invariably polished to be of optical quality and free from surface defects. They come as plates or wafers cut from bulk single crystals. Ideally, one would prefer 100% lattice matching between the substrate and the films to be deposited, but such is not always possible; so one settles for the best match. Lattice constant match between the substrate and the film is a stringent requirement, and a very careful selection of substrate materials become imperative if epitaxial films are desired.

2.3.5 Methods for Film Growth

There are many methods for film growth. The selection depends on many factors that include the nature of the film desired and the availability of instrumental resources in an academic or industrial laboratory. As far as electroceramics are concerned, there are five principal methods that are customarily used for film growth. They are (i) Solgel or chemical deposition method; (ii) Pulsed laser deposition (PLD) method which is also known as laser ablation method (LAM); (iii) Molecular beam epitaxy (MBE) method; (iv) Radio frequency magnetron sputtering method (RF Magnetron Method); and (v) Liquid phase epitaxy (LPE) method. Methods (i)–(iv) are the most used methods for deposition of oxide materials, whereas the LPE method is not so common because of the many unknown factors associated with the chemical reactions taking place during the process at elevated temperatures. In the next sections, we will outline the basic principles of these methods, and the relative merits and demerits for different methods for film growth.

2.3.5.1 Solgel Method

This is one of the simplest methods for film growth. It is efficient, less time-consuming, and instrumentation requirements are simple and less expensive compared to other film growth methods. It is an ancient process dating back to the mid-1880s, but it was not until around 1990 that its importance to the growth of ceramics, especially oxides, was appreciated. The process takes place in four distinct steps as outlined in Figure 2.9.

The most important and crucial step involves the preparation of the precursor consisting of an organic metal alkoxide in which fine ceramic particles are suspended. The alkoxide undergoes hydrolysis and polycondensation reactions to form colloids. The term *colloid* is used to describe a broad range of liquid–solid or liquid–liquid mixtures. The colloidal precursor must be prepared with great care, preferably by an organic chemist, who is well trained in this art. If the precursor is not well prepared, good oxide films will not precipitate.

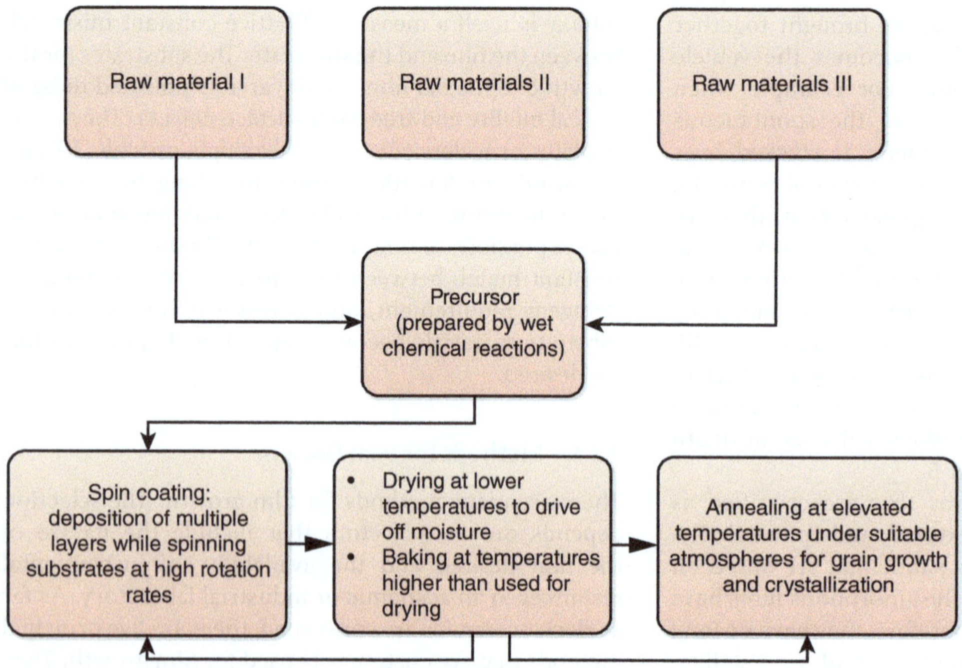

Figure 2.9 A block diagram outlining steps involved in sol–gel process of films.

The word *solgel* is the abbreviations combining two chemical concepts which are solvent (sol) and gel-like mixture containing the solid and liquid phases concurrently. In such a colloidal system the volume fraction of solid particles is so low that the two phases cannot be distinguished by naked eyes. Therefore the precursor looks like a dark liquid. It has a short life span varying from a few days to a few weeks. The rule of thumb is "the fresher the precursor, the better the quality of film produced."

The solgel process involves spin coating the substrate with the precursor using a high-speed centrifuge. The growth chamber is covered to avoid contamination from dust particles. The colloidal solution is injected dropwise on the rotating substrates, resulting in a uniform distribution of the liquid. Subsequently, the substrate with the film is first baked at low temperatures to remove excess solvent. Then additional layers may be added by repeating the process. Once a satisfactory thickness is achieved, the solid film is annealed at higher temperatures for grain growth and crystallinity. The annealing temperature for these films is much lower and its duration much shorter than for bulk ceramic samples. The quality of films grown by the solgel method are dense and unusually of high quality. These films are typically polycrystalline with good texturing. Solgel method is applicable also to growth of ceramic fibers and polymer films. It is the most used method for the growth of textured ferroelectric and piezoelectric films that can then be integrated with other materials.

2.3.5.2 Pulsed Laser Deposition (PLD) Method

PLD method is a sophisticated technique for the growth of simple and complex oxides. Broadly defined, it is a vapor phase epitaxy technique. Laser-assisted film growth of electronic materials began soon after the discovery of the laser in 1960 at Bell Laboratories. From then until the late 1980s, PLD remained impractical because of its many drawbacks. The picture changed drastically with the discovery of high temperature superconductivity in oxide systems in 1987. A high-power excimer laser was employed to deposit films of high-temperature superconducting oxides on a variety of substrates. Excimer lasers emit UV radiation with wavelengths varying from 126 to 351 nm. For PLD growth typically a KrF excimer laser emitting UV with 248 nm wavelength is used. During the growth process, the repetition rate of the laser is usually maintained at about 100 Hz with a pulse duration of ~10 ns. Over the years, many refinements have been made, and today, PLD has become one of the most used film growth methods for oxides. The growth mechanism is straightforward and simple. A high-power laser beam is focused on a ceramic target in relatively high vacuum resulting in the production of a very high temperature energetic plasma carrying the atoms and molecules ablated from the target surface. By directing the plasma to a substrate located a few centimeters away from the target, which is heated by a heater embedded in the substrate assembly, the vapor settles on the substrate. The heater is usually maintained at around 600–800 °C. During the experiment, one can use partial pressure of oxygen to compensate for the loss of oxygen in the process from the target. The substrate heater serves the double purpose. Once the growth is completed, the substrate with the film is left in the growth chamber so that in situ annealing can take place. The resulting film is of high quality and epitaxial. Another very important feature of these films is that the chemical stoichiometry is maintained between the substrate and the film. This is a very important attribute of the PLD system. Since the materials are removed from the surface of the target by the process of ablation, the PLD method is also called the laser ablation deposition (LAD) method. The growth chamber consists of many ports to accommodate a significant number of devices that can add to sophistication of the growth method.

We show an overview of a PLD chamber in Figure 2.10. The schematic is self-explanatory with one exception. We see that the chamber is fitted with a RHEED gun and a RHEED screen. The acronym refers to "reflection high energy electron diffraction." The chamber outfitted with a RHEED device enables the operator to monitor the electron diffraction pattern generated by the film surface in progress. The degree of epitaxy and therefore, the quality of film is monitored in situ that is a very desirable outcome of the growth process.

In Figure 2.11, the general setup of a PLD system is shown in the left side, whereas on the right side, the system under operation is depicted. Here one can easily see the plasma plume generated by the excimer laser with its cone in focus with the substrate.

The growth mechanism is simple and quite obvious, but the physical principle behind the PLD method is rather complex and still not well understood. What, exactly, is happening during the process? First, the laser energy is absorbed by the target. The conversion of the absorbed energy takes place in the form of electronic excitation, and then into thermal, mechanical, and chemical energies. To develop a model explaining such a complex mechanism is not a trivial enterprise. Simply to set up the thermodynamic equilibriums with the multiple facets of the phenomenon is mathematically a daunting task.

PLD is a valuable research tool in the hands of an experimentalist. It is a versatile method for film growth and is easily adoptable for the growth of nitrides, metallic multilayers, and superlattices. The process has many desirable attributes, but there are some drawbacks too. The film can contain islands of particles clustered together, and the process is limited to growth on a small surface area. These are the two handicaps that work against the PLD technique being adopted by the microelectronic industry.

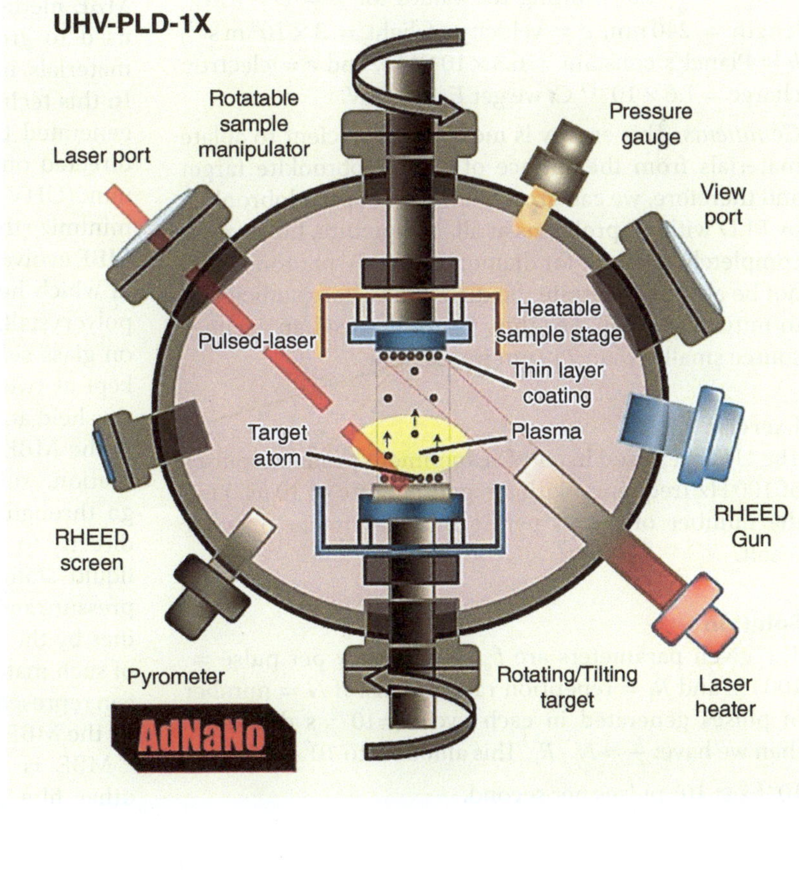

Figure 2.10 Schematic of a sophisticated PLD growth chamber. Source: http://www.adnanotek.com/uploads/1/6/5/3/16533032/4966021_orig.jpg?157.

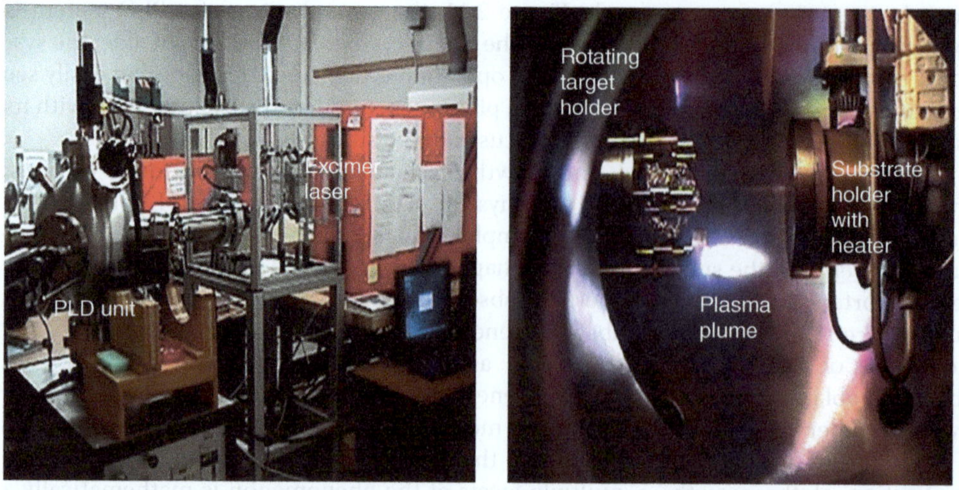

Figure 2.11 A PLD unit in operation. Source: Courtesy of Electronic Materials Laboratory at the University of Alabama, Tuscaloosa, AL, for the growth of ilmenite–hematite magnetic semiconductor.

Exercise 2.7
Find the photon energy associated with the UV radiation produced by an excimer laser of 240 nm wavelength. Would you use this laser for the growth of diamond ($E_g \approx 5.47$ eV) and pseudobrookite, Fe_2TiO_5 ($E_g \approx 2.18$ eV) films? Comment on your results.

Solution
According to Planck's law of radiation, E_p = Photon energy = $\frac{ch}{\lambda}$. Substituting the values for λ = UV wavelength = 240 nm, c = velocity of light = 3×10^8 m s^{-1}, h = Planck's constant = 6.3×10^{-34} J s, and e = electron charge = 1.6×10^{-19} C, we get $E_p \approx 5.2$ eV.

Comments: This energy is more than sufficient to ablate materials from the surface of a pseudobrookite target and therefore, we can grow a thin film of pseudobrookite by PLD with no problems at all. The picture, however, is completely different for diamond. The UV photon would not be able to penetrate the diamond surface sufficiently to initiate ablation. For this, we would need an excimer source smaller than 240 nm wavelength.

Exercise 2.8
The UV laser used in a PLD experiment produces pulses of 100 Hz frequency with a repetition rate of 10 ns. Find the number of pulses per second. Comment on your result.

Solution
The given parameters are f_p = frequency per pulse = 100 Hz and R_t = repetition rate = 10 ns. If N = number of pulses generated in each cycle (=10^{-2} s duration), then we have: $\frac{1}{f_p} = N \cdot R_t$. This amounts to 10^7 pulses per 10^{-2} s $\cong 10^9$ pulses per second.

Comments: From Exercise 2.7, we know that the single UV photon energy = 5.2 eV. Therefore, the energy penetrating the target surface $\approx 5.2 \times 10^9$ eV s^{-1}. This enormous amount of energy will raise the target temperature so high that the ablated particles would vaporize almost instantaneously. This will also then result in recrystallization of the surface layer of the target making it very smooth and shiny.

2.3.5.3 Molecular Beam Epitaxy (MBE) Method

MBE method is an evaporation technique that can be used to grow thin films of high-quality single crystal materials, including semiconductors, metals, and oxides. In this technique, neutral atomic and molecular beams, generated thermally from heated Knudsen sources are directed onto a heated substrate under ultra-high vacuum (UHV) conditions. UHV conditions are essential to minimize the incorporation of impurities during growth. MBE evolved from the work of Günther during the 1950s in which he used a three-temperature method to grow polycrystalline but stoichiometric films of InAs and InSb on glass substrates [2]. In this method, the sources are kept at two different temperatures, while the substrate was held at an intermediate temperature.

The MBE method is based on the principle of evaporation, sublimation, and condensation. Some solids go through the phase transition from the solid phase directly to the vapor phase without going through the liquid state, occurring at the critical combination of pressure and temperature. The vapor subsequently solidifies by the process of condensation. Classical examples of such materials are dry ice (CO_2) and naphthalene. We can represent the process taking place during film growth by the MBE technique with the help of Figure 2.12.

MBE is a highly sophisticated technology and no other film growth method comes even close to it in

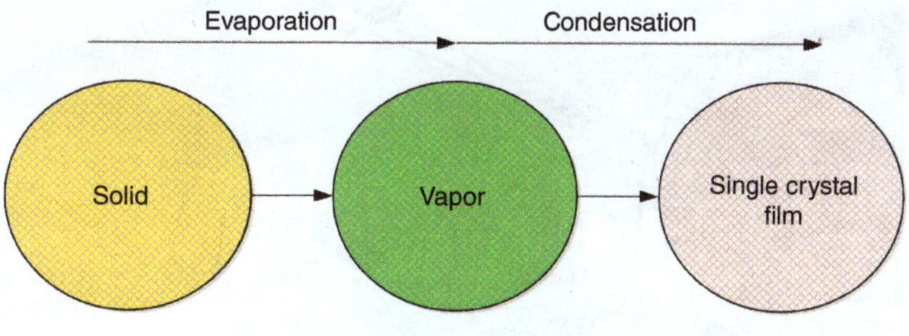

Figure 2.12 Solid-sublimation-condensation process applied to MBE film growth technology.

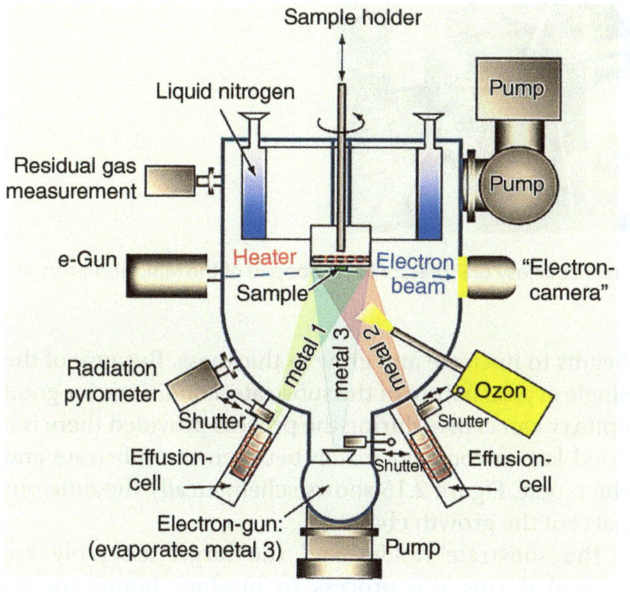

Figure 2.13 Schematic of a MBE system. Source: https://www.google.com/url?sa=i&rct=j&q=&esrc=s&source=images&cd=&ved=0ahUKEwi0-rXkj4vTAhUBVmMKHYuSDuEQjRwIBw&url=https%3A%2F%2Fwww.fkf.mpg.de%2F273938%2F30_Oxide_MBE_Lab&psig=AFQjCNHekD116tBBdx8ycrQ1zEo6LVkk1Q&ust=1491406359975668.

sophistication and quality of resulting films. But the system comes at a steep financial outlay. On the basis of initial investment and daily maintenance PLD is a "poor man's MBE."

Figure 2.13 shows a schematic of the main growth chamber of a MBE system. It is outfitted with a number of special devices such as e-gun, RHEED, and effusion cells. The chamber also has the provision for introducing oxygen, ozone, and nitrogen for the growth of oxides and nitrides. The substrate assembly is rotated for uniformity of film.

Figure 2.14 is a photograph of an operating MBE system for growth of oxides.

MBE technology as a practical film growth technique was developed in early 1970s at Bell Laboratories and since then it has become the major source of producing single-crystal thin films by epitaxy for various device applications. For a very long time, it was mostly used for the growth of epitaxial semiconductor films, especially III–V compounds including gallium arsenide, GaAs. It is of advancements that MBE has been modified to include the growth of oxide films. The method has also been very effective in the growth of high-quality quantum dots, quantum wells, and superlattices. Ever since its invention, the MBE technique has been effectively used not only for thin films but also for the growth of integrated structures and semiconductor devices such as diodes and transistors. It is considered to be the most reliable method for producing NSMs and for the development of nanotechnologies. The whole process takes place under an UHV environment (pressure $\approx$ 100 nPa $\approx 10^{-9}$ Torr). Vacuum is maintained at a very high level by a combination of ion pumps, cryo-pumps, and turbo molecular pumps. MBE systems also have cryo-panels that must be chilled to 77 K by using liquid nitrogen for additional pumping. For the growth of oxide materials, oxygen must be introduced in the chamber usually through a high-precision leak valve or a mass flow controller. Even films of organic materials can be grown by the MBE technique.

The principal objective of this technology is to produce chemically ultrapure materials as thin-film single crystals. To achieve this goal, high-purity raw materials are first heated until evaporation to produce atomic or molecular beams that impinges on a heated substrate where it ultimately nucleates as a single crystal. During the nucleation stage, the different species of the source materials undergo chemical reactions. At the end, they combine to form as perfectly epitaxial thin layers with true single crystal attributes. Different effusion cells are required to heat and evaporate different materials (as indicated as metal 1, metal 2, and metal 3 in the schematic of Figure 2.13. Heating is achieved by Knudsen effusion cells and electron beam evaporators for materials that have low vapor pressures. The temperature is controlled precisely by a temperature controller and the temperature of the substrate can be measured by radiation pyrometer. The quality of the layers growing epitaxially on the wafer is monitored in real time by RHEED.

2.3.5.4 RF Magnetron Sputtering Method

The sputtering method is in fact a physical vapor deposition (PVD) method for thin film growth of a large

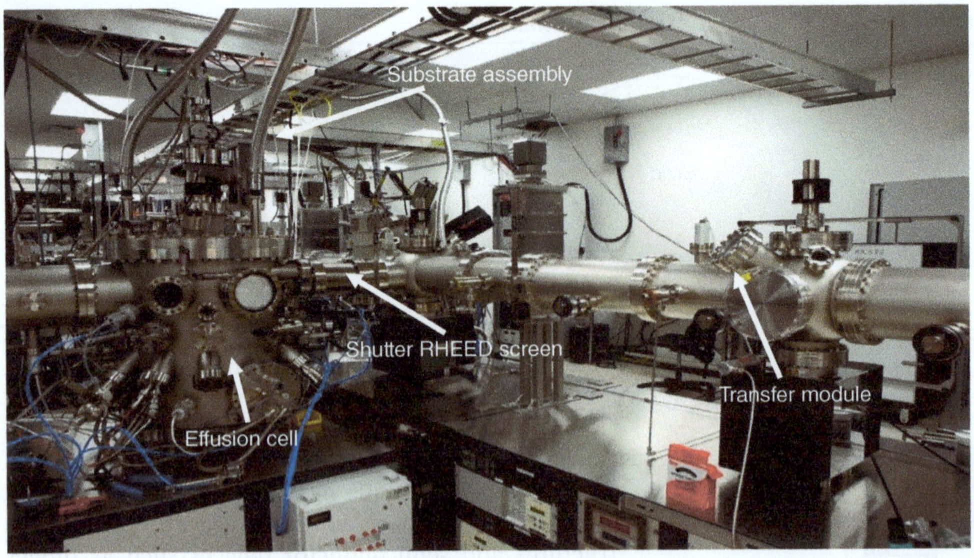

Figure 2.14 A sophisticated MBE unit in operation for oxide growth. Source: Courtesy of Professor Ravi Droopad of the MBE laboratory at Texas State University, San Marcos, TX.

number of materials. The mechanism involved is simple and straightforward. The target is bombarded by high energy ions or atoms with a wide range of energy. The bombardment at very high frequency knocks out atoms from a target carrying with them large energies. These energies can be of the order of tens of electronvolt, and the sputtered atoms can acquire temperature of the order of 10^5 K. As a consequence, a plasma plume is generated that reaches the proximity of the single crystal substrate placed directly above the target? Once the sputtered particles in vapor form settle on the substrate, thin film begins to nucleate and grow in thickness. Because of the single crystal nature of the substrate, conditions for good epitaxy can evolve during the process provided there is a good lattice constant match between the substrate and the target. Figure 2.15 shows schematically the different parts of the growth chamber.

The substrate holder and the target assembly are rotated during the process to produce homogeneous films with minimal compositional variations. In order to avoid charge build-up on an insulating target, the system is powered by a DC/RF power supply. The

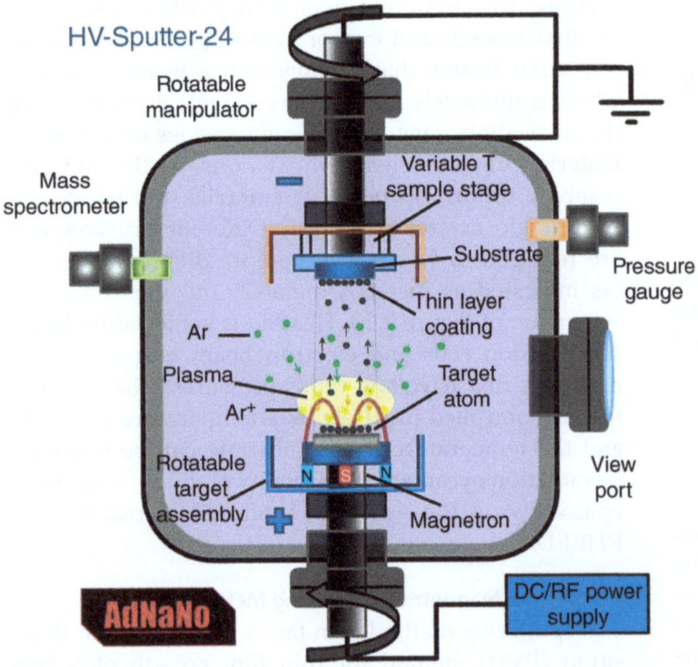

Figure 2.15 Schematic of a sputtering growth chamber. Source: http://www.adnano-tek.com/uploads/1/6/5/3/16533032/3126982_orig.jpg.

Figure 2.16 RF magnetron sputtering system with its associated attachments. Source: http://surface.iphy.ac.cn/sf03/images/DSCN4944.jpg.

radio-frequency source is commonly maintained at about 14 MHz leading to change of polarity between the cathode and anode bias at a rapid rate. Furthermore, the target assembly also includes a magnetron device whose purpose is to contain the plasma particles close to the target by making use of the electric and magnetic fields. The vaporization of the target induces ionization of its surface layers. The ejected ions of different masses travel toward the substrate in the presence of an electric and a magnetic field following a curved trajectory. The mass spectrometer installed close to the target serves in separating the ions according to their masses. This organized mass system becomes the signature of the sputtered film. Figure 2.16 shows a picture of the magnetron sputtering system.

RF magnetron sputtering is only one type of sputtering system. Other examples are ion beam sputtering (IBS); ion-assisted sputtering (IAS), and high utilization sputtering (HUS). They all work on the same basic principles of sputtering. Atoms are knocked out from the target surface which after high energy collision vaporize and then solidify as thin layers on a substrate.

2.3.5.5 Liquid Phase Epitaxy (LPE) Method

LPE method for film growth is based on a melt method and is applicable for the film growth of materials melting either congruently or incongruently. This method can be adopted for the film growth of metals, semiconductors, binary, and pseudo binary systems. As far as oxides are concerned one would prefer to use this growth method if they can be represented by either of the two phase diagrams depicted in Figures 2.3 and 2.4. This will help in optimizing the growth parameters, resulting in deposition of high-quality epitaxial layers.

The LPE method is also suitable for the growth of thick films. The same protocol as used for the solidification of materials using the two phase diagrams of Figures 2.3 and 2.4 can be adopted with minor modification for the LPE method. A growth chamber resembling that used for the Czochralski crystal growth method can also be employed for LPE method. Rotation of both the seed and the melt produces homogeneous, smooth, and stress-free films.

Both homoepitaxial and heteroepitaxial films can be grown using suitable substrates. But growing on substrates of the same material as the film (homoepitaxy) is advantageous because of their perfect lattice constant match and similar response to high temperature chemical reactions. The ideal situation involves systems with phase diagrams that possess a complete solid solution. For example, potassium tantalate ($KTaO_3$) and potassium niobate ($KNbO_3$) form a complete solid solution, allowing one to crystallize a ferroelectric potassium tantalate niobate (KTN) compound. Therefore, highly epitaxial layers of KTN can be precipitated on KTN substrate by following the protocol of slow cooling without being burdened by possible unfavorable chemical reactions. Another advantage of this approach would be the growth of films of any composition of KTN simply by adjusting the growth conditions. A large number of oxide films have been grown by this method. Magnetic GGG (Gd-Ga-garnet) film on GGG substrate is a classic example of the LPE growth technique. In 1980s, GGG film was a highly sought-after material for the development of magnetic bubble memory technology. For heteroepitaxy, caution should be exercised in the selection of substrates to form high-quality compounds.

2.3.6 Single Crystal Growth Methods for Ceramics

We have already learned about the different methods for the processing of bulk ceramics and the processing of thin films. The third processing method that we need to cover deals with the growth of bulk single crystals especially those applicable to the growth of electroceramic single crystals.

Crystal growth is a fascinating field of technology and is so important to science and technology that no other processing method can substitute for it. It requires a great degree of patience, diligence, and intuition on the part of its practitioners who must be technically competent and esthetically endowed. On top of these attributes, good luck is an important credential that only the Superior Power can grant. Crystal growth is both a science and an art. Nature is the Master of all masters in crystal growth. Diamond and scores of gems and other countless minerals are its handiwork.

The science and the art of crystal growth is an ancient field, and there are hundreds of materials with their own unique peculiarities and idiosyncrasies. The first synthetic gemstones were reported in the 1800s. Over the centuries, numerous crystal growth methods have

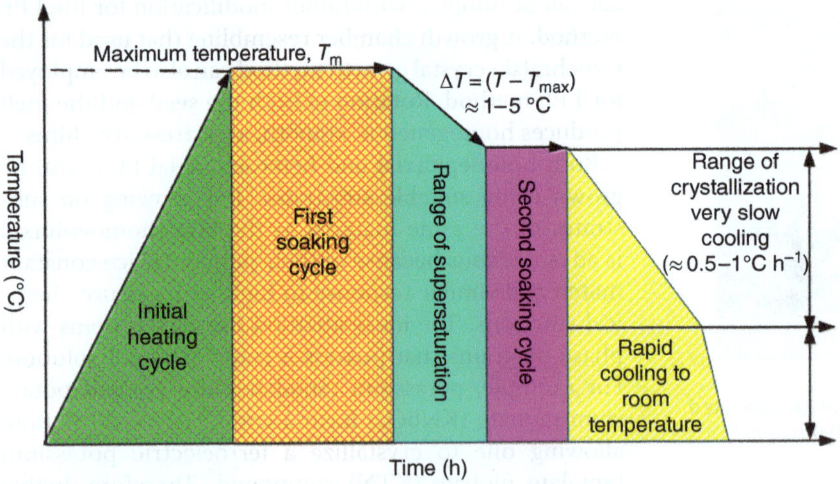

Figure 2.17 Multiple steps of heating and cooling involved in high temperature solid solution growth of crystals.

evolved, and it is not within the framework of this chapter to do justice to each of them. Keeping this in mind, we will limit ourselves to the growth of oxides that are the lead members of the electroceramic group for fundamental studies and its applications that are already established and evolving at a rapid rate. Readers interested in advanced knowledge of this field are referred to the many standard books on the subject.

The science of crystal growth for oxides can be broadly classified in three principal categories which are the following: (i) Growth from the vapor; (ii) Growth from the melt; and (iii) Growth from fluxes. In sections that follow, we will discuss each of them in some detail. Selection of one or another method will depend first on the nature of the melting points of solids and on their chemistry at high temperatures. It should be realized though that the chemistry of materials at high temperatures is not well documented, and so we will have to depend upon our intuition to guide us.

There are two types of melting points: one is congruent, and the other incongruent. Solids for which the melting and solidification takes place at the same temperature are said to have a *congruent* melting point. Some examples are elements such as pure metals. Those solids that melt at one temperature and solidify at a different temperature have an *incongruent* melting point. Examples of this group are binary and pseudobinary oxides. Very few oxides melt congruently, in fact, such oxides are rare. The majority of oxides melt incongruently. The discussions on the three most used methods for crystal growth of oxides now follows.

2.3.6.1 High Temperature Solution Growth (HTSG) Method or Flux Growth Method

This method is used to grow crystals of materials that melt incongruently at very high temperatures. In order to synthesize them as single crystals, first the materials must be dissolved in high-temperature solvents (also called fluxes) and then cooled very slowly for crystals to nucleate. In Figure 2.17, we outline a protocol for the growth of crystals by the HTSG or flux method.

The charge for crystal growth mixed with a flux is packed in a crucible usually of platinum. The growth takes place in air using resistive heaters. Platinum crucibles are employed because platinum is chemically inert and is not supposed to react with aggressive melts consisting of flux and the charge. Nevertheless, some fluxes do attack platinum at high temperatures, but the damage is minimal and does not render the crucibles useless.

The crucible with charge and flux must be maintained very clean and therefore needs to be contained in a covered tall alumina crucible during the growth process. There are a number of fluxes with the most used and prominent ones being a mixture of lead oxide (PbO) and vanadium oxide (V_2O_5); PbO and lead fluoride ($PbO{-}PbF_2$; PbO and boron oxide (B_2O_3), and barium borate ($BaO{-}B_2O_3$). At high temperatures, PbO has the tendency to lose oxygen that can reduce PbO to metallic Pb that is very aggressive in molten state. To compensate for this loss, sometime PbO_2 is usually added.

According to the protocol described in Figure 2.17, we notice that the charge is rapidly heated to the maximum temperature (T_m). It is held there for number of hours for soaking. Once the first soaking cycle is completed, the temperature is reduced by a very small amount. This is done to initiate supersaturation of the dissolved charge in the molten flux. After a short wait, the second-soaking period commences and continues for a few hours. This helps in achieving a homogeneous supersaturated melt. In the final stage of the process, a very slow cooling is initiated, usually with the rate of 0.5–1 °C h^{-1}. During this stage, a thermal range is achieved that is favorable to the nucleation of very tiny crystallites. These crystallites float in the turbulent melt and begin to grow in size. The very slow cooling cycle may last for days or even weeks at a time. On the completion of this cycle, the furnace may be shut off and allowed to reach room temperature rapidly. Knowledge of the phase can be advantageous because this can enable one to fix the various temperatures that are used for different thermal cycles. In the absence of such a phase diagram, the temperatures can be guessed intuitively and by trial and error, one can find the most useful temperature ranges that are needed.

The platinum crucible with the frozen mass is retrieved for harvesting the crystals varying in size from 1 to 12 mm in length. The harvesting is done by dissolving the solid mass in an acid bath and then filtering out the crystals. This process alone can take days to complete.

This is a very slow process, but if this growth method is the only option to produce some crystals for research and device development, then it is a good option. By this method, one can control the size of resulting crystals that can vary in size from nm to mm scale. The resulting crystals may contain flux as impurity and may not be defect free. In spite of these drawbacks, the high temperature solution growth (HTSG) method is widely used for growth of oxides for which no other method might be applicable. Famous oxide crystals such as magnetic garnets, YIG ($Y_3Fe_5O_{12}$), piezoelectric PZN–PT (lead-zirconate-titanate and lead titanate solid solution), antiferroelectric lead-titanate ($PbTiO_3$), ferroelectric barium titanate ($BaTiO_3$), and semiconductor pseudobrookite (Fe_2TiO_5) have been grown by the HTSG method.

The size of crystals can be enhanced by introducing crucible rotation, melt stirring, and a seed crystal during the growth process. Size enhancement can also be achieved by introducing some nutrients (tiny crystallites of the material obtained from previous growth runs) in the melt. As soon as the nucleation begins during the very slow cooling cycle these tiny crystallites, which may even be mm size, will prefer to adhere to the seeds due to their chemical affinity, and ultimately grow in bigger sizes.

2.3.6.2 Czochralski Growth Method

Czochralski Growth Method is named after Jan Czochralski of Poland who invented this technique in 1916 for crystal growth. This method was an accidental discovery (as many of them happen to be). Czochralski was experimenting with molten tin when he accidentally dropped his pen in the melt. He immediately pulled it out and to his dismay found that a thin fiber of solidified metallic tin was hanging from the nib. He replaced the nib with pointed objects and repeated the experiments. Each time the result was the same. On analysis, he found that these solidified fibers were indeed single crystals.

This technology got its modern-day importance after the semiconductor germanium crystals were grown at Bell Laboratories by Gordon Teale and his colleagues in 1948 after introducing multiple refinements to the instrumentation used by Czochralski. This led to the discovery of the transistor, also at Bell Labs; within a year after Ge single crystals became available. They were used to produce first the *point contact transistors* followed by the invention *of bipolar junction transistors*. These groundbreaking inventions gave a real push to the Czochralski crystal growth technique. In the meantime, this growth technique has been used for many oxides, some of which are needed for the fabrication of high-power lasers and electro-optical devices.

The growth takes place from the melt of congruent materials, so materials with congruent melting points are candidates for crystal growth by the Czochralski method (Cz-method). Nucleation is initiated on a seed by bringing it in contact with the melt. Once the initial growth is established and conditions are favorable, the crystal can be left growing in size and diameter until all the melt is exhausted in the crucible. A successful Cz-growth experiment can produce large crystal boules having the same exact crystal orientation as the seed. For example, 12-in. diameter and many feet long boules of silicon are routinely grown on an industrial scale these days. A good size silicon boule can easily weigh many kilograms. In Figure 2.18, a silicon boule is shown that just a part of the original boule.

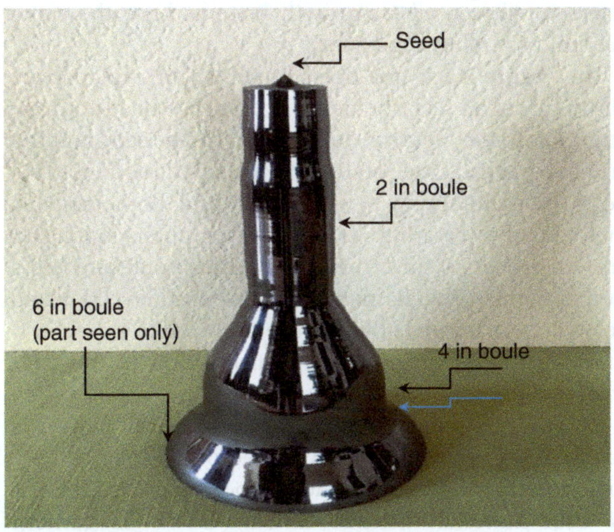

Figure 2.18 Part of a large silicon single crystal boule grown by the Cz-technique. Source: Courtesy of Texas Instruments, Inc., Dallas, TX.

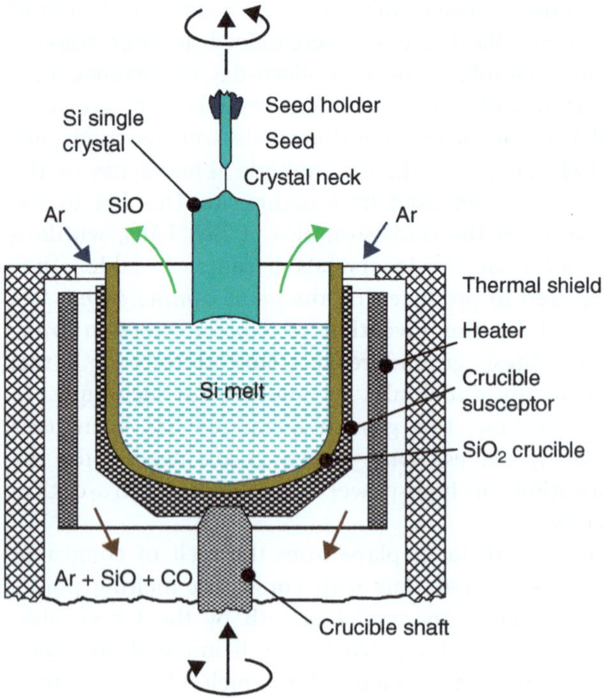

Figure 2.19 A sketch showing different parts of a Cz-growth chamber. Source: https://www.tf.uni-kiel.de/matwis/amat/elmat_en/kap_6/illustr/cz_si_growth.gif.

Here, we find that the boule diameter is changing from 2 to 4 in. and then finally to 6 in. This step-wise growth is done purposely to slowly establish the right thermal and mechanical conditions to produce a boule with perfect orientation and with minimal defects. The general instrumental arrangement for the Cz-growth method is shown in Figure 2.19. Although this figure refers to silicon, all Cz-growth systems have the same general type of arrangement.

This figure does not need much of an explanation; all the parts are clearly identified. But we need to draw attention to the fact that during growth, the seed and the crucible pedestal are rotated in opposite directions. This helps in maintaining a well-mixed and homogeneous melt in the rotating crucible. Computer-controlled growth process assures the quality of the boule and helps in maintaining the diameter at the desired level. During growth, pure argon is bled in the chamber to expel SiO_2 that can form from the chemical reaction of the melt with the quartz crucible.

Oxides identified as laser, electro-optical materials are routinely grown by the Cz-method so long as they melt congruently and have low vapor pressure. The list of laser materials include GSGG (Gd-Sc-Ga garnet), GSAG (Gd-Sc-Al garnet), YSGG (Y-Sc-Ga garnet), YAG (Y-Al-garnet), and GGG (Gd-Ga-garnet). Lithium niobate ($LiNbO_3$) and lithium tantalate ($LiTaO_3$) are established ferroelectric materials that exhibit excellent electrooptic properties. Another famous electrooptical material is Sr-Ba-niobate (SBN) which is also grown by the Cz-technique. These crystals are also used as substrate materials for thin film growth. For the growth of oxide materials, the crucible has to be either platinum or iridium. In the growth chamber, the temperature is high, and the melt is chemically aggressive. Neither the temperature nor the melt adversely affect the inert metals crucibles. Quartz crucibles, which are used for silicon growth, are chemically incompatible with oxide melts.

2.3.6.3 Top Seeded Solution Growth (TSSG) Method

This growth method is applicable to incongruently melting materials. It is a combination of the Cz-growth method and the HTSG method. A good knowledge of the phase diagram of the system of interest is essential to developing a working protocol for the top seeded solution growth (TSSG) technique. Leading ferroelectric materials such as barium titanate (BT) and potassium-tantalate-niobate (KTN) are routinely grown by this method to produce optical quality large crystals. Also many relaxor ferroelectrics can be grown as large single crystals using this method.

The crystals can be grown using a growth chamber similar to the one used in Cz-growth. Its thermal management is dictated by the phase diagram of the system. First, the ceramic charge is melted, and after thermal saturation is achieved a seed crystal is introduced below the melt (unlike in Cz-growth where the seed contact is made at the melt surface). Both the seed and crucible are rotated just like in the Cz-method. The seed becomes chemically etched by the molten flux that promotes nucleation. Then the slow cooling process begins and the growth temperature is monitored using an optical pyrometer, while a pre-programmed growth protocol is implemented by a computer. The growth usually continues for days without any interruptions until a good size crystal grows.

How would we know the range of cooling and its rate? This is where a good and reliable phase diagram becomes valuable. For the sake of discussion, let us look once again at the phase diagram of Figure 2.3. Let us say that we want the final composition of the crystal to be equivalent to $A_{0.2}B_{0.8}$, for this composition to crystallize, we might prepare a charge having a composition close to $A_{0.8}B_{0.2}$. Then by slow cooling between the temperature of T_i and T_1, we can grow a crystal with the desired composition. To clarify this point, let us consider the case of KTN as a practical example. We would use the phase diagram

Figure 2.20 KTN crystal grown by TSSG method. Source: Goeking et al. [3]. Reproduced with the permission of Taylor & Francis.

of potassium niobate ($KNbO_3$) and potassium tantalate ($KTaO_3$). With the solid solution nature of the phase diagram and logic used for the growth of a composition of interest (such as $A_{0.2}B_{0.8}$) we could grow KTN crystals.

The slow cooling of the melt may be at the rate of $1\,°C\,h^{-1}$ or even less. Once a good size crystal grows, one can slowly pull it out from the melt and rotate it vigorously to remove the liquid adhering to it. Then the chamber can be brought to room temperature slowly to avoid cracking of the crystal due to thermal stress. A large crystal of KTN with well-developed faces and steps signifying step-wise growth has been grown by this method and is shown in Figure 2.20 [3].

The method is also applicable to systems that do not form complete solid solution as KTN but must be dissolved in a flux to crystallize by slow cooling.

2.3.6.4 Hydrothermal Growth

This growth method is applicable to materials that can be dissolved in aqueous solution at high temperatures and pressures. There are two categories of this method, one in which the charge is soluble in water at atmospheric pressure and low temperatures, and the other in which the charge is soluble in water only at high pressures and high temperatures. Here high temperature refers to temperatures mostly in the range of 500–600 °C that can be far below the melting point of many oxides. As far as the nature of melting point is concerned, it can be either congruent or incongruent. The basic requirement is that the charge can be dissolved in aqueous solution under specific conditions of temperature and pressure.

Low temperature growth: For low temperature growth, usually between room temperature and 100 °C, supersaturated solution is first produced in a glass container that can be outfitted with temperature control and a stirring mechanism. The temperature is very slowly lowered resulting in rejection of excess charge from the supersaturated solution. These minute particles can be of nanometer size and act as nucleation centers for large crystals to grow. In the absence of a seed, a large number of highly transparent small crystals that are still big enough to handle and study, settle down at the bottom of the growth vessel. A typical example of this scenario is recrystallized sugar crystals or rock salt crystal grown from aqueous solution at room temperature. By introducing a seed crystal, larger crystals can be grown by this method.

High temperature and high pressure growth: The growth principle is similar to the one described above. However, it requires autoclaves that can withstand high pressures and temperatures. Precautions must be taken to prevent explosion, otherwise, the consequences can be catastrophic. The growth chamber is usually in the form of a cylinder in which the charge and water are contained. The pressure and temperature are slowly raised that accelerates the dissolution of charge. Once equilibrium is attained both temperature and pressure are lowered slowly. This promotes nucleation and ultimately growth of rather large crystals. The growth rate is fast, the resulting crystals have few defects and are of rather very high quality. This method has been used for the growth of quartz (SiO_2) since 1851. Because the growth takes place in a closed environment, the atmosphere may be controlled to produce oxidizing or reducing as desired.

Autoclaves of various designs are commercially available. Quartz or Pyrex glass vessels are satisfactory for growth at low pressure (≈10 atm) and low temperature (≈300 °C). Steel autoclaves work well up to 500 atm and 400 °C. For higher pressure and temperature, they must be welded and specially designed. Some examples of crystals grown by this method include pyroelectric TGS (triglycine sulfate), Al_2O_3 (sapphire), magnetic magnetite (Fe_3O_4), semiconductors, ZnO and ZnS, magnetic nickel ferrite ($NiFe_2O_4$), and piezoelectric aluminum phosphate ($AlPO_4$).

2.3.6.5 Some Other Methods of Crystal Growth

As already stated, there are as many different methods of crystal growth as there are different categories of materials. Some sublime, some melt, some dissociate on melting, some easily vaporize, some have low melting points, some have high melting points, and yet there are some that cannot be included in these general and broad schemes of crystallization. Table 2.3 outlines a few other methods used for crystal growth.

Table 2.3 Some other methods used for crystal growth of a variety of materials.

Method	Description	Materials grown
Bridgman–Stockbarger method	Growth under temperature gradient conditions	BGO ($Bi_4Ge_3O_{12}$), CaF_2, CeF_2, NaI:Tl, and LiF
Zone melting method or crystallization by zone refining	A polycrystalline rod is transported horizontally very slowly in a boat through a narrow zone with the temperature greater than the melting point of the rod. The molten zone is pulled very slowly out of the hot zone. The melt begins to form a single crystal	Applicable to semiconductor materials such as Si, Ge, PbTe, and PbSe This method is also used for producing highly purified ingots of semiconductor materials
Verneuil method; also called flame fusion method	Controlled melting of a powder charge in a flame and then allowing droplets to fall through a long column on a cold finger of seed crystal promotes growth of a larger crystal	Gem quality crystals such as sapphire and ruby, as well as quartz Growth takes place under nonequilibrium conditions that introduces defects in the crystal
Vapor phase method	Evaporation of a charge and guiding it to a seed crystal	Applicable to the growth of high-vapor pressure materials such as $CuSO_{4.5}$ H_2O. Difficult to maintain evaporation speed leading to growth fluctuations. Crystals are of poor quality and include clusters

Glossary

Ablation Particles removed from a solid material by evaporation.

Annealing A thermal process widely used in materials processing especially in ceramic technology and metallurgy. The process involves heat treating a sample at elevated temperatures but below its melting point, holding it there for an extended period and then slow cooling to room temperature. The process promotes grain growth and can produce samples with well-defined crystalline grains with chemical homogeneity. This is a step that must be used to produce high-quality samples of bulk ceramic. The process is also used for producing stress-free films and single crystals with high degree of crystallinity. The annealing can be done either in air, or oxidizing or reducing environments depending upon the desired chemical nature of the end product.

Ball milling A milling process for breaking down a mixture of raw chemicals to smaller particle sizes. For milling, small ceramic or agate balls are used to do the milling, while the chemicals are contained in a jar and rotated at high speeds. Thus the name ball milling.

Calcination This is also a thermal process, but it is distinctly different than annealing. This process is done to drive off moisture and for the decomposition of carbonates and other materials from the raw materials. The temperature is kept below the melting point of the raw chemicals during this process and cooling to room temperature can be rapid.

Concentration Molecular or weight percentage of one chemical into another chemical either in liquid or in solid phases. Its value can vary among $0 < x < 1$ where x is the concentration.

Czochralski growth A technique for the growth of bulk single crystal boules of materials. It is the most effective method for the growth of silicon and many other materials having congruent melting points. Named after Jan Czochralski of Poland who invented this technique in 1916 for crystal growth.

Epitaxy The origin of this word is Greek. With reference to film growth, it means the film that is deposited on the single crystal substrate has the same crystalline orientation as the host substrate.

Heteroepitaxy When the epitaxial process is conducted in which the single crystal substrate and the deposited film are of different materials.

Homoepitaxy When the single crystal substrate and the epitaxial film grown on it are of the same material and of the same crystal orientation.

HTSG growth This is the acronym for "high temperature solution growth" technique for growing bulk single crystals, mainly oxides.

LPE technique An acronym for "liquid phase epitaxy." In the LPE method, epitaxial films are grown using the liquid phase of the system.

Magnetron A device that can confine charged plasma particles close to the surface of the target by utilizing strong electric and magnetic fields.

MBE technique The acronym stands for "molecular beam epitaxy" that is a very powerful method to produce single crystal films of semiconductors and many oxides.

Phase The different physical states of a material at a thermodynamic equilibrium. Solid, liquid, and vapor are classical examples of phases in which a solid can exist physically. In solid-state physics and materials science, the word *phase* is used to represent well-defined states of a material that need not go through a change in its physical state.

PLD A film growth technique, PLD is an acronym for "pulsed laser deposition." It is also known as laser ablation deposition method or simply LAD.

RF sputtering A process through which charge buildup on insulating targets can be prevented during film growth. At a RF frequency (e.g.13.56 MHz), the sign of the cathode and anode bias varies so rapidly that the charge buildup cannot occur.

RHEED A device capable of monitoring the electron diffraction of the crystal surface during the film growth process in an enclosed chamber. The acronym, RHEED, stands for "reflection high energy electron diffraction."

Sintering A thermal processing step in ceramic technology. The mixed powders of constituent raw chemicals are subjected to sintering at temperatures below the melting points of the individual powders into a solid mass of single phase. This process can be done either in air or in special environments to produce a homogeneous mass, but without intentional grain growth the cooling to room temperature is usually done rather rapidly.

Sputtering A process by which atoms are dislodged from the surface of a material by collision with high-energy particles.

VPE technique Growth of epitaxial films on a substrate by utilizing the vapor phase of the target produced by different growth techniques.

Problems

2.1 Describe the Gibbs phase rule defining the terms phase, degree of freedom, number of compositional independent variables, and number of noncompositional variables.

2.2 Apply the Gibbs phase rule to determine the number of phases that coexist if $C = 2$, $N = 2$, and $F = 2$. Comment on your result.

2.3 What is the significance of the triple point? Describe in as much detail as you can an arbitrary pressure–temperature phase diagram.

2.4 Construct a concentration–temperature phase diagram of two materials, A and B, which are totally miscible in each other. A has the molecular weight of 100 and melts at 1000 °C, B has the molecular weight 200 and melts at 1500 °C. Describe how you would proceed to grow single crystal of the composition $A_{0.2}B_{0.8}$?
   (a) What is the molecular weight of $A_{0.2}B_{0.8}$?
   (b) What is the weight ratio between liquid at L and solid at S?

2.5 Describe the various methods for the growth of thin films of oxides emphasizing the merits and demerits of each method. If you were to grow thin film of an unknown oxide material which method you would prefer and why.

2.6 Briefly describe the different methods for the growth of single crystals of materials? Describe the most effective method for the growth of a large single crystal of an oxide with a noncongruently melting point.

2.7 What are the different methods for the processing of ceramics? Describe one method in detail that you would use to produce high-quality dense ceramic for your own research.

2.8 The supercapacitor CCTO ($CaCu_3Ti_4O_{12}$) is a mixture of perovskite $CaTiO_3$ and perovskite $CuTiO_3$. Calculate the amount of raw materials consisting of $CaCO_3$, CuO, and TiO_2 that would yield 100 g powder of CCTO after thermal treatments.

2.9 Describe how you would prepare the powder of CCTO to be of uniformly in nm particle size before proceeding to press ceramic discs? What advantages you would get by using powder of nanometer-sized particles instead of using a powder mixture consisting of raw materials without ball milling?

References

1 Nazarov, A.A. and Mulyukov, R.R. (2003). Nanostructured materials. In: *Handbook of Nanoscience, Engineering and Technology* (ed. W.A. Goddard, D.W. Brenner, S.E. Lyshevski and G.J. Iagrate). CRC Press.

2 Günther, K.G. (1968). *Z. Naturforsch.* 3 (a): 1081.

3 Goeking, K.W., Pandey, R.K., Squattrito, P.J., and Beratan, H.R. (1989). *Ferroelectrics* 92, 89.

3

Methods for Materials Characterization

CHAPTER MENU

Insanity is doing the same thing over and over, and expecting a different result.

Albert Einstein

3.1 Introduction

In the previous chapter, we discussed various methods of processing ceramic materials as bulk, thin film, and single crystals. Once the processing is completed, one needs to know the quality of the material produced. Electrical, magnetic, and optical properties, which form the basis for ascertaining potential applications of a material, is highly dependent upon the structural and surface properties of any material, be it in bulk ceramic form, film, or single crystal. For best performance, the highest quality of a substrate is a requirement, and until the structural and surface integrity are guaranteed, the physical properties would invariably vary from substrate to substrate. Such materials would not be useful to a device engineer nor will it be studied by any respectable scientist. No one wastes his or her time in working with a bad material. This obviously puts a heavy burden on a materials processing scientist. His or her superior trade is indispensable to the world of science and technology. Figure 3.1 adequately illustrates the importance of the interdependence of processing, characterization, property, and performance of materials. We will once again come back to this point in Figure 4.2.

In this chapter, we will discuss primarily the various methods employed for the surface and structural characterization of materials. Physical properties and their determinations are discussed in other chapters.

3.2 Methods for Surface and Structural Characterization

There are large numbers of imaging techniques that are routinely used for materials characterization. They fall into two broad categories: one being conventional in which the imaging sources are light, X-rays, and electrons. Neutrons are also used for imaging but mostly for magnetic materials. The nonconventional methods produce images of a surface by scanning it with a probe referred to as scanning probe microscopy (SPM).

The traditional microscopes use diffraction, deflection or scattering of light, X-rays, or electron waves to produce images. The SPM, on the other hand, uses the interaction between a tip and the sample surface for producing images. Examples of traditional microscopes are optical, X-ray diffraction (XRD), scanning electron microscopy (SEM), energy dispersive X-ray analysis (EDAX), transmission electron microscopy (TEM), and X-ray photoelectron spectroscopy (XPS), which is also known in some circles as electron spectroscopy of chemical analysis (ESCA) and Rutherford back scattering spectroscopy (RBS). Scanning probe microscope generates images by detecting the force existing between the surface structure and a scanning probe. Atomic force microscopy (AFM), magnetic force microscopy (MFM), and piezoelectric force microscopy (PFM) are prominent examples of force microscopes.

Out of these many methods, some are more applicable and essential for bulk ceramics, some for films, and some for single crystals. Of course, one would like to evaluate a new material with as many methods as possible, but investment in time, labor, and cost require prudence. Therefore, each type of structure is usually characterized in practice by three or four different methods that give

Fundamentals of Electroceramics: Materials, Devices, and Applications, First Edition. R. K. Pandey.

Companion website: www.wiley.com/go/Pandey/Fundamentals_Electroceramics

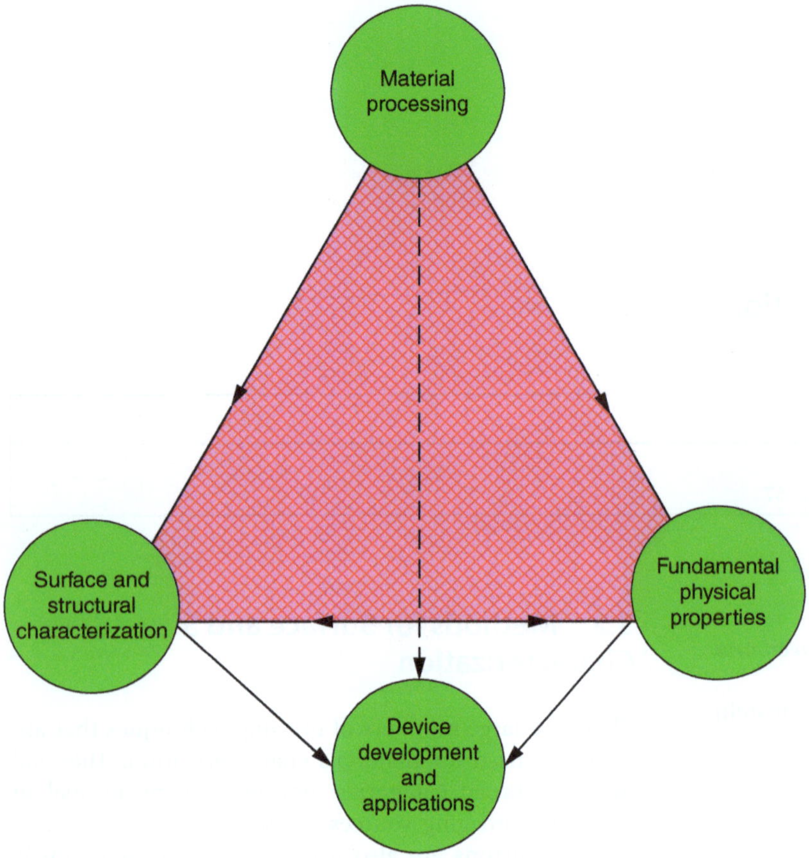

Figure 3.1 Interdependence of materials processing-characterization-property-performance.

a good insight into the quality of the material. In the next sections, we will amplify some of these techniques already mentioned.

Photons (light or X-rays), neutrons, and electrons are the common sources for producing images to study the properties of crystalline materials. The nature of diffracted waves from the crystal lattice with embedded information is the result of the interactions between the crystal structure and the wavelength of radiation. In Table 3.1, a comparative picture of X-rays, electrons, and neutrons with respect to their wavelength and the energy are given.

The most important characteristic of a microscope is its resolving power. The traditional microscopes come with the resolution in the range of $10^{-10} < d < 10^{-6}$ m. They are particularly good for X–Y lateral imaging. The force microscopes can have very high resolution among $10^{-10} < d < 10^{-9}$ m. They can profile depth and therefore are good for topography imaging. It is important to remember that both categories of microscopes are needed for materials characterization. They complement each other and are not in competition for superior or inferior performance.

3.2.1 Optical Microscopes

The word *microscopy* is of Greek origin and translated simply means, "observation of small objects." In Greek, the word for small is *mikros* and for "to observe," it is *skopien*. Optical microscopes use visible light for imaging very tiny features, which are not observable by human eyes, through a set of optical lenses. The configuration of the lens assembly is such that features are magnified manyfold to produce images for close inspection. It is an ancient tool of invaluable importance to scientists, biologists, geologists, and pathologists just to name a few disciplines. It is a piece of equipment whose presence is ubiquitous in all types of laboratories throughout the world. In the hands of ceramists and metallurgists, it is a powerful tool to examine quickly the surface structures of samples and identify whether the samples are free of voids and microcracks.

The history of optical microscope goes back to the thirteenth century. Many claims have been made, but the real inventor of the compound microscope has not been identified. But it is certain that the great astronomer and scientist, Galileo Galilei, is the inventor of the telescope. Around 1610, he explored the sky with this new found

Table 3.1 Energies of X-rays, neutrons, and electrons suitable for crystallography.

Source	Wavelength	Basis	Comments
X-rays	$\lambda_{\text{X-ray}} \approx \frac{1.24}{E\,(\text{eV})} \approx$ 0.154 18 nm for $E \approx 8$ keV This is the wavelength of Cu K_α radiation. This is the most used wavelength for crystallography. Because of its high melting point and excellent heat conduction copper makes an excellent target for generating X-ray	Planck's law of radiation $\lambda_{\text{photon}} = \frac{ch}{E}$	Energy up to 50 keV is used for crystallography Because of very high energy, X-rays can penetrate below the surface
Neutrons	$\lambda_n \approx \frac{0.028}{[E(\text{eV})]^{\frac{1}{2}}}$ $\lambda_n \approx 0.1$ nm for $E \approx 0.08$ eV	de Broglie's relationship $\lambda_n \approx \frac{h}{(2m_n E)^{\frac{1}{2}}}$ $m_n \approx 1.675 \times 10^{-27}$ kg	Neutrons of $E \approx 0.08$ eV are called thermal neutrons. They can be used for crystal diffraction because the wavelength is compatible to lattice parameters Neutron diffraction is used for structural studies of magnetic crystals because neutrons have magnetic moments and therefore can interact with electrons
Electrons	$\lambda_e \approx \frac{0.122}{[E(\text{keV})]^{\frac{1}{2}}}$ For $E = 10$ keV, $\lambda_e \approx 0.4$ nm	de Broglie's relationship same as above equation $m_e \approx 9.11 \times 10^{-31}$ kg	They can interact with solids because they are charge particles. Their penetration depth is small compared to X-rays. Therefore, they are commonly used for surface studies such as to determine bonds, defects, energy states

tool to observe stars and other celestial bodies and in the process made many discoveries.

Until very recently, visible light remained the only source for imaging using an optical microscope with limited resolution. The picture changed drastically when imaging was created by fluorescent lint from molecules opening the world of nano dimensions to optical microscopy. In 2014, the Nobel Prize in Chemistry was awarded for this paradigm-shifting discovery. The recipients were Eric Betzig of USA, Stefan W. Hell of Germany, and William E. Morener of USA. With this discovery, the optical microscope became a nanoscope almost instantaneously.

Visible light microscopes are limited in their ability for magnifying images because very high magnifying lenses produce fuzzy images when viewed with transmitted light. Magnifying power and resolution are the two most important characteristics of a microscope. Magnification enlarges the object with respect to its original size such that a small object appears to be large. However, the resolving power refers to the capacity of a microscope for resolving extremely tiny features as distinguishable objects. The resolution, R, of a microscope is directly proportional to the wavelength of light, λ, and inversely proportional to the numerical aperture, NA, of the objective lens as expressed in Eq. (3.1).

$$R = \frac{\lambda}{2\text{NA}} \tag{3.1}$$

The numerical aperture, NA, is a design parameter and has its own inherent limitation. The highest value obtained for NA in air medium is 0.95, which increases to 1.5 for oil as the medium. Using, for example, green light of $\lambda \approx 550$ nm and NA = 0.95 for air, we get $R \approx 290$ nm that can be increased $\approx$180 nm by replacing air with oil.

Exercise 3.1

If the source of light is a mercury fluorescent lamp emitting in UV of 185 nm wavelength calculate the resolution, R, of the optical microscope having the numerical aperture (NA) equal to 1.5. Comment on the advantage of using the mercury source instead of a visible light.

Solution

Using Eq. (3.1), we get $R \approx 60$ nm which is an improvement of 300% compared to the best resolving power using green light. This is a big advantage of replacing the source of light from visible to UV. But we cannot use UV source directly because its wavelength is too small for human eye to see. Adding Hg to the UV source makes it fluorescent, which we can then see.

Today's optical microscopes are vastly improved instruments compared to its humble beginning long ago. Besides transmission microscope, there are a number of sophisticated optical microscopes that use polarized light as the source. They can produce color images with

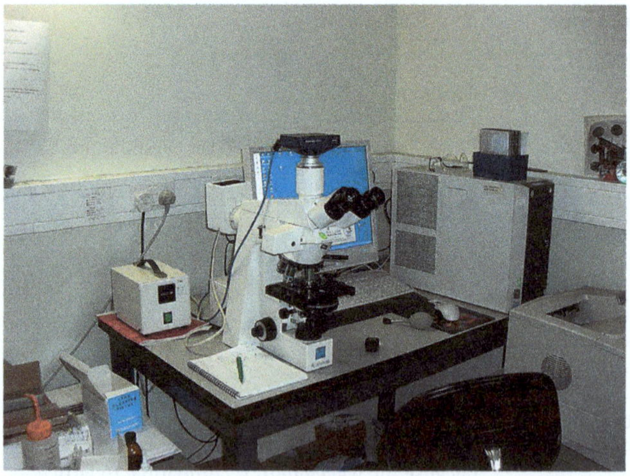

Figure 3.2 Picture of an optical microscope for fluorescent microscopy. Source: https://en.wikipedia.org/wiki/File:Microscope.

good resolution. For many applications such as in optics and geosciences, a polarization microscope is considered to be an indispensable research tool. A commercially available modern microscope with fluorescent light source outfitted with a digital camera and a desktop computer is depicted in Figure 3.2. So far as cost is concerned, the optical microscopes cost just a fraction of what is needed to acquire all other scopes.

3.2.2 X-ray Diffraction Analysis (XRD)

The XRD method of characterizing materials is the most used nondestructive tool available to us. It is not only an indispensable tool for materials characterization and scholarly research rather also an asset to industry for quality control. Using short wavelength X-rays XRD can analyze all sorts of materials ranging from fluids to powders and crystals. Chemists, geologists, crystallographers, solid-state scientists, and engineers would be vastly inhibited without this techniques being available to them for identifying new materials and determining their crystal structures and symmetry.

X-ray was discovered in 1895 by Wilhelm C. Röntgen of Germany, and he was the first recipient of Nobel Prize in Physics awarded in 1901. While experimenting with electromagnetic radiations, he accidentally discovered radiations of extremely small wavelengths. Not knowing what these radiations were he named them X-ray, and this name is still in vogue. However, in some countries, X-ray is also known as Röntgen-ray.

Since its discovery, X-ray has been in service of humanity with its applications covering a very wide field of spectrum from basic crystallography to medical diagnostics to industrial quality control. During World War I, Madame Curie of France, who discovered radioactivity for which she shared Physics Nobel Prize in 1903 with her husband, Pierre Curie, used X-ray for the diagnosis and treatment of wounded soldiers. For crystallographers, materials scientists, and geologists, it is indispensable representing the premier characterization tool for determining crystal structures, phase identification, and symmetry.

3.2.2.1 XRD Diffractometer: Intensity vs. 2θ Plot

In earlier days, samples had to be crushed to a fine powder for phase analysis and determination of crystal structure. The method was called "Powder Diffraction Method" or "Debye–Scherrer Method." With the advancement in electronics and computation, modern-day diffractometers have evolved, and destructive testing is rarely done. Newer versions of diffractometers are versatile with the capabilities for standard powder diffraction, glancing incidence diffraction, in-plane diffraction, high-resolution diffraction, X-ray reflectivity, and small angle X-ray scattering.

There is a wide range of wavelengths associated with X-rays. They can vary from 10^{-2} to 10^{1} nm with the equivalent frequencies of 3×10^{16} to 3×10^{9} Hz. On the basis of wavelengths, X-rays are shorter than UV but larger than γ rays. They carry very large energies with them in the range of 10^{2}–10^{5} eV. Obviously, they should be handled with great care; otherwise, serious damages can result including the onset of certain type of cancer. On the one hand, X-ray is a boon in the hands of professionals but can be deadly if handled poorly.

X-ray diffractometers operate on the principle of the Bragg's law. It was proposed by W.L. Bragg and his father, W.H. Bragg of England, for which the father and son pair were awarded the Physics Nobel Prize in 1925. Since then this law has become the very foundation of X-ray crystallography. The Braggs assumed that atoms in a crystal lattice are tightly packed and are uniformly distributed in parallel planes having a fixed separation between them. They further assumed that X-rays also like light can give rise to diffraction phenomenon. The sketch shown in Figure 3.3 represents two-dimensional arrays of atoms in parallel planes of a crystal.

In order to have to an expression for Bragg's law, let us examine this figure. We see that the incident wave glances at the first atomic plane at an angle θ. Once it hits an atom, there it bounces back with angle θ as well. The parallel planes are separated by a distance d. Now consider the second incident wave that strikes an atom on the second plane. It bounces off there too with the same glancing angle θ. Using simple geometry we can find the difference in the path length, Δ, between these two consecutive waves. It amounts to be equal to the sum of the points

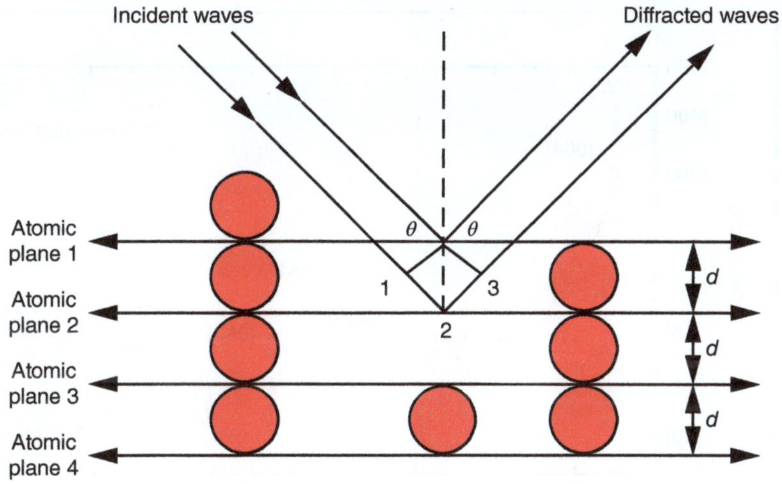

Figure 3.3 A two-dimensional representation of a crystal lattice.

(1–2) and (2–3). Equation (3.2) gives the mathematical expression for this.

$$\Delta = 2\,d\,\sin\,\theta \tag{3.2}$$

Furthermore, if the two planes are to scatter in phase, the path difference between them must also be equal to the product, $n\lambda$ where $n = 1, 2, 3, \ldots$. Here 1 represents the first-order diffraction, 2 the second-order diffraction, and so forth. This leads us to Eq. (3.3).

$$2d\,\sin\,\theta = n\lambda \tag{3.3}$$

This is the famous Bragg's law of XRD. Alternatively, we can also express it as in Eq. (3.4).

$$\theta = \sin^{-1}\left(\frac{n\lambda}{2d}\right) \tag{3.4}$$

The significance of knowing θ is that it allows us to determine the lattice spacing, d. Once we know the lattice spacing, we can then find the crystal structure of the sample. In practice, it is common to plot intensity vs. 2θ to determine the planes involved in the diffraction process. From that information, one can then deduce the lattice constant and the crystal structure. We will discuss this in greater depth in the next chapter while dealing with crystal structures and crystal symmetries. We can still go a little bit deeper in the significance of the Bragg's law. Bragg's law can also be written as Eq. (3.5) simply by substituting wavelength λ with its wave number k $\left(=\frac{2\pi}{\lambda}\right)$in Eq. (3.3).

$$k\,\sin\,\theta = n\left(\frac{\pi}{d}\right) = \frac{C}{d} = d^{*} \tag{3.5}$$

Here d^{*} is the reciprocal lattice and C a constant. The Bragg's law also enables us to determine directly the magnitude of the reciprocal lattice that is fundamental to crystallography.

In Figure 3.4, a typical XRD pattern of a polycrystalline material is plotted [1]. The pattern was generated by the method using a high-angle X-ray diffraction (HAXRD). Each major peak represents a particular crystal plane and their random orientations point to the polycrystalline nature of the sample.

In comparison, Figure 3.5 is the XRD pattern of a film grown epitaxially on a single crystal GaAs substrate [2]. The peaks of bismuth ferrite (BFO) represent the planes with the indices of $n(001)$ with n being integers 1, 2, 3, etc. Such symmetry means that these planes are parallel to each other pointing to the single crystal nature of the epitaxial film. The sharp peak at 60° is an artifact from the X-ray tool.

Also notice the presence of $n(002)$ GaAs planes. An additional evidence of high degree of epitaxy is provided by the inset. The inset is the full width at half maximum (FWHM) equal to 0.27° of (001) plane. In crystallography, the FWHM is a measure of the quality of a single crystal.

3.2.2.2 Laue X-ray Diffraction Method

In 1912, Max von Laue of Germany proposed that a crystal lattice with its atomic periodicity and symmetry should act as a diffraction grating for X-rays because their small wavelengths may be comparable its interplanar spacing. This was a brilliant idea that completely changed the world of crystallography and opened up new avenues for physicists, chemists, and crystallographers. Laue encouraged his assistants Walter Friedrich and Paul Knipping to perform an experiment using a copper sulfate crystal. They followed through under Laue's guidance. Laue's hunch paid off, and the result showed a highly symmetrical distribution of large number of diffraction spots on a flat photographic plate. This provided an experimental evidence for his theory paving his way to the Physics Nobel Prize in 1914.

The experimental set up for Laue diffraction is rather simple. First a beam of white X-rays of continuous wavelength is allowed to pass through a pinhole that

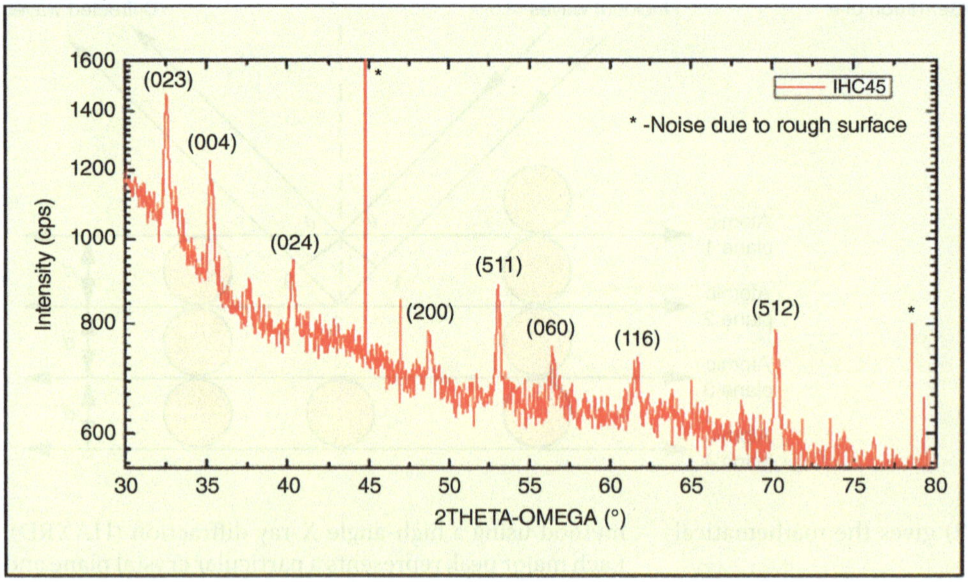

Figure 3.4 XRD pattern of an ilmenite–hematite 45 (HC 45) bulk ceramic sample.

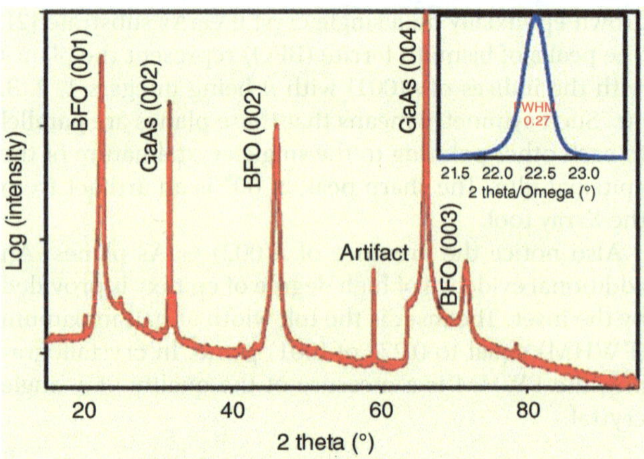

Figure 3.5 XRD pattern of bismuth ferrite (BFO) single crystal film grown by MBE method on a strontium titanate (STO)-buffered gallium arsenide (GaAs) substrate. The inset is the FWHM of (001) plane.

produces a collimated beam. If a copper target is used to produce white X-ray, then it may include the wavelength range of 0.02–0.22 nm. A stationary crystal is irradiated by the collimated beam. The beams of white X-ray are diffracted from crystal planes for which the Bragg's condition is satisfied. Consequently, a series of pronounced diffracted spots are generated on the photographic film. The resulting diffraction pattern is called the Laue pattern. It exhibits the internal structure and symmetry with reference to the central beam. Using Bragg's law, the interplanar separation (d) and crystal symmetry can be determined. A typical example of the Laue pattern is given in Figure 3.6. Notice the high degree of symmetry with which the spots are distributed in the figure. With reference to the central beam, they are showing a fourfold symmetry. Two sets of diffraction spots showing fourfold symmetry about a [001] direction are identified here with arrowed crosses. It appears that the sample was perhaps a cubic crystal, though it could also be either a tetragonal or an orthorhombic crystal.

Over the years, as one would expect, Laue's original experiment has gone through multiple refinements. Laue patterns are now routinely generated by rotating a crystal through 360°, which then gives a large number

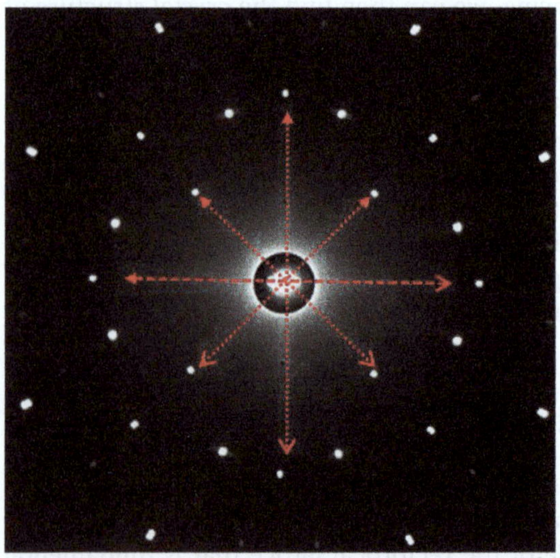

Figure 3.6 Laue diffraction pattern of a bulk single crystal; fourfold symmetry is pronounced. Source: http://www.xtal.iqfr.csic.es/Cristalografia/archivos_06/laue1.jpg.

diffracted spots in a wide range of scattering angles with which one can work. The Laue method is ideally suited to orient a crystal quickly for number of experiments of interest to crystallographer and solid-state scientists. It has also a very practical application. It is the method of choice for orienting a seed crystal for Czochralski growth such as for the growth of long silicon crystal boules with one unique orientation. Once crystal boules are grown with one unique orientation, they are then sliced into thin wafers, each being parallel to the other. The result is that the orientation of each wafer follows that of the big boule itself. Laue-like patterns with a high degree of symmetry can also be generated by using neutrons and electrons.

3.2.3 Electron Microscopes

Based on the wavelengths given in Table 3.1 for X-rays (≈0.154 nm), neutrons (≈0.08 nm), and electrons (≈0.04 nm), we can conclude that the magnifying power of a microscope using any of these three sources for imaging would be far superior to that of an optical microscope. Compared to an optical microscope with visible light of 550 nm (green) as its source, an electron beam-based microscope would be in a position to magnify an image by approximately 14-fold. It has been established experimentally that the magnification produced by an electron microscope is 200 times greater than that of an optical microscope.

The questions now to answer are the following: First, what is an electron microscope? And second, how does it work? The answer to the first question is provided by the de Broglie hypothesis that particles are waves and waves are particles. Its experimental verification paved the way for the construction of the electron microscope. The answer to the second question comes, of course, from the combined genius of an experimental physicist and an engineer.

The first electron microscope was built by Ernst Ruska of Germany nine years following the discovery of de Broglie in 1924. The Nobel Prize for Physics went to him in 1986 for this wonder microscope. Today, the electron microscope and its variants are routinely used in research and development in physics, microelectronics, materials science, chemistry, biology, and medicine. How an electron interacts with a solid surface and what results from such an encounter are summarized in Figure 3.7. Many useful characterization tools have been developed using the transmitted and scattered beams. The transmitted electrons are the basis for the transmission electron microscope (TEM), whereas the Auger surface analyzer and backscattered analyzers are based on backscattered electrons. Scanning electron microscope (SEM) uses the backscattered electrons for its operation.

When electrons strike a surface some are absorbed, some are transmitted, and yet some are absorbed. Electrons are generated with very high energy. Absorption of high energy electron beam heats up the solid. For transmission to occur, the sample thickness must be very small because of the limited penetration depth of electrons. The transmitted electrons are used for imaging and are the foundation for TEM. The backscattered secondary

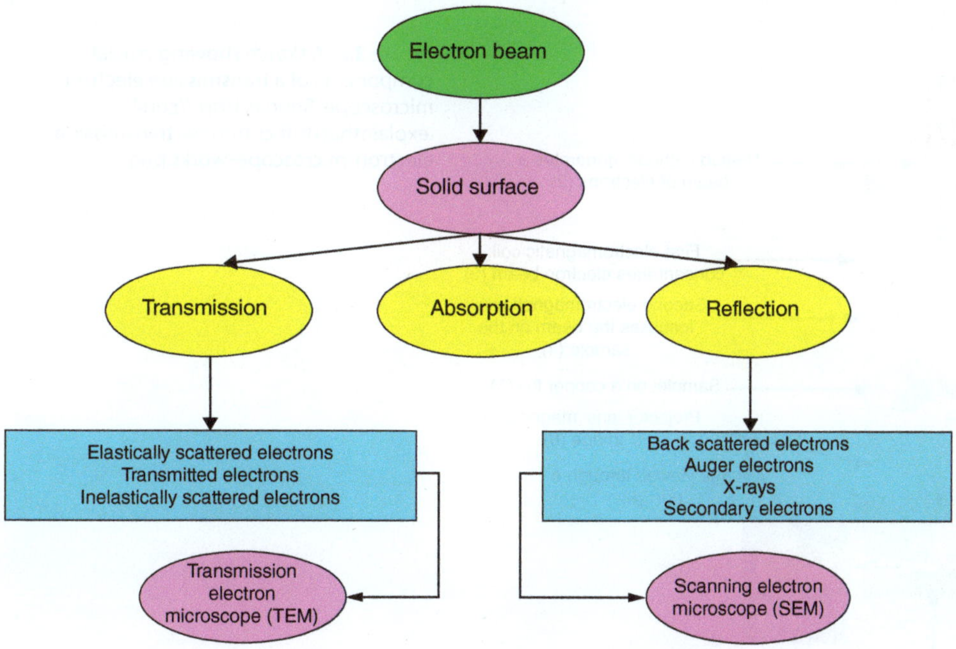

Figure 3.7 Electron–solid interactions and resulting phenomena and devices.

electrons are also used for imaging. For example, Auger electrons are used for surface analysis, backscattered electrons for imaging, and X-rays for chemical analysis.

Electron microscopes fall into two principal categories that are TEM and SEM. TEM is based on transmitted electrons, whereas the other is on backscattered electrons. Both these tools are very important for materials characterization, and we need to learn more about them. Scanning transmission electron microscope (STEM) is another variant of the TEM. It is powerful hybrid between a TEM and a SEM microscope. In a STEM microscope, the electron beam is focused as a point and then scanned over the sample on a raster.

There is one more electron microscope, which is called the "scanning tunneling microscope" (STM). It is based on the electron tunneling effect of quantum mechanics. It was developed in 1981 by Gerd Binning and Heinrich Rohrer at the IBM laboratories at Zürich, Switzerland. They were awarded Nobel Prize in Physics in 1986 for their discovery. STM can resolve objects at the atomic scale. Therefore, it can manipulate individual atoms and generate their images. It is a boon for nanotechnology and is widely used for research in nanomaterials and devices.

3.2.3.1 Transmission Electron Microscope (TEM)

The fundamental nature of electron is that its transport can be manipulated by an electric or magnetic field. By proper configuration of these two fields, a diverging electron beam can be brought to a focus. In other words, one can build an electromagnetic loop that would work as a lens for an electron beam.

In an electron microscope, the transmitted electrons are guided through a series of electromagnetic coils to focus on a solid surface. If the sample is made very thin, the electrons can penetrate and can exit from the other side as a transmitted beam carrying with it the signature of the atomic distribution of the solid material. This information-rich beam is then projected to a lens that magnifies the image and focusses it on a fluorescent screen. The image becomes visible and then can be monitored through a viewing port. Finally, the image is recorded on a photographic plate or as an image on a TV screen. The TEM images are called *electron micrographs*. In Figure 3.8, we show the different components of a TEM and how images are produced and recorded.

Since the kinetic energy of electrons can be controlled with the help of applying accelerating voltages of varying strengths, electrons with different wavelengths can be generated. Usually, they lie between 50 and 300 kV. These wavelengths, as we know from Table 3.1, will be smaller than UV light. For example, an electron beam of 1 nm wavelength is approximately 4×10^3 times shorter than the UV light. Therefore, a TEM can magnify an object thousands of time more than the most efficient optical microscope.

A TEM has many attractive attributes, and because of that it is an indispensable tool for materials characterization. It can easily resolve features as small as 0.2 nm, can generate direct imaging of crystalline lattices, and identify defects in samples. However, since the penetration depth of an electron wave is very small, the samples must be made very thin in order for the beam to transmit through it. The very first TEM was built in 1933 by

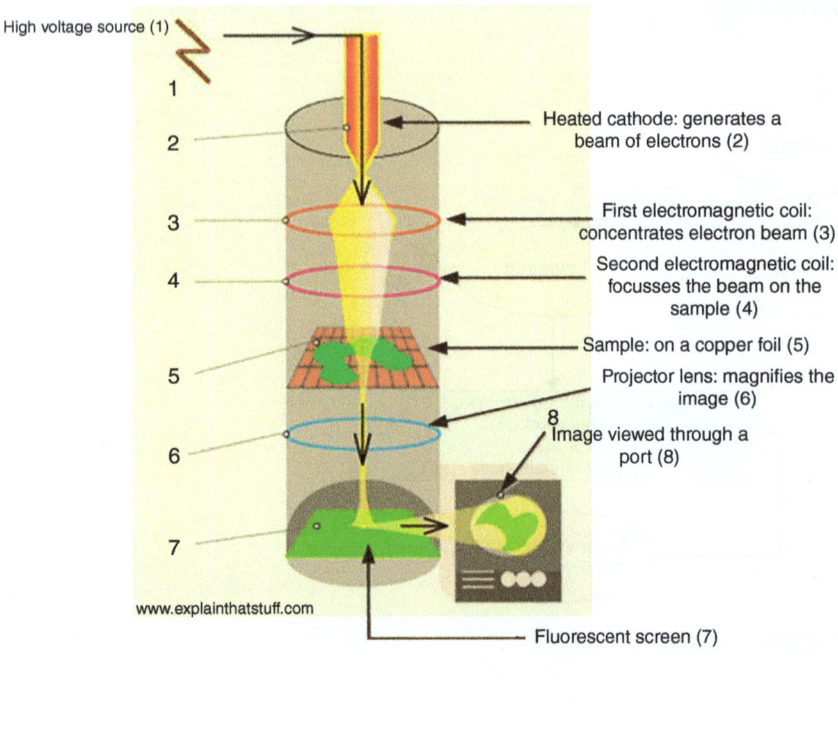

Figure 3.8 A sketch showing crucial components of a transmission electron microscope. Source: http://cdn4.explainthatstuff.com/how-transmission-electron-microscope-works.png.

Figure 3.9 First TEM microscope built in 1933 by Ruska, and it is now in Deutsches Museum in Munich, Germany. Source: https://commons.wikimedia.org/w/index.php?curid=5309032.

Ruska which is now housed in the Deutsches Museum in Munich, Germany, Figure 3.9.

Exercise 3.2

An optical microscope uses the green light as its source with the wavelength of 0.55 μm and a TEM of an electron beam of 0.04 nm wavelength. Find the magnification of the TEM microscope assuming for a moment that the numerical aperture of both the microscopes is 1.5 (which is not true, the numerical aperture of a TEM is much smaller than of the optical microscope).

Solution

From Eq. (3.1), we have $R_{\text{green}} \approx 185$ and for $R_{\text{electron}} \approx 0.013$. $\frac{R_{\text{green}}}{R_{\text{electron}}} \approx \frac{185}{0.013} \approx 14\,230$. The TEM is in the position to magnify the image by approximately14 000 times more than the optical microscope. *Comment:* In microscopy, resolution and magnification are related. We know that resolution is the capacity of the microscope to distinguish between very tiny features and the magnification is its power to make these features to appear large.

3.2.3.2 Scanning Electron Microscope (SEM)

Both conducting and nonconducting samples can be examined by a SEM, and three-dimensional images are generated by rotating the sample. The resolving power of a SEM is just a few tens of a nanometer, which is superior to that of a TEM. In Figure 3.10, we find the schematic of a typical SEM. The accelerating voltage for a SEM is in the range of 20–30 kV. The electrons are contained and focused using a set of electromagnetic coils, just like it is done in TEM. Subsequently, they are guided to a sample mounted on a rotating stage. The secondary electrons emitted from the specimen are detected and guided to scan the sample surface.

In contrast to the TEM, backscattered electrons are used for obtaining the sample characteristics in a SEM. The samples are exposed to a narrow beam of backscattered electrons. The beam is rapidly scanned across the specimen surface. As the beam penetrates the surface, emission of electrons or photons occurs carrying with them the salient features of the sample. The images are recorded representing the sample characteristics.

3.2.3.3 Scanning Transmission Electron Microscope (STEM)

This is a powerful tool for studying materials. It combines all the functions of a TEM and a SEM into one single microscope. STEMs can produce three-dimensional topography revealing the internal and external structure of samples. STEMs based on electrons with accelerated voltages of 5–25 kV are called low-voltage STEMs. These microscopes combine, in one single unit, the functionalities of a TEM, SEM, and STEM. That makes them extremely valuable tools for material characterization and device development.

We know that when a solid surface is irradiated by an electron beam, X-rays are also emitted which can be detected by an X-ray spectrometer. A STEM microscope outfitted with an X-ray diffractometer can be used for EDAX. This added feature makes a STEM microscope

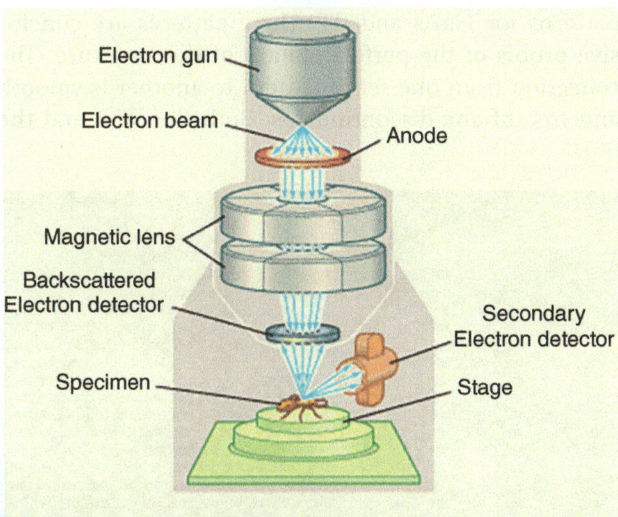

Figure 3.10 Schematic of a scanning electron microscope. Source: Courtesy of Encyclopedia Britannica, Inc. [3].

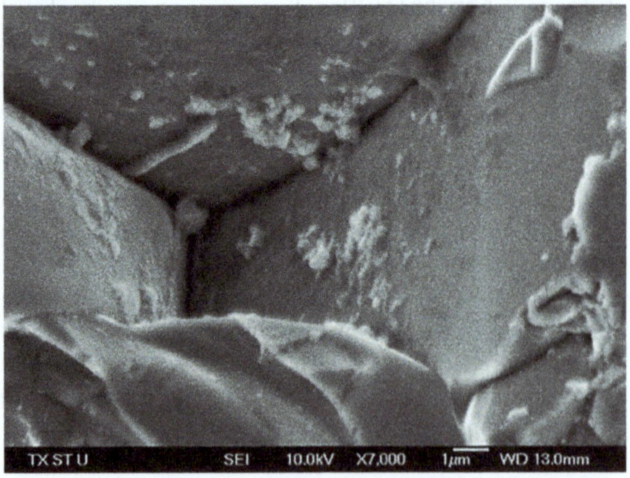

Figure 3.11 A SEM image showing the microstructure of a Ca–Cu–Ti–O_3 supercapacitor bulk ceramic sample.

an indispensable tool for solid-state research and development. Whereas a STEM provides us with information about the internal structure of a solid, the EDAX gives us the chemical analysis of the same sample.

Figure 3.11 depicts the microstructure of a ceramic sample of CCTO (Ca–Cu–Ti–O_3) that can be used for a supercapacitor [4]. The image was generated by a SEM operating at 100 kV and with the magnification of 7000. We can clearly identify the grains and grain boundaries separating them. In Figure 3.12, we show the diffraction pattern obtained for an oxide-semiconductor multilayer structure consisting of gallium arsenide (GaAs)–strontium titanate (STO)–barium titanate (BT). The STO interface sandwiched between GaAs and BT clearly shows its single crystal nature and so do the patterns for GaAs and BT. These patterns are conclusive proofs of the perfect epitaxy of the structure. The transition from one set of pattern to another is smooth and free of any discontinuities. Both the STO and the BT crystal lattice possess a 45° rotation in the plane of the surface to accommodate the lattice mismatch. Closer examination of the lattice constant and with the 45° rotation, the BT layer has a minimal lattice mismatch, whereas the STO layer has a 2.3% mismatch. Since the thickness of the STO is less than the critical thickness, this layer is tensile strained.

3.2.3.4 X-ray Photoelectron Spectroscopy (XPS)

This is another very useful and powerful materials analysis method. Its principle is based on the photoelectric effect, which we have discussed in Chapter 1. Instead of using photons of light, X-ray is used in the XPS method to eject electrons from a solid surface. XPS is also known as ESCA that stands for "electron spectroscopy of chemical analysis."

The materials are examined by this method under ultrahigh vacuum conditions with pressure in the chamber maintained $\leq 10^{-9}$ mbar. XPS can provide all types of information about surface chemistry. Information such as elemental composition as parts per thousand, empirical formula, chemical, and electronic state of the elements is routinely determined by XPS. Its sensitivity is such that it can detect a whole lot of elements with atomic number, $Z > 3$. Not only the elements but also the XPS method can characterize all types of materials such as ceramics, inorganic compounds, semiconductors, metals, and glass. Since it can determine the binding energy of elements, it is customary to determine the relationship of peak intensity with the binding energy of elements present at the surface. The magnitude of the binding is determined using the Rutherford equation which is given in Eq. (3.6).

$$E_{\text{binding}} \approx E_{\text{photon}} - (E_{\text{kinetic}} + \phi) \tag{3.6}$$

where ϕ is the work function.

X-rays used as the source in XPS can penetrate up to 10 nm of the surface being examined, and it can

Figure 3.12 GaAs–STO–BT super-lattice imaging produced by a STEM microscope. Source: Courtesy of Professor Ravi Droopad of Texas State University, San Marcos, TX.

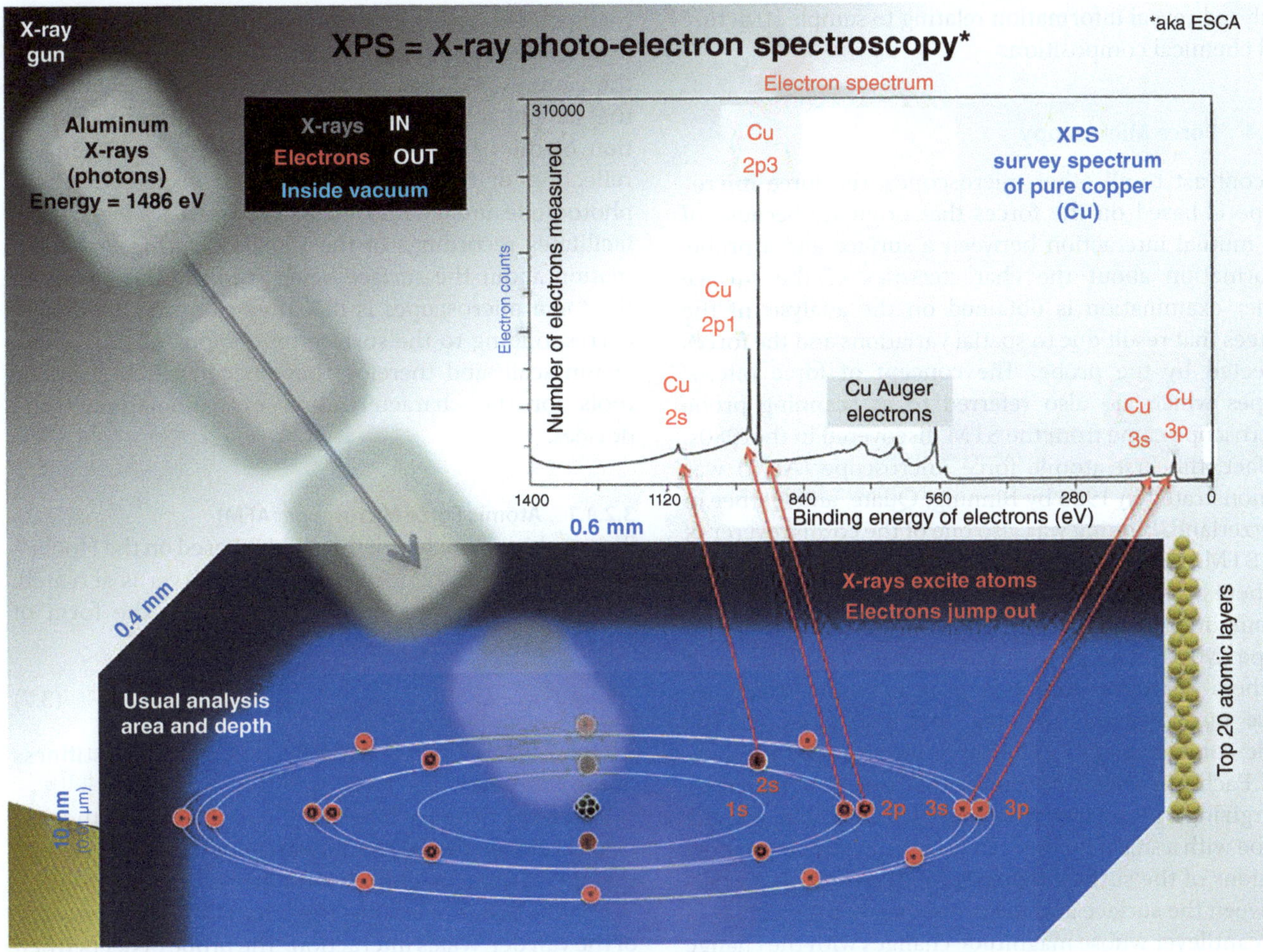

Figure 3.13 A schematic showing the experimental arrangements of XPS and its results for copper. Source: https://commons.wikimedia.org/w/index.php?curid=17267647.

generate a large number of information that can be analyzed for understanding the chemical nature of materials.

The experimental setup for XPS is shown in Figure 3.13. In the bottom part of the figure, an atom of copper is shown with its different orbits. In Chapter 1, we have discussed the meaning of electron configurations of elements and rules that govern it. For example we know that the atomic number of Cu is 29, and therefore, its electronic configuration is [*A*] $3d^{10}4s$. The quantum states such as 1s, 2s, 2p, 3s, and 3p are identified in this figure. The plot in the upper part of the figure shows the binding energy of these states with respect to the number of emitted electrons.

In Figure 3.14, we find the spectrum for two states of Bi, namely, $Bi4f_{5/2}$ and $Bi4f_{7/2}$. The well-centered peak at the binding energy values of 158 and 164 eV, respectively, are shown here for the BFO film [2].

Another characterization tool similar to XPS is called RBS. This technique is also used to determine the structure and elemental analysis of the sample. Instead of

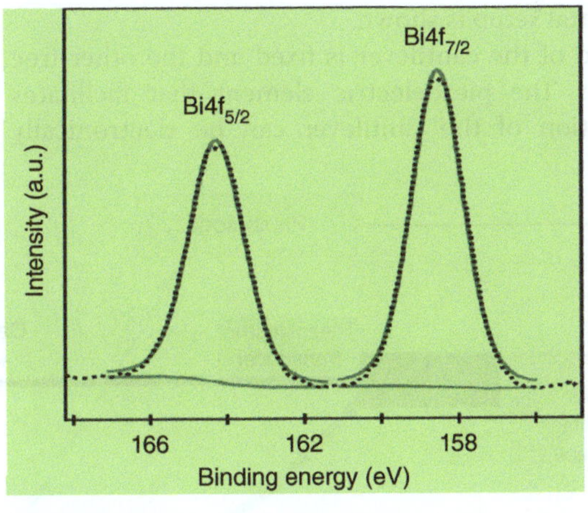

Figure 3.14 High-resolution XPS spectrum of Bi4f of Bismuth ferrite film.

X-rays, high-energy ions (protons or alpha particles) are allowed to backscatter from the solid surface that carries the structural and chemical signatures of the sample. Large numbers of data are generated, and when analyzed reveal information relating to sample structure and chemical composition.

3.2.4 Force Microscopy

In contrast to all other microscopes, the force microscope is based on the forces that originate because of the mutual interaction between a surface and a probe. Information about the characteristics of the surface under examination is obtained on the analysis of the images that result due to spatial variations and the forces detected by the probe. The concept of force microscopes which are also referred to as scanning probe microscope came from the STM discovered in the1980s. In fact, the first atomic force microscope (AFM) was demonstrated in 1986 by Binning, Quate, and Gerber in Switzerland. Binning was also one of the co-discoverer of the STM.

There are three principal classes of force microscopes of our interest. They are AFM, magnetic force microscope (MFM), and piezo-force microscope (PFM). Each of them has three components in common which are force measurement, imaging, and manipulation. The basic configurations of the three microscopes are similar. Each one uses a cantilever that can be actuated by energizing a piezoelectric element attached to it, and the probe with a sharp tip that can detect the minutest deformations of the surface topography. The force that exists between the surface and the probe causes a deflection of the cantilever whose magnitude changes with the change in the surface topography. This translates into a generation of electrical signals, which in turn are detected and analyzed. In Figure 3.15, the basic configuration of the experimental setup is shown.

One end of the cantilever is fixed and the other free to vibrate. The piezoelectric element that facilitates the actuation of the cantilever can be electronically manipulated to produce deflections in nm range. The probe with a pointed tip is affixed near the open end of the cantilever. A *xyz*-drive facilitates the movement of the sample in three dimensions (3D). When a deflection of the cantilever occurs, it induces a shift in the reflection of the laser beam, which is detected by a photodiode and fed into a detector. Feedback electronics facilitates recordings of the signals carrying the information about the surface being studied. The beauty of the force microscopes is that they can produce images corresponding to the surface topography of the sample at nm scale and therefore have become indispensable tools for the characterization of nanomaterials and devices.

3.2.4.1 Atomic Force Microscope (AFM)

The working principle of the AFM is based on the Hooke's law that defines the force when a cantilever is actuated. Mathematically expressed, this law takes the form of Eq. (3.7).

$$F = -kZ \tag{3.7}$$

where F is the force, Z the deflection, and k the stiffness of the cantilever.

When the distance between the probe tip and the surface is very small, the Columbian interaction is repulsive that increases in magnitude with the decrease in distance. When the distance is large, the force is attractive because of the van der Waals interaction. The probe deflects when it senses either of these two forces. But there is no deflection at all, when it is faraway from the surface because the probe does not experience any force. The AFM can generate images in 3D with a high degree of resolution. Resolution on the order of a nanometer can be achieved by AFM, which is superior by a factor 10^3 compared to the resolution of an optical microscope.

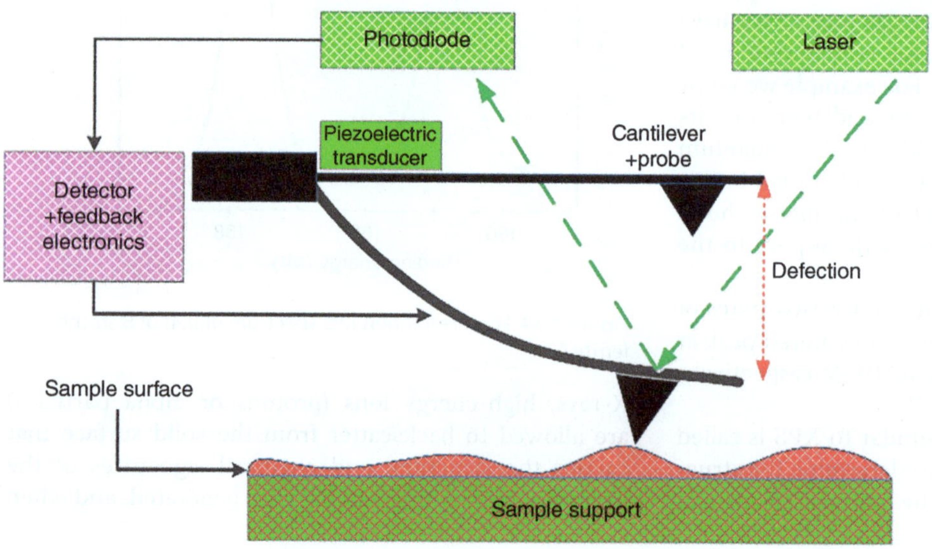

Figure 3.15 Basic experimental configuration of scanning force microscopes.

The design of the probe and its tip requires sophistication. Insulating tips are commonly used for AFM imaging. These tips are microfabricated highly precise units made of silicon (Si), silicon nitride (Si_3N_4), or diamond. An example of AFM imaging is given in Figure 3.16. It is the surface morphology of a glass surface, and we can clearly see the 3D nature of the image that the AFM can produce.

3.2.4.2 Magnetic Force Microscope (MFM)

AFM is its precursor and was developed in 1987. This tool is used for characterizing the topography of a magnetic surface. Here a magnetic tip is used to scan the surface that gives rise to a force because of the interaction between the surface and the probe. The AFM probe with the coating of a magnetic material is used for MFM. Cobalt chromium coating is done to make the tip magnetic. Information carrying the signature of a magnetic surface with its topography is generated and stored. The MFM can measure many types of magnetic interactions including the magnetic dipole–dipole interaction. The force existing between the surface and the probe is given by Eq. (3.8).

$$\vec{F} = \mu_0(\vec{M} \cdot \nabla)\vec{H} \tag{3.8}$$

Here μ_0 is the magnetic permeability of free space, M the magnetic moment of the probe, and H the stray magnetic field which is inherent with the surface. Alternatively, the Eq. (3.8) can also be rewritten in the form of Eq. (3.9).

$$F = M_x\left(\frac{\partial^2 H_x}{\partial z^2}\right) + M_y\left(\frac{\partial^2 H_y}{\partial z^2}\right) + M_z\left(\frac{\partial^2 H_z}{\partial z^2}\right) \tag{3.9}$$

Here M_x, M_y, and M_z are the components of moment M in x, y, and z directions, and similarly H_x, H_y, and H_z are the components of the stray magnetic field in x, y, and z-directions.

The second-order derivatives of the stray magnetic field suggest that the contrast should be pronounced in images when magnetization changes abruptly. Abrupt change of magnetization is well documented in magnetic domains and therefore, the MFM microscope reveals the domain patterns with clear contrast. Resolution of 30 nm can be routinely achieved by a MFM with a good tip profile, and it is possible to improve it to 10–20 nm with some elaborate improvements in the system. A MFM image of a magnetic hard-disk drive is depicted in Figure 3.17. The resolution of magnetic domains and their contrast is impressive.

3.2.4.3 Piezoelectric Force Microscope (PFM)

This is the third member of the force microscope, and it was developed in Germany by Güthner and Dransfeld in 1992 [5]. It is used for characterizing the surface topography of a ferroelectric or piezoelectric material. The PFM probes are conductive. The insulating tips of AFM are coated with metals to make them conductive. Usually, gold and platinum are used for this purpose. Piezoelectricity is found in certain classes of materials with noncentro-symmetry. Some materials develop electric polarization when stressed because of the direct piezoelectric effect. Conversely, the same group of materials can exhibit converse piezoelectric effect when subjected to an electric field. This effect is the basis for the operation of the PFM. We will discuss in-depth piezoelectric and ferroelectric phenomena and materials

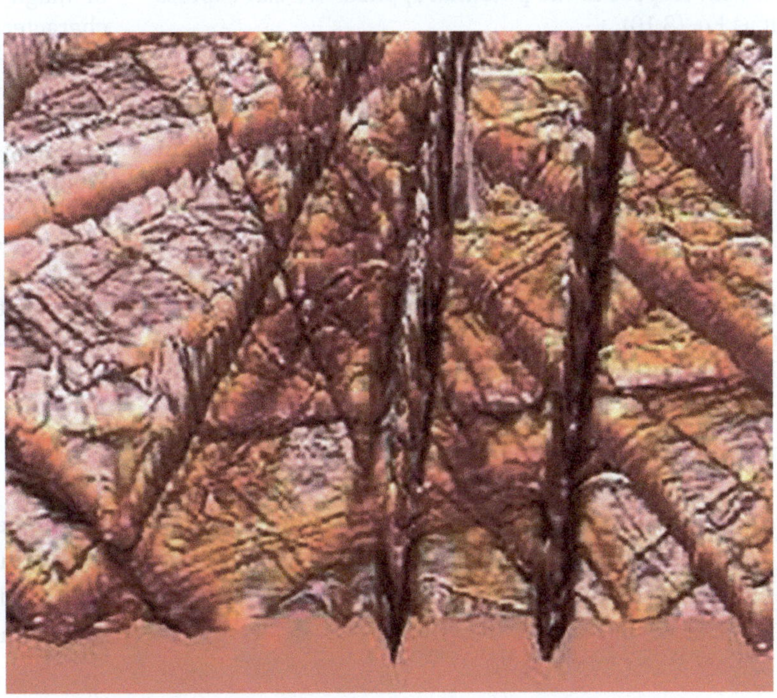

Figure 3.16 AFM image of a glass surface. Source: https://upload.wikimedia.org/wikipedia/commons/e/e8/AFMimageRoughGlass20x20.

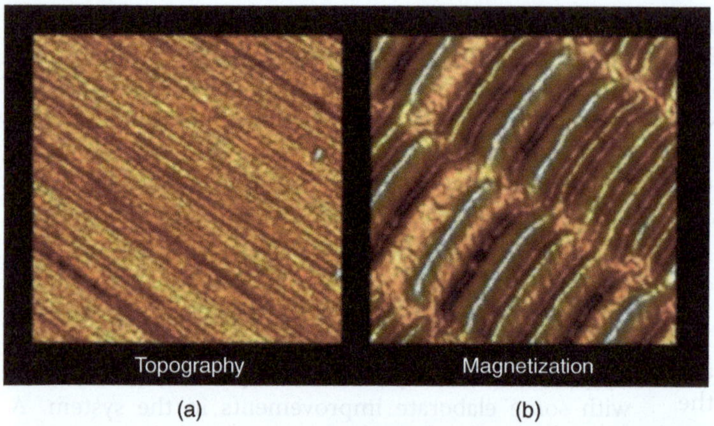

Figure 3.17 MFM-generated image of a magnetic hard-disk drive: (a) the surface topograph; (b) magnetic domains showing the magnetization distribution of the surface. Source: https://www.nanoandmore.com/uploads/gallery/Hard-disk-drive-HDD-platter-surface-20120828090836_small_320_200.png.

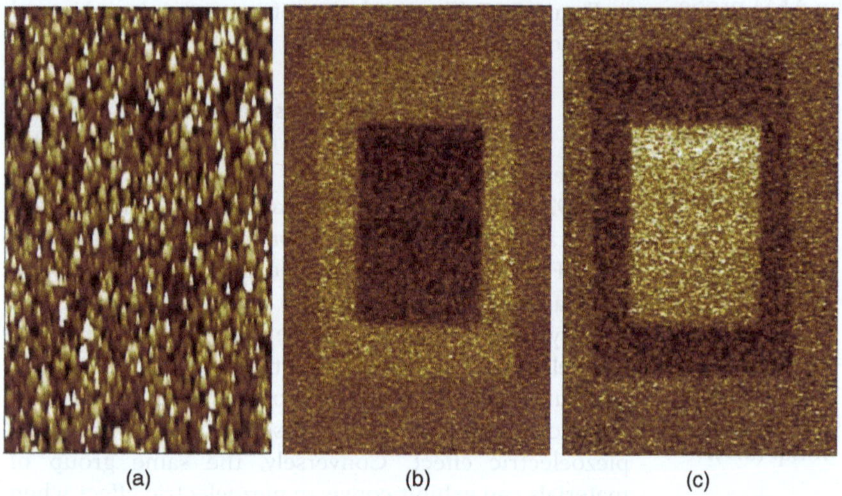

Figure 3.18 $12 \times 12\,\mu m^2$ PFM micrograph of bismuth ferrite surface: (a) general topography; (b) out of plane phase, and (c) out of plane amplitude.

in Chapters 5 and 6, respectively. Because of the strain experienced by the sample there is change in its length, L, with respect to the potential applied. We can express it as Eq. (3.10).

$$\int dV \approx d \int dL \tag{3.10}$$

The coefficient d in the equation plays a major role in piezoelectricity and is defined as charge coefficient. The piezoelectric transducer in Figure 3.15 is powered by applying an AC potential that because of the converse piezoelectric effect promotes a deflection in the cantilever. By raster scanning the probe tip, it detects the surface topography of the sample in 3Ds. The information gathered are stored and analyzed. The result is the image showing the spatial distribution of polarization just as a MFM displays the magnetic structures of magnetic surface. The PFM is now widely used in characterizing piezoelectric–ferroelectric and surfaces to detect the polarization variations at nanometer scale. The domains present in ferroelectric materials are now routinely observed by PFM. Their images are colored with excellent contrast and sharpness. Similarly, the magneto-electric materials are studied using PFM that reveals their fine structures. Like AFM and MFM, it is a must-have tool for characterizing nanomaterials. Its resolution is in the range of 10–20 nm. A PFM micrograph of a BFO film is shown in Figure 3.18 [2]. Notice the clear distinction in the images of (b) and (c).

Glossary

AFM Atomic force microscope. In this method of materials characterization, spatial distribution of weak surface forces is plotted by a probe and displayed on a screen.

Binding energy The energy that binds the nucleus to its constituents that are atoms, ions, and molecules. It is also defined as the energy that is required to separate an atom, an ion, or a molecule from its nucleus.

EDAX Energy dispersive X-ray analysis. This method is applied for the determination of elements and their percentages in ceramics, semiconductors, and other materials present in a sample surface.

ESCA Electron spectroscopy for chemical analysis. ECSA is another name for XPS.

Magnification The capacity of a microscope to magnify the original image by manyfold.

MFM Magnetic force ,microscope. Used for detecting the magnetic topography of a surface.

PFM Piezoelectric force microscope. A PFM detects the polarization distribution of a piezoelectric surface by using the converse piezoelectric effect.

Piezoelectric probe The probe used in a PFM microscope. It is coated with a conductive material just like it is for the atomic force microscope.

Resolution The capacity of a microscope to distinguish between extremely tiny features of a material.

SEM Scanning electron microscope. A variant of the electron microscope. The images are produced by raster scanning the sample surface.

STM Scanning tunneling microscope. This imaging tool is based on the quantum mechanical principles of tunneling. It is the precursor for the atomic force microscope.

TEM Transmission electron microscope. This microscope is based on the transmitted electrons from an ultrathin sample. Its resolving power is orders of magnitude higher than of the most powerful optical microscope.

XPS X-ray photoelectric spectroscopy. A technique used to detect and analyze the presence of elements in a solid sample. It can measure the binding energy of elements present and their percentages. It can also display the binding energy of quantum states of electrons with respect to intensity of emitted electrons.

Problems

3.1 Assume that a NaCl single crystal when radiated with X-ray of wavelength 0.155 nm satisfies the condition for Bragg diffraction. Consider that the Bragg diffraction angle is 30° and the order of diffraction is 1. Find the interplanar distance and the phase difference.

3.2 If the X-ray used in the above experiment is generated at 30 keV, find the associated photon energy.

3.3 If neutrons of 0.08 eV instead of X-rays are used to analyze the NaCl in Problem 3.1, find the diffraction angle, θ assuming that the Bragg condition is satisfied. Comment on your result.

3.4 Discuss the fundamental difference between a Laue diffraction pattern and a typical XRD pattern. When should you use Laue diffraction instead of typical X-ray diffraction? Justify your answer.

3.5 Discuss the various types of electron microscope emphasizing the basic differences between them.

3.6 How does an optical microscope work? What are the advantages of using a TEM instead of an optical microscope in analyzing a sample?

3.7 Discuss the mechanisms in the operation of different type of force microscopes. How do they differ from each other so far as their applications are concerned?

References

1 Pandey, R.K., Stapleton, W.A., Sutanto, I. et al. (2014). Properties and applications of varistor–transistor hybrid devices. *J. Electron. Mater.* 43 (5): 1307–1316.

2 Shafiqur Rahman, Md. (2018) Integration of Multi-functional oxide thin film heterostructure with III–V semiconductors. Dissertation. Texas State University, San Marcos, TX.

3 Encyclopedia Britannica, Inc. (2008). https://www.coursehero.com/file/10164387/Lecture-3-conventional-Microscope/.

4 Pandey, R.K., Stapleton, W.A., Tate, J. et al. (2013). *AIP Adv.* 3: 062126.

5 Güthner, P. and Dransfeld, K. (1992). Local poling of ferroelectric polymers by scanning force microscopy. *Appl. Phys. Lett.* 61 (9): 1137.

4

Binding Forces in Solids and Essential Elements of Crystallography

CHAPTER MENU

"If I have seen further than others, it is by standing on the shoulders of giants."
"I can calculate the motion of heavenly bodies, but not the madness of people."

Sir Isaac Newton

4.1 Introduction

In this chapter, we will study how the solids are held together in a structure that can be either polycrystalline or single crystal, how solids are classified based on their governing binding forces, the relationship between structures and the properties they acquire when subjected to an external agent such as an electric or a magnetic field, and, finally, how these properties help engineers and scientists develop devices and their applications. We will also study the nature and properties of single crystals and how their unique properties help us in understanding their importance to science and technology.

4.2 Binding Forces in Solids

We can easily visualize the presence of a strong attractive force existing between the negatively charged electrons and positively charged protons making the atom stable. The nature of this attractive force, F, is, of course, the famous Coulombian force of attraction first postulated by Coulomb in 1784 (Charles Augustin de Coulomb was a French physicist). Coulomb's law states that the force interacting between two oppositely charged particles obeys the following relationship.

$$F = k\frac{q_1 \times q_2}{r^2} \approx \frac{q_1 \times q_2}{4\pi\epsilon_0 r^2} \tag{4.1}$$

here k is Coulomb's constant, q_1 and q_2 the electric charges with opposite polarity, r the distance separating the two charges, and ϵ_0, as always, is the permittivity of vacuum. Coulomb's constant is numerically equal to $9 \times 10^9\ \mathrm{N\,m^2\,C^{-2}}$.

From the above simple equation, we can conclude that as the distance r between the two charges increases, the force decreases. Conversely, it becomes very large if the separation is very small, which is the case at the atomic scale. This becomes obvious once we consider the simple example of Exercise 4.1.

Exercise 4.1
Compute the Coulomb's attractive forces between the two charges of opposite polarity that are 1.6×10^{-19} and

Fundamentals of Electroceramics: Materials, Devices, and Applications, First Edition. R. K. Pandey.

Companion website: www.wiley.com/go/Pandey/Fundamentals_Electroceramics

3.2×10^{-19} C, respectively, when they are 1 nm apart and 1 mm apart. Use 8.854×10^{-12} F m^{-1} for dielectric constant of free space.

Solution
Using Eq. (4.1), we get force, F_1, for 1 mm separation $= 4.6 \times 10^{-3}$ N. The force, F_2, corresponding to 1 nm separation $= 4.6 \times 10^{9}$ N. That is $F_2 = 10^{12} F_1$. The force increases as the separation between the two charges changes and decreases.

Life would have been simple if only a few atoms were present in a real solid. Because of large number of atoms, molecules, and ions, the actual picture is more complicated, but fortunately well described by fundamental physics. Without going into the details of the background, it is sufficient for us to learn that there are two principal categories of *bonds* present in a solid. One is the principal bond, and the other the secondary bond.

Principal bonds are of three types that are ionic bond, metallic bond, and covalent bond. Similarly, the secondary bonds are van der Waals bond, polar molecule-induced dipole bonds, and permanent dipole bonds. Familiarity with these bonds would facilitate our understanding of solids and crystals.

4.2.1 Ionic Bonding

As the name suggests, this type of bond holds the ionic solids together. A classic example of an ionic solid is the sodium chloride (NaCl) crystal. Ionic bonding forms when in a molecule one of the atoms is negatively charged because it has gained an electron, and the other positively charged because it has lost an electron. This is the strongest bond that exists in nature. In the case of a NaCl molecule, Cl-ions are negatively charged because it has gained the electron that Na-atom has lost. When ions of opposite polarity approach each other, each ion tends to neutralize its charge by surrounding itself with the ions of opposite polarity. This leads to the formation of periodic lattices in which the environment of all similar atoms is the same. In such a lattice, the sum of all the positive and negative ions equals to zero, making the ionic solids stable and electrically neutral. Besides NaCl crystals alkali metals and halides are also good examples of solids having ionic bonds.

4.2.2 Covalent Bonding

An atom with unsaturated valency is energetically unstable. In order for the solid to acquire stability, the atoms present in the solid must somehow reach saturation in its valence state. This is consistent with the fundamental law of physics that states that the state of equilibrium of a system corresponds to its minimum energy. Once it attains this status, it is held together by a covalent bond in which the electrons are shared between the molecules. A covalent bond is formed when two electrons with antiparallel spins participate. It is a strong bond and is comparable to an ionic bond so far as the bond strength is concerned. It is a dominant bond of organic molecules. The classic example of a molecule in which the covalent bond dominates is our hydrogen molecule (H_2). We know that hydrogen is the lightest element and therefore, the first member of the periodic table. Its electronic configuration is H^1 meaning that only one electron is present in its outermost orbit. These are, as we studied earlier, called the s-electrons, and there can be maximum of only two s-electrons to saturate the s-state. Once a hydrogen atom acquires two electrons, its valency reaches the saturated state and as a result acquires stability. Silicon, too, has covalent bonding, and in fact, it is considered to be an example of giant covalent bonding. Silicon being tetravalent (4) bonds with another four silicon atoms to reach the state of equilibrium. Nature knows many covalently bonded atoms and molecules. A simplified example of covalent bond for silicon and hydrogen are shown in Figure 4.1.

Other examples are silicon dioxide (SiO_2) and the natural gas, methane (CH_4). Like silicon, carbon too is tetravalent and resides in the same group (Group IV) as silicon. Examples of two covalent bonded H_2 and Si are shown in the figure below. The covalent bonds are represented by double arrowed two parallel lines. However, it is not necessary to represent the covalent bonds with double lines. Many authors prefer only one line because of its convenience.

4.2.3 Metallic Bonding

A defining attribute of a metal is its capacity to conduct electricity. We know institutively that copper is a better

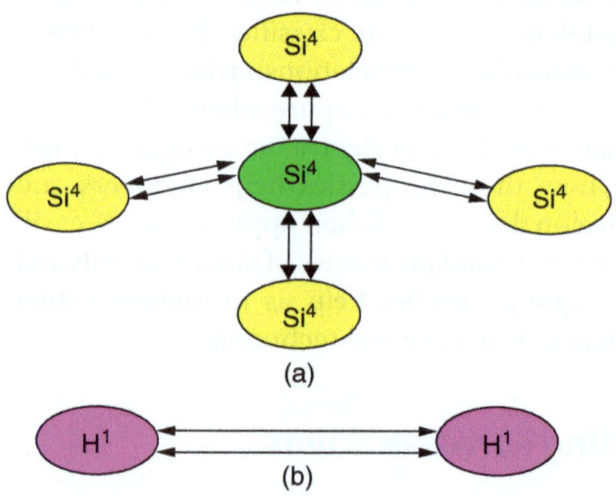

Figure 4.1 Representation of a covalent bond in (a) Si and (b) H_2.

conductor of electricity than, for example, an insulator such as a piece of alumina (Al_2O_3). The reason being that metals are built of atoms in which the valence electrons are free to move around. We can also define metals as home of free electrons with the luxury of aligning themselves along the direction of the applied electric field facilitating the conduction of electricity. Atoms in metals are ionized resulting in the formation of an electron sea leading to the binding of charged nuclei. We can visualize it as the binding between two H atoms resulting in the existence of an H_2 molecule. Alkali metals (Na, K, Cs, etc.) that reside in Group IA of the periodic table are good examples of array of positive charges embedded in the sea of a negatively charged environment. The characteristic feature of a metallic bond is the lowering of the energy of the valence electrons. The metallic bonds are not as strong as the ionic and covalent bonds. Metals tend to crystallize in close-packed structures such as hexagonal close-packed (hcp) or face-centered cubic (fcc) unlike a loosely packed structure such as diamond.

4.2.4 Van der Waals Bonding

The inert gases (He, Ne, Ar, Kr, Xe, and Rn) typically exhibit van der Waals bonding. We define inert gases as those whose electronic configurations are represented by completely filled quantum mechanical states with absolutely no vacancies. Since they consist of closed shells, they are relatively immune to chemical reactions. Therefore, they are called *inert gases*. It is postulated that moving charges in the molecules of the inert gases give rise to the formation of localized electric dipoles, which can interact with each other. The interactive force is relatively weak and is inversely proportional to the seventh power of the intermolecular separation. This force is rather large for large molecules and small for smaller molecules. This concept helps us in distinguishing between gases, liquids, and solids. At room temperature, the small molecules tend to be gases, medium-sized molecules result is liquids, whereas the larger molecules form solids. Organic solids are examples of solids formed because of large interactive forces. They are typically soft and have low melting points.

4.2.5 Polar-molecule-induced Dipole Bonds

This is a very special type of bond. A polar-molecule can induce a dipole effect in an atom that comes in its close vicinity. This can give rise to a bonding. For example, water (H_2O) is a polar molecule in which H-ions are positively charged and the O-ions are negatively charged. The H_2O molecule can induce a dipole in a nearby atom resulting in the formation of a bond.

4.2.6 Permanent Dipole Bonding

This also is a special class of bonding. The H-end of the hydrogen molecule can bond with the negatively charged O-end of another dipole molecule. Consequently, a permanent dipole bonding results. Hydrogen bond in ice is a good example of this type of bonding.

4.3 Structure–Property Relationship

Solids come in all types of forms and structures. We can define solids as amorphous, polycrystalline, single crystals, and thin and thick films. Amorphous solids are those in which long-range order of a crystal is absent and only short-range order exists. They do not have well-defined crystal structure, and as such they cannot be treated on the basis of crystal properties and symmetry. Yet amorphous solids offer scientific challenges and are studied widely. They also have been identified as good materials for certain applications. Glass is a typical example of an amorphous solid. However, for unlocking the potential of solids with the objective of finding novel and unique applications, particularly in electronics, ceramics (polycrystalline), and single crystal are preferred. Thin and thick films are other examples of solids widely used in applications. They too can be structurally either polycrystalline or epitaxial (which is equivalent to having single crystal material). Materials by themselves are passive. They become active only after we subject them to certain external agents such as an electric field.

The simple block diagram presented in Figure 4.2 shows that an external agent induces certain properties in crystalline solids. Manipulation and optimization of these properties form the backbone of novel applications enabling our technological horizon to expand. For example, the application of an electric field (E) gives rise to the generation of an electric current (I) in a conductor and polarization (P) in an insulator. The relationship between the polarization and the electric field can be either linear or nonlinear. Both types are used in applications.

Magnetic field (H) induces magnetization (M) in a solid. For a nonferromagnetic material, the relationship between M and H is linear, whereas it is nonlinear for a ferromagnetic material.

The nonlinearity of polarization manifests itself as a hysteresis loop in a ferroelectric material (such as found in barium titanate, $BaTiO_3$), whereas nonlinearity of M vs. H is the basis for the formation of a hysteresis loop in a ferromagnetic material (an example is iron), which is the key for success of ferromagnetic materials in technology. We will discuss these issues in depth in subsequent chapters.

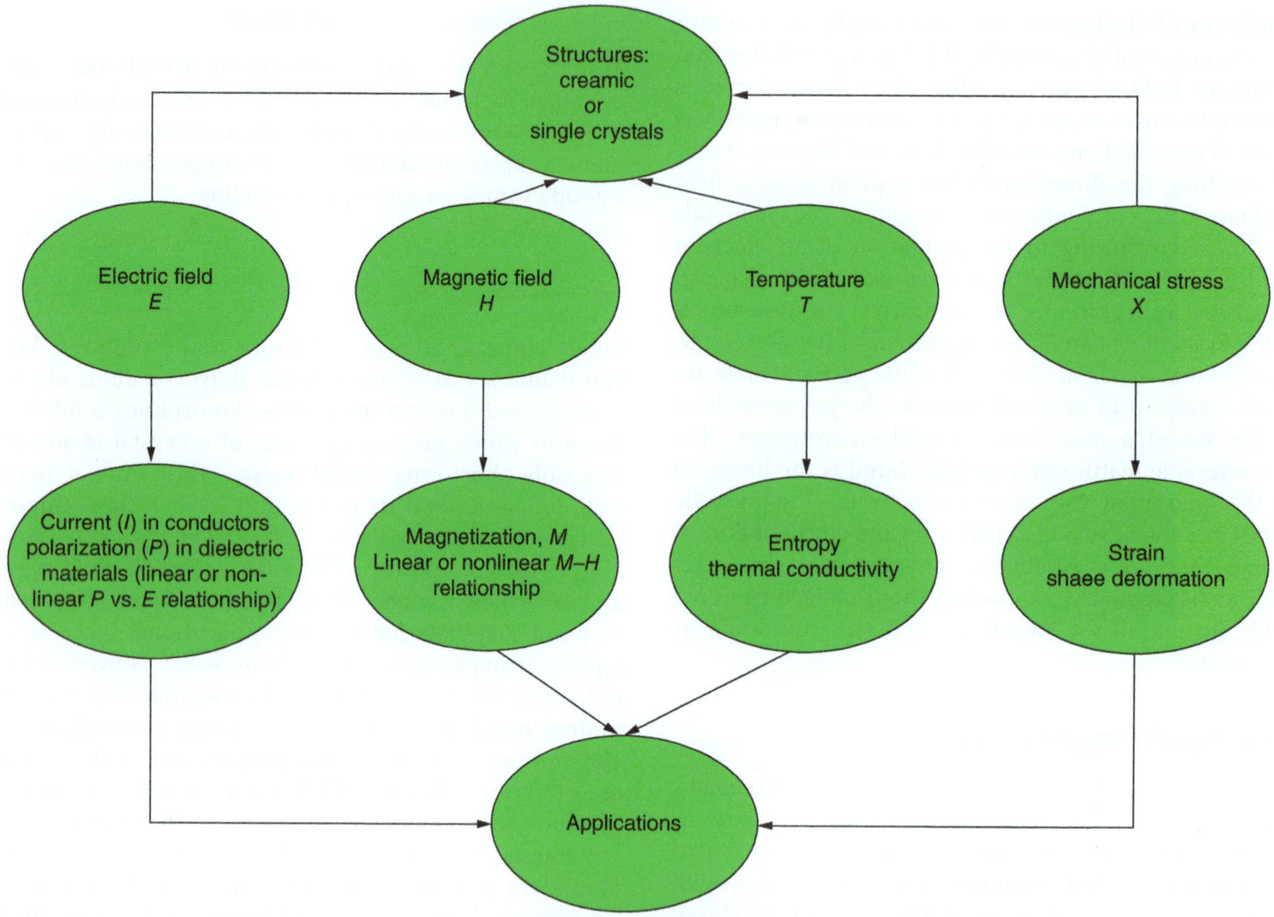

Figure 4.2 Agents induced properties in solids leading to their applications. External agents are electric field (E), magnetic field (H), temperature (T), and stress (X).

Temperature induces entropy (S) and thermal conductivity (κ). A mechanical stress (X) such as the application of pressure can cause a strain (x) in a material resulting in shape deformation that too can be exploited to develop applications. These simple examples show that there is a direct relationship between structure and property. This is an active field of research, and we will see in this and subsequent chapters, how properties can be optimized by manipulating the structure of the solid itself which can meet the requirements of many evolving technology.

Different solids respond differently when subjected to an external agent. For example, in a ceramic, the physical properties are isotropic, meaning that it is the same no matter in which direction we monitor them. The situation is, however, very different if the structure happens to be a single crystal. In general, a single crystal, displays strong anisotropy in the same property. Anisotropy is defined as that characteristic of a crystal that causes a property to vary significantly in magnitude from one crystal direction to another. Therefore, a great care is to be taken in selecting structures to meet the requirements of new devices and technology. Knowledge of atomic radii, nature of bonds, anisotropic characteristics, and electronic band structure and symmetry elements are some of the attributes of materials that must be considered to identify the suitability of a structure for device design and applications.

We will learn in this chapter in some detail about the symmetry elements of a single crystal and the dependence of physical properties on them. While considering the symmetry elements, one ought to keep in mind also the symmetry of external agents, symmetry of resulting parameters, and symmetry of any physical changes that is the result of applied external agent.

We can define the structure–property relationship in two broad categories. One being the *primary* interaction, and the other the *secondary* interaction. Induced properties outlined in Figure 4.2 are some examples of the *primary interaction*. On the other hand, if a physical phenomenon evolves as a result of structural response to two agents applied concurrently to a structure, then we call it the *secondary* interaction. In Figure 4.3, we show some

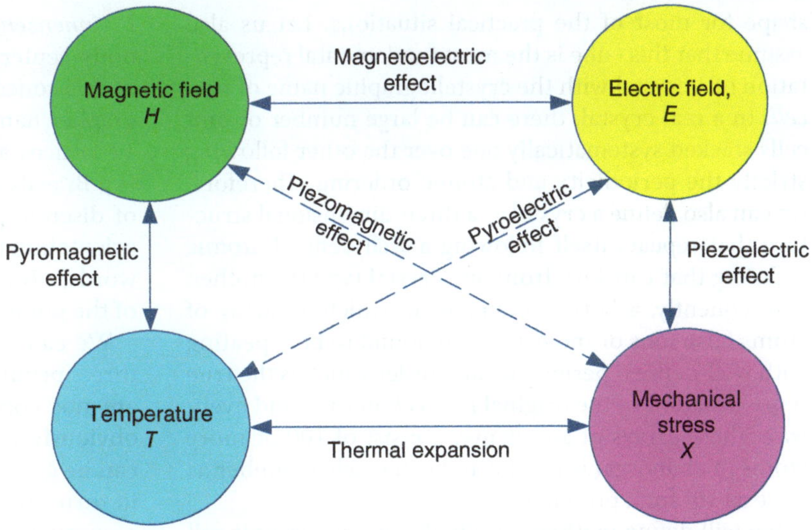

Figure 4.3 Representation of coupled properties in solids giving rise to some unique physical phenomena.

examples of *secondary* interactions originating from the application of two agents concurrently.

For example, a magnetic field and an electric field can combine together to give rise to the magnetoelectric effect. This effect has been known for a very long time, but it is only recently that new materials have been found in which large magnetoelectric effect has been reported. As a result, this has become a very active field of research because its technical importance is projected to be enormous. We will discuss in some detail this effect and its impact on data storage and memory elements in one of the chapters that will follow. Interaction between an electric field and mechanical stress gives rise to the piezoelectric effect in some solids. Piezoelectricity has become a big field by its own right, and we will cover this subject along with its application in Chapter 6. Other examples of secondary effects are pyroelectricity, piezomagnetism, pyromagnetism, and thermal expansion effect.

4.4 Basic Crystal Structures

The building blocks of solids are atoms, ions, and molecules. Structurally, they can be randomly oriented as is the case in a ceramic or they can be distributed in three-dimensional (3d) space with a high degree of order and symmetry. When a structure is highly ordered and possesses symmetry, we call them *single crystals*. In a ceramic structure, the grains (which are very small crystallites) are randomly distributed in space and separated from each other by very thin and narrow walls called the *grain boundaries*. In comparison, a single crystal does not have *grain boundaries*; the whole structure ideally consists of one single crystallographic orientation. This is the fundamental difference between a ceramic and a single crystal. We will learn subsequently in this chapter about the specific attributes of a single crystal including the crystallographic orientation. In Chapter 3, we learned in some detail about the differences in ceramic, single crystals, and films, and how they are processed.

Now the question arises: how do we define a single crystal? There are many different approaches to this problem but let us stay with the most accepted argument of them all. An ideal crystal is built by infinitely repeating identical basic structural unit in three-dimensional space. In other words, a crystal is a three-dimensional structure that repeats itself in space without undergoing any change in its external shape or internal atomic distribution. Spatial periodicity and high degree of atomic ordering are its attributes that distinguishes a crystal from any other structure.

A crystal is customarily represented as a cube as shown in Figure 4.4. Let us assume that each corner of the cube has one atom, which we understand to be spherical in

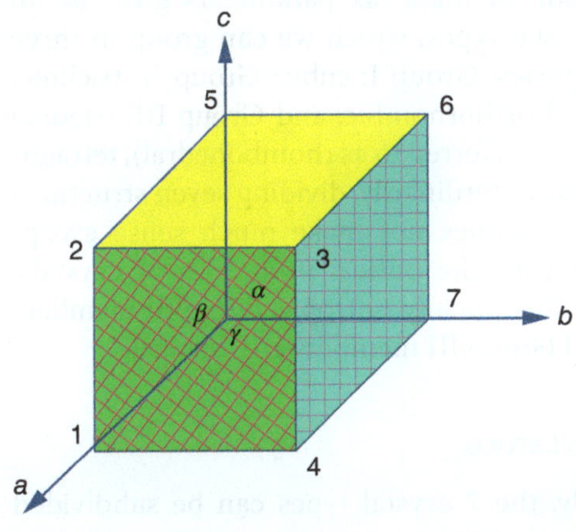

Figure 4.4 A cubic crystal with its defining parameters.

shape for most of the practical situations. Let us also assume that this cube is the most fundamental representation of a crystal with the crystallographic name of *unit cell*. In a real crystal, there can be large number of unit cells stacked systematically one over the other following strictly the periodicity and atomic ordering. Therefore, we can also define a crystal as a three-dimensional structure that repeats itself following a well-defined atomic ordering that can vary from one crystal type to another. Consequently, a lattice originates in which an array of atoms (also ions or molecules) are found to be repeating with well-defined periodicity and order which is the true representation of the original *unit cell* in each and every way. These arrays of atoms may consist of 100 or more atoms in an inorganic crystal and can reach a number as large as 10^3 for a protein crystal.

We will define mathematically the atoms per unit cell later, but for the time being let us accept the fact that a unit cell can have one or more atoms per unit cell. However, when it consists of only one atom per unit cell, it has a special name and meaning in crystallography. It is called the *primitive cell*. The distinction between a *primitive cell* and a *unit cell* is subtle; the distinguishing feature being that a *primitive cell* cannot have more than one atom per cell, whereas a *unit cell* must contain more than one atom per cell. Having resolved this important issue of definition, we can now move on to other defining attributes of a crystal.

The geometry of a unit cell is defined by three principal directions called axes with the labels of *a*, *b*, and *c*. The physical dimensions of the axes *a*, *b*, and *c* are called *lattice constants or lattice parameters*, and they are characteristic of a particular material. For example, it is 0.3567 nm for graphite (C), whereas it is 0.5431 nm for silicon (Si) even though both these materials crystallize in diamond structure. The angle opposite to the a-axis is given the label of α, whereas it is β for the angle opposite to the b-axis, and γ opposite to the vertical c-axis. The combination of these six parameters give rise to seven basic crystal types, which we can group in three principal categories: Group I: cubic; Group II: triclinic, monoclinic, and orthorhombic; and Group III: trigonal (in older literature referred to as rhombohedral), tetragonal, and hexagonal. Ordinarily, dividing seven structures into three groups does not make much sense except when we deal with the optical properties of crystals. Cubic crystals are optically isotropic; Group II members are biaxial; and Group III members are uniaxial.

4.4.1 Bravais Lattice

Mathematically, the 7 crystal types can be subdivided into 14 different lattices grouped into 4 cell types identified by the letters *P* (primitive), *C* (base-centered), *I* (*innenzentriat* which is a German word meaning body-centered), and *F* (face-centered). The foundation of this concept was given by a brilliant French crystallographer named Auguste Bravais in 1850. Therefore, the 14 lattices are called Bravais lattices. But what exactly is a Bravais lattice? It can be defined as an infinite array of discrete point with such a unique arrangement and orientation that is invariant under a translation. In other words, a Bravais lattice appears the same from whichever of the points the array is viewed.

We can describe a primitive lattice with the help of three primitive lattice vectors: a_1, a_2, and a_3 which are not coplanar. These vectors are 3d-vectors which obviously depend on the geometry. The infinite lattice can be generated by infinite array of discrete points that in turn can be generated by a set of discrete translation operations. Making use of these concepts, we can now represent a 3d-lattice by Eq. (4.2).

$$R_n = n_1a_1 + n_2a_2 + n_3a_3 \tag{4.2}$$

where R_n is the position vector that generates the infinite lattice points; and n_i are integers. We have already introduced its concept in Chapter 1 (Eq. (1.61)). It is important to appreciate the fact that a lattice is just a geometrical construction and not a crystal structure. Only when a lattice is occupied by atoms at well-defined sites a lattice is a crystal structure. The atoms can occupy the corners of a lattice or any points. The atomic distribution in a lattice is called the *basis*. We define a crystal structure as follows:

$$\text{lattice} + \text{basis} = \text{crystal structure} \tag{4.3}$$

The 14 Bravais lattices are reproduced in Figure 4.5 [1]. Notice the construction of each cell with atoms (open circles) located in their specific positions. For example, the primitive cell has one atom in each of the eight corners of a three-dimensional structure. Here only the corners are occupied, and there is no other atom anywhere else. Besides at the corners, atoms can also occupy the body-centered or face-centered or base-centered if the Bravais lattice so allows. The body-centered atoms are located at the intersection of the body diagonals whereas the face-centered atoms find themselves at the intersection of face diagonals. Similarly, the base-centered (C) atoms occupy the intersection of face diagonals only at the top and bottom faces of the unit cell.

The 14 Bravais lattices also point to the structural symmetry of a crystal cell. However, much more is involved in the symmetry of a crystal than just the atomic distribution in cell grouped into four different categories. We will come to this point later when we will discuss different types of symmetry elements distinguishing one type of crystal from another. Let us just say at this

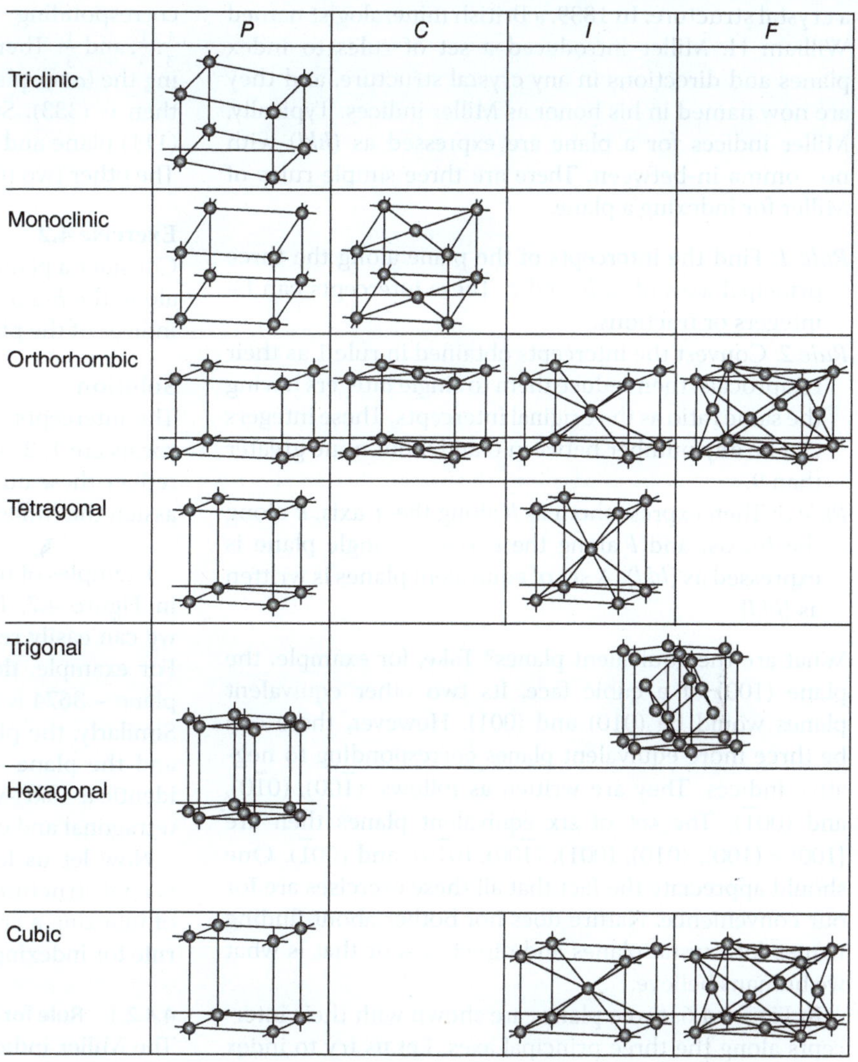

Figure 4.5 Seven crystal structures and their 14 Bravais lattices. Cubic has the highest symmetry and triclinic the lowest. Source: Borchardt-Ott 1995 [1]. Reproduced with permission of Springer.

point that the number 7 is equivalent to the highest symmetry (for cubic structure) on the scale of 1–7 and 1 the lowest (triclinic). The order of symmetry follows the descending order from cubic to triclinic. In Table 4.1, the typical characteristics of the seven crystal types are given.

4.4.2 Miller Indices for Planes and Directions

We can imagine a large number of planes and directions that can exist in a given unit cell. Unless we follow some rules to name them, it would be impossible to distinguish between the infinite number of planes and directions in

Table 4.1 Axial and angular-based definitions of crystal classes.

Crystal type	Lattice constants	Angular relationship	Type of Bravais lattice	Order of symmetry
Cubic	$a = b = c = a_0$	$\alpha = \beta = \gamma = 90°$	*P*, *I*, and *F*	7 (highest)
Trigonal	$a = b = c$	$\alpha = \beta = \gamma < 120°$	*P* (*R*)	6
Orthorhombic	$a \neq b \neq c$	$\alpha = \beta = \gamma = 90°$	*P*, *C*, *I*, and *F*	5
Monoclinic	$a \neq b \neq c$	$\alpha = \gamma = 90° \neq \beta$	*P* and *C*	4
Hexagonal	$a = b \neq c$	$\alpha = \beta = 90°$ but $\gamma = 120°$	*P*	3
Tetragonal	$a = b \neq c$	$\alpha = \beta = \gamma = 90°$	*P* and *I*	2
Triclinic	$a \neq b \neq c$	$\alpha \neq \beta \neq \gamma$	*P*	1 (lowest)

a crystal structure. In 1839, a British mineralogist named William H. Miller introduced a set of rules to index planes and directions in any crystal structure, and they are now named in his honor as Miller indices. Typically, Miller indices for a plane are expressed as (*hkl*) with no comma in-between. There are three simple rules of Miller for indexing a plane.

Rule 1: Find the intercepts of the plane along the three principal axes of *a*, *b*, and *c*. These intercepts can be integers or fractions.

Rule 2: Convert the intercepts obtained in rule 1 as their reciprocals, then reduce them to single integers having the same ratio as the original intercepts. These integers can be any number between 0 and 9 but none greater than 9.

Rule 3: Then express them as *h* along the *a*-axis, *k* along the *b*-axis, and *l* along the *c*-axis. A single plane is expressed as (*hkl*). A set of equivalent planes is written as {*hkl*}.

What are the equivalent planes? Take, for example, the plane (100) of a cubic face. Its two other equivalent planes would be (010) and (001). However, there can be three more equivalent planes corresponding to negative indices. They are written as follows: $(\bar{1}00)$, $(0\bar{1}0)$, and $(00\bar{1})$. The set of six equivalent planes then are $\{100\} \approx (100), (010), (001), (\bar{1}00), (0\bar{1}0)$, and $(00\bar{1})$. One should appreciate the fact that all these exercises are for our convenience. Nature does not bother about finding names for crystal planes and directions, or that is what we humans believe.

In Figure 4.6, three planes are shown with their intercepts along the three principal axes. Let us try to index them using the rules of Miller indices that we have just learned.

First consider the plane with the intercepts of 3 along *a*-axis, 3 along *b*-axis, and 3 along *c*-axis. Their corresponding reciprocal numbers are the fractions: $\frac{1}{3}, \frac{1}{3}$, and $\frac{1}{3}$. Therefore, the three Miller indices identifying the (*abc*)-plane are $h = 3$, $k = 3$, and $l = 3$. The plane then is (333). Such a plane is obviously parallel to the (111) plane and therefore, can also be written as 3(111). The other two planes are 2(111) and (111).

Exercise 4.2

Consider a plane with intercepts of 1 along the *a*-axis, ½ along the *b*-axis and ¼ along the *c*-axis. Find the Miller indices of the plane.

Solution

The intercepts are 1, ½, and ¼. Therefore, their reciprocals are 1, 2, and 4, respectively. Obviously, we cannot reduce these numbers to anything smaller. Therefore, we assign the Miller indices to this plane as (124).

Examples of major planes in cubic structure are shown in Figure 4.7. Following the Miller rules of indexing, we can easily recognize the frontal planes in Figure 4.7. For example, the front plane – 1234 is (100), the side plane – 3674 is (010), and the top plane – 2563 is (001). Similarly, the plane – 152 is indexed as the (111) plane and the plane – 1267 as the (110) plane. Planes with identical (*hkl*)-values can also be identified easily for tetragonal and orthorhombic structures.

Now let us learn about indexing the directions in a crystal structure. This task is even simpler than the task of indexing a crystal plane. We simply follow the Miller rule for indexing crystal directions.

4.4.2.1 Rule for Indexing a Crystal Direction

The Miller indices for a direction is expressed as [*uvw*]. The numerical values for *u*, *v*, and *w* are the set of smallest integers that maintain same ratio among each other as

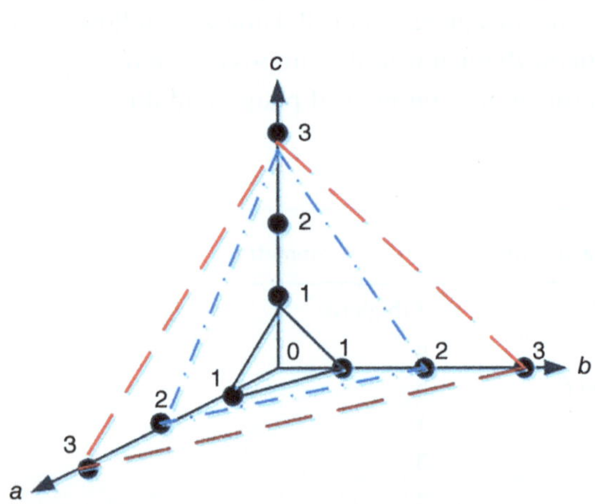

Figure 4.6 Three planes with intercepts in *a*, *b*, and *c* axes.

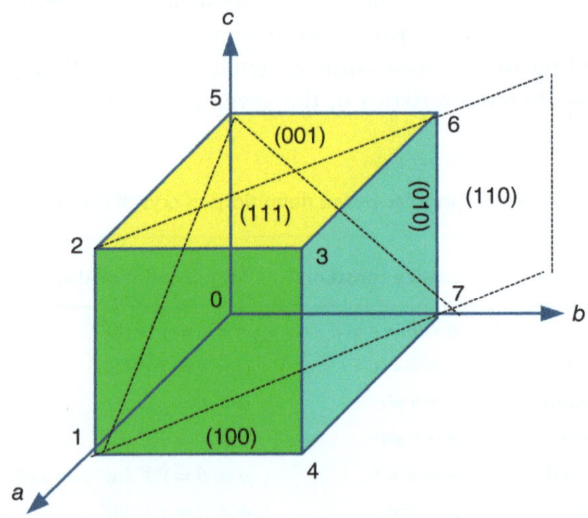

Figure 4.7 A cubic crystal with its prominent planes.

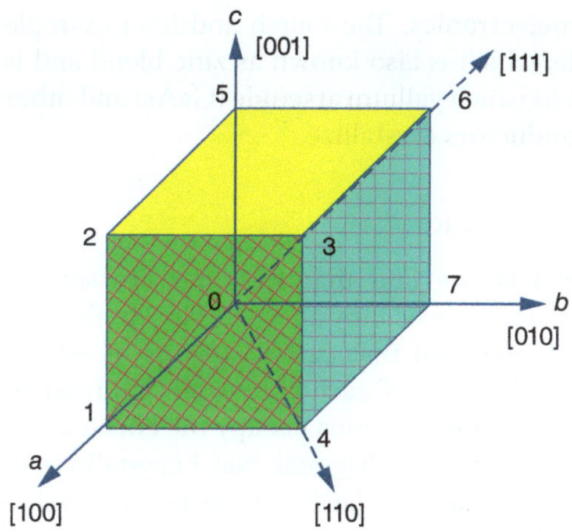

Figure 4.8 Prominent directions in a cubic crystal.

the components of a vector in the desired direction. The direction along the a-axis is referred to as [100]. Similarly, it is [010] along the b-direction and [001] along the c-direction. These directions are identified in Figure 4.8 for a cubic crystal.

The other two directions, namely, [110] and [011] are also identified in this figure. These are the most prominent directions of a crystal. The same indices we can assign for the equivalent directions for the tetragonal and orthorhombic structures.

Hexagonal and trigonal structures can also be indexed according to Miller's rules of indexing but with a small exception. Four indices are used that are ($hkil$). Here h, k, and l indices are identical to the indices discussed already. The index i is new, and the relationship between h, k, and i must obey the following condition.

$$h + k = -i \tag{4.4}$$

This condition makes the i-index redundant because it can be easily expressed as the sum of $(h + k)$ with sign reversed. Therefore, it is customary to express the hexagonal planes and direction only using the hkl-indices.

In a hexagonal structure, the ab-plane is its basal plane, and the vertical c-axis is perpendicular to it. In its primitive cell, $a = b$ and the angle between a and b equals 120°. Hexagonal structures also come as closed-pack, which is abbreviated as hcp. In an ideal hcp structure $c = 1.633a$. Many metals crystallize with hcp structure. Some examples are Be, Mg, Ti, Zr, etc. Another example of closed-packing is the face-centered-cubic crystal.

4.5 Reciprocal Lattice

There are two types of crystal lattices. One is the *direct lattice* and the other the *reciprocal lattice*. So far we have been dealing only with the direct lattice in this chapter. We can define it as a periodic function in real space. The reciprocal lattice, on the other hand, exists in reciprocal space which is referred to as *momentum space* in solid-state physics. Another name is K-space which is also a common vocabulary of solid-state physics. The Fourier transform becomes a valuable tool in dealing with reciprocal lattice. Therefore, the reciprocal lattice of a reciprocal lattice is the original direct lattice. The reciprocal lattice is of fundamental importance in understanding the neutron and X-ray diffraction. The concept of reciprocal lattice, or more precisely of momentum-space, is widely used in the interpretation of Brillouin zones in solid-state physics.

The concept of reciprocal lattice was first advanced by Paul Peter Ewald (1888–1925), a German Physicist, for explaining the X-ray diffraction patterns for crystal structure. One can find a reciprocal lattice corresponding to any direct lattice simply by drawing a normal from the origin for each set of lattice planes (hkl) and assigning it a length of d^* which must satisfy the following condition.

$$d^* = \frac{C}{d_{hkl}} \tag{4.5}$$

where C is a constant, and d_{hkl} the inter-planar distance between a set of parallel planes with the Miller indices of (hkl).

For an excellent treatment of reciprocal lattices, we recommend reference [2].

4.6 Relationship Between d^* and Miller Indices for Selected Crystal Systems

For a cubic crystal with the lattice parameter of a, Eq. (4.5) is the most used standard relation for inter-planar distance, d, which exists between the parallel planes having (hkl) indices.

$$d_{hkl} = \frac{a}{\sqrt{h^2 + k^2 + l^2}} \tag{4.6}$$

The corresponding reciprocal lattice, d^* is given then by Eq. (4.6).

$$d^{*^2}_{hkl} = \frac{1}{d^2_{hkl}} = \frac{h^2 + k^2 + l^2}{a^2} \tag{4.7}$$

For a tetragonal structure with the lattice constants of a and c, d^* is

$$d^{*^2}_{hkl} = \frac{1}{d^2_{hkl}} = \frac{(h^2 + k^2)}{a^2} + \frac{l^2}{c^2} \tag{4.8}$$

For a orthorhombic structure with the lattice constants of a, b and c, d^* is given by Eq. (4.9).

$$d^{*^2}_{hkl} = \frac{1}{d^2_{hkl}} = \frac{h^2}{a^2} + \frac{k^2}{b^2} + \frac{l^2}{c^2} \tag{4.9}$$

For a hexagonal structure with the lattice constants a and c, d^* is given by Eq. (4.10).

$$d^{*2}_{hkl} = \frac{1}{d^2_{hkl}} = \frac{4}{3}\left(\frac{h^2 + hk + k^2}{a^2}\right) + \frac{l^2}{c^2} \tag{4.10}$$

Exercise 4.3

Find the inter-planar distances between the planes (100), (110), and (111) for silicon with the lattice constant of 0.5431 nm. Comment on the results.

Solution

To find the values of inter-planar distances, d_{hkl}, we can use Eq. (4.4). We get: $d_{100} = 0.5431$ nm; $d_{110} = \frac{0.5431}{\sqrt{2}} = 0.3840$ nm; and $d_{111} = \frac{0.5431}{\sqrt{3}} = 0.3136$ nm.

Comment: (i) Higher values of (hkl) result in reduced spacing between the parallel planes and (ii) $d_{100} : d_{110} : d_{111} = 1.73 : 1.22 : 1$. That is the d-spacing is 73% more for (100) plane and 22% more for (110) plane with respect to the spacing of the (111) planes.

4.7 Typical Examples of Crystal Structures

In this section, we will discuss the examples of four materials which are of our prime concern. The first example is of sodium chloride (NaCl), which is also commonly known as table salt and is the model representation of an ionic bonding. The second example is calcium titanate ($CaTiO_3$), which is the prototype of pervoskite structure in which ferroelectric barium titanate ($BaTiO_3$) crystallizes. The third example is diamond structure in which silicon crystallizes. We know that silicon dominates the field of microelectronics. The fourth and final example is zinc sulfide, which is also known as zinc blend and is the structure in which gallium arsenide (GaAs) and other III–V semiconductors crystallize.

4.7.1 Sodium Chloride, NaCl

From Figure 4.9a, we find that Na^+-ions are smaller than Cl^- ions, and they occupy alternate sites in the unit cell. The Na-ions (blue) that carry a charge of +1 are located at the midpoint of each cube edge sandwiched between two Cl^--ions. Cl^--ions occupy the corners and midpoint sites of the face diagonal. NaCl crystallizes in fcc structure. There are four lattice points per unit cell of NaCl. We will learn later how to calculate this. There are also six nearest neighbors; each type of atom having six nearest neighbor of the other kind. Its lattice constant is 0.563 nm. As already discussed before, NaCl is the typical example of ionic bonding. Potassium chloride (KCl) with the lattice constant of 0.629 nm, lead sulfide (PbS) with the lattice constant of 0.592 nm, and magnesium oxide (MgO) with the lattice constant of 0.420 nm are other prominent examples of the fcc structure. Lead sulfide is a semiconductor material and was used as such long before silicon and germanium were discovered. Magnesium oxide is an important substrate material for the growth of thin films of many oxides.

4.7.2 Perovskite Calcium Titanate

Its unit cell is shown in Figure 4.9b. Calcium titanate, as we already know, is the most prominent member of the perovskite family represented by the general chemical formula of ABO_3. Ca atoms (blue) occupy the corners of the orthorhombic structure, whereas oxygen (red)

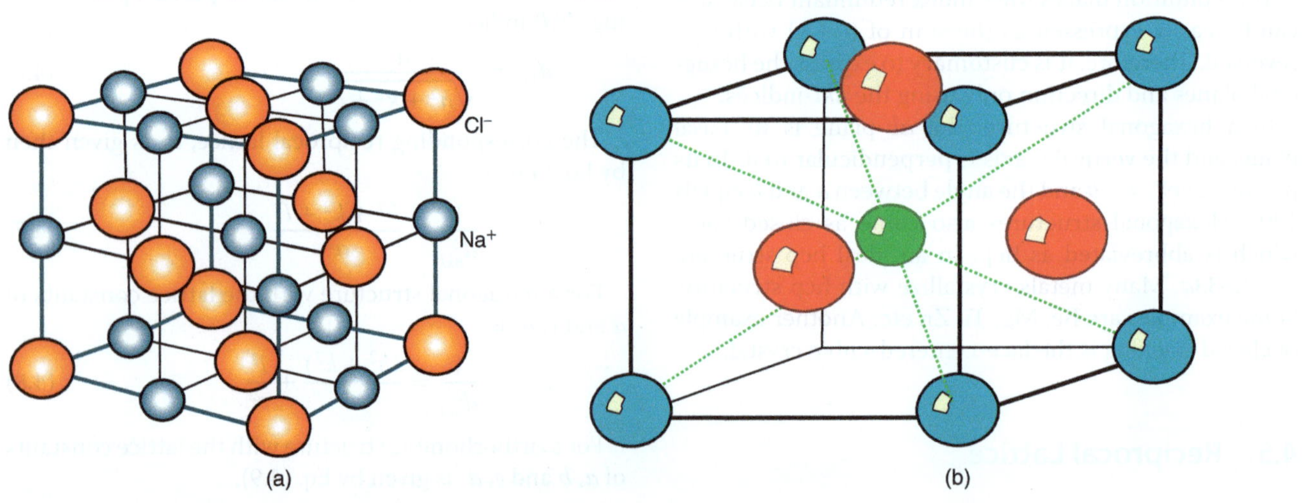

Figure 4.9 Examples of some common crystals: (a) NaCl (rock salt) and (b) $CaTiO_3$ (perovskite). http://www.tf.uni-kiel.de/matwis/amat/mw_for_et/kap_3/backbone/r3_2_2.html.

the midpoint of the face diagonals and titanium (green) the body-centered site (which is equivalent to the midpoint of the body diagonals or volume diagonals). If we eliminate the central titanium atoms, the structure becomes a face-centered cell with Ca and O as its building blocks. The presence of the titanium atom gives the uniqueness to this structure, which we can consider as the combination of a body-centered and a face-centered structures. The lattice points per unit cell is 5 (4 from its face-centered basis and 1 from the central Ti atom which is not shared by any other neighboring units). Usually, it comes as an orthorhombic structure with the lattice parameters of $a = 0.538$ nm, $b = 0.5443$ nm, and $c = 0.7645$ nm. In comparison, ferroelectric $BaTiO_3$ crystallizes in tetragonal perovskite structure with the lattice constants of $a = 0.3994$ nm and $c = 0.4034$ nm. This is a structure that has gained importance in technology ever since it was discovered that the famous example of high temperature superconductor, YBaCu-oxide (123 YBCO) has the structure of perovskite.

4.7.3 Diamond Structure

It is an intriguing structure having the internal symmetry that is a tetrahedral and external that of a face-centered cubic crystal. We introduced the tetrahedral structure in Figure 4.1, in which each silicon atom is covalently bonded by the other four silicon atoms. The same pattern will hold for any tetrahedral structure irrespective of the material.

Figure 4.10 represents a diamond structure that is the same structure in which famous semiconductors silicon and germanium of Group IV synthesize [3].[1] The structure got this name because indeed natural diamonds (C_x) crystallizes with this structure. The center of the tetrahedral structure is identified as numbers 1–4 in this figure. Following these centers, we can easily see a repeating pattern that culminate in a diamond structure. Each tetrahedral central atom is covalently attached to one corner atom and three face-centered atoms. Each of them is located at a distance of $\frac{a\sqrt{3}}{4}$ (a being the lattice constant) from one corner atom of the body diagonal and $\frac{3a\sqrt{3}}{4}$ from the other corner atom. The distance equal to $\frac{a\sqrt{3}}{4}$ is called the bond length. There are eight lattice points per unit cell, which is twice the value found for the fcc structure. However, it is pretty much a vacant structure. Only 34% of its total available space is occupied by atoms, which is exactly $\frac{1}{2}$ of the value found for a bcc structure. The nearest neighbor distance is just equal to $\frac{a\sqrt{3}}{4}$, which is the same as the bond length of the diamond structure.

The atomic distribution for the diamond structure is shown in Figure 4.11 projected on a cube face of the diamond structure. The four corners represent a coordinate of 0. Beginning from any one of them, we will find atoms at sites given by fractional numbers. These are the fractions of the lattice parameter, a. Notice the fractions $\frac{1}{4}$ and $\frac{3}{4}$. They refer to the sites occupied by the tetrahedral centers. The fraction $\frac{1}{2}$ is the site occupied

1 http://iuc.twien.ac.at

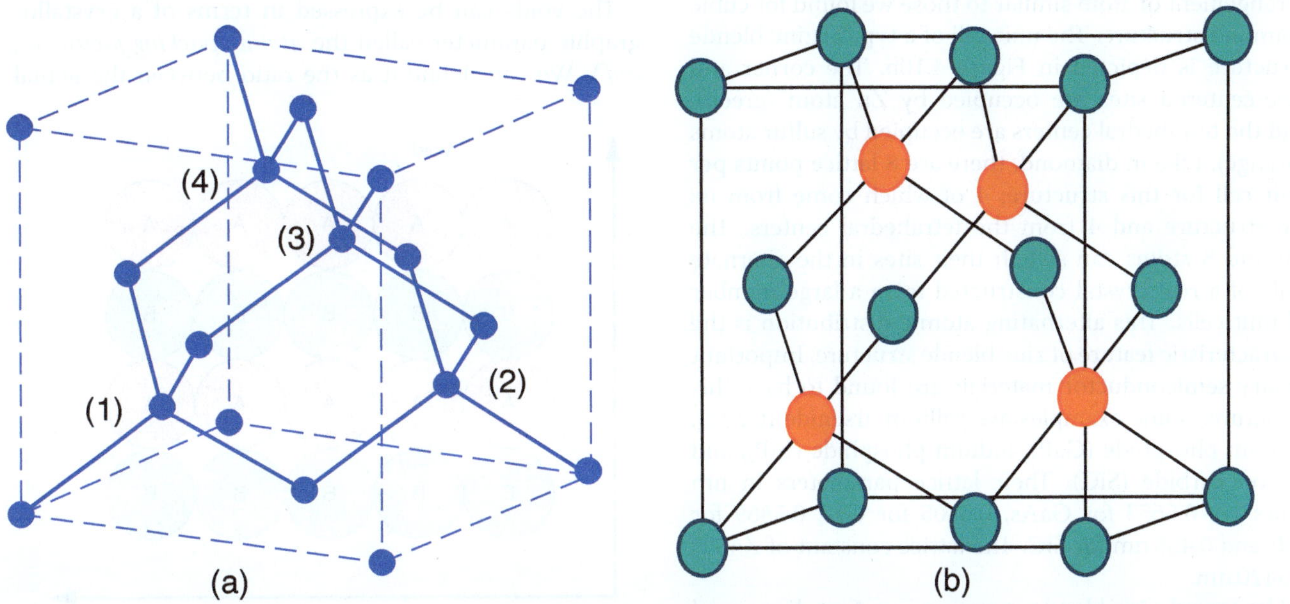

Figure 4.10 (a) Diamond (http://iuc.tuwien.ac.at) and (b) zinc blende (ZnS, http://www.tf.uni-kiel.de/matwis/amat/mw_for_et/kap_3/backbone/r3_2_2.html) structures.

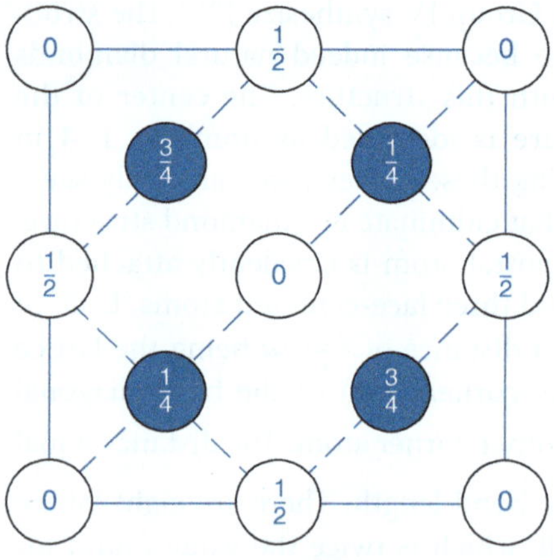

Figure 4.11 Atomic distribution of a diamond structure. Source: Kittel 1976 [2]. Reproduced with permission of John Wiley and Sons.

by an atom at the midpoint of the cube edge (see NaCl figure for clarification if need be).

Diamond (C_x), silicon (Si), germanium (Ge), and tin (Sn), all of which are elements of the Group IV of the Periodic Table, crystallized with the diamond structure. Their lattice constants in nm units are 0.3567 for C_x, 0.5431 of Si, 0.5658 for Ge, and 0.4249 for Sn. Diamond, silicon, and germanium are semiconductors, and tin a superconductor at low temperatures.

4.7.4 Zinc Blende (Also Wurtzite)

Zinc blende comes in two structures: cubic and hexagonal. Cubic zinc blende (or ZnS) has the tetrahedral arrangement of atom similar to those we found for cubic diamond structure. The unit cell of a typical zinc blende structure is depicted in Figure 4.10b. The corner and face-centered sites are occupied by Zn atom (green), and the tetrahedral centers are occupied by sulfur atoms (orange). Like in diamond, there are 8 lattice points per unit cell for this structure; 4 of which come from its fcc structure and 4 from the tetrahedral centers. The Zn and S atoms can switch their sites in the alternate cells of a real crystal constructed from a large number of unit cells. This alternating atomic distribution is the characteristic feature of zinc blende structure. Important binary semiconductor materials are found to have this structure. Some examples are gallium arsenide (GaAs), gallium phosphide (GaP), indium phosphide (InP), and silicon carbide (SiC). Their lattice parameters in nm units are 0.5653 for GaAs, 0.5405 for GaP, 0.5869 for InP, and 0.435 nm for SiC. The lattice constant of ZnS is 0.5420 nm.

Hexagonal zinc blende structure was first discovered in meteorites. It has been synthesized in laboratory. The cubic and hexagonal forms of zinc blende are related and have the identical values of density. The two forms can be simulated using carbon as the atoms. This structure is also known as wurtzite, named after the mineral wurtzite. As expected, the tetrahedral structure is also the building block of this structure. Some important members of this group with their respective lattice constants in nm scale are ZnO; with $a = 0.325$ and $c = 0.51$; ZnS with $a = 0.381$ and $c = 0.623$; SiC with $a = 0.325$ and $c = 0.521$; cadmium sulfide (CdS) with $a = 4.13$ and $c = 6.75$; hexagonal diamond with $a = 2.52$ and $c = 4.12$; and gallium nitride (GaN) with $a = 0.3189$ and $c = 0.5184$. GaN is the lead material for optoelectronics.

4.8 Origin of Voids and Atomic Packing Factor (apf)

We know that atoms can be assumed to be spherical in shape and also that they are stacked in a certain order occupying the specified sites in the unit cell. During the process of sequential stacking, atoms can touch each other, but not overlap, as shown if Figure 4.12. Notice that a void is produced when four atoms come together in a two-dimensional space. There can be many different ways in which the atoms can be sequentially distributed in three dimensions, giving rise to a large numbers of voids. For simplicity, we are considering only the two-dimensional sequence of atoms A (red) and atoms (B). The two rows of A and B atoms alternate each other. The voids can be of two principal configurations that are octahedral and tetrahedral. They play a very important role dealing with defects in structures.

The voids can be expressed in terms of a crystallographic parameter called the *atomic packing factor* (or, *apf*). We can define it as the ratio between the actual

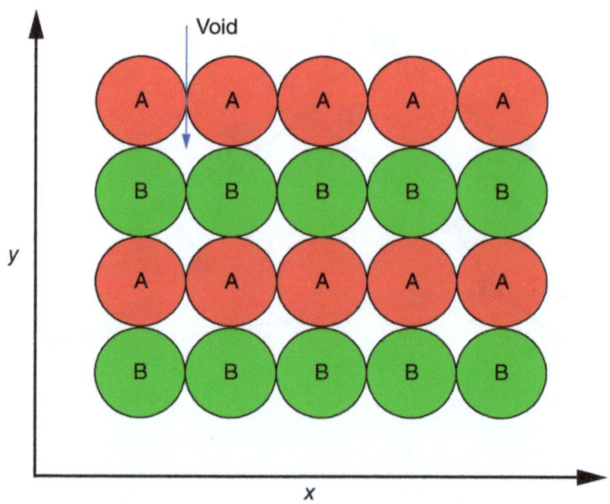

Figure 4.12 Atoms stacked over each other forming voids.

volume occupied (V_{oc}) and the theoretical volume (V_{th}) available in an unit cell with the assumption that atoms are spheres, and they touch each other along the nearest neighbor distance. Mathematically expressed, we get Eq. (4.11).

$$\text{apf} = \frac{\text{actual volume occupied}}{\text{theoretical volume}} = \frac{V_{oc}}{V_{th}} = n\left(\frac{V_{atom}}{V_c}\right) \tag{4.11}$$

here n = number of atoms per unit cell, V_{atom} = volume of one atom, and V_c the unit cell volume. Equation (4.11) tells us that the concept of apf is another example of how the crystal structure is related to the physical properties of a crystal structure. Let us now calculate the apf for cubic crystal with primitive Bravais lattice. This will guide us to calculate apf for other structures.

4.8.1 apf for a Primitive Cubic Structure (P)

In this structure, as we might recall, we have a cell with just eight corner atoms and nothing else in-between. Its $n = 1$. The nearest neighbor distance is equivalent to the lattice constant itself. We can therefore consider that the lattice constant $a = 2r$ (distance between center of one atom to the center of another atom) where r = atomic radius. Using Eq. (4.10) we get

$$\text{apf} = \frac{1\left(\frac{4\pi}{3}\right)r^3}{a^3} = \frac{1\left(\frac{4\pi}{3}\right)\left(\frac{a}{2}\right)^3}{a^3} = 0.52 = (52\%)$$

From this example, we conclude that only 52% of the available space is actually occupied by the atoms in a simple cubic structure. Void occupies the 48%.

Exercise 4.4
Calculate the apf for a bcc, fcc, and diamond structure. Comment on your result.

Hint: Students are encouraged to calculate the values for apf for the three unit cells. If correctly done, the answers would be the apf for the bcc structure = 0.68, it is 0.74 for the fcc structure, and 0.34 for the diamond structure.

Comments: Among all these structures, the fcc appears to have the maximum percentage of occupancy. Therefore, this is also called the cubic closed-packed (ccp) unit cell. Diamond, on the other hand, has the lowest occupancy among all these cubic unit cells and yet has an intriguing internal arrangement of atoms. It is in fact only 50% of the apf of a bcc structure. No wonder it is so rare and so costly because Nature appears to exhibit its mastery in arts and sciences in producing diamond crystals!

We tabulate the most important properties of cubic crystals in Table 4.2. This should serve as a ready reference material for students while solving homework problems.

4.9 Hexagonal and Cubic Close-packed Structures

In the preceding section, we discussed the cases of some important structures with importance to technology. In all these cases, we were not concerned about the packing of the atoms. Now we need to discuss two special cases that are close-packed hexagonal structures and close-packed cubic structures. Many metals crystallize with these structures.

Nature, in producing close-packed structures, tries to minimize the voids so that the resulting structures can have the maximum possible space occupied by atoms. Mathematically, it can be achieved by using two methods of atomic stacking order. In one, a hexagonal structure results with the apf of 0.74. In another stacking order, the

Table 4.2 Basic crystallographic properties of cubic crystals with the lattice constant being a.

Crystal type	Volume per cell	Lattice points per cell	Number of nearest neighbors	Nearest neighbor distance	Atomic packing fraction
Simple	a^3	1	6	a	$\frac{\pi}{6}$(= 52%)
Body-centered	a^3	2	8	$\frac{a\sqrt{3}}{2}$	$\frac{\pi\sqrt{3}}{8}$ (= 68%)
Face-centered	a^3	4	12	$\frac{a}{\sqrt{2}}$	$\frac{\pi\sqrt{2}}{6}$(= 74%)
Diamond (C_n)	a^3	8	4	$\frac{a\sqrt{3}}{4}$	$\frac{\pi\sqrt{3}}{16}$(= 34%)

Source: Kittel 1976 [2]. Reproduced with permission of John Wiley and Sons.

Table 4.3 Some metals and their crystal structures.

Structures	Metals
bcc	Cr, Fe, Mo, Ta, and W
fcc	Al, Cu, Au, Pb, Ni, and Pt
hcp	Cd, Co, Mg, Ti, and Zn

structure produced is the face-centered cubic structure which too has the apf of 0.74. In the case of closed-packed hexagonal structure (hcp), the atoms are distributed in alternating A and B layers. The sequence that follows is A-B-A-B. In the case of a fcc (or ccp) structure, the atoms follow the sequence of ABC. In Table 4.3, we list the metals with their crystal structures.

4.10 Predictive Nature of Crystal Structure

By now, we should be convinced that atoms, which are the building blocks of all materials, can be arranged in many different ordered structures by Nature that we call crystal structures. Therefore, it is no wonder that crystal structures of materials are intimately related to the predictive nature of physical properties. One such example is the density, which is the most fundamental property of all solids. By definition, the theoretical density (ρ_{th}) is nothing else than the ratio between the mass of a solid and the volume associated with it. We can also express this simple ratio on the basis of the chemical composition of a solid and its crystal structure as expressed in Eq. (4.11)

$$\rho_{th} = \frac{\text{cell mass}}{\text{cell volume}} = \frac{nA}{N_A V_c} \quad (4.12)$$

here A is the atomic weight, N_A the Avogadro's number = (6.023×10^{23} molecules/mol) and V the unit cell volume, and n the number of atoms per unit cell. In Eq. (4.12), we can readily know the value of all identities on the right-hand side except the number of atoms per unit cell (n). You might recall that we have already made use of this concept in our previous discussions without knowing how we came to the numerical values. Let us now try to find out how to determine this quantity. Consider Figure 4.13 and the argument that follows.

In the above figure, we have shown a set of four cube faces labeled I, II, III, and IV. Each corner has one red atom, and the central atom is green, and all are identical atoms. The central atom is located at the intersection of the four cube faces. Now imagine that we place an identical structure on top of this structure. In that event, central atom is shared by eight equivalent cubes. Only $\frac{1}{8}$ of this atom could be considered belonging to one cube. We can extend this concept to all other atoms, remembering that any one of them could be at the intersecting point of eight cubes. Therefore, in a cube with 8 corner atoms, there will be only 1 atom per cell that can belong exclusively to the cell. This would be the case of a primitive cell, which can have only 1 lattice point per cell according to the rules of Bravais lattices. It is worth remembering that the number of atoms per unit cell is the same as lattice points per cell.

Exercise 4.5

Find the number of atoms per unit cell for the bcc and fcc structures.

Solution

Case 1. Consider first the bcc structure: From Figure 4.5, we can determine that the bcc (*I*) structure consists of 8 corner atoms plus 1 atom at the center of the body diagonals. The corners will contribute 1 atom per unit cell, but the body-centered atom belongs exclusively to this structure and cannot be shared by any other cube. This gives us 2 atoms (or 2 lattice points) per unit cell for the bcc structure.

Case 2. Fcc (F) structure: Here too, we will get just 1 corner atom per unit cell. But each face-centered atom can be concurrently shared by two cubes (top and bottom). There are six faces to the cubic cell, making the contribution to be 3 per unit cell. Then for an fcc structure, we get 4 lattice points per unit cell. Now let us put this theoretical argument to test. In order to do this, let us calculate the density of a structure, say, for example, NaCl, using Eq. (4.12).

We have for NaCl, $A = 54.45$ and $n = 4$. Its lattice constant, $a = 0.564$ nm ($= 4.64 \times 10^{-8}$ cm). Substituting these

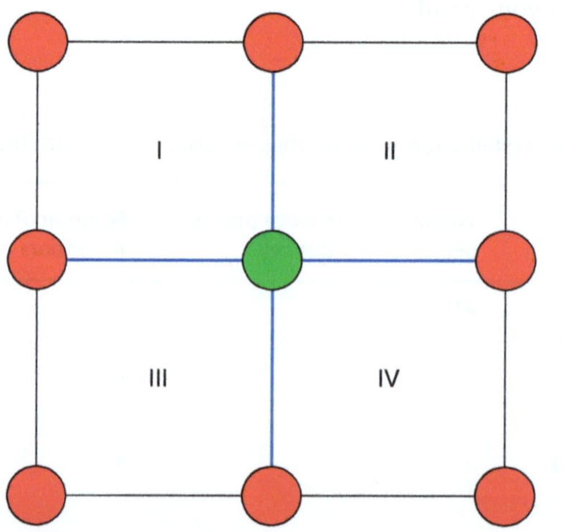

Figure 4.13 Two-dimensional atomic distribution among four identical cubic faces.

values in Eq. (4.12), we get

$$\rho_{th} = \frac{nA}{N_A V_c} = \frac{4 \times 54.45}{6.023 \times 10^{23} \times 4.64^3 \times 10^{-24}}$$

$$= \frac{233.8}{6.023 \times 14.94} = \frac{233.8}{108} = 2.16\ \mathrm{g\,cm^{-3}}$$

The experimental value of density for NaCl is $2.15\ \mathrm{g\,cm^{-3}}$. The agreement is excellent with the theoretical value validating the assumptions made in determining the values of lattice points per unit cell. This simple example should serve to convince the readers of the intimate relationship that exists between a crystal structure and its physical properties of solids. This is just one example. In the subsequent sections, we will learn how the coupled physical phenomena, which are fascinating and very important for a host of applications, are linked intimately with the symmetry elements of crystal structures.

4.11 Hypothetical Models of Centrosymmetric and Noncentrosymmetric Crystals

We have defined a crystal structure with respect to spatial periodicity of atoms in a 3d-space that keeps on repeating itself and is the exact replica of the smallest representation of the larger structure. A quick inspection of the 14 Bravais lattices given in Figure 4.5 will convince us that each structure displays a high degree of external symmetry. For a better understanding, let us consider the case of a body-centered cubic crystal. We can cut it in half either vertically or horizontally with respect to the body-centered atom, and will find that the two halves are identical in each and every respect. The result will be the same even if decide to halve it in any other way. This is what is meant by external symmetry. But this is just one type of symmetry. For a complete understanding to the nature of crystal symmetry, external and internal, we will have to consider certain types of operations called symmetry elements. We will elaborate on these points in the next section. But before we can understand the meaning of symmetry elements, we need to accept the fact that symmetry operations can produce, in general, two types of lattices; one being centrosymmetric and the other noncentrosymmetric.

What exactly is meant by these terms? An explanation of Figure 4.14 might be helpful. In this figure, we have the representation of two hypothetical models labeled (a) and (b).

In a real crystal, the positive and negative ions share an inversion point, which is electrostatically neutral. This is a true statement and is valid for all crystal types only if the inversion point is at the lattice points and midway between any two of them. For example, in a P lattice, the inversion points are located at the following coordinates: $0, 0, 0; \frac{1}{2}, 0, 0; 0, \frac{1}{2}, 0; 0, 0, \frac{1}{2}; \frac{1}{2}, \frac{1}{2}0; \frac{1}{2}, 0, \frac{1}{2}$, and $\frac{1}{2}, \frac{1}{2}, \frac{1}{2}$.

The inversion point is also referred to as the center of symmetry. When an inversion point is present in a lattice, the crystal structure is called centrosymmetric. This situation does not change even under a stress, which might squeeze the crystal, but will fail to destroy its inversion point. The model of Figure 4.14a represents the case of a crystal in which an inversion point is present and it is invariant. However, there are a large number of crystals in which the inversion point is absent because the atomic distribution is such that an internal stress is present which causes slight distortion in the symmetry within the lattice. The distorted structure gives rise to dipoles. Such a crystal is called noncentrosymmetric crystals.

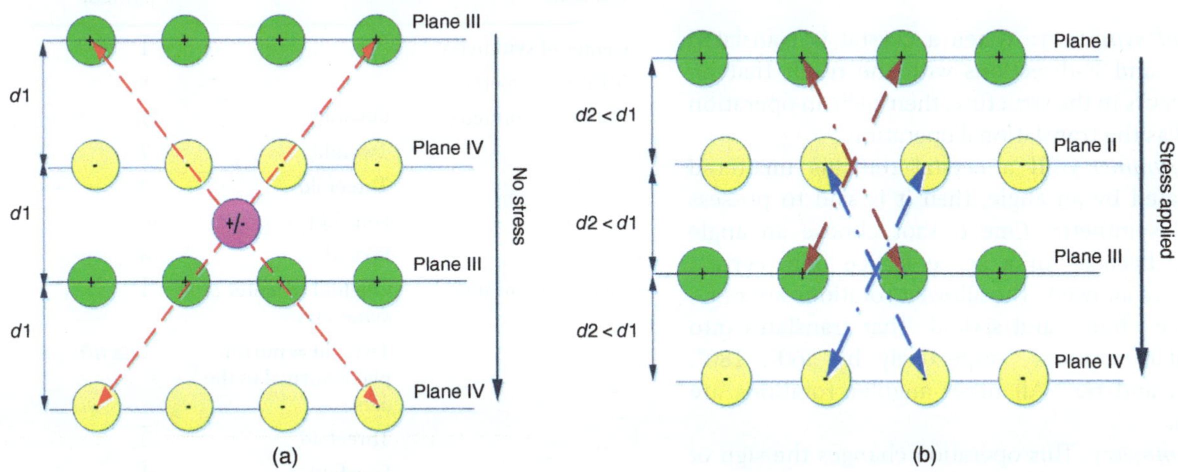

Figure 4.14 Models of (a) centrosymmetry and (b) noncentrosymmetry in a crystal.

Now consider the part (b) of the figure. For a moment, let us assume that the symmetry of the lattice in Figure 4.14a can be destroyed by the application of an external stress. As expected the structure will suffer a squeeze along the direction of the applied stress causing a relative shift in the equilibrium positions of the positive and negative ions. Also the interlunar distance $d1$ of Figure 4.14a will become $d2$ of Figure 4.14b such that $d2 < d1$. The new structure will lose the original inversion point and dipoles will form as shown in Figure 4.14b. The absence of an inversion point gives rise to some very unique physical phenomena such as ferroelectricity, piezoelectricity, and pyroelectricity with enormous appeal to technology. These phenomena cannot be observed in centro-symmetric crystals. A good example of a noncentrosymmetric crystal is perovskite barium titanate.

4.12 Symmetry Elements

There are four types of symmetry operations, each of which is based on different mathematical manipulations. It is also common in literature to refer to the symmetry operations as symmetry elements. The four symmetry elements are translational symmetry, rotational symmetry, inversion symmetry, and mirror symmetry. The selection of the origin for the symmetry operations and coordinate system must be done with the sole objective of achieving the highest possible symmetry. Usually, the Cartesian coordinates are chosen for symmetry operations. Under these conditions, the cubic crystals have the highest symmetry and the triclinic the lowest (see also Table 4.1). We have expressed them on the relative scale of 1–7, 7 being the highest and 1 the lowest. Let us now try to understand the four categories of symmetry elements.

Translational symmetry: When a crystal is translated in X-, Y-, and Z-directions with the result that no change occurs in the structure, then such an operation is defined as the translational operation.

Rotational symmetry: If a crystal remains unaltered when rotated by an angle, then it is said to possess rotational symmetry. One cannot choose an angle arbitrarily because in a space lattice only certain rotations are allowed. The allowed rotations are one-, two-, three-, four-, and sixfold. That translates into the angular rotations, respectively by 360°, 180°, 120°, 90°, and 60°. All other angular rotations are forbidden.

Inversion symmetry: This operation changes the sign of the coordinates of each point with respect to the inversion point.

Mirror symmetry: When reflection across a plane restores the original structure, then it is said to possess mirror symmetry.

These four symmetry elements are mostly used in practice. However, an addition symmetry element is sometimes used which is called *improper rotation*. It is simply an operation in which both rotation and inversion operations are involved.

When these symmetry operations are applied to the 14 Bravais lattices, a total of 32 point groups distributed over the 7 crystal structures emerge. In order to distinguish between these 32 point groups, two principal nomenclature methods are used in practice. One is called the Schönflies method, and the other the international method or the Hermann–Mauguin method. (History tells us that Carl Hermann was a German crystallographer who first devised the international method of identifying the symmetry elements. However, in 1931, a French mineralogist modified Hermann's nomenclature scheme, and since then the international method bears the name of both of these scientists.)

Different symbols are used for the same symmetry elements by the followers of Schönflies School and the international school. The Schönflies method is preferred by spectroscopists, whereas the international method is preferred by crystallographers. This nomenclature was adopted as standard by the International Table for Crystallographers in 1934. We will, therefore, use the international symbols for identifying a given symmetry element. In Table 4.4, all the symmetry elements with their international symbols are tabulated.

Table 4.4 Types of symmetry elements and their respective international symbols.

Symmetry elements		International symbol
Center of symmetry		$\bar{1}$
Mirror symmetry		m
Rotation symmetry	Onefold	1
	Twofold	2
	Threefold	3
	Fourfold	4
	Sixfold	6
Inversion symmetry	Onefold = center of symmetry	$\bar{1}$
	Twofold = mirror plane normal to the axis	$\bar{2}$ ($\cong m$)
	Threefold	$\bar{3}$
	Fourfold	$\bar{4}$
	Sixfold	$\bar{6}$ ($\cong 3/m$)

Table 4.5 Lattice type generated by symmetry elements.

Symmetry elements	Generated lattice type
1	Parallelogram
m	Rectangle
2	Parallelogram
$2m = mm$	Rectangle, rhombus
3, $3m$	Hexagonal
4, $4m$	Square
6, $6m$	Hexagonal

We find that the four principal symmetry elements give rise to 10 international symbols. In Table 4.5, their geometrical representations are given. We also find from this table that each symmetry element can generate only one particular type of lattice. We hope that this table will help our readers in associating the geometrical shape represented by each of the 10 international symbols. This, we also hope, will help in visualizing the difference between one symmetry element and the other.

4.13 Classification of Dielectric Materials: Polar and Nonpolar Groups

The point groups are indispensable tools in understanding the nature of dielectric materials especially the oxides which form the largest group of electroceramic materials. Many of the oxides crystallize in two categories of centrosymmetric and noncentrosymmetric structures. Among the noncentrosymmetric group we find furthermore *polar* and *nonpolar* groups which form the basis for large number of applications based on the unique properties of such oxides. The 32 point groups define all 14 Bravais lattices irrespective of the fact whether they are centrosymmetric or noncentrosymmetric. In Figure 4.15, we show how we can divide the 32 point groups in different categories, resulting in the basis of piezoelectricity, pyroelectricity, and ferroelectricity.

First we find that out of 32 point groups, 11 are centrosymmetric and 21 noncentrosymmetric. Then again out of the 21 noncentrosymmetric point groups, 20 are classified as the piezoelectric point groups. The one that is left over is the point group 432 in the cubic system which is not piezoelectric and is therefore not included in the discussion. The 20 piezoelectric point groups are furthermore divided into the two categories: one being polar having 10 point groups and other nonpolar also having another set of 10 point groups.

Polar materials are those that exhibit nonreversible spontaneous polarization, and the nonpolar materials lack this type of polarization. Pyroelectric materials are a typical example of polar materials with irreversible spontaneous polarization. From Figure 4.15, we can also conclude that piezoelectricity and pyroelectricity can coexist in polar materials. An example being the ferroelectric materials. On the other hand, pyroelectricity cannot exist in nonpolar materials though they can be piezoelectric. Some examples of polar piezoelectric materials are barium titanate ($BaTiO_3$), lead-zirconate-titanate (PZT), and cadmium telluride

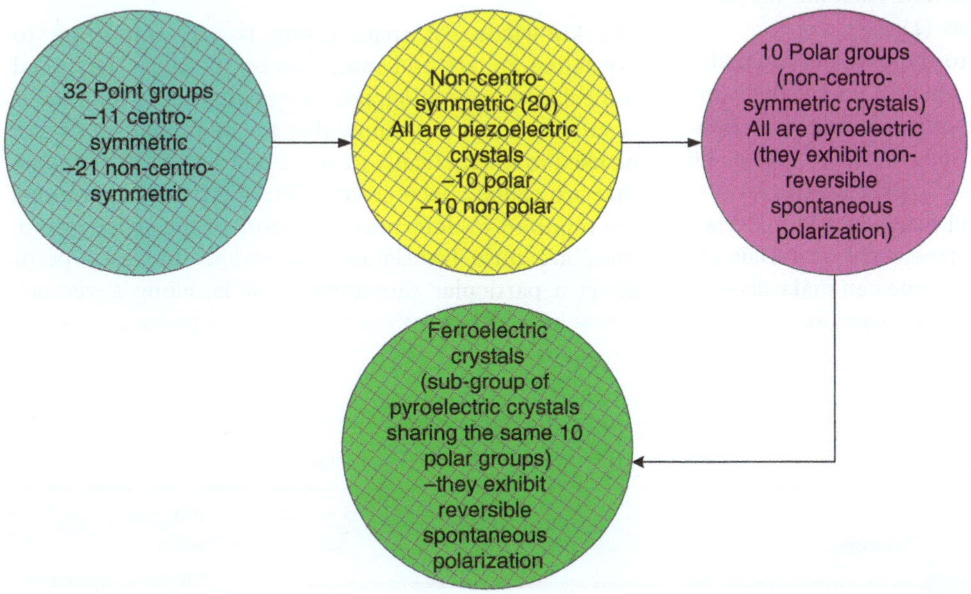

Figure 4.15 Schematic of 32 point groups distributed between polar and nonpolar groups.

Table 4.6 Symmetry elements for piezoelectric materials (20).

Triclinic (1)	Tetragonal (5)	Hexagonal (5)	Monoclinic (2)	Orthorhombic (2)	Trigonal (3)	Cubic (2)
1	$4, \bar{4}, 422, 4mm$	$6, \bar{6}, 622$	2	222	3	23
	$\bar{4}2m$	$6mm$	m	$mm2$	32	$\bar{4}3m$
		$\bar{6}m2$			$3m$	

Table 4.7 Symmetry elements for pyroelectric (and ferroelectric) materials (10).

Triclinic (1)	Tetragonal (2)	Hexagonal (2)	Monoclinic (2)	Orthorhombic (1)	Trigonal (2)
1	4	6	2		3
	$4mm$	$6mm$	m	$mm2$	$3m$

(CdTe). Similarly, some examples of nonpolar piezoelectric materials include quartz (α-SiO_2), beta zinc sulfide (β-ZnS), and tellurium oxide (TeO_2).

There are a large number of materials with the identical point groups as the pyroelectric materials but with reversible spontaneous polarization. They are called ferroelectric materials. By definition, a ferroelectric material is both a piezoelectric material and a pyroelectric material. We will learn a lot about these three types of materials which will be covered in subsequent chapters.

Table 4.6 lists the 20-point groups specific to of piezoelectric materials. They are distributed between the seven crystal structures. Notice the population to be rather high for tetragonal (5) and hexagonal structures (5). Then comes the trigonal group with (3) point groups, monoclinic, cubic, and orthorhombic each has (2) point groups, whereas triclinic has only (1).

The point groups belonging to the pyroelectric materials are given in Table 4.7. As already stated, the same point groups are shared with ferroelectric materials. Notice the common point groups between Table 4.6 (piezoelectric group) and Table 4.7 (pyroelectric group). It is a well-cited proverb that all piezoelectric materials are ferroelectrics, but not all ferroelectric materials are piezoelectric. The same statement one can make also for pyroelectric materials. Notice the conspicuous absence of cubic structure in Table 4.7. The 10 pyroelectric point groups are distributed as follows: 2 each for tetragonal, hexagonal, monoclinic, and orthorhombic and 1 each for triclinic and orthorhombic.

The centrosymmetric point groups are given in Table 4.8. The 11 of them are distributed as 2 each for tetragonal, hexagonal, trigonal, and cubic crystals and 1 each for triclinic, monoclinic, and orthorhombic structures. Notice the symmetry arrow going from the cubic side to the triclinic side.

All 32 point groups covered in Tables 4.6–4.8 are given once again in the modified composite Table 4.9 [3].

4.14 Space Groups

The knowledge of space group is very important to crystallographers and mineralogists because they need to deal with crystallography in greater detail than us. It is sufficient for us to know that the 32 point groups can be used mathematically to arrive at a set of 230 elements that are called space groups. Three types of operations are required to determine elements of the space group. They are pure translation that simply moves a point along a particular directions (that is, along a vector); screw axes that involves rotation of a point around an

Table 4.8 Symmetry elements of centrosymmetric group (11).

Triclinic (1)	Tetragonal (2)	Hexagonal (2)	Monoclinic (1)	Orthorhombic (1)	Trigonal (2)	Cubic (2)
$\bar{1}$	$4/m$	$6/m$	$2/m$	mmm	$\bar{3}$	$m3$
	$4/mmm$	$6/mmm$			$\bar{3}m$	$m3m$
Lowest symmetry	←					Highest symmetry

Table 4.9 Crystal systems and their point groups (32) in international notations.

Crystal system	Number of symmetry elements	International symmetry symbol	Pyroelectric	Piezoelectric	Centrosymmetric
Triclinic	2	1	Yes	Yes	
		$\bar{1}$			Yes
Tetragonal	7	4	Yes	Yes	
		$\bar{4}$		Yes	
		$4/m$			Yes
		422		Yes	
		$4mm$	Yes	Yes	
		$\bar{4}2m$		Yes	
		$4/mmm$			Yes
Hexagonal	7	6	Yes	Yes	
		$\bar{6}$		Yes	
		$6/m$			Yes
		622		Yes	
		$6mm$	Yes	Yes	
		$\bar{6}m2$		Yes	
		$6/mmm$			Yes
Monoclinic	3	2	Yes	Yes	
		m	Yes	Yes	
		$2/m$			Yes
Orthorhombic	3	222		Yes	
		$mm2$	Yes	Yes	
		mmm			Yes
Trigonal	5	3	Yes	Yes	
		$\bar{3}$			Yes
		32		Yes	
		$3m$	Yes	Yes	
		$\bar{3}m$			Yes
Cubic	5	23		Yes	
		$m3$			Yes
		432	No	No	No
		$\bar{4}3m$		Yes	
		$m3m$			Yes

Source: Lines and Glass 1977 [3]. Reproduced with permission of Oxford University Press.

axis while simultaneously translating it parallel to the axis and glide plane which includes reflection of a point through a plane while translating it parallel to the plane. We can accurately describe physical phenomena of our interest without restoring the complex nature of space groups.

Glossary

Bravais lattice The 7 basis crystal structures can be mathematically organized in a group of 14 separate lattices called collectively as Bravais lattices.

Crystal structure A crystal structure consist of a basis and a lattice. There are seven fundamental crystal structures which are cubic, tetragonal, trigonal, hexagonal, triclinic, monoclinic, and orthorhombic.

Inversion point A crystal is said to have an inversion point when the structure is centrosymmetric with respect to a given point in the lattice. Inversion points are normally electrically neutral. Crystal structures come as centrosymmetric and noncentrosymmetric.

Inversion symmetry A centrosymmetric crystal is said to have an inversion symmetry. Lattice Any three-dimensional structures can be called a lattice. It is a geometric concept. Unless the lattice points are occupied by atoms (or ions or molecules), it is not a crystal structure.

Lattice points Each site that can be occupied by an atom (or ions and molecules) in a lattice is a lattice point.

Nonpolar materials Materials lacking the spontaneous polarization are called the nonpolar materials. Any material will get polarized when subjected to an electric field. But it will disappear as soon as the field is withdrawn. This is not the case for polar materials.

Point groups 14 Bravais lattices can be mathematically manipulated by a set of four operations to arrive at a group of 32 distinct points. Collectively, they are called point groups.

Polar materials These are the group of materials such as pyroelectrics and ferroelectrics that are spontaneously polarized even in the absence of an electric field.

Primitive cell When a crystal has only one lattice point per cell, it is called a primitive cell. There is at least one primitive cell in each of the seven fundamental crystal structures.

Symmetry elements The four mathematical operations that can produce 32 point groups are also called symmetry elements.

Unit cell When a crystal structure has more than 1 lattice point per cell, then it is said to have a unit cell in contrast to a primitive cell which can have only 1 lattice point per cell.

Problems

4.1 Using a diagram show the structure–property relationship when the external agents are electric field, magnetic field, and mechanical stress.

4.2 Distinguish between the covalent bonding and ionic bonding giving some examples for each of the two bondings.

4.3 Define magnetoelectric, piezoelectric, pyroelectric, and piezomagnetic effects. Show with the help of a block diagram the interdependence of these coupled phenomena.

4.4 Name the 14 Bravais lattices and distinguish between the three Bravais lattices found in a cubic crystal using figures representing them.

4.5 Find d_{hkl} and d^*_{hkl} of a cubic crystal with $a_0 = 3$ nm and having the following Miller indices: $h = 1$, $k = 2$, and $l = 2$. Comment on your result.

4.6 Find d_{hkl} and d^*_{hkl} if the crystal is tetragonal assuming that the Miller indices are $h = 2$, $k = 2$, and $l = 1$ and the lattice constants are $a = 2$ nm and $c = 5$ nm.

4.7 Find the atomic packing factor for Ge and $CaTiO_3$. The lattice constant of Ge is 0.57 nm. For $CaTiO_3$ lattice constant are $a = 0.538$ nm, $b = 0.5443$ nm, and $c = 0.7645$ nm. For this problem, assume that the Ca, Ti, and O atoms are of equal size. Comment on the results.

4.8 Calculate the theoretical density for Si with the lattice constant of $a_0 = 0.543$ nm. Compare it with the actual density and then comment.

4.9 Prove that the apf of diamond structure is exactly ½ of the apf for the bcc structure.

4.10 Calculate the ratio between the atomic radii for Si and Ge considering the lattice constant for Si to be 0.543 nm and that for Ge it is 0.57 nm.

4.11 Using the value for the atomic radius for Si found in the previous problem prove that the bond length for Si is 0.235 nm.

4.12 So far as the symmetry is concerned crystals come as centrosymmetric and noncentrosymmetric. Discuss the difference between them and give some examples of the two categories of the crystal symmetry.

4.13 Name the symmetry elements and the international symbols to represent them.

4.14 Divide the 32 point groups in polar and nonpolar groups identifying the physical phenomena they represent.

References

1 Borchardt-Ott, W. (1995). *Crystallography*, 2e. Springer.
2 Kittel, C. (1976). *Solid State Physics*, 5e. John Wiley and Sons.
3 Lines, M.E. and Glass, A.M. (1977). *Principles and Applications of Ferroelectrics and Related Materials*. Clarendon Press.

Further Reading

Newnham, R.E. (2013). *Properties of Materials: Anisotropy, Symmetry and Structures*. Oxford University Press.
Nye, J.F. (1985). *Physical Properties of Crystals: Their Representation by Tensors and Matrices*. New York: Oxford University Press.

5

Dominant Forces and Effects in Electroceramics

CHAPTER MENU

Nature seems to take advantage of the simple mathematical representations of the symmetry laws. When one pauses to consider the elegance and the beautiful perfection of the mathematical reasoning involved and contrast it with the complex and far-reaching physical consequences, a deep sense of respect for the power of the symmetry laws never fails to develop.

Chen Ning Yang, Nobel Lecture

5.1 Introduction

In this chapter, we will learn about the dominant forces that are responsible for inducing both linear and nonlinear effects in materials resulting in the origin of some interesting physical properties. We will try to understand the mechanism behind the two types of effects and how we can make use of the nonlinear properties to advance the field of electroceramics technology. The external agents of interest are electric field, magnetic field, temperature, and mechanical stress. It is our challenge to understand how they interact in collaboration with each other to give rise to unique and coupled phenomena that distinguishes linear dielectrics from nonlinear dielectrics.

5.2 Agent–Property Relationship

The structure–property relationships are vital in understanding the nature of materials because they form the basis for engineering applications. We have discussed this subject in reference to crystal structure and symmetry. In this chapter, we will investigate how solids respond to the applications of external agents leading to the origin of novel effects that are exploited in a number of ways to develop new devices and their applications. We can define two types of agent–effect relationship: the first being the interactions arising from electric field (E), mechanical stress (X), and temperature (T); and second consisting of E, X, and magnetic field (H). Both categories form the backbone for a variety of applications. For example, the E–X–T relationship gives rise to multiple interesting phenomena having significant impacts on the electronic and mechanical technology, sensor and detectors, and harvesting of green energy. We will then cover the issues related to the other set of trio consisting of E–X–H. They are responsible for the emergence of a hybrid system of materials commonly known as multiferroics. Multiferroics are oxides displaying coupled properties that are the foundations for the next generation of storage media, memory, and sensors.

Fundamentals of Electroceramics: Materials, Devices, and Applications, First Edition. R. K. Pandey.

Companion website: www.wiley.com/go/Pandey/Fundamentals_Electroceramics

5.3 Electric Field (E), Mechanical Stress (X), and Temperature (T) Diagram: Heckmann Diagram

The interaction of solids with each of these three agents can manifest itself in variety of ways. They are the foundations for the origin of primary and coupled effects exhibiting some unique electronic, thermal, and mechanical properties. There are three principal effects and six associated coupled effects. It is not a trivial task to present them in a unified and consistent way. This problem was solved in 1925 by G. Heckmann of Germany who invented a diagram, which now bears his name, representing all the effects that can occur in solid when subjected to the interactive forces of $E–X–T$ [1].

Ever since the diagram became known, it has been used by scientists and engineers to understand the nature of solids and the effects they can display. We give a simplified version of the Heckmann diagram in Figure 5.1.

The three corners of the larger triangle are occupied by the agents E, X, and T, respectively. The smaller triangle has its corners labeled as electrical displacement (D), mechanical strain (x), and entropy (S). These six parameters form three groups that are (i) between electric field (E) and displacement (D); (ii) between stress (X) and strain (x); and (iii) between temperature (T) and entropy (S). They are called the three *principal effects.*

Principal effect No. 1: Permittivity (ϵ) is a fundamental property of all solids, but it plays a very special role in the field of dielectric materials like an oxide. We know from Chapter 2 that the relationship between D and E for linear dielectric materials is given by the following equation.

$$D = \epsilon E \tag{5.1}$$

Since both D and E are first rank tensors, Eq. (5.1) can also be expressed as in Eq. (5.2).

$$D_i = \epsilon_{ij} E_j \tag{5.2}$$

The permittivity (also called dielectric constant) now gets a new meaning. It is no more an isotropic property. For a noncubic single crystal, it will depend upon

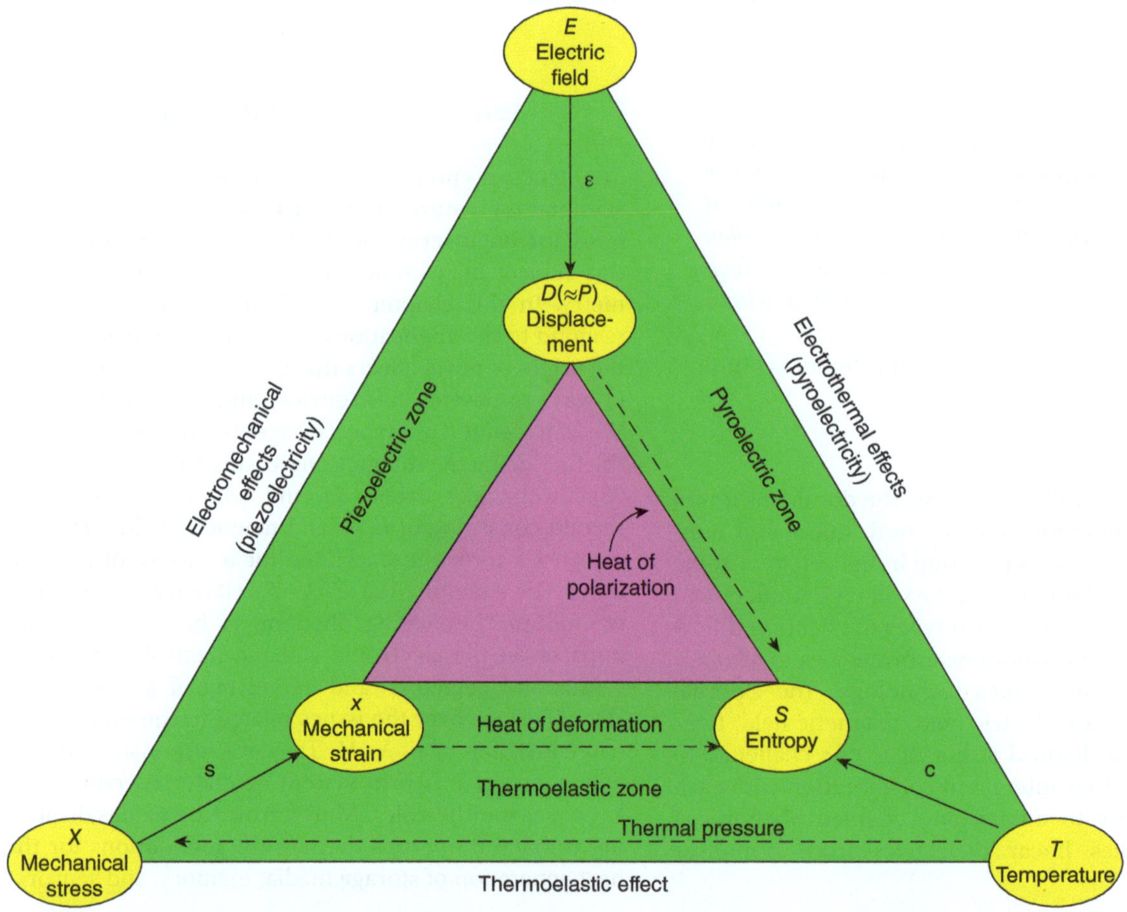

Figure 5.1 Modified electric field, mechanical stress, and temperature diagram. ϵ, s, and c are dielectric constant, elastic compliance, and heat capacity, respectively. Source: Heckmann 1925 [1].

the crystal axes and therefore, will be an anisotropic parameter.

It should be noted, however, that ordinarily we do not have to express physical parameters such as D, E, and ϵ in tensor forms, but for properties requiring advanced formulation for explanation, it is convenient to express them in tensor forms. Assuming that many of our readers may not be comfortable with advanced understanding of tensors, we will use the tensor forms when it is necessary; otherwise, we will stay with the simple formulations without tensor notations. For advanced readers, we recommend two excellent resource materials for getting a deeper understanding of tensor analysis relevant to physical properties of materials [2, 3].

Principal effect No. 2: When a material experiences a mechanical stress (X) such as pressure it suffers strain (x). Stress is defined as force per unit area with the unit of (N m^{-2}). Strain is defined as change in length per unit length, $\left(\frac{\Delta L}{L}\right)$ where L is the original length. It is obviously a unitless parameter. The coupling between stress and strain is given by the linear relationship as expressed in Eq. (5.3).

$$X = sx \tag{5.3}$$

Equation (5.3) is the formulation of the Hooke's law that describes the response of a linear elastic material to the application of mechanical stress (X). The coefficient s is called the elastic compliance. The tensor form of this equation is given in Eq. (5.4).

$$X_{ij} = s_{ijkl}x_{kl} \tag{5.4}$$

We notice in the above equation that both stress and strain are second-rank tensors that makes the compliance, s, a fourth rank tensor. We will make use of Eqs. (5.2) and (5.4) while discussing piezoelectricity later in this chapter.

Principal effect No. 3: This effect constitutes temperature (T), entropy (S), and heat capacity (c). Entropy is a thermodynamic parameter and can be understood as a measure of molecular disorder within a microscopic system. In a reversible process for a unit volume, the change in temperature, dT, produces a change in entropy, dS, so that

$$\mathrm{d}S = \left(\frac{c}{T}\right)\mathrm{d}T \tag{5.5}$$

The relationship between displacement (D) and entropy (S) is called heat of polarization and similarly between the strain (s) and entropy (S) it is called the heat of deformation.

Besides these three *principal effects*, there are six secondary effects that form the group of *coupled effects*. They reside in the three zones identified in Figure 5.1 as the pyroelectric zone, the piezoelectric zone, and the thermoelastic zone. Each of these zones contains two *coupled effects*. We discuss below the characteristics of each of these three zones.

5.3.1 Piezoelectric Zone

Figure 5.2 is the detailed version of piezoelectric zone. We see here that there are two types of piezoelectric effects: one being the direct piezoelectric effect, and the other the converse piezoelectric effect. These effects can occur only in noncentrosymmetric crystals that are divided into two groups of polar and nonpolar materials.

The direct piezoelectric effect occurs when a material gets electrically polarized because of being subjected to a mechanical stress. The polarization, P, is defined as charge per unit area with the unit of (C m^{-2}). The converse piezoelectric effect, on the other hand, occurs when an electric field causes a mechanical strain to develop in a material. Both these types of piezoelectric effects are the basis for many novel and practical devices. We will study this topic in greater detail later in this chapter.

5.3.2 Pyroelectric Zone

The pyroelectric zone consists of the two coupled effects that are the pyroelectric effect and the electrocaloric effect. They are shown in Figure 5.3. The topic of

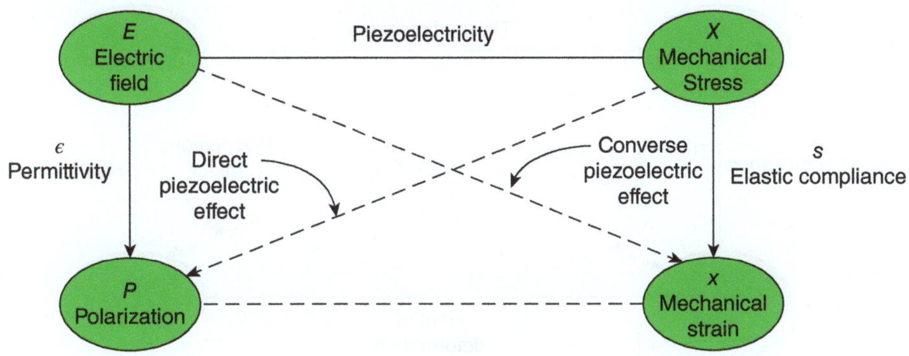

Figure 5.2 Expanded piezoelectric zone showing the origins of direct and converse piezoelectric effects.

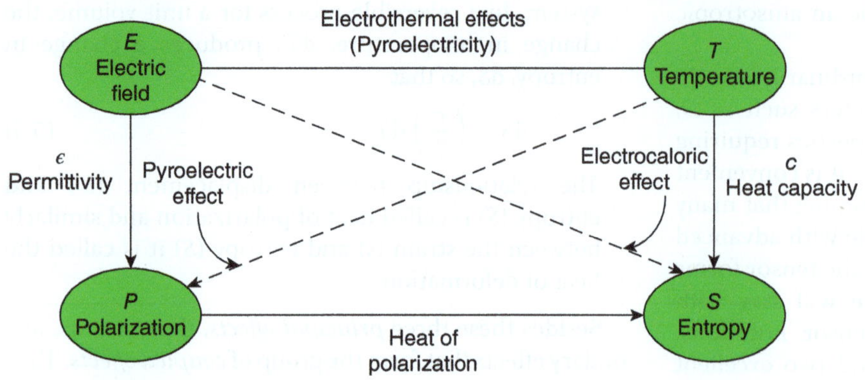

Figure 5.3 Expanded pyroelectric zone showing pyroelectric and electrocaloric effects.

pyroelectricity will be discussed in depth in Chapter 9. Therefore, here we will cover it only briefly.

Some materials acquire electrical charge (or polarization) when heated. Semiprecious crystal tourmaline is the classical example of a pyroelectric material. For this effect to take place, the material must be noncentrosymmetric and belong to the group of polar materials. In contrast to piezoelectricity, this effect does not take place in nonpolar materials. In pyroelectric materials, a change in temperature can induce a change in the electrical polarization. The relationship between the temperature gradient ($\mathrm{d}T$) and the change in resulting polarization ($\mathrm{d}P$) is linear with the coefficient being the pyroelectric coefficient, $P(T)$. Mathematically expressed, it is given by Eq. (5.6).

$$\mathrm{d}P_i = P_i(T)\mathrm{d}T \tag{5.6}$$

Both polarization (P) and pyroelectric coefficient $P_i(T)$ are vectors, and therefore, shown in Eq. (5.6) as first-rank tensors. The second effect shown in the diagram is the electrocaloric effect. The coupling between electric field and entropy leads to the electrocaloric effect. This effect is governed by Maxwell relation of thermodynamics:

$$\left(\frac{\delta S}{\delta E}\right)_T = \left(\frac{\delta P}{\delta T}\right)_E \tag{5.7}$$

Thermodynamically, the electrocaloric effect may also be understood as the process through which an electric field induces a reversible temperature change in a polarizable material. This obviously can be exploited for cooling. Because of the importance of green energy as an alternative source of energy, materials exhibiting electrocaloric effect are actively investigated for refrigeration technology.

5.3.3 Thermoelastic Zone

The details of the third zone identified as thermoelastic zone in Figure 5.1 are given in Figure 5.4. The two coupled effects here are thermal expansion effect and piezocaloric effect.

The thermal expansion is related to temperature and strain. It is a common experience that all materials when heated undergo some change in their shapes which is because of the mechanical strain they experience. Consider a sample of length L, which expands when heated. Let us assume that the change in length is $\mathrm{d}L$ for a temperature change of $\mathrm{d}T$. The relationship between $\mathrm{d}L$ and $\mathrm{d}T$ is linear with the coefficient being the *coefficient of thermal expansion*. Normally, it is represented by the symbol of α_L. Mathematically, it is defined by Eq. (5.8).

$$\alpha_L = \frac{1}{L}\left(\frac{\mathrm{d}L}{\mathrm{d}T}\right) \tag{5.8}$$

The temperature can induce not only the linear change but can also cause changes in both area and volume as well. Following the logic applied to Eq. (5.8), we can

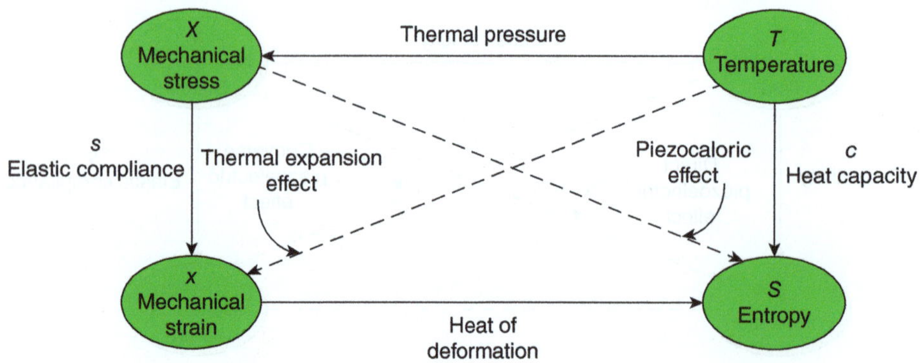

Figure 5.4 Thermoelastic zone details: X, x, T, and S coupling.

define the *coefficient of area expansion* (α_A) and *coefficient of volume expansion* (α_V) as expressed in Eqs. (5.9) and (5.10).

$$\alpha_A = \frac{1}{A}\left(\frac{dA}{dT}\right) \tag{5.9}$$

$$\alpha_V = \frac{1}{V}\left(\frac{dV}{dT}\right) \tag{5.10}$$

Obviously, in the above two equations A is the area and V the volume. The unit of all these three coefficients is the same as °C^{-1} or K^{-1}.

The second effect in this zone is the piezocaloric effect that develops in a material because of the coupling between mechanical stress (X) and entropy (S). Like the electrocaloric effect, the piezocaloric effect can also contribute to the green energy technology. We will address this issue in some detail in Chapter 10. This effect is related to thermal expansion and can be formulated as in Eq. (5.11).

$$\left(\frac{\delta S}{\delta X}\right)_T = \left(\frac{\delta x}{\delta T}\right)_X \tag{5.11}$$

5.4 Electric Field, Mechanical Stress, and Magnetic Field Diagram

Similar to the Heckmann diagram, there is another diagram that describes the interactive forces and their resulting effects when the external agents are electric field (E), mechanical stress (X), and magnetic field (H). That is, this new diagram simply replaces the temperature (T) of Heckmann diagram with the magnetic field (H). It was proposed by Spaldin and Fiebig [4]. This has become a good tool for understanding the forces responsible for multiferroics phenomena just like the Heckmann diagram is for understanding the phenomena of ferroelectricity, pyroelectricity, and piezoelectricity.

We can see the similarity between the Heckmann diagram and this diagram. Both are based on interactions between the external agents that give rise to physical properties in a solid. Two of the three primary linear effects are common in both the diagrams. They are the outcome of the interactions between electric field (E) and polarization (P), and between stress (X) and the resulting strain (x). Here the third primary effect is different. It is given by the linear relationship between the magnetic field H and the resulting magnetization M. We saw that Eqs. (5.1) and (5.4) describe the linear relationship between E and P, and between X and x, respectively. Similarly, the relationship between the magnetic flux, B, and magnetic field, H is given by

$$B = \mu H \tag{5.12}$$

Here μ is the permeability which, as we know, is the magnetic counterpart of permittivity, ϵ. The magnetic flux, B, in the above equation can also be replaced with the magnetic moment (or, magnetization), M. In the same way, we replaced the electric displacement, D, with polarization, P. Many authors prefer to use the magnetization, M, instead of the magnetic flux, B. But, for the time being, we will stay with B and replace it with M later while discussing the topic of magnetoelectric (ME) effect.

For an isotropic medium such as a polycrystalline solid or a cubic unit cell with center of symmetry, the permeability (μ) is a scalar quantity. Otherwise, it is tensor as is the permittivity (ϵ) discussed before. Since both B and H are first-rank tensors, the magnetic permeability, μ, is a second-rank tensor. In tensor notation Eq. (5.12) becomes

$$B_i = \mu_{ij} H_j \tag{5.13}$$

In Figure 5.5, we also encounter six coupled effects. The first set of two are the direct piezoelectric effect and converse piezoelectric effect (exactly the same thing as in Figure 5.1). Then there are two piezomagnetic effects, which are direct piezomagnetic effect and converse piezomagnetic effect (obviously, these are analogous to the direct and converse piezoelectric effects). The third set of two coupled effects is ME effects defined by the interactions between the electric field (E) and magnetization (M), and between electric polarization (P) and magnetic field (H). We will appreciate the scientific importance and potential technical impact of ME efforts when we consider them while discussing multiferroic materials.

In principle, both direct and converse piezomagnetic effects can coexist only in certain classes of magnetocrystalline symmetry. Both are linear phenomena: the interaction between stress and magnetic moment is linear and so is the interaction between strain and magnetic field. The origin of piezomagnetism lies in the transition of an antiferromagnetic state to the weakly ferromagnetic state caused by the application of a mechanical stress that induces changes in the crystal symmetry. This effect is rather rare and has been found to exist only in a few antiferromagnetic materials. Antiferromagnetic materials are also magnetically polarized materials like the ferromagnetic materials, but unlike in a ferromagnetic material, the elementary magnetic moments (also called magnetic spins) are in antiparallel arrangement in an antiferromagnetic material. This results in a zero spontaneous magnetic moment. More detailed description of antiferromagnetism and ferromagnetism will be discussed in Chapter 11. Because of the emergence of ME technology, new composites are being developed in which the ME effect is induced by

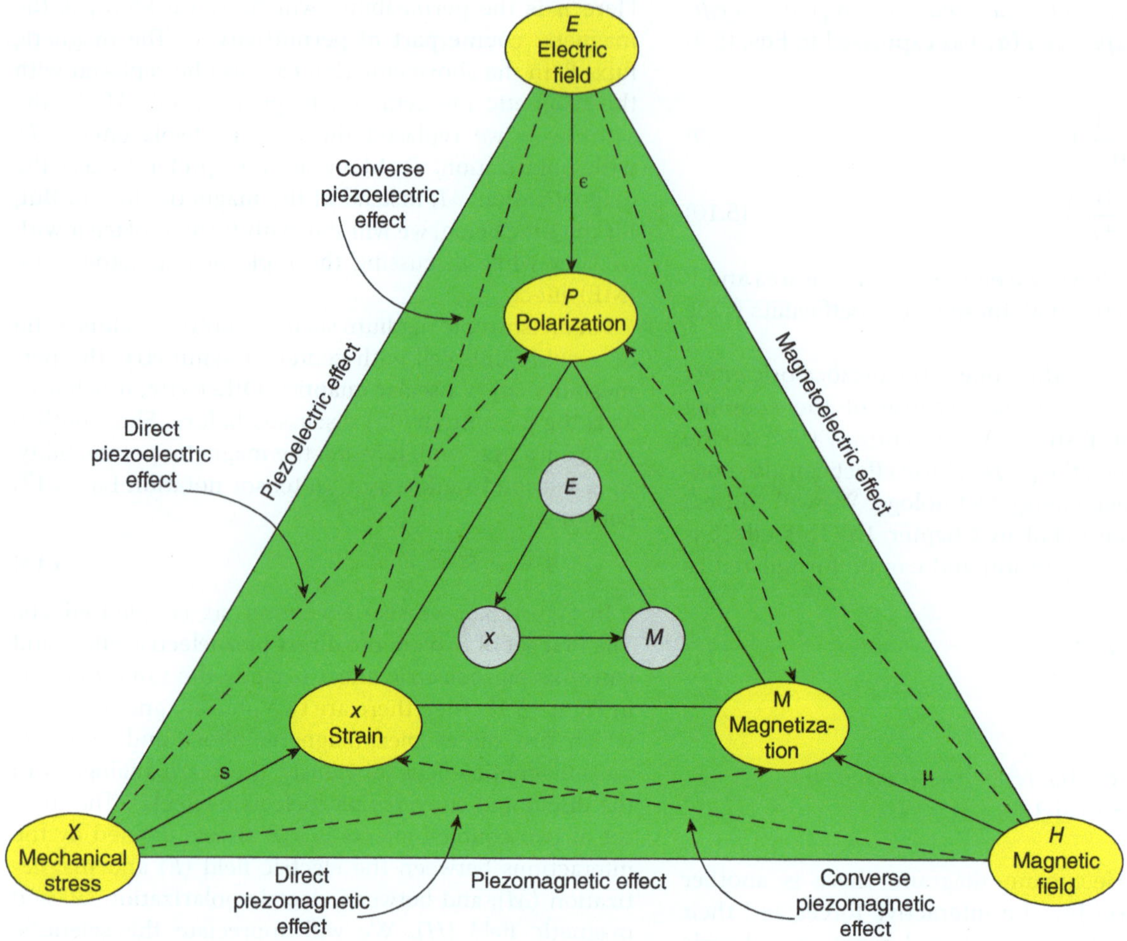

Figure 5.5 Modified electric field, mechanical stress, and magnetic field diagram: *E*–*X*–*H* coupling. Source: Spaldin and Fiebig 2005 [4].

piezomagnetism. It is also known as magneto-elastic effect or magneto-mechanical effect, and it is analogous to the well-known piezoelectric effect but not as common.

We can express the direct piezomagnetic effect between stress (X) and the resulting magnetization (M) as follows:

$$M_i \approx \eta_{ijk} X_{jk} \tag{5.14}$$

Here, η is the coefficient of the direct piezomagnetic effect, and it is a third-rank tensor. Its unit is $\mathrm{Wb\,N^{-1}}$. The converse piezomagnetic effect follows the relationship given in Eq. (5.15). Here the coefficient is once again a third-rank tensor with the symbol of γ. Its unit is $\mathrm{m\,A^{-1}}$.

$$x_{ij} \approx \gamma_{ijk} H_k \tag{5.15}$$

The smallest triangle with its corners as E, X, and M in Figure 5.5 points to the possibility of influencing the magnetic state of a ferromagnetic material with the help of an electric field (E). In other words, it means that one could use the converse piezoelectric effect to induce changes in the magnetic state, and thereby indirectly achieve coupling between an electric field and magnetic polarization. This is just one proposed way to achieve this goal. However, there are some other ways in which this goal could be attained.

Since the discovery of ferroelectricity in 1921, it has been the dream of scientists to combine the phenomena of ferroelectricity and ferromagnetism in the single-phase material. The driving force behind this dream is first of all the ambition to add a new dimension to solid-state physics, and second, the hope of unlocking the enormous technical potentials such materials might have.

Numerous groups throughout the world have worked on this problem for decades, but the success remained elusive until Schmid reported about multiferroic MEs in 1994 [5]. The new field of multiferroics appeared on the scene, and today, it is one of the most active fields of basic research and electronic device technology. The underlying principle behind the ME effect is rather simple and straightforward. Conceptually, one should be able to alter the magnetization, M, of a ferromagnetic material by the application of an electric field, E; and similarly, the polarization, P, of a ferroelectric material could be modified with the help of a magnetic field, H. It turns out that, in reality, it is not so simple because of entirely different physical basis that give rise to ferromagnetism and ferroelectricity.

5.5 Multiferroics Phenomena and Materials

The name multiferroics employs a system of materials in which at least two ferroic phenomena couple. Examples of ferroic materials are ferroelectrics, ferromagnetics (also, ferrimagnetics), and ferroelastics. Each of these three materials displays hysteresis loops that are their signature properties and also the foundation for a large number of applications. The idealized loops for these three ferroics are shown in Figure 5.6.

Ferroelectricity is a macroscopic phenomenon and has many different origins, but relative displacement between the positive and negative ions in the lattice unit cell is considered to be the principal reason for its origin. In comparison, ferromagnetism is a quantum mechanical phenomenon that is caused because of the partial vacancy in the d-orbitals of an atom, but more precisely, in the unfilled states of transition metals (examples being Fe, Co, and Ni) in their d-shells.

We are showing the hysteresis loop of the ferromagnetic material as M–H loop and not as the B–H loop, which is a common practice among scientists and engineers. For most practical purposes, they are interchangeable. We prefer the M–H loop instead of the B–H loop because by definition for a ferromagnetic material $B \approx M$. The magnetization, M, is a very important parameter that we use in gaining a deeper insight into the fundamental nature of a ferromagnetic material, which is the classical example of a nonlinear material. Different units are used for M, which are (emu cm^{-3}), (emu g^{-1}), or (G cm^{3}). However, the standard internationally accepted unit is (A m^{-1}).

Notice that in Figure 5.6, we have also added ferrimagnetic material in the ferromagnetic group. There is a good reason to do it. But what are ferrimagnetic materials? In short, they are not the same as ferromagnetic materials. There is a subtle difference between a ferromagnetic material and a ferrimagnetic material. Ferromagnetism is associated with either the metallic elements such as Fe, Co, and Ni or their alloys, whereas ferrimagnetism is the phenomena exhibited exclusively by insulating magnetic oxides such as magnetite, Fe_3O_4. Magnetite is the principal ingredient in all magnetic oxides in which ferrimagnetism dominates. In technical language, magnetic oxides are called ferrites that are the backbone of magnetic devices having multiple unique applications.

On the fundamental level, the internal structure of a ferromagnet consists of elemental magnetic moments of equal strength, and they are aligned parallel to each

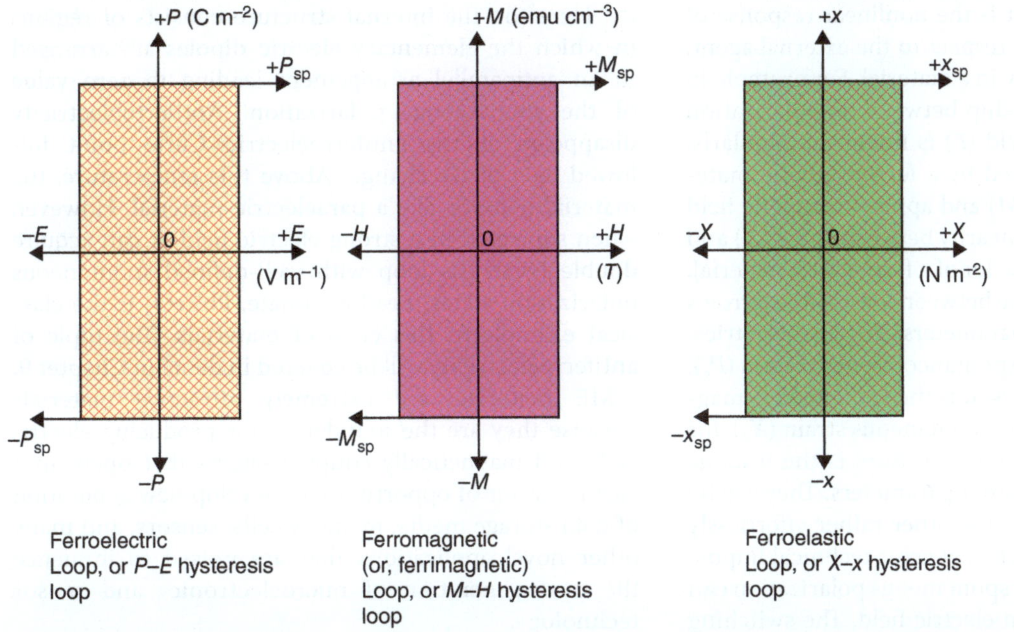

Figure 5.6 Idealized hysteresis loops of the three primary ferroic materials. The spontaneous polarization, magnetization, and strain are labeled as P_{sp}, M_{sp}, and x_{sp}, respectively.

other giving rise to a large spontaneous magnetization. In comparison, the magnetic moments in a ferrimagnetic structure are of two different strengths; one set being larger than the other set. The two sets of magnetic moments are considered to be localized on two sites classified as site A and site B. In each of these two sites, the magnetic moments are parallel to each other, but considered jointly, they form antiparallel groups of spins. In other words, if all magnetic spins are parallel in site A, then all spins are antiparallel in site B. Since they are of unequal magnitude, the net result is that also a ferrimagnetic material is spontaneously magnetized, but its strength is normally weaker than that of a ferromagnetic.

Both types of magnetic materials go through a phase change at the Curie point, T_C, above which they completely lose their spontaneous magnetization and exhibit paramagnetism. It is common to lump these two phenomena in one and collectively refer both of them as ferromagnetism. But so far as the multiferroics are concerned, we inherently mean ferrimagnetism and not ferromagnetism because we deal with oxide magnetic materials. However, we will not emphasize their subtle distinctions any more while discussing the topic of multiferroics. We will discuss magnetic materials in general and ferrites in particular in greater detail in Chapter 10.

The physical mechanism, as explained already, behind the origin of ferromagnetism and ferroelectricity are so different that one cannot imagine them to be reconciled to give rise to the multiferroic effect. Yet, it is possible, and we now discuss some of the ways to overcome the inherent difficulties. We begin our discussions by examining the hysteresis loops of Figure 5.6. They have three things in common. First, it is the nonlinear response of the physical property with respect to the external agent, which induces the property in a material. For example in ferroelectrics, the relationship between its polarization (P) and applied electric field (E) is nonlinear. Similarly, the nonlinearity is displayed by a ferromagnetic material in its magnetization (M) and applied magnetic field (H) relationship. The nonlinearity between stress (X) and strain (x) is the typical behavior of a ferroelastic material. The second common feature between these three ferroics is the presence of order parameters. For ferroelectrics, the order parameter is the spontaneous polarization (P_s). Similarly, for ferromagnetics, it is the spontaneous magnetization (M_s), and it is the spontaneous strain (X_s), for ferroelastics. The third common feature is the bistable states of each of the three-order parameters. They can be switched from one state to the other rather effortlessly and that is the feature which is of great technical importance. In ferroelectrics, the spontaneous polarization can be switched by applying an electric field. The switching force for ferromagnetics is a magnetic field, whereas it is a mechanical stress for ferroelastics. We will discuss the salient features and the technical importance for the hysteresis loops exhibited by ferroelectric and ferromagnetic materials in Chapters 9 and 11, respectively.

By definition, multiferroics are those materials in which at least two ferroic orders are present. For example the coupling between ferroelectricity and ferromagnetism gives rise to a special type of multiferroics that are called MEs. The other examples of two ferroic order multiferroics are coupling between ferroelectricity and ferroelasticity, and between ferroelasticity and ferromagnetism.

The three ferroic phenomena of ferroelectricity, ferromagnetism, and ferroelasticity form the primary ferroic group. The definition of multiferroics could be extended to include also two secondary ferroic phenomena, which are antiferromagnetism and antiferroelectricity. Both these order parameters might be combined to produce ME effect.

Antiferromagnetic materials are those in which the internal structure consists of antiparallel configuration of magnetic spins. As such in the absence of a magnetic field, the net spontaneous magnetization is zero. These materials go through the phase change at a well-defined temperature called the Néel point (in honor of the French Physicist, Louis Néel) with the symbol of T_N. It is analogous to the Curie point, T_c, of a ferromagnetic material. The phase change that takes place at the Néel point is from the antiferromagnetic state to paramagnetic state. Antiferromagnetism occurs mostly in oxides such as FeO, MnO, MnTe, and CuO.

Analogous to antiferromagnetic materials, there are some dielectric materials that exhibit antiferroelectricity. Here too the internal structure consists of regions in which the elementary electric dipoles are arranged in an antiparallel arrangement leading to zero value of the spontaneous polarization. Antiferroelectricity disappears at the antiferroelectric Curie point followed by a phase change. Above this temperature, the material behaves like a paraelectric material. However, when subjected to a strong electric field, it can acquire double hysteresis loop with well-defined spontaneous polarization states. Lead zirconate, $PbZrO_3$ is the classical example of this class of materials. The topic of antiferroelectricity will be covered in detail in Chapter 9.

ME materials are extremely attractive materials because they are the foundation for producing electrically and magnetically coupled effects that open up a vast reservoir of opportunity to develop new generation of data storage media, memory cells, sensors, and many other novel applications that are poised to influence the next generation of microelectronics and sensor technology.

ME effect is not new. It has a long and interesting history associated with the giants of physics. Already in

1888, Wilhelm Conrad Röntgen, who later discovered X-rays in 1895, discovered that a moving dielectric became magnetized when placed in an electric field. Six years later in 1894, the great French physicist, Pierre Currie, predicted the existence of ME effect in single crystals on the basis of symmetry considerations. The term *ME* was coined by American physicist Peter J. Debye (born in the Netherlands) who was one of the greatest minds of the twentieth century. In 1960, the ME effect became fully established when it was experimentally demonstrated in a single crystal of antiferromagnetic Cr_2O_3. But the effect was rather small and therefore not of any practical value. The field gets a new emphasis after the discovery of multiferroics magnetoelectronics in 1995. It took another 10–12 years for the field to go through a major revival.

The status quo changed drastically when in 2007 Ramesh Ramamoorthy of Berkeley successfully grew an epitaxial film of $BiFeO_3$ in which his group could clearly demonstrate the switching of magnetic domains by applying an electric field [6]. This was the first time that a strong coupling between ferroelectricity and antiferromagnetism was clearly established at room temperature. This was the breakthrough that paved the way for a worldwide race for discovering new and novel materials exhibiting strong ME effect at room temperature. $TbMn_2O_3$ and perovskite $TbMnO_3$ were found to be magnetoelectronic materials in 2004 and 2007, respectively [7, 8]. But the effect is present only at low temperatures. That leaves $BiFeO_3$ so far as the only room temperature magnetoelectronic material.

Bismuth ferrite, $BiFeO_3$, is an extremely interesting material. It is the lead ME material and has all three primary ferroic orders (see Figure 5.6). It is both ferroelectric and ferroelastic with the ferroelectric Curie point of 1103 K, and an antiferromagnetic with the Néel point, T_N, of 643 K. It is interesting to note that in bulk $BiFeO_3$ the multiferroic effects are rather weak, but in epitaxial films, they are greatly enhanced [6]. The multiferroic coupling occurs both in single phase and composite materials. Thin-film forms are the desired structures because of their importance to device development and integration with other materials. In Table 5.1, we present a list of multiferroic materials.

5.6 Magnetoelectric (ME) Effect and Associated Issues

5.6.1 Basic Formulations Governing the ME Effect

ME effect manifests itself when the electrical polarization (*P*) is altered by the application of a magnetic field (*H*) as well as by inducing changes in magnetization (*M*) by the application of an electric field (*E*). Mathematically, these two couplings can be expressed in the form of Eqs. (5.16) and (5.17).

$$P_i = \sum a_{ij}H_j + \sum b_{ij}H_jH_k \pm \cdots \tag{5.16}$$

$$M_i = \sum a_{ij}E_j + \sum b_{ij}E_jE_k \pm \cdots \tag{5.17}$$

Table 5.1 Selected single-phase multiferroic materials.

Group	Materials	Comments
Oxide, not perovskite	Cr_2O_3	First magnetoelectric (ME) effect discovered (1960)
	$TbMn_2O_3$	Discovered in (2004)
Perovskite group: ABO_3	$Pb(Fe_{0.}5Nb_{0.5})O_3$ (PFN)	Exhibit ferroelectric, ferromagnetic, or weakly ferromagnetic effects
	$Pb(Fe_{0.5}Ta_{0.5})O_3$ (PFT)	All are Pb-based materials
	$Pb(Fe_{0.67}W_{0.33})O_3$ (PFW)	Weakly anti-ferromagnetic and ferroelectric; $BiFeO_3$ is the most active material for magnetoelectric studies
	$BiMnO_3$, and $BiFeO_3$	Both are Bi-based and considered environmentally compatible
	$ErMnO_3$, $YbMnO_3$, $YMnO_3$, $TbMnO_3$ and $LuMnO_3$	Antiferromagnetic and ferroelectric effects They belong to the rare-earth group
Mixed perovskite solid solutions	PFW-$PbTiO_3$, PFW-$Pb(Mg_{0.5}W_{0.33})O_3$, PFN-$Pb(Zr_{0.2}Ti_{0.8})O_3$	Ferroelectric and ferromagnetic effect; Pb-ions responsible of polarization
Nonoxides	$BaMnF_4$, $BaNiF_4$, $BaFeF_4$	Ba-based fluorides
Spinel chalcogenides	$ZnCr_2Se_4$, $CsCrwS_4$	Low temperature multiferroics; also having colossal magnetocapacitance

Note: Except Cr_2O_3 and $TbMn_2O_3$ all other materials in the table have been adapted from a table given in Ref. [9].
Source: Ortega et al. 2015 [9].

Coefficients a_{ij} is the linear ME susceptibility and b_{ij} the nonlinear susceptibility. For a detailed discussion of these equations, the interested readers might wish to review the excellent paper by Fiebig [10].

5.6.2 Composite ME Materials

In single-phase materials such as Cr_2O_3 and $BiFeO_3$, the ME effect is the coupling between the electric and magnetic order parameters. The picture is, however, different in composite ME materials. Composites are by definition a combination of two dissimilar bulk materials A and B exhibiting the cross coupled properties of both A and B. That is, they consist of at least two phases and may even be mixtures of many phases. Most of the ME composites are two phase materials in which magnetostrictive and electrostrictive properties are fused together to produce the resulting material with the properties acquired from the parent materials. An example would be the combination of a ferrimagnetic material such as cobalt ferrite, $CoFe_2O_4$ (CFO) and a piezoelectric material such as lead zirconate titanate, $PbZr_{1-x}Ti_xO_3$. There are by now available a good number of composite ME materials with excellent properties. Some other examples of composite ME materials are BTO/CFO, BTO/LMO, CFO–BFO, etc. Here BTO stands for ferroelectric barium titanate, LMO for lanthanum manganite, $LaMnO_3$, and BFO for bismuth ferrite, $BiFeO_3$.

5.6.3 ME Integrated Structures

Bulk composites are usually not well suited for the development of novel electronic devices that can be integrated with silicon technology. As such materials in thin-film configurations with excellent ME properties are actively pursued with the objective of exploiting them for fabrication of novel devices that can advance the emerging ME technology. The access to highly sophisticated tools for producing integrated thin-film structures with a variety of configurations has been the key for the success of heterostructured or integrated structured ME devices. The ME effect is achieved by integrating magnetostrictive and piezoelectric components. Magnetostriction is the property of ferromagnetic materials and exhibits itself by going through a dimensional change under the influence of a magnetic field. We have already defined piezoelectricity as the effect that arises because of the coupling between stress and electric field. For producing the ME effect, a thin film of magnetostrictive material is first grown epitaxially on a single-crystal piezoelectric/ferroelectric substrate. The structure when placed in a magnetic field develops strain in the magnetostrictive layer that subsequently propagates to the piezoelectric layer via the interface inducing polarization. The entire process taking place in the integrated structured configuration leading to the onset of the ME effect can be qualitatively represented by Eqs. (5.18) and (5.19) [11].

$$\text{ME effect} = \left(\frac{\text{electrical}}{\text{mechanical}}\right) \times \left(\frac{\text{mechanical}}{\text{magnetic}}\right) \quad (5.18)$$

$$\text{ME effect} = \left(\frac{\text{magnetic}}{\text{mechanical}}\right) \times \left(\frac{\text{mechanical}}{\text{electrical}}\right) \quad (5.19)$$

5.6.4 Experimental Determination

There has been a proliferation of efforts in the past few years in producing thin films and composites exhibiting large ME effects. For the growth of integrated film structures, the most used methods are the molecular beam epitaxy (MBE), pulsed laser deposition (PLD), and RF sputtering. In Figure 5.7, a general schematic of a heterostructure consisting of ME ferroic films is shown. The PLD is currently the most used method for producing such structures because of its capacity to retain the stoichiometry of the original targets and the ease with which it can be operated.

A heterostructure consisting of ferroelectric/piezoelectric lead–zirconate–titanate, PZT, and lanthanum–strontium–magnetite, LSMO, which is a well-known colossal magnetoresistive (CMR) material was investigated to establish the coupling between polarization and magnetization [9]. The structure was grown by the PLD method on the insulating lanthanum aluminate, $LaAlO_3$, substrate which remains passive and insulated during the process of inducing ME effect. The top metallic electrodes are needed for accessing the structure electrically. Platinum and gold are the two most-preferred electrode metals for such investigations.

When an electric field, E, is applied the PZT ferroelectric film gets polarized. This leads to the development of the hysteresis loop as depicted in Figure 5.8a,b [9]. We find that the ferroelectric loop of Figure 5.8a shows a small effect of applied magnetic field of 0.3 T (=3000 G). Even though the effect is very small, yet it does show

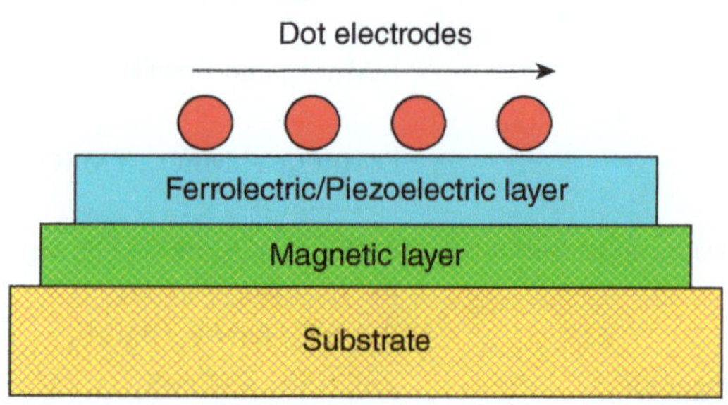

Figure 5.7 Multiferroic hybrid structure consisting of a ferroelectric/piezoelectric layer integrated with a magnetic layer grown on a substrate.

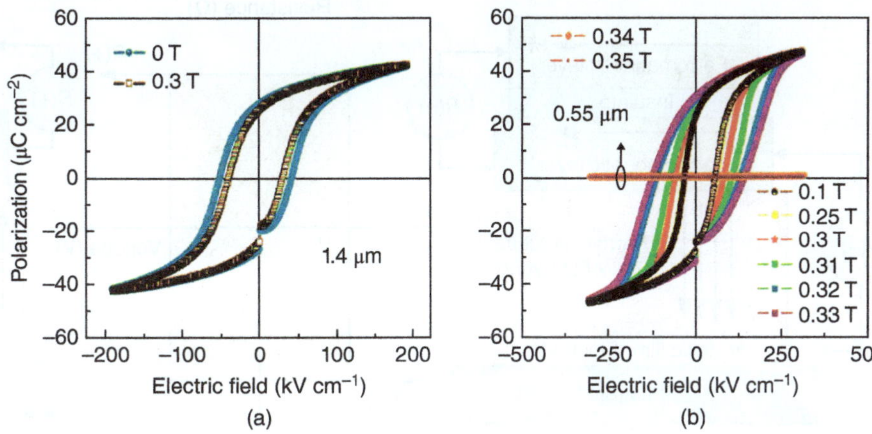

Figure 5.8 Polarization vs. electric field of PZT films as function of applied magnetic fields (a) 1.4 μm thick and (b) 0.55 μm thick. Source: Ortega et al. 2015 [9].

the onset of ME effect in the integrated structure. This becomes more pronounced when the film thickness is reduced from 1.4 μm of film (a) to 0.55 μm of film (b). It is remarkable that now the effect of magnetic field is substantial and the changes in the ferroelectric loop can be induced even by relatively small fields of $1000 < H < 3300$ G. The mechanism for the onset of the ME effect in the structure is obviously by the process defined by Eq. (5.18).

5.7 Applications of Multiferroics

It should be obvious by now that the field of multiferroics holds great potentials in developing a host of new applications and devices because of the cross coupling between electric and magnetic fields. Some potential applications are nonvolatile computer memory consisting of two bistable polarization states and two bistable magnetizations states; attenuators, filters, transducers converting the electrical energy into mechanical energy and vice versa, field probe, and ferroelectric/multiferroic tunnel junctions (MTJ).

5.7.1 Ferroelectric and Ferromagnetic Coupled Memory

Currently, memory technology is based on two principal categories: one being the ferroelectric random access memory (FeRAM) and the other the magnetic random access memory (MRAM). Historically, it is important to remember that in the beginning of the memory technology, just a few decades ago only the ferrite core memory was around. The FeRAM is a development of 1980s. In each of these two types of memory, the manipulation of the hysteresis loops of ferroelectrics and magnetics play crucial roles. How they are manipulated to achieve the goal of data write-in and read-out are described in some detail in Chapter 9 for FeRAM and in Chapter 11 for MRAM. Here it should suffice to understand that the hysteresis loops of ferroelectrics provide us with two bistable states of polarizations, $\pm P$ at electric field $E = 0$, which are equivalent, and since they are there even at zero electric field, they are called nonvolatile. The nonvolatility implies the information stored in the memory cell does not disappear when power fails. In contrast, a semiconductor random access memory (RAM) is considered volatile because all stored information can be lost in the event of a power failure. On the other hand, read only memory (ROM) and Not-And (NAND) flash memories are nonvolatile.

The polarization, P, at $E = 0$ is called the remnant polarization, P_r. It is bistable with the values of $\pm P_r$. Similarly, we have bistable and nonvolatile $\pm M_r$ states in a ferromagnetic hysteresis loop. The remnant parameters, $\pm P_r$ and $\pm M_r$, of the two types of loops form the heart and soul of FeRAM and MRAM, respectively. The multiferroics materials now offer us the opportunity of developing quadruple-state memory instead of the bistable conventional memories. In such a memory system, the data storage capability will be enormous. We can, therefore, hope that the next generation of memory, which may be called magnetoelectric random access memory (MERAM), will be built on the cross coupling of electric and magnetic fields resulting in ultrafast and ultrahigh density memory elements. The possibility is certainly there, but its realization can only happen when all the other technical challenges are overcome. One of the most important challenges is the integration of ferroelectric and ferromagnetic film with excellent interface properties so that cross-coupled parameters can evolve without being inhibited by any sort of perturbations. Another nontrivial issue is how to achieve seamless integration of ME heterostructure with conventional silicon technology. The schematic of Figure 5.9 is an example of a modified MERAM memory proposed conceptually [12].

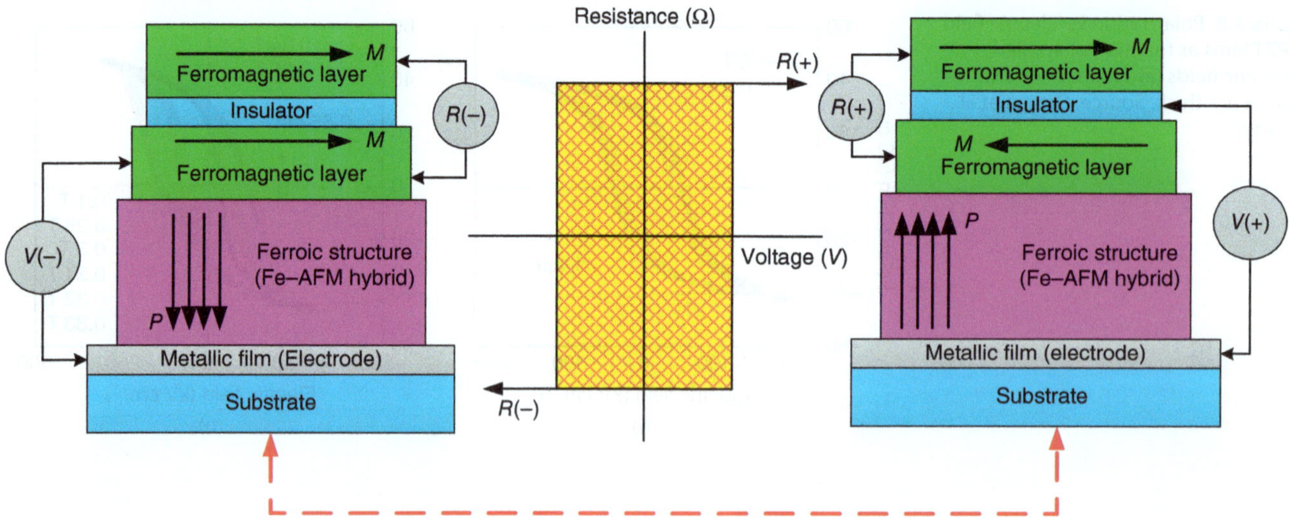

Figure 5.9 Modified model of a resistive MERAM memory consisting of ferromagnetic and ferroic (ferroelectric–antiferromagnetic) hybrid structures.

In Figure 5.9, the blue blocks are ferromagnetic layers separated by a very thin insulating layer so that spin tunneling can take place. The green block is the ferroelectric layer that is also antiferromagnetic (FE-AFM). No special effort is made to induce antiferromagnetism in the ferroelectric film. Nature provides this magnetic property in almost all ferroelectrics. We simply use it. The basic principle of operation behind this device is unique and novel. It combines the concept of spin exchange and the ME coupling in one. In other words, it integrates the ME coupling with the interfacial coupling between a multiferroic layer and a ferromagnetic layer by applying a potential. The ME coupling enables the electric field to control the spin exchange coupling at the interface. To begin with, let us assume that the magnetization in the ferromagnetic films (top right) are antiparallel with the resistance equal to R_{ap}. The ferroelectric layer is polarized by the applied voltage, $V+$ which results in the development of a ferroelectric polarization vectors, P, pointing up. The next step is to apply a negative voltage, $V-$ between the bottom blue layer and the FE-AFM layer (bottom left). The polarization, P, reverses its orientation resulting in the new resistance (R_p) between the two ferromagnetic layers. Now the magnetization, M, in the two magnetic layers is parallel. The magnetic states of the structure are expressed as a function of resistance, R_{ap} and R_p. These resistances are obviously dependent upon the magnitude and polarity of the voltage applied. As such a resistance vs. voltage hysteresis loop evolves, as seen from Figure 5.9, with well-defined bistable and nonvolatile states representing "on" and "off" states. If the concept could be translated into practice, then undoubtedly, it will pave the way for the advancement of MERAM technology.

5.7.2 Multiferroic Tunnel Junctions (MTJ)

Another novel device with many applications that could be developed is the MTJ. For tunneling effect to occur, the films must be ultrathin; between 3 and 5 nm thick. The structure consisting of such ultrathin films based on two different ferroic materials must be separated by an insulating barrier that must also be just a few nanometer thick. The tunneling induced by applying an external voltage will lead to the development of a resistance which is called the tunnel electroresistance (TER). It is easy to visualize how the MTJ device based on room temperature multiferroic materials would be a boon to highly sophisticated sensors and ultrafast memory technology.

5.8 Magnetostriction and Electrostriction

Both these phenomena are encountered in normal materials irrespective of their crystal symmetry. They are different from the piezoelectric and piezomagnetic effects and yet they bear some resemblance to them. These effects can be important for inducing changes in the fundamental properties of multiferroic materials and thereby can be useful for tuning multiferroic devices.

5.8.1 Magnetostriction

When a ferromagnetic material is magnetized by the application of a magnetic field a dimensional change occurs. The length of the sample can change which is usually very small and is mostly of the order of a few parts per million. This effect was discovered by James Joule in

1842. The strain (x) that is induced by the magnetic field is proportional to the square of the magnetization (M).

$$x = nM^2 \quad (5.20)$$

The constant, n, is called the magnetostriction coefficient. It is a fourth-rank tensor. Equation (5.20) when expressed in its tensor form becomes Eq. (5.21).

$$x_{ij} = n_{ijkl}M_kM_l \quad (5.21)$$

The quadratic magnetostrictive effect is usually larger than the linear piezomagnetic effect. In practice, for applications the magnetostrictive constant, λ, is more used than the strain coefficient, n. The constant, λ, is, equivalent to $\frac{dl}{l}$, where l is the original length and dl the corresponding change. The typical plot of magnetostriction for a ferromagnetic material is shown in Figure 5.10 and values of magnetostriction coefficient for some materials are given in Table 5.2.

Using the magnetostrictive effect, it is possible to transform the magnetic energy into kinetic energy, or conversely, the kinetic energy into magnetic energy. Magnetostriction effect can be exploited for building actuators for providing precise movement and control of a system such as a stage of a scanning electron microscope (SEM) or semiconductor mass aligner. The magnetostrictive effect can also be useful in sensor technology.

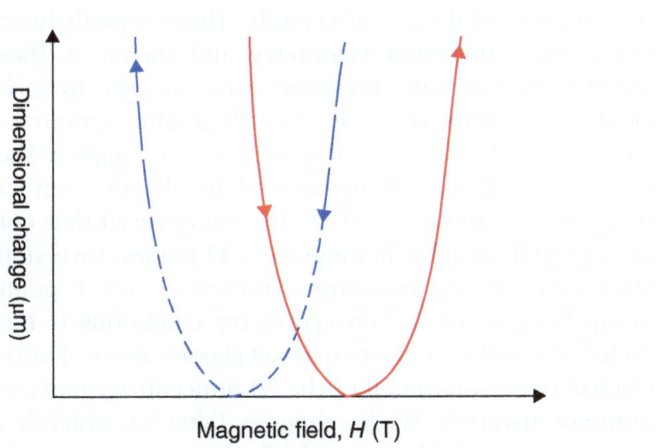

Figure 5.10 Magnetic field vs. dimensional change in a ferromagnetic material. Red curve: Magnetic field, H, increasing; Blue dashed curve: Magnetic field, H, decreasing.

Table 5.2 Magnetostriction constant, λ, for some magnetic materials.

Materials	$\lambda \times 10^{-6}$	Single crystal, orientation
Iron, Fe	21	[100]
Cobalt, Co	46	[100]
Nickel, Ni	120	[100]
Magnetite, Fe_3O_4	−33	[111]
Cobalt ferrite, $Co_{0.8}Fe_{2.2}O_4$	120	[111]

Source: Newnham 2013 [3].

5.8.2 Electrostriction

Electrostriction is analogous to magnetostriction. A very small change in shape takes place when an electric field is applied to a dielectric material. The positive ions move in the directions of the applied field, whereas the negative ions move in the opposite direction. This causes a strain to develop within the sample. As we know strain refers to the fractional change in the shape of the sample and is a dimensionless property of materials.

It should not be confused with the converse piezoelectric effect that is the linear relationship between the strain induced and the electric field. In comparison, the electrostriction effect is a nonlinear quadratic effect as represented by Eq. (5.22). Here q is the constant of proportionality and is also fourth-rank tensor. Its dimension is $\left(\frac{m^4}{C^2}\right)$.

$$x = qP^2 \quad (5.22)$$

In tensor formulation, it becomes Eq. (5.23).

$$x_{ij} = q_{ijkl}P_kP_l \quad (5.23)$$

Figure 5.11 depicts the strain vs. electric field relationship for a typical ferroelectric material [13]. The technical name for q is electrostriction coefficient and it is the figure-of-merit of the material. The larger its value, the more attractive, the material becomes for applications.

The butterfly loop is the signature of the nonlinear electrostrictive effect found in ferroelectric materials like barium titanate, $BaTiO_3$. The curve A is the initial path followed by the x–E relationship. While generating

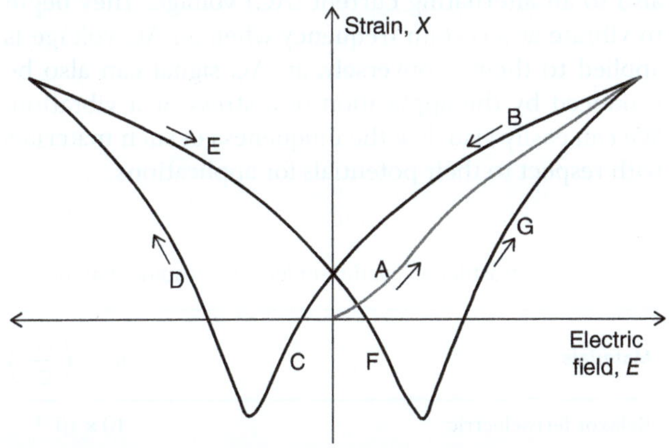

Figure 5.11 Strain vs. electric field plot of ferroelectric material [13].

the butterfly loop, it traverses through the path indicated by A →B→ C → D → E →F→ G.

Relaxor ferroelectrics exhibit the largest values of the electrostrictive constants. The prominent materials with large values of q are lead magnesium niobate (PMN), composite of lead magnesium niobate–lead titanate (PMN-PT), and lanthanum-substituted lead zirconate titanate (PLZT).

Electrostriction effect is used for developing SONAR (sound navigation and ranging) for submarines and surface vessels. Some of the other applications of this effect include micropositioner elements with high degree of precision (actuators) for applications in electron microscopy, high-resolution optical microscopy, scanning electron microscopy, translation and alignment for laser mirrors, microelectronic probe poisoners, micromachining tool controls, and X-ray lithography position controls. In Table 5.3, values for the electrostriction coefficient q is given for some materials [14].

Quartz is widely used in the industry for applications of electrostriction because of its high value of electrostriction coefficient and frequency response.

5.9 Piezoelectricity

Piezoelectricity is a fascinating phenomenon with a very long history of successful applications in technology and basic science. It was discovered by the famous French physicist brothers, Jacques and Pierre Curie in 1880. The name is, as is common in science, derived from the Greek word *piezo*, which means mechanical stress such as pressure. There are a large number of devices and applications that are based on piezoelectricity. We will discuss some of them later in this chapter. The piezoelectric effects are not limited only to direct current (DC) signals. The piezoelectric materials respond well also to an alternating current (AC) voltage. They begin to vibrate at a certain frequency when an AC voltage is applied to them. Conversely, an AC signal can also be produced by the application of a stress or a vibration. We can easily visualize the uniqueness of such materials with respect to their potentials for applications.

Table 5.3 Electrostriction coefficient for some leading materials.

Materials	$q_{1111}\left(\frac{m^4}{C^2}\right)$
Relaxor ferroelectric	10×10^{-3}
Quartz	10
PVDF copolymer	−2 to −2/5

The applications of piezoelectricity cover a wide range of topics, some of which are generation of high voltages, production and detection of sound, energy harvesting, fabrication of transducers, precision mass aligners, frequency generators, microbalances, surface acoustic wave (SAW) devices, acoustic amplifiers, and ultrasonic diagnostic tools. Valuable scientific instruments such as scanning tunneling microscope (STM), atomic force microscope (AFM), SONAR, and piezoelectric force microscope (PFM) are also based on piezoelectric effects. The general-purpose consumer gadgets using piezoelectricity are quartz watches, lighters, and recreational toys.

5.9.1 Crystallographic Considerations for Piezoelectricity

In order for the piezoelectric effects to occur, the materials (more precisely a crystal) must meet certain crystallographic conditions. It is not a universal phenomenon like electrostriction or conductivity or even thermal conduction. Rather piezoelectricity is the result of the direct interaction between a mechanical stress and the electrical polarization that it can cause to originate in certain special class of crystals. These crystals must not possess inversion symmetry and therefore, they are classified as noncentrosymmetric crystals. In such crystals, rigorous rules of crystallographic symmetry elements apply. We study this topic in this chapter. The rules of symmetry elements lead to classification of noncentrosymmetric crystals. We might recall that out of 32 crystallographic point groups, 11 possess inversion symmetry, 20 are noncentrosymmetric, and 1 point group because of certain symmetry conditions is not included in either of the two broad classifications. Piezoelectricity is found in all of the 20 noncentrosymmetric group of materials. We list them in Table 5.4, which is a combination of Tables 4.6 and 4.7.

Here the 20 point groups for piezoelectric crystals have been divided into two groups of polar and nonpolar elements with their respective symmetry elements identified. There are 10 members in each of these groups. All members of Group I are nonpolar. The symmetry elements of this group allow piezoelectricity to exist but forbid pyroelectricity. In contrast, Group II point groups allow both pyroelectricity and piezoelectricity to coexist simultaneously. By definition, ferroelectricity is the subset of pyroelectricity and therefore, they share the symmetry elements of Group II with pyroelectricity. In summary, not all piezoelectric materials are polar. They are found both in polar and nonpolar groups.

Now let us revisit Figure 5.2 once again. While discussing the piezoelectric zone outlined in this figure, we defined two types of piezoelectric effects: one being

Table 5.4 Symmetry elements for piezoelectricity: polar and nonpolar.

	Cubic	Triclinic	Tetragonal	Hexagonal	Monoclinic	Orthorhombic	Trigonal
Group I (10)	23	None	$\bar{4}$	$\bar{6}$	None	222	32
Non-polar	$\bar{4}3m$		422	622			
			$\bar{4}2m$	$\bar{6}m2$			
Group II (10)	None	1	4	6	2	$mm2$	3
Polar			$4mm$	$6mm$	m		$3m$

the direct effect and the other the converse effect. These are reversible effects and are therefore the basis for a large number of applications. In order to gain some insight into these two important physical effects, we need to acquire a certain degree of familiarity with their mathematical representations.

5.9.2 Mathematical Representation of Piezoelectric Effects

Before we can consider the piezoelectric effect in any depth, we need to refresh our memory about tensors. Piezoelectricity is one area in which its various concepts and definitions become more comprehensible if one has some knowledge of tensor analysis. All physical properties can be broadly categorized either as a scalar such as mass of a body or a vector such as velocity with which the mass can move. Tensor is the third way to describe physical properties and is an elegant mathematical tool that theoretical physicists and mathematicians love to use. On the other hand, the experimentalists and practicing engineers have found ways not to depend too heavily on tensors. We will examine later how it can be done.

Let us first consider Eq. (5.2), which expresses the relationship between the electrical displacement, D (which is another name for electrical polarization) and electric field, E.

Suppose we are now dealing with a cubic crystal with the principal axes as 1, 2, and 3, respectively (see Figure 5.12). If the electric field is applied along the axis 1 and displacement D measured also along the same axis, then Eq. (5.2) reduces to Eq. (5.24)

$$D_1 = \epsilon_{11} E_1 \tag{5.24}$$

Similarly, we can write $D_2 = \epsilon_{22}E_2$ and $D_3 = \epsilon_{33}E_3$. When both the subscripts of a coefficient have the same values for which they are employed that the parameters representing cause and effect are parallel to each other. Information obtained under such a condition is only partial. To get the whole picture, the nonparallel components must also be considered. Then Eq. (5.2) can be

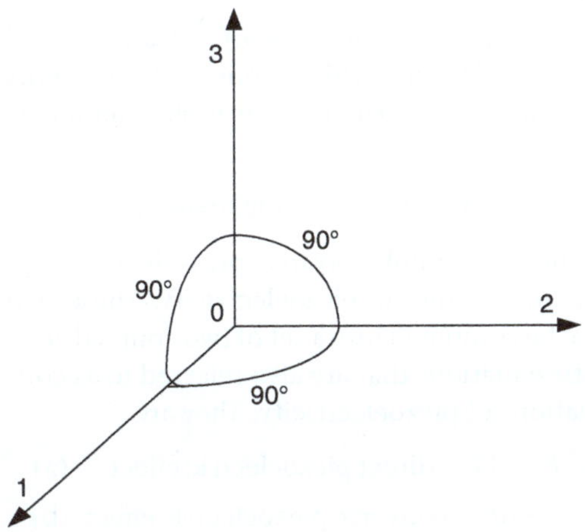

Figure 5.12 Representation of the rectangular coordinates of a cubic crystal.

written as a set of three equations as follows.

$$\begin{aligned} D_1 &= \epsilon_{11}E_1 + \epsilon_{12}E_2 + \epsilon_{13}E_3 \quad \text{(a)} \\ D_2 &= \epsilon_{21}E_1 + \epsilon_{22}E_2 + \epsilon_{23}E_3 \quad \text{(b)} \\ D_3 &= \epsilon_{31}E_1 + \epsilon_{32}E_2 + \epsilon_{33}E_3 \quad \text{(c)} \end{aligned} \tag{5.25}$$

We now notice that the permittivity (ϵ) has three components corresponding to each value of the displacement. We know that by definition, the permittivity is a second-rank tensor that can now be expressed more elegantly in a tensor form as in Eq. (5.25).

$$\epsilon \equiv \begin{pmatrix} \epsilon_{11} & \epsilon_{12} & \epsilon_{13} \\ \epsilon_{21} & \epsilon_{22} & \epsilon_{23} \\ \epsilon_{31} & \epsilon_{32} & \epsilon_{33} \end{pmatrix} \tag{5.26}$$

It is important to understand what the subscripts mean. The first one refers to the row and the second to the column. This rule must be followed when expressing a physical property in its tensor form. A second-rank (or second-order) tensor is expressed as a 3×3 matrix amounting to 9 elements in total. As a general rule, we can state that the number of elements of any tensor is equal to 3^n, where n = rank of the tensor. Then the first-rank tensor will have three elements represented by a matrix of 3×1. For the second-rank tensor, it is $3^2 = 9$, for the third rank, it is $3^3 = 27$, for the fourth rank, it is $3^4 = 81$, and so on. Incidentally, a scalar is considered a zero-rank tensor and a vector a first-rank tensor. Imagine now how many tensor elements we will have to write if we want to express Eq. (5.4), which deals with the relationship between the mechanical stress (X) and the resulting strain (x) each of which is a second-rank tensor.

Simply said, we would need 81 ($=3^4$) tensor elements just to express the elastic compliances, in its tensor form, which would not be a trivial enterprise to accomplish.

Exercise 5.1
Write the tensor elements for stress (X), strain (x), and magnetic permeability (μ). (This exercise is for students to work independently. Instructors may wish to monitor.)

5.9.3 Constitutive Equations for Piezoelectricity

Fortunately, there is a simple way to express the Eqs. (5.2) and (5.4) for understanding piezoelectricity. These two equations can be combined into a set of two coupled electric and elastic equations that are also referred to as constitutive equations of piezoelectricity. They are

$$D = \epsilon^X E + \mathrm{d}X : \text{direct piezoelectric effect} \quad \text{(a)}$$
$$x = s^E X + \mathrm{d}E : \text{converse piezoelectric effect} \quad \text{(b)} \tag{5.27}$$

The superscript E indicates zero or constant electric field. Similarly X refers to zero or constant stress. In these equations, d is a piezoelectric coefficient, and s the materials compliance, which is the inverse of materials stiffness. If $X = 0$, from Eq. (5.27) (a) we get $D = \epsilon E$ which is the standard formula defining the very basic nature of a dielectric material, and if $E = 0$, we get from Eq. (5.27) (b) $x = sX$ which is nothing else than the famous Hooke's law. This simple analysis justifies the validity of using the coupled equations of Eq. (5.27) to describe the two types of piezoelectric effects.

There are four different types of piezoelectric coefficients that are identified by symbols d, g, e, and h. Piezoelectric applications are classified based on these four parameters and therefore, they play a crucial role in identifying suitable materials that could meet the needs of specific applications. They can be defined with the help of Eqs. (5.28) through (5.35).

- Piezoelectric charge (or strain) coefficient, d ($\mathrm{C\,N^{-1}}$) or ($\mathrm{m\,V^{-1}}$)

$$d = \left(\frac{D}{X}\right) = \left(\frac{x}{E}\right) \tag{5.28}$$

Notice that the coefficient d is identical for both the direct and converse piezoelectric effect with the unit of ($\mathrm{C\,N^{-1}}$) for the direct piezoelectric effect and ($\mathrm{m\,V^{-1}}$) for the converse effect. Materials with large values of d are found suitable for actuators and sound-generating devices.

- Piezoelectric voltage coefficient, g ($\mathrm{V\ m\ N^{-1}}$) or ($\mathrm{m^2\ C^{-1}}$)

$$g = \left(\frac{E}{X}\right) = \left(\frac{x}{D}\right) \approx \frac{d}{\epsilon_r \epsilon_0} \tag{5.29}$$

Materials with large values of g are suitable for the generation of voltage in response to a mechanical stress. Classic examples are gas lighters and phonograph pickups. Obviously, they could be considered also for producing electrical energy of small magnitude. Equations (5.30) and (5.31) are the formal representations of the coefficients d and g. This form is mostly used for theoretical work.

$$d = \left(\frac{\delta D}{\delta X}\right)_E = \left(\frac{\delta x}{\delta E}\right)_X \tag{5.30}$$

$$g = -\left(\frac{\delta E}{\delta X}\right)_D = \left(\frac{\delta x}{\delta D}\right)_X \tag{5.31}$$

- Piezoelectric stress coefficient, e ($\mathrm{C\,m^{-2}}$); this is equivalent to the inverse of voltage coefficient.

$$e = -\left(\frac{X}{E}\right) \approx -\left(\frac{1}{g}\right) \tag{5.32}$$

- Piezoelectric coefficient, h ($\mathrm{V\ m^{-1}}$) or ($\mathrm{N\ C^{-1}}$); this is equivalent to the inverse of the charge coefficient.

$$h = -\left(\frac{E}{x}\right) \approx -\left(\frac{1}{d}\right) \tag{5.33}$$

These two coefficients in their differential forms can be written as in Eqs. (5.34) and (5.35), respectively.

$$e = -\left(\frac{\delta X}{\delta E}\right)_x = \left(\frac{\delta D}{\delta x}\right)_E \tag{5.34}$$

$$h = -\left(\frac{\delta E}{\delta x}\right)_D = -\left(\frac{\delta X}{\delta D}\right)_x \tag{5.35}$$

By convention followed in practice, the piezoelectric coefficients are identified with two subscript such as d_{31} or d_{33}. There is a good reason for this. From the definitions of the piezoelectric effects, we gather that they evolve only after two parameters couple, for example, stress and polarization (for direct effect) or electric field and strain (converse effect). Both these parameters must be identified for defining the coefficients precisely. The first subscript refers to the direction of the electric field or displacement; and the second subscript as the direction of the stress or strain. The convention may be easily understood by considering the rectangular coordinates of Figure 5.12 with the three orthogonal axes being 1, 2, and 3.

Case I: Let us assume that in a crystal, we are measuring the polarization P as a result of the stress X applied (direct piezoelectric effect) in a single crystal. If both of them are parallel to each other, then the resulting piezoelectric coefficients should be d_{11} or d_{22} or d_{33}. However, these measurements are usually made in poled crystals in which conventionally the polar axis is assigned to be the *c*-axis. Since axis-3 of Figure 5.12 corresponds to *c*-axis of a crystal, the correct way to express this *d* coefficient would be d_{33}.

Case II: If the polarization *P* is along the axis 3 and stress *X* along the axes 1 or 2, then the resulting piezoelectric coefficients will be given by d_{31} or d_{32}.

Exercise 5.2

Write all *d* coefficients using two-subscript convention for the converse piezoelectric effect in a single crystal. You can make use of the coordinates shown in Figure 5.12. (This exercise is for students to work independently. Instructors may wish to monitor.)

We have so far built our concepts of the piezoelectric effects considering the anisotropic nature of single crystals. But industry produces piezoelectric sensors and devices based on polycrystalline electroceramics. How are the piezoelectric coefficients to be determined in such devices? It is simple, and there is a good solution. All we have to do is to electrically pole the ceramic sample. By poling, it is meant that the electrical polarization is induced in the direction of the applied electric field while changing the sample temperature from a higher value to room temperature without withdrawing the electric filed. By poling, it is possible to lock the polarization vectors in one unique direction which by convention is set to be along the along the direction identified as 3 in Figure 5.12. The poling mechanism will be discussed in detail in Chapter 6. It is a good practice to measure piezoelectric parameters only after poling a sample irrespective of the fact whether it is a single crystal, an epitaxial, or a textured film or simply a polycrystalline electroceramic. The values of piezoelectric coefficients show strong dependence on the strength of the electric field and temperature used for poling. By trial and error, one can find out the optimal conditions for poling. A short list of leading piezoelectric materials is given in Table 5.5.

Compared to other materials, we find from this table that the piezoelectric coefficient, *d*, is rather large for both barium titanate and PZT making them suitable for many piezoelectric applications. Quartz, on the other hand, is one of the most-used piezoelectric materials by industry and is widely used for precise control of frequency such as in watches. Good-quality single crystals of quartz are abundantly available commercially. Its another benefit is its excellent thermal stability. Similarly, lithium niobate and lithium tantalate crystals are also used on commercial scale because of their superb thermal properties and availability of high-quality crystals. These are the materials of choice for electro-optic devices and SAW devices that find their applications in signal processing and electronic filters.

5.10 Experimental Determination of Piezoelectric Coefficients

5.10.1 Charge Coefficient, *d*

In Eq. (5.28), we defined the piezoelectric charge coefficient, *d*, as the ratio between the displacement, *D*, and stress, *X*. That is it can also be expressed in terms of charge, *Q* and force, *F*. By definition, we know that the displacement $D \approx P$ (polarization) $\approx Q/A$ (charge/area). Therefore,

$$d = \left(\frac{\Delta Q}{\Delta F}\right) \tag{5.36}$$

Also, in Eq. (5.28), we used the converse piezoelectric effect to define the charge coefficient, *d* (=*x*/*E*). In other words, a piezoelectric material undergoes a change in sample length (*L*) when a voltage (*V*) is applied. Therefore, we can also express the strain coefficient, *d*, as described by Eq. (5.37).

$$d = \left(\frac{\Delta L}{\Delta V}\right) \tag{5.37}$$

These two alternative definitions of the coefficient, *d*, have been used for a very long time to measure the longitudinal charge coefficient, d_{33}, and transverse charge coefficient, d_{31} in bulk sample having a well-defined geometry. We found earlier that the magnitude of the charge coefficient remains the same whether we determine it by using the direct or converse piezoelectric effects. Only they should be identified by using the correct unit.

Table 5.5 Selected piezoelectric materials and their piezoelectric coefficient (*d*).

Material	Silicon dioxide (Quartz)	Barium titanate	Lithium niobate	Lithium tantalate	Lead zirconate titanate (PZT)	Polyvinyl chloride
Chemical formula	SiO_2	$BaTiO_3$	$LiNbO_3$	$LiTaO_3$	$Pb_{1-x}Zr_xTiO_3$	PVC
d (ncm V^{-1})	0.225	19.0	0.085	0.3	23.4	70

Source: Kao 2004 [15]. Reproduced with the permission of Elsevier.

Though it looks simple and straightforward, it is not that easy to measure the very small dimensional change, ΔL, in the sample that occurs when it sees a voltage across itself. Usually, the dimensional change is in the order of nm. Apart from that it is assumed that the sample must not have any inconsistency in its geometry, otherwise, the result would not be accurate. That demands that the sample must be prepared with great care not to have any discrepancy in its length, width, and thickness. Only then can the value for d be determined rather accurately by measuring the small dimensional change using either an AFM or a sensitive photonic sensor such as a laser sensor.

5.10.2 Stress Coefficient, *e*

Accurate measurements of very small dimensions become even more challenging in film samples that are in demand for many of the newer applications of piezoelectricity. It is very difficult to measure the charge coefficient, d, in thin film samples. The film samples are universally integrated to a bulk substrate and therefore, they are clamped and not free to bend or undergo displacement. The solution for this dilemma is provided by measuring the stress coefficient, e, as defined by Eq. (5.30). Bending of the substrate causes stress to be induced in the film capacitor giving rise to the e_{31} response. That is

$$\Delta D = -e_{31} \cdot \Delta x \tag{5.38}$$

We can also rewrite this equation in terms of the equivalent change in the film capacitor charge, Q, when the lateral dimension, L, of the film changes because of the bending effect.

$$\Delta Q = -e_{31} \cdot \Delta L \tag{5.39}$$

The advantage of this method is that the both the charge and the force to bend the sample can be measured accurately. By using the mechanical model of the substrate, one can then obtain the values of the d coefficient from the measured e coefficient. We describe now a method for measuring the stress coefficient, e, and other associated parameters for an integrated film of a piezoelectric material. These films have gained importance because they are compatible to CMOS processing of integrated circuits as well as being suitable for MEMS (microelectromechanical systems) applications. In Figure 5.13, a layout for the experimental setup using a laser sensor is outlined [16]. The integrated structure consists of a piezoelectric film (red) of PZT grown on platinized silicon. This structure acts as a cantilever that is actuated by the bulk PZT actuator when an AC signal is applied. Consequently, it induces a deflection in the PZT film that is sensed by the laser sensor. Also a charge is generated on the platinum electrode. After going through a charge amplifier and a lock-in-amplifier, these two parameters are monitored by the oscilloscope.

In Figure 5.14, the dependence of the cantilever deflection on the driving voltage of the actuator is plotted [17]. An AC voltage of 10 Hz was applied to the PZT cantilever either on side I or side II for actuating it and the resulting deflection at 38.8 mm monitored by the laser sensor. The total length of the cantilever was 43.85 mm. Notice the magnitude of the deflection that is just in nm range. The deflection of the film sample resulted in the generation of charge that was collected on side II (or I).

We see in Figure 5.15 that the PZT cantilever responds well to varying frequency and exhibits its characteristic resonant frequency at 260 Hz. The deflection magnitude increases rapidly with the increasing frequency reaching

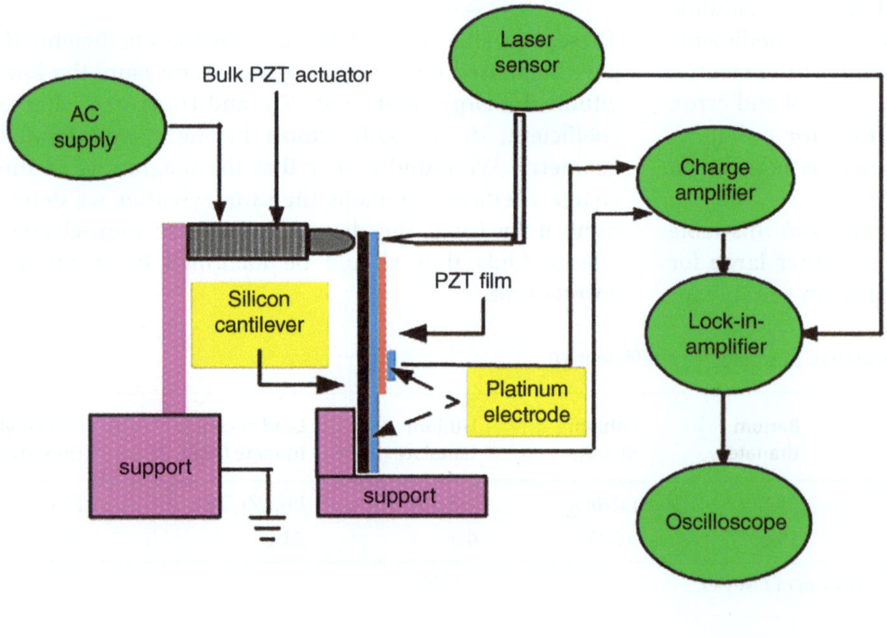

Figure 5.13 Modified schematic of an experimental setup for determination of the charge coefficient, e_{31} of an integrated film structure. Source: Zhong 2008 [16].

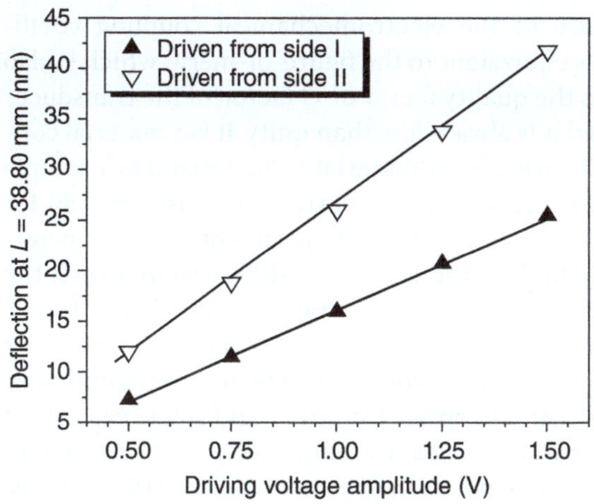

Figure 5.14 Cantilever deflection as a function of applied driving voltage for PZT double sided film. Source: Zhong et al. 2007 [17]. Reproduced with the permission of Taylor & Francis.

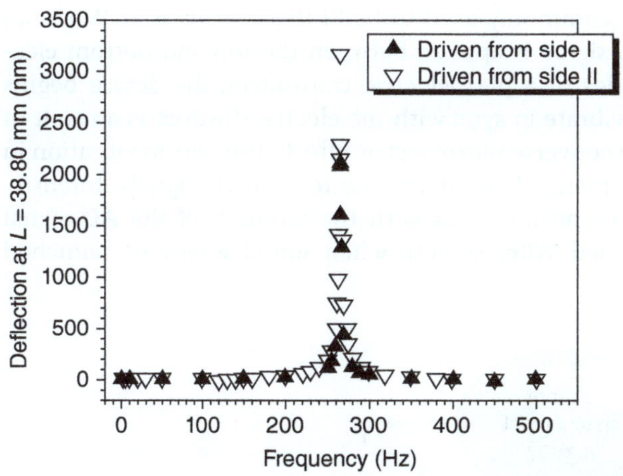

Figure 5.15 Cantilever deflection as a function of frequency for a PZT double sided film. Source: Zhong et al. 2007 [17]. Reproduced with the permission of Taylor & Francis.

its maximum at the resonant frequency and then drops off to its original value around 500 Hz.

In Figure 5.16, the effect of varying static strain on the stress coefficient, e_{31}, for the PZT cantilever is plotted. By driving the actuator with an AC signal biased with a DC potential static strain is induced in the structure. The static strain is varied by adjusting the distance between the cantilever and the actuator. The stress coefficient

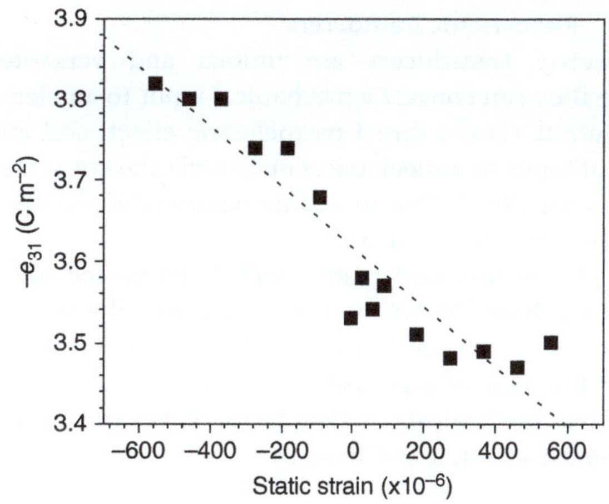

Figure 5.16 Stress coefficient, $e_{31,}$ of a PZT film cantilever as a function of static strain. Source: Zhong et al. 2007 [17]. Reproduced with the permission of Taylor & Francis.

decreases linearly with increasing strain with a slope of $-3.69\,\mathrm{C\,m^{-2}}$.

In Table 5.6, the values for piezoelectric coefficients, *d* and *e*, for ZnO, AlN, and PZT are given. Both ZnO and AlN are used extensively along with PZT for the fabrication of integrated structures for MEMS applications. From the table, we find that both the coefficients are much larger for PZT than for ZnO and AlN. Therefore, it is the materials of choice for fabrication of piezoelectric transducers for large number of applications.

5.10.3 Piezoelectric Devices and Applications

We indicated in Section 5.1 that because of the unique properties of piezoelectric materials, they are used in a large number of applications. The distinctive characteristics of a piezoelectric material are the direct piezoelectric effect and the converse piezoelectric effect. Both these properties are exploited in technology to fabricate piezoelectric transducers that are the backbone of almost all piezoelectric applications. What is a transducer? By definition a transducer is a device that can transform one type of input to another type of output whose physical characteristics are dissimilar. Some examples of transducer devices are an electric motor that converts mechanical energy into electrical energy; a solar cell that converts light into electricity; and light bulb that converts an electrical input to a photonic output.

Table 5.6 Piezoelectric charge and stress coefficients for some materials.

	Zinc oxide, ZnO	Aluminum nitride, AlN	Lead-zirconate-titanate, PZT (1–3 μm)	PZT double-sided film
Charge coefficient, d_{33} (pm V^{-1})	5.9	3.9	60–130	—
Stress coefficient, $-e_{31}$ (C m^{-2})	1.0	1.05	8–12	3.4–3.9

5.10.3.1 Piezoelectric Transducers

Piezoelectric transducers are unique and versatile because they can convert a mechanical input to an electrical output via the direct piezoelectric effect, and an electrical input to a mechanical output via the converse piezoelectric effect. The reversible nature of these two effects is shown in Figure 5.17.

As we know, thermodynamics forbids 100% efficiency for any machine. Therefore, we can assume the transfer function of a transducer must be less than 1. The transfer function of a piezoelectric transducer is called the electro-mechanical coupling factor, k. For the direct piezoelectric effect, it is defined by

$$k^2 = \frac{E_{\text{output}}}{M_{\text{input}}} = \frac{\text{converted electrical energy}}{\text{total input of mechanical energy}} =< 1 \quad (5.40)$$

Similarly, for the converse effect, it is defined by Eq. (5.41).

$$k^2 \approx \frac{M_{\text{output}}}{E_{\text{input}}} \approx \frac{\text{converted mechanical energy}}{\text{total input of electrical energy}} =< 1 \quad (5.41)$$

The square of the electromechanical coupling coefficient, k, is equivalent to the figure-of-merit, which is also known as the quality factor or Q factor, of the transducer device and it is always less than unity. It is a material constant and varies from material to material. For example it is about 0.12 for SiO_2 (quartz), whereas it is 0.35 for $BaTiO_3$ and 0.72 for PZT. Experimentally, it is determined by finding the frequency difference between the maximum impedance and minimum impedance. Piezoelectric transducers are common features of ultrasonic devices, intrusion alarms and detectors, microphones, earphones, and beepers. Ceramics and single crystals of piezoelectric materials are employed in fabrications of oscillators for generation of radio frequency (RF) signals.

5.10.3.2 Generation of Sound and an AC Signal

A piezoelectric transducer, shown in Figure 5.18, is used to produce sound waves or an AC signal. Integrated films are commonly used to build these structures. When an AC signal is applied between the top and bottom electrodes of a piezoelectric transducer, the device begins to vibrate in sync with the electrical waves as a result of the converse piezoelectric effect. This sets in vibration in the metal plates giving rise to acoustic signals. Intensity of sound increases with the strength of the AC signal applied. Alternatively, when sound waves are launched

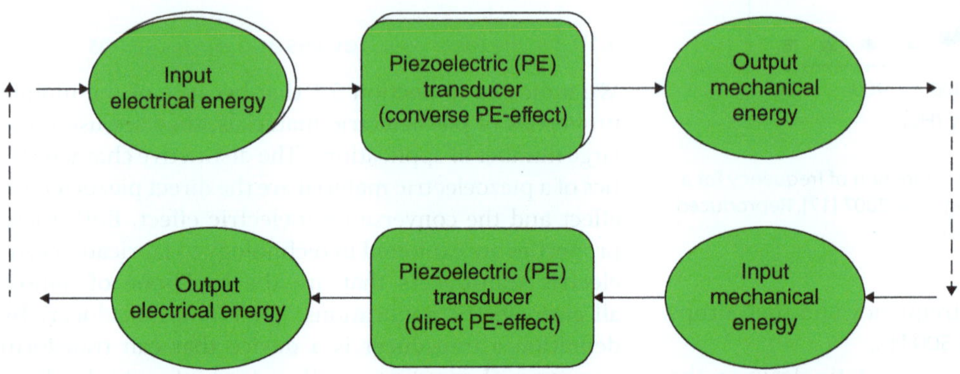

Figure 5.17 Reversible nature of a piezoelectric transducer.

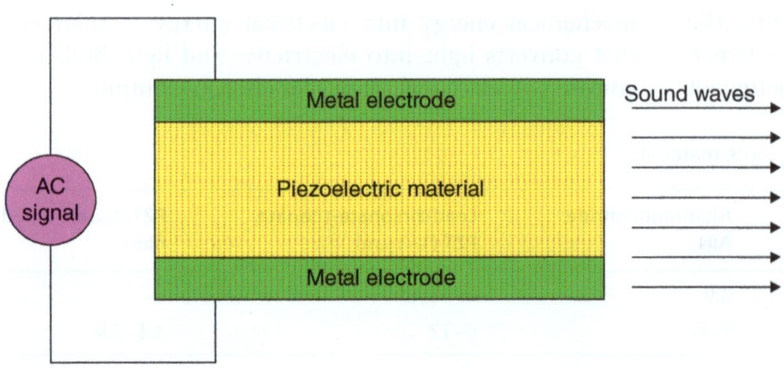

Figure 5.18 Schematic of a piezoelectric capacitor for signal conversion.

on one or both metal electrodes vibration sets in. The piezoelectric medium picks up the vibration that is then converted into an AC signal because of the direct piezoelectric effect. This principle of conversion of electrical signals into sound and vice versa is the basis for many useful devices including ultrasonic generators.

Sound waves in the range of 20 Hz–20 kHz are audible to human auditory system. We humans cannot detect sound below 20 Hz nor can we detect those above 20 kHz. Sound waves with frequency less than 20 Hz are called infrasonic, and those above 20 kHz are called ultrasonic. The range of ultrasonic extends from 20 kHz to about 10^4 GHz.

Ultrasonics is used in a large number of applications and is an indispensable tool for the medical diagnosis technology. They are used in trackless surgery in which no incision is required. An example is to destroy kidney stones with shock waves formed by bursts of focused ultrasonic waves. Focused ultrasound is used also in the treatment of Parkinson's disease. Navigation and guidance is another area in which ultrasound is used. Speed monitoring of a moving object such as an automobile or a submarine is another important application of ultrasound. This is based on the Doppler effect according to which if the moving object approaches the monitoring device, the frequency of the reflected waves from the moving object is increased. Similarly, if the object moves away from the monitoring device the frequency is decreased. High-intensity ultrasound is also used for cleaning of objects large and small. It is also routinely used for nondestructive testing of materials. Its another use is for detecting the integrity of industrial parts by identifying portions with microcracks and other faults.

5.10.3.3 Surface Acoustic Wave (SAW) Device

This device has the distinction of being conceived theoretically in 1856 by none other than by Lord Rayleigh himself who was one of the most celebrated English physicists of the nineteenth century. Its experimental counterpart had to wait more than a century. In 1965, interdigitated transducers (IDT) were invented that paved the way for the fabrication of SAW devices on a commercial scale. A SAW configuration is depicted in Figure 5.19.

Two sets of IDTs are located on each end of the piezoelectric substrate. When an AC signal is launched at the IDT on the left end, the electrical waves begin to travel on the surface that deforms as a consequence of the converse piezoelectric effect. As a consequence, these electrical signals are converted into acoustic waves that reach the IDT on the other end. If the pitch of the IDT and SAWs match, an electrical signal is generated that can then be monitored across the load at the output. The frequency of the generated waves can be modified by controlling the pitch of the IDT. SAW devices are extensively used in signal processing as filters and delay lines. SAW filters are also used in cordless, cellular phones, and pagers. Further uses include in broadcasting technology such as in TV broadcasting, radio broadcasting, and satellite broadcasting. As filters, they find their applications in navigation devices such as in GPS, telematics, and automobile navigation. Various types of waves are used in SAW devices; their selection is dictated by intended applications. Their typical characteristics include propagation velocity of waves, electromechanical coupling coefficient, temperature coefficient, and delay time. The commonly used SAW substrates are single crystals of piezoelectric

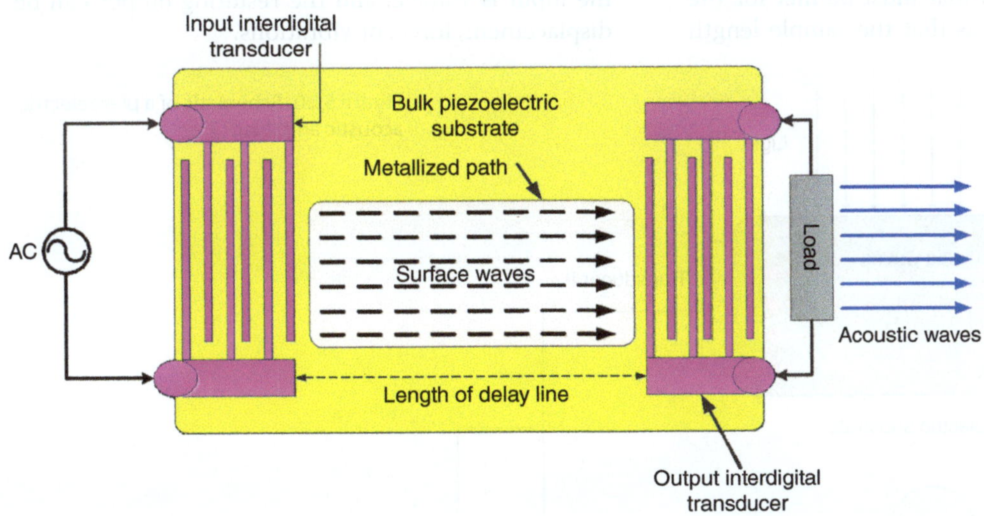

Figure 5.19 Configuration of a surface acoustic wave, SAW, device.

quartz (SiO_2), ferroelectric-piezoelectric lithium niobate (LiNbO3), and ferroelectric-piezoelectric lithium tantalate ($LiTaO_3$).

5.10.3.4 Piezoelectric Acoustic Amplifier

A piezoelectric acoustic amplifier is an interesting application of converse piezoelectric effect. The schematic shown in Figure 5.20 is a representative of this device. An AC signal applied between the two transducers sets up acoustic waves on the surface of a piezoelectric semiconductor substrate (such as cadmium sulfide, CdS). Signal amplification is achieved by irradiating the surface of the transducer with photons. When light waves strike the sample surface, the photons interact with the acoustic waves. Because the photons have a higher velocity of propagation than the acoustic waves, the energy is transferred from the photons to the acoustic waves setting in amplification of the output signal. The main advantage of such as device is its compactness compared to an audio amplifier based on electromagnetic waves. The wavelength of sound is smaller by five orders of magnitude than the wavelength of electromagnetic waves. Therefore, one can build a very compact piezoelectric acoustic amplifier compared to its electromagnetic counterpart.

5.10.3.5 Piezoelectric Frequency Oscillator

One of the most important applications of piezoelectricity is the frequency oscillator. Single crystal quartz has been used as the material of choice for building these devices for a very long time. These oscillators have impressive degrees of accuracy which amounts to about 1part per 10^8 or even 1 part per billion (10^9). The principle of operation is straightforward. We know that all capacitors have a resonant frequency of vibration at which their response to an external source is amplified manyfold. The condition that must be met for the resonant frequency to occur is that the sample length (L) must be an odd number of the half wavelength (λ) (that is, $\frac{\lambda}{2}, \frac{3\lambda}{2}, \frac{5\lambda}{2}$, etc.). Because the acoustic wave in quartz would travel with the velocity of sound, v_s, its associated frequency, f, would satisfy the condition of $f = \frac{V_s}{2L}$ for the first resonance mode. One can easily visualize that simply by adjusting the length of the oscillator one can produce a number of resonant frequencies. Since both the speed of sound and the length are very stable parameters at a constant temperature, there is no reason why the resonant frequencies should also not be very stable. Piezoelectric oscillators are used in keeping a radio station on a fixed wavelength. By the same token, they also provide accuracy to time-keeping devices such as clocks and watches. Wherever high degree of reliability of frequency is a necessary condition quartz oscillators are used. High-quality single crystals of quartz with different cuts are used for piezoelectric frequency oscillators.

5.10.4 MEMS Actuator

An actuator is the generic term for any device that is capable of producing and controlling mechanical motions. There is an array of actuators available to technology. They are categorized as electrical, mechanical, thermal, hydraulic, and pneumatic. Each of them is based on a different physical principle. The piezoelectric actuators are based on the principles of direct and converse piezoelectric effects. Their individual output is dictated by the nature of the input it receives. We can distinguish between the two classes of piezoelectric actuators as:

- Actuators based on direct piezoelectric effect: If the input is pressure, impact, or sound waves, then the output is voltage.
- Actuators based on converse piezoelectric effect: Here the input is voltage, and the resulting output can be displacement, force, or vibrations.

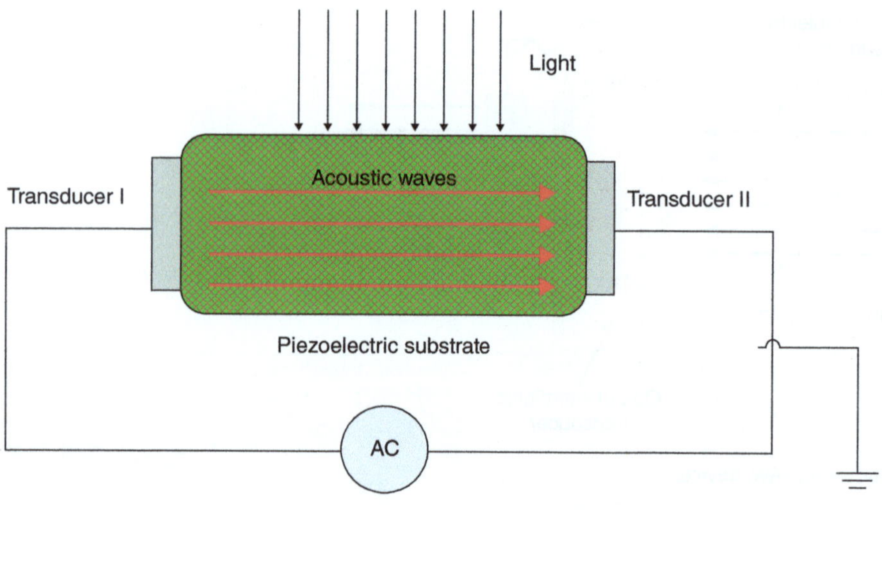

Figure 5.20 Schematic of a piezoelectric acoustic amplifier.

The most efficient piezoelectric actuators are built using the MEMS cantilever as shown in Figure 5.21. These cantilevers are produced by using micromachined silicon substrates that provide support for thin metallized piezoelectric element (PZT). From the configuration shown in this figure, we can conclude that the PZT cantilever is anchored on one end and free to bend on the other end. The actuator is capable of producing displacement when the structure is excited by an electrical voltage (V). Following the converse piezoelectric effect, the cantilever bends providing displacement whose amplitude increases with increasing voltage. The displacement is usually in the nm range and can be controlled with great precision.

PZT actuators have a large number of applications ranging from control of mass aligners for semiconductor processing to control of translation of sample stage of highly sophisticated microscopes such as AFM, SEM, and transmission electron microscope (TEM). The list for the applications of piezoelectric actuator is long. New applications are being invented at a rapid rate. In principle, these actuators can be fabricated by any piezoelectric material, but PZT is the material of choice because of its excellent piezoelectric properties compared to other potential materials.

So far we have described only one application of the actuator. Such cantilevers can also be used for producing electrical energy that can be harvested. How can it be done? It is not difficult. Consider that the piezoelectric cantilever is basically a capacitor whose capacitance is inversely proportional to its thickness and directly proportional to its area. By making the cantilever very thin and long, as can be done by micromatching, one can produce a capacitor whose capacitance (C) is large. Then even a modest level of force as input should generate sufficient charge to power miniature electronic components. We describe this in the next section.

Exercise 5.3

A quartz cantilever is 100 μm long, 10 μm wide, and 1 μm thick. It is anchored at one end. Consider that the relative permittivity of SiO_2 is 3.9 and of free space it is 8.85×10^{-12} F m^{-1} and also that the cantilever is metallized completely on its top and bottom surfaces. Calculate the capacitance of the cantilever.

Solution

The structure is similar to that of a parallel plate capacitor. Therefore, its capacitance C is given by: $C = \epsilon_r \cdot \epsilon_0 \cdot (A/d)$ where A = area of the metallized surface and d = the thickness of the cantilever. Therefore, C of the cantilever $= (3.9 \times 8.85 \times 10^{-12}) \times (10^3 \times 10^{-12}) \times (1/10^{-6}) = 34.52 \times 10^3 \times 10^{-18} \approx 3.45 \times 10^{-14}$ F.

Exercise 5.4

Now assume that 100 V is applied between the top and bottom electrodes of the SiO_2 cantilever of the previous example. Consider that the charge coefficient, d, of quartz is 2·3 pm V^{-1} and its electromechanical coupling coefficient, k, is 0.12. Calculate the following parameters: polarization, P, beam deflection, L, and the figure-of-merit of the device. Comment on the usefulness of quartz cantilever for energy harvesting.

Solution

The polarization, P, of the cantilever = charge/area = $CV/(lw) = (4 \times 10^{-14} \times 100)/(10^3 \times 10^{-12}) = 4 \times 10^{-3}$ C m^{-2} = 4 mC m^{-2}. By definition, the piezoelectric charge coefficient, d, is given by $d = L/V$ (from Eq. (5.37)) where L = deflection and V = voltage applied. Therefore,

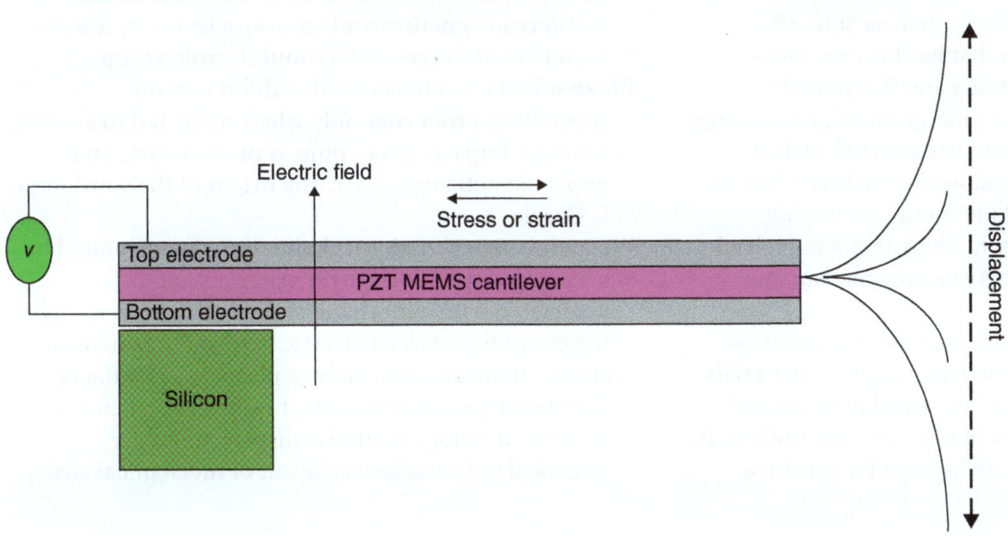

Figure 5.21 Configuration of a micromachined PZT actuator.

$L = d{\cdot}V = 0.23$ nm. The electromechanical coupling coefficient $k = 0.12$. By definition, the figure of merit is k^2 (see Eq. (5.41)). Therefore, the efficiency of the SiO_2 cantilever device is just 0·0144 (=1.4%). This means that the energy conversion capacity of a quartz cantilever is very small. In comparison, k^2 for barium titanate ($k = 0.35$) is ≈0·13 (=13%) and for PZT ($k = 0{\cdot}72$) it is ≈ 0·52 (52%). It is now obvious why PZT is the material of choice for many piezoelectric applications including for energy harvesting.

Glossary

Actuator An actuator is the generic term for any device that is capable of producing and controlling mechanical motion. They come in a variety of different types.

Cantilever A mechanical structure whose aspect ratio (length/width) is high and is clamped on one end enabling the other end to bend. Some cantilevers can be clamped on both ends with the bending occurring in the middle.

Doppler effect It states that when two moving objects are approaching each other, the monitoring device will record an increase in frequency, whereas a decrease in frequency is noticed when they are moving away from each other. This effect is of profound importance to communication and navigation technologies. The effect is also called the Doppler shift. It was discovered in 1842 and is named after the Austrian Physicist Christian Doppler.

Electrostriction When an electric field is applied to materials, a very small change in their dimensions occurs. This effect is different from the converse piezoelectric effect. The electrostriction effect is a nonlinear quadratic effect, whereas the converse piezoelectric effect is a linear effect. This effect manifests itself as a butterfly loop for ferroelectrics. Though the effect is small yet suitable for designing precise sensors and actuator. It is also used for navigation using sound waves such as SONARs.

Energy harvesting A term that has become more prevalent in the last few years and has gained importance because of the emergence of green energy. Any device that can be used to extract electrical energy from a nonelectrical source such as a force or impact is said to be good for energy harvesting.

Magnetoelectronic The technology that has evolved because of the coupling between magnetism and ferroelectricity.

Magnetostriction It is analogous to electrostriction but can be observed only in ferromagnetic materials. These materials undergo a very small dimensional change that is of the order of few parts per million. It is a quadratic effect and can be used for sensitive sensors and actuators. Also, it can be used for conversion of energy from magnetic to kinetic and vice versa.

MEMS This is the abbreviation for microelectromechanical system. This technology is rather of recent origin and is based on micromachined structures. The MEMS devices are both electrical and mechanical simultaneously and as such they impact on sensor and detector technology.

Micromachining It is a term reserved for producing highly intricate and miniature three-dimensional structures that are the basis for the MEMS technology. Now there are many different specialized tools available to scientists and engineers for micromachining a material for production of MEMS structures.

Multiferroics A group of nonlinear oxide materials in which at least two-order parameters couple. One example being coupling between ferromagnetism and ferroelectricity; and the other ferromagnetism and ferroelasticity. Ferromagnetism, ferroelectricity, and ferroelasticity are the principal members of the ferroic group; each of them have one-order parameter that is spontaneous magnetic moment in ferromagnetic (also ferrite) materials, spontaneous polarization in ferroelectric materials, and spontaneous strain in ferroelastic materials. Antiferromagnetism and antiferroelectricity are the members of the secondary multiferroic group.

Piezo effects Some materials exhibit certain specialized properties only when subjected to a stress, force, or impact. They come as piezoelectric and piezomagnetic materials. The origin of the word *piezo* is Greek.

Piezoelectricity A physical phenomenon exhibited by certain groups of dielectric materials and it is associated with a mechanical force. Piezoelectricity is the coupling between electric field and mechanical stress. There are two types of piezoelectric effects. The direct piezoelectric effect originates when a material develops electrical displacement (or electrical polarization) as result of mechanical stress.

On the other hand, when an electric field gives rise to a mechanical strain (shape distortion or deformation), then this is called the converse piezoelectric effect.

SAW devices It is the abbreviation for the *surface-acoustic-wave* devices in which an electrical signal is transformed as acoustic waves that travel from one end to the other end of the device structure. At the other end, it is converted back as an electrical output. These are very important electronic devices and are used in communications, signal processing, delay lines, and in many other applications.

Stress It is defined as the mechanical force per unit area and has the unit of (N m^{-2}).

Strain It is defined as the change in length per unit length $\left(\frac{\Delta L}{L}\right)$. It is obviously a unitless quantity.

Transducer A device that is used to transform one form of energy into another. Piezoelectric transducers can convert a mechanical force into electrical voltage (polarization) or can induce displacement (or other forms of strain) in a material by the application of an electrical potential.

Problems

5.1 With the help of the modified Heckmann diagram identify the piezoelectric, pyroelectric and thermoelectric effect in a dielectric material and discuss the agent–property relationships giving rise to each of these three physical phenomena.

5.2 Represent the stress (X) and strain (x) and the magnetic permeability (μ) in tensor format.

5.3 Distinguish between the direct and converse piezoelectric effect. Give two examples of their respective applications.

5.4 Piezoelectric coefficients are defined as piezoelectric voltage coefficient (g) and piezoelectric charge coefficient (d). Define these parameters and discuss at least one application based on each of the two coefficients.

5.5 Explain how the subscripts are assigned to d coefficient for a single crystal? Can you use the same concept for writing d coefficient for a ceramic sample?

5.6 What is a transducer? Explain how a piezoelectric transducer operates and what is the maximum conversion efficiency we can expect from such a transducer?

5.7 Consider a piezoelectric spark generator. Its electric cylinder is 15 mm long with a diameter of 5 mm. The piezoelectric material used for fabricating this device has a dielectric constant of 2000 and the piezoelectric coefficient $d = 500 \times 10^{-12}$ m V^{-1}. Assume that its breakdown voltage is 5 kV. Calculate the force that is required to create the spark.

5.8 There are many unique devices and applications based on piezoelectric effect. Describe in as much detail as you can the operation mechanisms of a SAW device and of a piezoelectric acoustic amplifier. Give a few examples of the applications of these two devices.

5.9 What is a MEMS actuator? Describe the configuration of this device. How are these devices used?

5.10 Consider a cantilever of the following dimensions in microns: length = 10, width = 2, and thickness = 0.01. It is completely covered with gold films on top and bottom. The material of the cantilever is quartz having the permittivity of 4. Calculate the charge that can accumulate when a field of 10 V m^{-1} is applied between the electrodes.

5.11 Using the conditions given in Problem 5.9, calculate the following parameters: polarization, P, beam deflection, L, and the figure-of-merit of the device. Assume that the charge coefficient (d) of quartz is 2.3 pm V^{-1} and the electromechanical coupling coefficient K is 0.15.

5.12 What is meant by the term *Ferroics?* Discuss their relevance in the emerging field of magnetoelectric devices.

5.13 Explain how a magnetoelectric device operates and how it is likely to impact the computer memory technology of the future.

References

1 Heckmann, G. (1925). Die Gittertheorie der festen Körper. In: *Ergebnisse der Exacten Naturwissenschaften*, vol. 4, 140. Germany: Springer.

2 Nye, J.F. (1985). *Physical Properties of Crystals: Their Representation by Tensors and Matrices*. New York: Oxford University Press.

3 Newnham, R.E. (2013). *Properties of Materials: Anisotropy, Symmetry and Structures*. Oxford University Press.

4 Spaldin, N.A. and Fiebig, M. (2005). The renaissance of magnetoelectric multiferroics. *Science* 309: 391–392.

5 Schmid, H. (1994). Multi-ferroic magnetoelectrics. *Ferroelectrics* 162: 317–338.

6 Ramesh, R. and Spaldin, N.A. (2007). Multiferroics: progress and prospects of thin films. *Nat. Mater.* 6: 21–29.

7 Hur, N., Park, S., Sharma, P.A. et al. (2004). Electric polarization reversal and memory in a multiferroic material induced by magnetic fields. *Nature* 429: 392–395.

8 Cheong, S.-W. and Mostovoy, M. (2007). Multiferroics: a magnetic twist for ferroelectricity. *Nat. Mater.* 6: 13–20.

9 Ortega, N., Kumar, A., Scotti, J.F., and Katiyar, R.S. (2015). Multifunctional magnetoelectric materials for device applications. *J. Phys. Condens. Mater.* 27: 1–23.

10 Fiebig, M. (2005). Topical review: review of the magnetoelectric effect. *J. Phys. D: Appl. Phys.* 38: R123–R152.

11 Nan, C.W. (1994). Magnetoelectric effect in composites of piezoelectric and piezomagnetic phases. *Phys. Rev. B* 50: 6082–6088.

12 Bibes, M. and Barthélémy, A. (2008). Toward a magnetoelelctric memory. *Nat. Mater.* 7: 425–426.

13 www.aerotech.com/productcatalog.

14 Allen, H. Meitzler, Susan, Trolier-McKinstry, John, Ballato et al. (2003). IEEE standard definitions of terms associated with ferroelectric and related materials (draft 16). *IEEE Trans. Ultrason. Ferroelectr. Freq. Control* 50 (12): 1–32.

15 Kao, K.C. (2004). *Dielectric Phenomena in Solids*. Elsevier Academic Press.

16 Zhong, J. (2008). Piezoelectric and ferroelectric properties of CSD-prepared PZT films and their applications. Dissertation, The University of Alabama, Tuscaloosa, AL, USA.

17 Zhong, J., Kotru, S., Han, H., and Pandey, R.K. (2007). Double-sided PZT films on platinized Si: film properties, structural and piezoelectric properties. *Integr. Ferroelectr.* 92: 123–134.

6

Coupled Nonlinear Effects in Electroceramics

CHAPTER MENU

An equation means nothing to me unless it expresses a thought of God.

Srinivas Ramanujan

In the simplest array of digits, [Ramanujan] detected wonderful properties: congruence, symmetries and relationships which had escaped the notice of even the outstandingly gifted theoreticians.

James Roy Newman, in *The World of Mathematics* (1956)

6.1 Introduction

In this chapter, we will study about ferroelectricity, pyroelectricity, and antiferroelectricity. All these three classes of materials belong to the group of polar and nonlinear dielectrics, and their origin lies in the crystal symmetry that we studied in Chapter 5. Furthermore, we found in Chapter 6 that when external agents such as electrical field (E), temperature (T), and mechanical stress (σ) are applied to a dielectric material, piezoelectricity and pyroelectricity evolve. Both are strongly coupled physical phenomena with large number of novel and unique applications that impact our daily lives positively. Ferroelectricity is a subset of pyroelectricity and is a class by itself. Along with ferrites (magnetic oxides), it forms a very interesting class of materials we called multiferroics and dealt with in Chapter 5. Multiferroics, as we know, hold great promise for advancing the data storage media and sensors in near future. In a ferroelectric material, the nonlinearity of polarization with respect to the applied electric field manifests itself as a hysteresis loop similar to the hysteresis loop exhibited by a ferromagnetic material. That is why ferroelectricity is considered as the electrical equivalence of ferromagnetism. Ferroelectric materials are also good examples of multifunctional materials as can be seen from Figure 6.1.

We notice in this figure that a ferroelectric material can also be a semiconductor. In fact, they come as both an n-type and p-type extrinsic wide bandgap semiconductors with bandgap greater than 2.00 eV. They are invariably good rad-hard materials (radiation-hardened materials) and have been found to be suitable for space electronics. Also, many piezoelectrics and pyroelectrics are found to be semiconductors as well.

We know from the discussions advanced in Chapter 5 while covering the 32 point groups that are based on symmetry elements, we can identify 20 noncentrosymmetric groups that give rise to piezoelectricity. Out of these 20 piezoelectric unit cells, 10 are concurrently pyroelectrics as well. That leaves 10 point groups that are piezoelectric, but not pyroelectric. Ferroelectrics share the point groups of pyroelectrics as shown in Table 6.1 (identical

Fundamentals of Electroceramics: Materials, Devices, and Applications, First Edition. R. K. Pandey.

Companion website: www.wiley.com/go/Pandey/Fundamentals_Electroceramics

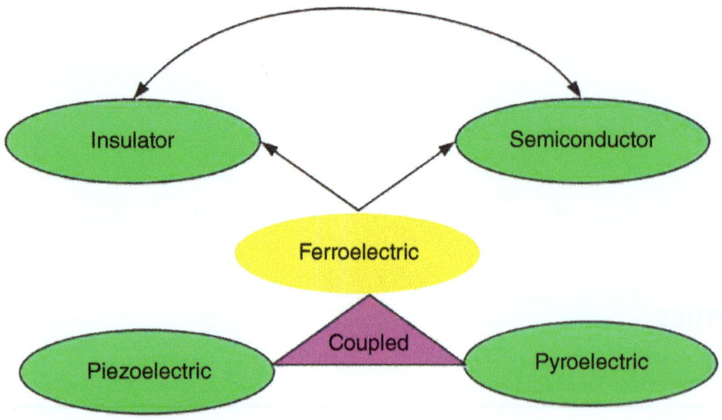

Figure 6.1 Multifunctional nature of ferroelectrics.

Table 6.1 Unit cells with their point groups (10) for pyroelectricity and ferroelectric single crystals.

Unit cell	Triclinic	Tetragonal	Hexagonal	Monoclinic	Orthorhombic	Trigonal
Point groups	1	4	6	2		3
		$4mm$	$6mm$	m	$mm2$	$3m$

to Table 4.7). All ferroelectrics are also piezoelectrics, but not all piezoelectrics are ferroelectrics.

The coupled nature of piezoelectricity, pyroelectricity, and ferroelectricity makes it possible to envision and develop a vast array of novel devices for different types of applications. We covered the applications based on piezoelectricity in Chapter 6. Here we will cover some of the applications of ferroelectricity and pyroelectricity.

Figure 6.2 shows a brief selection of potential applications of ferroelectric, antiferroelectric, and pyroelectric materials. Each of these three classes of materials offers enormous opportunities to develop novel devices to meet the needs of established and emerging technologies. While both pyroelectrics and antiferroelectrics form the foundation of very useful applications it is ferroelectricity that is dominating the field of data storage, energy

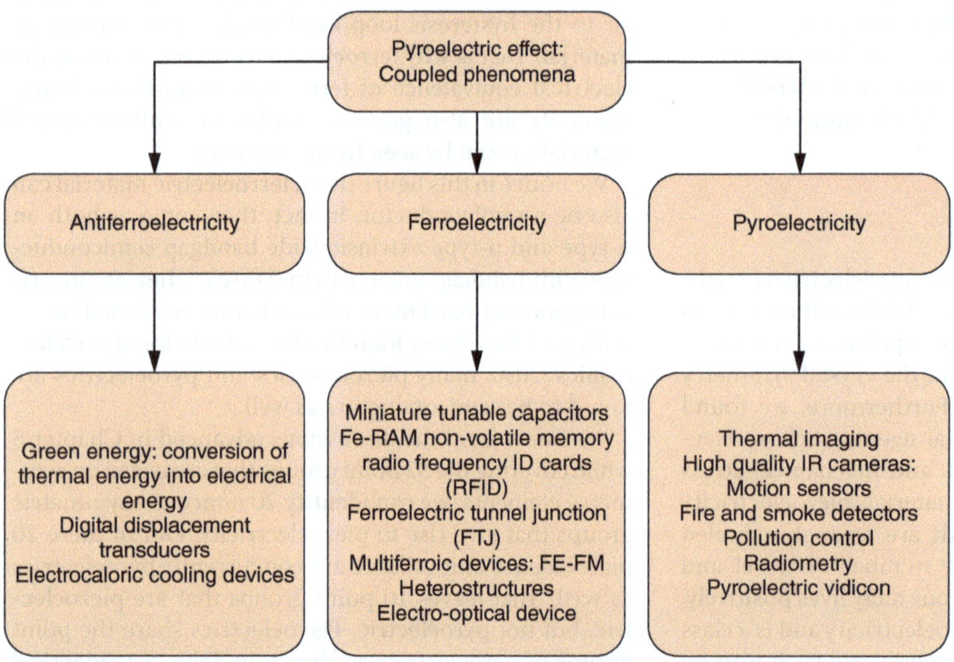

Figure 6.2 Some applications of ferroelectrics, antiferroelectrics, and pyroelectric materials.

storage, and play a vital role in the success of multiferroics. On the one hand, because of their high dielectric constant, ruggedness, and reliability with which the materials can be processed reproducibly in very large volumes, they are the materials of choice for the fabrication of capacitors varying in sizes from miniature, which are used in microelectronics, to large, which is used in power circuits. On the other extreme of the spectrum, they find applications in such modern technologies as electrooptics, nonvolatile memory, microelectromechanical system (MEMS), wireless communication, and emerging nanotechnology. The pyroelectric-based applications are also numerous impacting the technologies of thermal detectors, night vision, security systems, and many others. Antiferroelectrics are emerging as potential contributors to green energy development and electrocaloric cooling devices for many applications but particularly for flat panel display and flexible electronics. The brief background covered here should be sufficient to justify including this topic in our discussions.

6.2 Historical Perspective [1]

The phenomenon now universally known as ferroelectricity got its name more because of its phenomenological similarity with ferromagnetism than due to the underlying physics describing these two phenomena. The name is deceptive in some sense. Ferroelectricity, as the word employs, leads one to assume erroneously that it has something to do with iron (*ferum* in Latin). In reality, it has hardly anything to do with this magnetic metal. In fact, only a handful of materials having iron (or, other prominent members of the ferromagnetic family such as nickel and cobalt) have been reported to exhibit ferroelectricity. However, they are certainly not members of the mainstream ferroelectrics and rarely researched actively. Some examples are cadmium iron niobate, Cd_2FeNbO_6, some members of the barium-fluoride group (examples, barium iron fluoride, $BaFeF_4$; barium cobalt fluoride, $BaCoF_4$; and barium nickel fluoride, $BaNiF_4$), and antimony sulfide iodide type compounds.

Almost all well-known ferroelectric materials are manmade in contrast to leading ferromagnetic materials, which are found abundantly in nature. Classic examples are iron, cobalt, and nickel and their alloys. So far as ferroelectricity in naturally occurring minerals is concerned, some claims have been made that iron sulfide (FeS) and ilmenite ($FeTiO_3$) exhibit ferroelectricity. But this still remains to be confirmed.

Ferroelectricity was discovered by Joseph Valasek in 1921when he observed hysteresis effect between polarization and electric field in potassium sodium tartarate tetrahydrate ($KNaC_4H_4O_6 \cdot 6H_2O$), which is commonly known as Rochelle salt after its discoverer who lived in France in the seventeenth century [2]. From 1920 until 1945, only four or five ferroelectric materials were discovered, and all of them having H_2O group in common. In 1945, barium titanate ($BaTiO_3$) was discovered, and it was the first ferroelectric material which did not contain H_2O in its chemical formula. Ever since $BaTiO_3$ was discovered, it has remained the king of the ferroelectric materials similar to the undisputed status of silicon in microelectronics.

Now ferroelectrics number in hundreds enabling the emergence of novel technologies that exploit various properties commonly found in ferroelectrics. The range of applications of this class of materials is vast as already pointed out in the previous section. Ferroelectricity is a macroscopic phenomenon originating from the long-range interactions of the electric dipoles and the noncentro symmetry of the unit cell of the crystal. In comparison, ferromagnetism is a quantum mechanical phenomenon in which the magnetic state is associated with the two orientations of spins (that is, $\pm\frac{1}{2}$).

The phenomenological similarity between ferromagnetism and ferroelectricity goes beyond the presence of a hysteresis loop in each of these two materials. Like ferromagnetism, ferroelectricity also shows the existence of the Curie point at which it ceases to be ferroelectric and enters the nonpolar phase called the paraelectric state, analogous to paramagnetism. Furthermore, like antiferromagnetic materials, there are also antiferroelectric materials. Both ferromagnetic and ferroelectric materials have domains; in the case of ferromagnetism, they are to be understood with the help of magnetic order, and in the case of ferroelectrics, they are to be interpreted with the help of orientations of the electrical dipoles which is equivalent of dipole ordering.

6.3 Signature Properties of Ferroelectric Materials

A ferroelectric material is defined as a class of material in which the electric polarization exists even in the absence of an external electric field and it is said to be spontaneous. It is capable of existing in two equivalent states that are totally reversible, making ferroelectrics suitable for many applications.

The other unique properties of a ferroelectric material consist of (i) nonlinear relationship between electric polarization and electric field that manifests itself in the shape of a hysteresis loop. The nonvolatility is the result of such a loop; (ii) ferroelectrics exhibit high values of dielectric constant in the order of 10^3–10^4. They show strong frequency dependence and form the basis for applications in MW and RF ranges; (iii) the properties of a ferroelectric material are strongly temperature-dependent so much so that the value of the spontaneous polarization reduces to zero at the Curie point (also called, ferroelectric transition point) and the inverse of the dielectric constant tends to be infinity; (iv) the Curie point is that characteristic temperature at which a ferroelectric material suffers phase transition. Ferroelectricity dominates below the Curie point (its symbol being T_C), and it enters the regime of so-called paraelectric state in which a ferroelectric material becomes a normal linear dielectric material; and (v) the unit cell of a ferroelectric crystal lacks center of symmetry, which is the necessary crystallographic condition for the existence of polar material such as a ferroelectric material.

6.3.1 Hysteresis Loop: Its Nature and Technical Importance

We have already discussed the importance of a ferroelectric loop with respect to its uniqueness and applications. Let us now interpret a typical hysteresis loop as shown in Figure 6.3 in order to identify its salient parameters and their technical importance. In its virgin state at room temperature, ferroelectric materials exist in their depolarized state meaning thereby that its polarization is zero. Only when an electric field is applied, the material begins to acquire finite values of polarization.

Initially, the polarization increases with respect to increasing electric field, following the path shown as dashed curve in the figure. This curve is commonly

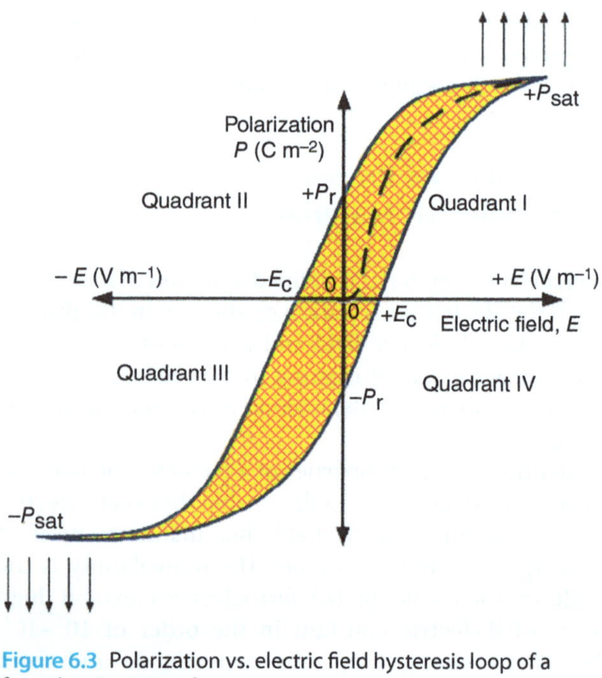

Figure 6.3 Polarization vs. electric field hysteresis loop of a ferroelectric material.

referred to as the virgin curve of the material. Once it reaches its maximum value identified as $+P_{sat}$ point, there is no further increase in its value irrespective of the increasing electric field. That is the state of saturation. The $+P_{sat}$ state is symbolically represented with the help of parallel arrows all pointing in the same upward direction. The message here is that once the maximum value of polarization is reached, each domain within a ferroelectric material is in a fully saturated state such that all available polarization vectors point in the same direction. We will provide the experimental proof of polarization domains later in this chapter.

It should be understood that we are calling the maximum value of polarization that can be achieved as the spontaneous polarization for technical convenience. Strictly speaking though this definition is not exactly correct because theoretically the spontaneous polarization is defined as that polarization that exists in a ferroelectric material at absolute zero (0 K) and at zero electric field. We also notice in the figure that the hysteresis loop has been divided into four quadrants labeled I thorough IV.

In quadrant I, when the field is reduced from its maximum value at P_{sat}, it does not retrace its original dotted path. Instead, it prefers a different path, following a solid nonlinear path that terminates at point, $+P_r$, corresponding to zero electric field. This is what remains of $+P_{sat}$, and therefore, it is called appropriately the rearmament polarization, P_r. The existence of remnant polarization even when there is no electric field present is the uniqueness of a ferroelectric material. The material acquires remnant polarization state only after it has overcome its virgin state.

Usually, the ratio of P_r:P_{sat} < 1. Ferroelectric materials, with this ration close to 1, display square loop, which is of great importance for data storage. On further reduction of the field in the negative direction, it brings us to quadrant II. We find now that the polarization, P, is zero at point $-E_c$. E_c is called the coercivity or coercive field. In quadrant II, it is that field which is needed to bring the remnant polarization to zero, making the material gets technically depolarized. Traversing further in the negative field direction, we enter the III quadrant where the P–E curve assumes a concave profile. Finally, we reach the point of maximum saturation in this quadrant that is identified as $-P_{sat}$ and symbolically represented by arrows pointing downward meaning that here all the polarization vectors have switched by 180° and are antiparallel to the vectors at $+P_{sat}$ state. Here the hysteresis curve has completed its half cycle. Now traversing in the opposite direction of the field, we reach the point $-P_r$. This is once again the value of the remnant polarization having its direction opposite to that of $+P_r$. The two states of polarization are equivalent, and they can switch by 180° going from one state to the other. The presence of the two stable equivalent states of the remnant polarization gives rise to the bistable states of polarization making the ferroelectric memory nonvolatile. Following the curve from $-P_r$ point to $+P_s$, we follow the solid path of quadrant IV reaching the point $+E_c$. This is equivalent to the $-E_c$ point and is appropriately called coercivity or coercive force in the positive field direction. Once again, like the $-E_c$ point, we can understand that the value of the electric field to depolarize the material in the positive side of the electric field is nothing other than the $+E_c$ field. By further increasing the field beyond the IV quadrant, we follow the path in region I and terminate our journey at point $+P_{sat}$. We have now completed the one full cycle and have generated a hysteresis loop. The hysteresis plot will keep on repeating itself no matter how many times we chose to go from $+P_{sat}$ state to the $-P_{sat}$ and in each subsequent full cycle, we will encounter the two states of remnant polarization and the two states of coercivity. The reproducibility of the hysteresis curve is an asset that can be exploited in many different ways from one application to another.

These four states have very important technical significance. We can artificially depolarize the material by the appropriate application of $\pm E_c$, and 180° switching of polarization is achieved simply by going from $+P_r$ state to $-P_r$ state. In a memory cell $+P_r$ state is referred to as "on" state or "write in," and $-P_r$ state as "off" state or "read out." The switching time between the two states of remnant polarization is determined by the method proposed in 1966 by Fatuzo and Merz [3]. Ultrafast switching speed of approximately 400 pS has been reported for PZT thin film. Like barium titanate, PZT is a very good ferroelectric material, and numerous applications are based on it. PZT stands for lead-zirconate-titanate which is a solid solution of lead zirconate and lead titanate.

6.3.2 Temperature Dependence of Ferroelectric Parameters

In Figure 6.1, we identified three sets of bistable parameters of a ferroelectric hysteresis curve. They are $\pm P_s, \pm P_r$, and $\pm E_c$. They all are also strongly temperature-dependent. As the temperature increases, their values begin to become smaller and smaller as the temperature reaches the Curie point. They become zero at and above the Curie point. The way they disappear is of great significance in physics; the transition at and close to the Curie point are defined as first order and second order and their mechanism can be understood with the knowledge of advanced thermodynamics. We will make an effort to understand it by discussing the simplified theory that we would cover later in this chapter. Each ferroelectric material, and there are hundreds of them, has to have the Curie point and as expected, they differ in value from material to material and for different chemical compositions.

The temperature at which the hysteresis loop collapses is also a measure of the Curie point. However, it gives only the approximate value. The exact value is to be determined by careful measurement of the spontaneous polarization as a function of temperature. Above the Curie point, ferroelectricity ceases to exist, and the material goes into the paraelectric state in which the field dependence-induced polarization increases linearly with the field. Once the field is removed, the polarization vanishes. This is the picture we find for linear dielectric or nonpolar dielectric materials. We can consider the Curie temperature as the dividing line between the polar and nonpolar states. Above the Curie point, the domain orientations are random, and as the temperature is lowered below the Curie point, the spontaneous polarization in domains begin to nucleate and their number keeps on increasing as the temperature is lowered. At absolute zero, they attain the highest order allowing the spontaneous polarization to reach its maximum possible value.

6.3.3 Temperature Dependence of Dielectric Constant

Compared to normal nonpolar dialectic material, the relative dielectric constant (ϵ_r) of a ferroelectric material is usually a large number. Value as high as 50 000 has been reported. On the other hand, some ferroelectric materials can be only of the order of a few hundreds. They too display a strong temperature-dependence; first increasing with the increasing temperature and reaching a peak in the vicinity of the Curie temperature. Above this point, it drops off rapidly as we can see from Figure 6.4. It follows the Curie–Weiss law above T_c, which is stated in Eq. (6.1)

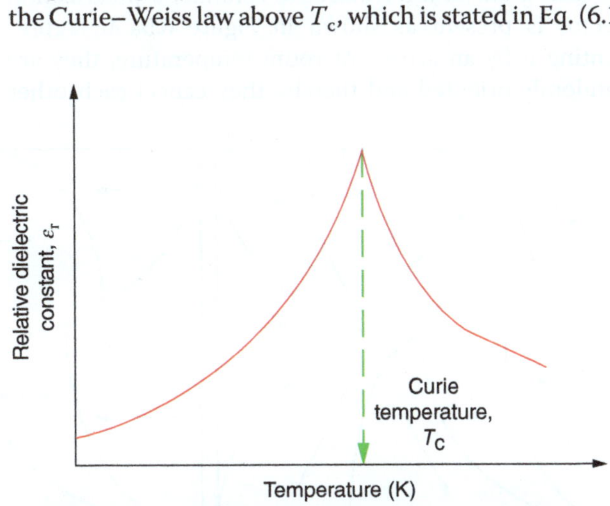

Figure 6.4 Temperature dependence of dielectric constant.

for all ferroelectrics. The parameter C in the equation is called the Curie constant, and it has the same dimension as the temperature.

$$\epsilon_r = \frac{C}{T - T_c} \tag{6.1}$$

According to this law, when $T = T_c$, $\epsilon_r \approx \infty$.

It should be pointed out for the sake of history that Curie–Weiss law was first developed for ferromagnetic materials. It predicts the behavior of magnetic susceptibility in the vicinity of the Curie temperature. The ferroelectric community simply borrowed it from magnetism to validate the behavior of dielectric constant when temperature approaches the Curie point.

6.3.4 Ferroelectric Domains

Ferroelectric crystals consist of a large number of *domains* in which the polarization is oriented in one unique direction. The neighboring domains are separated from each other by *domain walls*. All the polarization vectors point in one single direction and are parallel to each other, resulting in the maximum value of the spontaneous polarization for the bulk of the material. This picture is attainable by a ferroelectric material under two conditions: (i) close to the absolute zero temperature, and (ii) when it is subjected to a strong external electric field. Between the absolute zero and the Curie point, the domain structure goes from fully ordered state (maximum polarization) to fully disordered state (zero polarization).

A polycrystalline material, which is also called ceramic material, consists of large number of grains, and each grain is separated from the other by means of a grain boundary existing between the two neighboring grains. In each grain of a ferroelectric ceramic, a polarization vector is present as shown in Figure 6.5a by representing it by an arrow. At room temperature, they are randomly oriented and thereby they cancel each other resulting in net zero polarization. When an electric field is applied, these individual polarization vectors will follow the direction of the electric field and as a result, they end up being parallel to each other as shown in Figure 6.5b. When such a configuration is present, we should be able to measure the net polarization. We can actually see the ferroelectric domains visually by using a high-quality polarization microscope or by means of a piezoelectrics force microscope (PFM). We will discuss these experimental methods later in this chapter.

6.3.5 Electrets

There is a class of materials known as electrets that are often confused with ferroelectricity. These are also dielectric materials in which quasi-permanent real charges can reside on the surface. They can also find themselves in the bulk of the material, or as frozen-in aligned dipoles in the bulk. Electrets behave like a battery or acts like an electrical counterpart of a permanent magnet. However, they are not ferroelectrics. Ferroelectricity is distinctly different from them in two ways: (i) its polarization is spontaneous, which is retained even at zero electric field, and (ii) it is controlled by the crystal's symmetry. An exhaustive treatment of this topic is given in Ref. [4]. Interested readers may wish to refer to this excellent book.

6.3.6 Relaxor Ferroelectrics

Relaxor ferroelectrics are a special class of ferroelectric materials. They were discovered in 1950, and because of their complex chemical compositions and physical properties no unified theory has emerged so far to explain the mechanism involved in their origin. They are primarily found in perovskite group (ABO_3) which is also the leading group for a large number of conventional ferroelectrics such as barium titanate, $BaTiO_3$. The prominent members of relaxors are lead magnesium niobate (PMN),

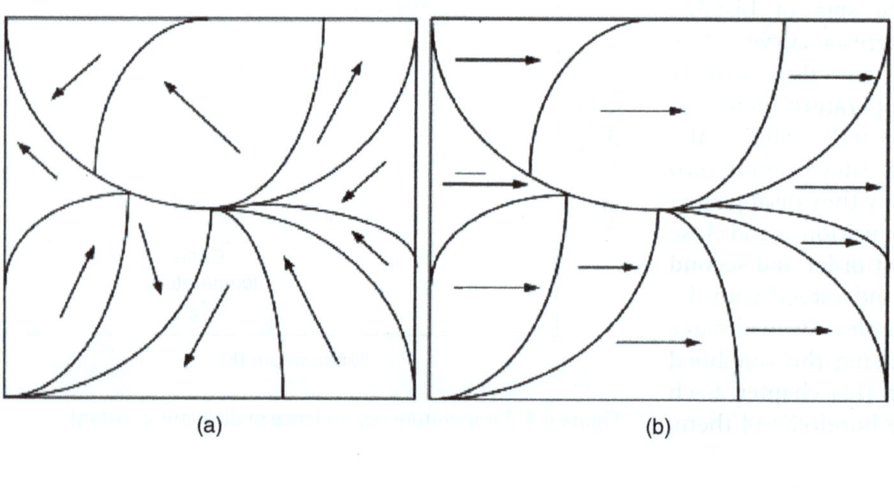

Figure 6.5 (a) Randomly oriented domains at room temperature and (b) oriented domains in a polycrystalline material.

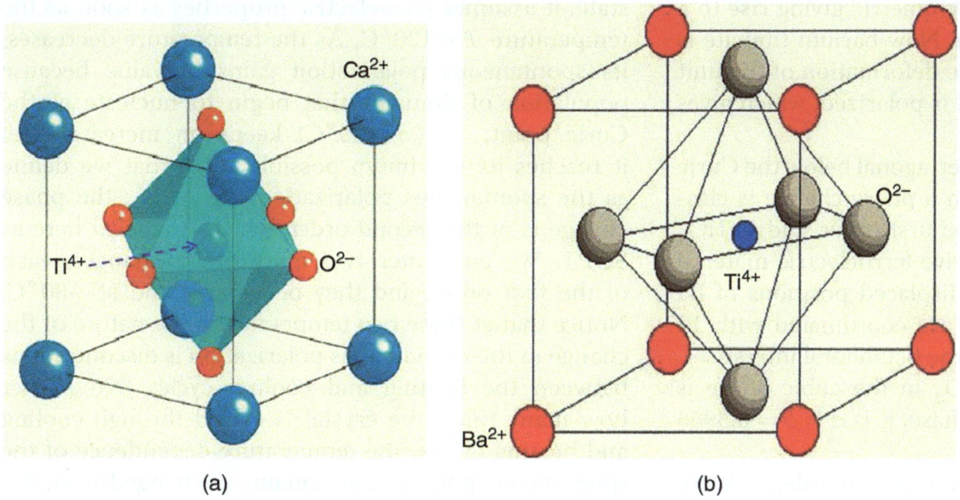

Figure 6.6 Cubic unit cells of (a) $CaTiO_3$ at room temperature and (b) $BaTiO_3$ above its Curie point at 120 °C.

lead magnesium niobate, lead titanate (PMN-PT), and lead lanthanum zirconate titanate (PLZT). Both PMN and PLZT are studied extensively and many applications identified.

The relaxors exhibit high electrostriction and strong frequency dependence of dielectric constant. Their frequency dispersion behavior can vary over a wide range of frequencies. Because of these two interesting properties of interest to technology, they are actively studied as new compositions are emerging. One of the signature properties of a relaxor ferroelectric is that the transition from perovskite to cubic phase is both temperature- and frequency-dependent. In PMN, with the chemical formulation of $Pb_3Mg_3Nb_2O_9$, one observes the presence of double peak in the temperature-dependence of dielectric constant that shifts both in its amplitude and location with respect to frequency. The same behavior is naturally also found for the temperature and frequency dependence of the loss tangent. Like ferroelectrics, relaxors also display hysteresis loops, but these are narrow loops consisting of relatively smaller area compared to ferroelectric loops.

6.4 Perovskite and Tungsten Bronze Structures

6.4.1 Perovskite Structure

From Table 6.1, we see that ferroelectricity (as well as pyroelectricity) can exist in all fundamental unit cells except the cubic. The only condition that must be met is the condition of noncentrosymmetry of the unit cell. There are specialized unit cells in which most of the leading ferroelectric materials are found; one is the perovskite structure and the other tungsten bronze (TB). We will study both these structures in this section. First, let us examine the two examples of perovskite structures shown in Figure 6.4.

Figure 6.6a is the unit cell of calcium titanate, $CaTiO_3$, which is the pro-type of the perovskite structure. Here Ca^{2+} ions occupy the corner of the cube, Ti^{4+} is at the intersection of the body diagonals site, and the O^{2-} ions occupy the intersection of the face diagonals that is also called the face-centered position. Its lattice constant is $a_0 \approx 0.379$ nm. As we can see, the net electrostatic charge is zero which is the necessary condition for the stability of the unit cell. We can visualize the $CaTiO_3$ unit cell as the one produced by nature as a combination of body-centered cubic and fact-centered cubic structures. The resulting unit cell then is perovskite. It is a fascinating unit cell in which many interesting physical phenomena are found. Besides ferroelectricity and its coupled phenomena also, the high temperature superconductivity was discovered in 1986 in a material with perovskite structure. After a close examination of the $CaTiO_3$ structure, we can easily determine that the body diagonal Ti^{4+} atom is centrosymmetric. Because of this reason, $CaTiO_3$ is not a ferroelectric material.

The $BaTiO_3$ unit cell above the Curie point is shown in Figure 6.6b. It is identical to the structure of $CaTiO_3$ except that the Ba^{2+} ions occupy the sites previously occupied by Ca^{2+}. The sites of the other two ions, Ti and O_2, are the same as in $CaTiO_3$. As we know that in the cubic state, $BaTiO_3$ is not ferroelectric. Below this temperature, the structure is tetragonal and the unit cell gets slightly deformed as a result of which both the Ba and Ti ions get displaced from their original sites with respect to the O^{2-} ion. In the process, then it is possible that the O^{2-} ions may move up or down. As a result, the structure becomes noncentrosymmetric giving rise to a structure-induced polarization. Now barium titanate is no more neutral, because of the deformation of the unit cell from cubic to tetragonal, it is polarized, which gives rise to ferroelectricity.

The structure changes from tetragonal below the Curie point to cubic above this. Such a phase change is classified thermodynamically of the first order, and $BaTiO_3$ itself is referred to be a displacive ferroelectric material because its origin lies in the displaced positions of Ba and Ti ions. The barium ions are coordinated with 12 oxygen ions and the Ti ions in the octahedral interstices. The lattice constants of $BaTiO_3$ in the cubic phase is 0.4 nm and in the tetragonal phase, it is $a = b = 0.3966$ and $c = 0.4035$ nm.

In Figure 6.7, we present the temperature-dependence of the spontaneous polarization of single crystal barium titanate. It was first reported by Merz in 1949 [5]. In this figure, four distinct regions have been identified, which are trigonal in the temperature range of −80 °C and below, above it is the range of orthorhombic, which exists between −80 and 0 °C, then comes the tetragonal range between 0 and +120 °C. Finally, comes the cubic structure in which barium titanate exists above its Curie point, T_c. Barium titanate is ferroelectric in the temperature (T) of $-80 < T < 120$ °C. We infer from this that the spontaneous polarization, P_s, is strongly temperature-dependent and as the temperature changes, so does the crystal structure of barium titanate. Structure changes occur at a specific temperature such as at 120 °C, the crystal switches from cubic to tetragonal structure. While in the cubic phase, it is in paraelectric state, it assumes ferroelectric properties as soon as the temperature $T < 120$ °C. As the temperature decreases, its spontaneous polarization gains in value because population of domains that begin to nucleate at the Curie point, T_c (≈+120 °C) keeps on increase until it reaches its maximum possible value that we define as the spontaneous polarization, P_s. At T_c, the phase change is of the second order, which is labeled here as SOPT. We encounter two more phase changes, each of the first order, and they occur at 0 and at −80 °C. Notice that at these two temperatures, the nature of the change in the spontaneous polarization is discontinuous between the heating and cooling cycles. No matter how many times the crystal is cycled through cooling and heating cycles, the temperature dependence of the spontaneous polarization remains unchanged from the picture depicted in Figure 6.7. The phase change at 0 °C takes place when the crystal structure changes from tetragonal to orthorhombic; and again at −80 °C for the change in crystal structure from orthorhombic to trigonal. These phase changes are labeled as first order phase transition (FOPT) in Figure 6.7. The spontaneous polarization for barium titanate crystal at room temperature is 26 μC cm^{-2}.

Like most of the physical properties of a crystal, the spontaneous polarization is anisotropic. It is directional dependent and differs in magnitude from one crystal axis to the other. For barium titanate in tetragonal phase, it resides along the c-axis. The value of the spontaneous polarization along the c-axis is the largest compared to the values found for axes corresponding to either a- or b-axis. The c-axis of the tetragonal structure is the

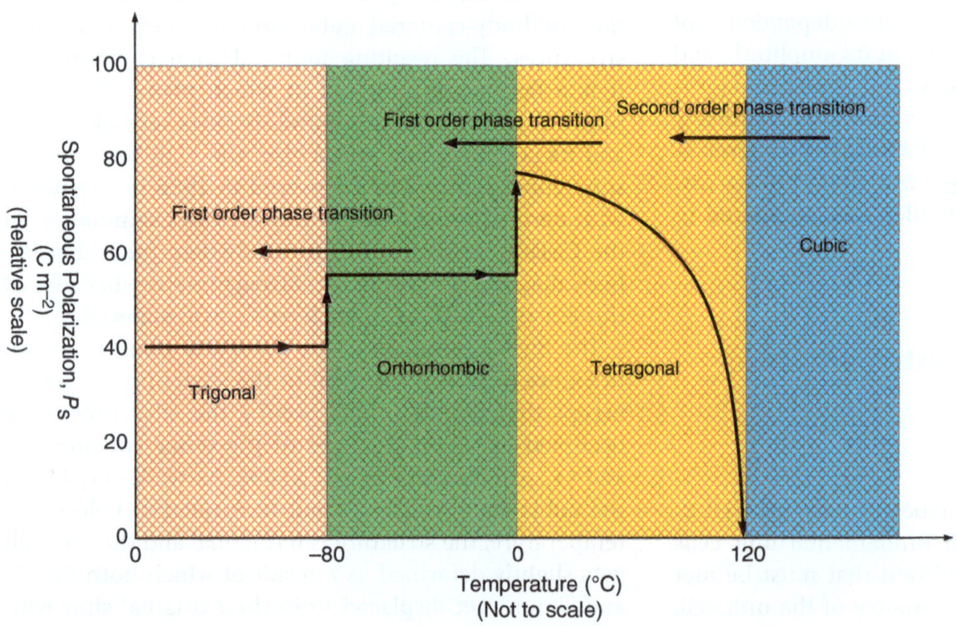

Figure 6.7 Spontaneous polarization as a function of temperature for single crystal barium titanate.

Table 6.2 Crystal structure dependence of polar axis in barium titanate single crystal.

Crystal structure	Temperature, T (°C)	Polar axis	Spontaneous polarization, P_s
Trigonal	$T < -80$	[001]	$P_s > 0$
Orthorhombic	$-80 < T < 0$	[011]	$P_s > 0$
Tetragonal	$0 < T < +120$	[111]	$P_s > 0$
Cubic	$T > +120$	None	$P_s \approx 0$

polar axis of barium titanate. Polar axes corresponding to each ferroelectric phase for barium titanate is given in Table 6.2. Notice that how the unit cell changes from trigonal to orthorhombic to tetragonal to cubic as the temperature increases from its low values to high value is pointing to the thermodynamic nature of phase changes.

The relative dielectric constant, ϵ_r, barium titanate also shows strong temperature-dependence, and it is presented in Figure 6.8. Here too, we find that it goes through major changes at the three Curie points identified above. It shows strong discontinuity at these temperatures, where the crystal goes through different phases. The maximum values of ϵ_r at each of the three Curie points manifest themselves as well-defined peaks that increase in amplitude as the crystal transforms from trigonal to cubic structures. The anisotropy of the dielectric constant is also well pronounced; its value is substantially greater along the a-axis than along the polar c-axis. At $T_c \approx 120\,°C$, it reaches a value of 10 000. This makes barium titanate a very attractive material for capacitors, and in fact, it dominates the tunable capacitors market. We will discuss this application in Chapter 10 while discussing the green energy.

The majority of the well-known ferroelectric materials synthesize in perovskite structure. Some examples of technologically important ferroelectric materials with this structure are potassium-niobate (KN), potassium-tantalate-niobate (KTN), PZT, etc. It is interesting to note that besides ferroelectrics, many other electronic materials of great scientific and technological importance can also crystallize in this structure. We have already discussed high-temperature superconductivity in Chapter 1. Other important materials crystallizing in perovskite structure are the colossal magneto-resistive materials of the type La-Ca-Cu-oxide and La-Sr-Cu-oxides. These recent discoveries have made the perovskite group of materials a subject of intensive research in the hope of discovering some novel phenomenon with the prospect of impacting next generation of electronic sensors and many other devices.

Exercise 6.1

Estimate the percentage change in volume when barium titanate goes from ferroelectric phase to nonferroelectric phase at the Curie point. Comment on the result.

Solution

In the ferroelectric phase, it is tetragonal with the lattice constant of $a = b = a = b = 0.3966$ and $c = 0.4035$ nm. And in the cubic phase, it is $a_0 = 0.4$ nm. We calculate that $(\text{Volume})_{\text{cubic}}/(\text{Volume})_{\text{tetragonal}} \approx 1016$. That means that barium titanate crystal expands by approximately 1.06% while going from the tetragonal phase to cubic phase. Though the percentage change in volume is small, yet its consequences are not. Not only that, drastic changes in the physical properties (P_s and ϵ_r) occur, and the very nature of the phase change manifests itself in a remarkable way, which we will appreciate once we have studied the mean field theory of ferroelectricity.

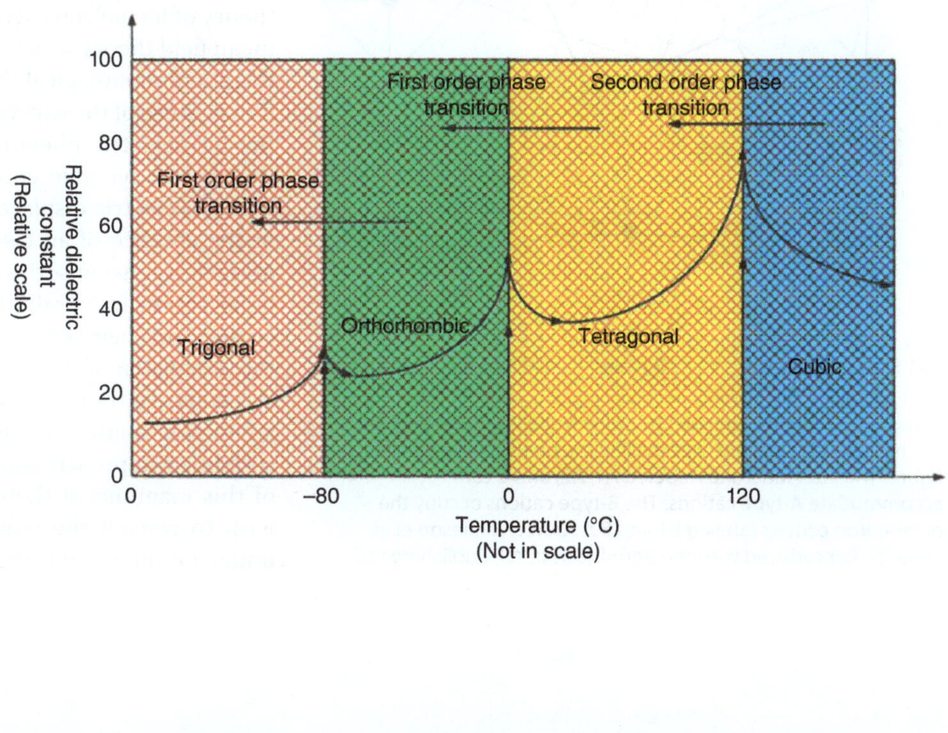

Figure 6.8 Dielectric constant as function of temperature for single crystal barium titanate.

6.4.2 Tungsten Bronze Structure

Another prominent ferroelectric group is found in tungsten-bronze (TB) structure which is essentially related with the perovskite structure and yet has its own unique crystallographic properties. Its basic structure is rather complicated and requires a sound knowledge of crystallography to understand. So far as we are concerned it is sufficient to say that it can be represented by the general formula of AB_2O_6. Both A and B ions can occupy two different sites identified by (A1, A2), and (B1 and B2) sites. C can occupy any of the available A sites. Such a distribution of ions makes the framework of TB structures quite challenging to comprehend and visualize. We show a cross-sectional view of tetragonal TB oxygen octahedral in Figure 6.9 [6].

TB ferroelectrics are mostly found in solid solutions such as between $BaNbO_3$ and $SrNbO_3$. Obviously the barium strontium niobate, Ba-Sr-niobate series can be represented by the general formula of $Ba_xSr_{5-x}Nb_{10}O_{30}$, with $0<x<1$. One of the members with the formula of $Ba_2Sr_3Nb_{10}O_{30}$ is an excellent electro-optical material and is widely used to fabricate optical delay lines. It is commonly referred to as SBN. Other prominent members of the TB family are $Ba_{4+x}Na_{2-2x}Nb_{10}O_{30}$ and $K_{6-x}Li_{4+x}Nb_{10}O_{30}$. Lead (meta) niobate ($PbNb_2O_6$), lead (meta) tantalate ($PbTa_2O_6$), and lead potassium niobate (PKN) having the formula of ($Pb_2KNb_5O_{15}$) are some of the other examples of TB ferroelectrics. PKN possesses large piezoelectric coupling coefficient and as such is an attractive material for many piezoelectric applications. But it is a very difficult material to synthesize as a single crystal because during the cooling process, it develops multiple cracks. Many members of the TB family are also relaxor ferroelectrics and possess interesting properties of interest to technology. With the possibility now for growing highly sophisticated films of complex compositions, it is hoped that many of these relaxors will become materials for the state-of-the-art devices. In Table 6.3, we present a list of important ferroelectrics with their properties and crystal structures.

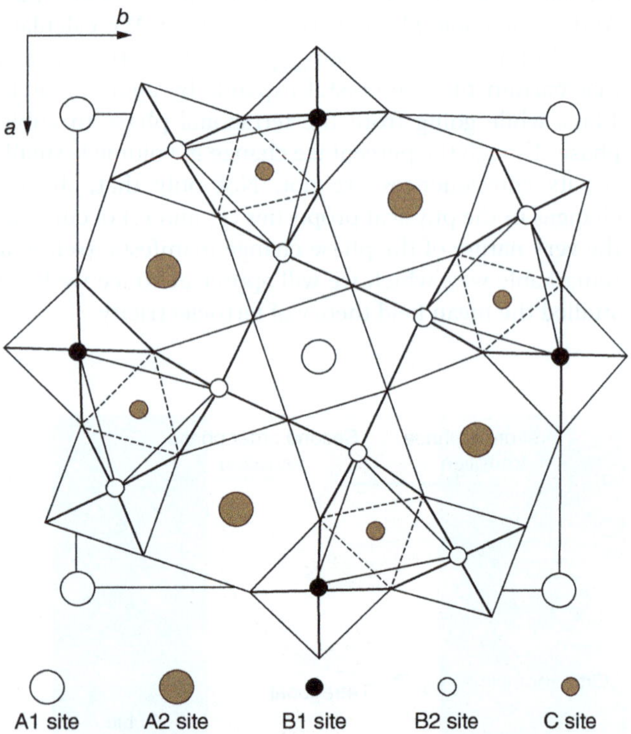

Figure 6.9 Tungsten Bronze Structure. Note: While looking down the tetragonal *c*-axis one would view the picture shown in this figure. The interstitial sites labeled A1, A2, and C can accommodate A-type cations. The B-type cations occupy the octahedron centers labeled B1 and B2. Source: Jamieson et al. 1968 [6]. Reproduced with the permission of AIP Publishing LLC.

6.5 Landau–Ginsberg–Devonshire Mean Field Theory of Ferroelectricity

The Landau–Ginsberg–Devonshire (LGD) mean field theory of ferroelectricity is a profound piece of theoretical work that successfully predicts the temperature-dependence of ferroelectric polarization and dielectric constant. It sets the mathematical conditions under which physics of phase transitions in a ferroelectric material becomes conceivable. First Landau's theory of second-order phase transition was developed in 1936. Based on this theory, Ginsburg in 1945 developed the first phenomenological theory of ferroelectricity. In 1954, Devonshire came forward with a more detailed theory of ferroelectricity that was based on Ginsburg's work. This became the most accepted theory of ferroelectricity and is appropriately called LGD mean field theory which is also referred to in literature as the phenomenological theory of ferroelectricity because it makes use of the experimentally established facts about the ferroelectric phenomenon. In essence, it combines the thermodynamic concept of Gibbs' free energy (G) with the electric displacement (D) and strain in a crystal in the absence of an electric field. The result is a polynomial expansion series consisting of all components of polarization and strain that is possible to exist in a polar crystal and their corresponding coefficients. As a result, the expansion series becomes formidable and requires the skills of a trained theoretical physicist to handle such a complex mathematical piece of work. We will take a simplified approach and try to understand the essence of this magnificent theory. The advanced students may wish to consult the references given here for a deeper understanding of this theory [7, 8].

Table 6.3 Some ferroelectric materials and their properties.

Name (abbreviation)	Chemical formula	Curie temperature T (°C)	Spontaneous polarization, P_s (μC m^{-2})	Crystal structure Above T_C	Below T_C
Barium titanate, BT	$BaTiO_3$	120	26.0	Cubic	Tetragonal
Lead titanate, LT	$PbTiO_3$	490	50.0	Cubic	Tetragonal
Potassium niobate, KN	$KNbO_3$	435	30.0 at 250 °C	Cubic	Tetragonal
Potassium dihydrogen phosphate (KDP)	KH_2PO_4	150	4.8 at −177 °C	Tetragonal	Orthorhombic
Triglycine sulfate (TGS)	$(NH_2CH_2COOH)_3$ H_2SO_4	49	2.8	Monoclinic (CS)	Monoclinic (NCS)
Potassium-sodium tartrate-tetrahydrate (Rochelle salt)	$KNaC_4H_4O_6 \cdot 4H_2O$	24	0.25 at 5 °C	Orthorhombic (CS)	Monoclinic (NCS)
Lead zirconium titanate (PZT)	$Pb(Zr,Ti)O_3$	200–480	Composition dependent	Cubic	Rhombohedral or tetragonal (depends on composition)
Lead niobium zirconium titanate (PNZT)	$Pb_{1.1}Nb_{0.04}Zr_{0.2}Ti_{0.8}O_3$	400	58	Cubic	Tetragonal
Barium strontium niobate (SBN)	$Ba_2Sr_3Nb_{10}O_{30}$	78	34	?	Tungsten bronze

CS, centrosymmetric; NCS, noncentrosymmetric; cubic by definition is always centrosymmetric. All values of spontaneous polarization refer to at room temperature unless identified otherwise.
Source: Partially modified Table 4.2 of Ref. [4].

Let us write an expansion series representing the Gibbs' free energy (G) only as a function of polarization (P), assuming the origin of energy to be zero for the free unpolarized, unrestrained crystal. Equation (6.2) expresses such a relationship.

$$G = \frac{1}{2}\alpha P^2 + \frac{1}{4}\beta P^4 + \frac{1}{6}\gamma P^6 - EP \tag{6.2}$$

In this equation, α, β, and γ are temperature-dependent coefficients and E the electric field. The Gibbs' free energy is defined as the maximum work that is reversible and can be performed under isothermal (constant temperature) and isobaric (constant pressure) by a thermodynamic system. What the concept of potential energy is to a mechanical system, so is the Gibbs' free energy to a thermodynamic system. The concept of G was advanced by Willard Gibbs in 1873; it is a powerful concept and is widely used in solving theoretical problems involving temperature in physics, chemistry, mechanical engineering, chemical engineering, and materials science. Ever since it was conceived, it has remained an unchallenged and unaltered concept of science. In literature, we often encounter other names given to Gibbs' free energy that are Gibbs function or free enthalpy.

So far as we are concerned, we can simply borrow from thermodynamics, the physical importance of Gibbs' free energy without going into the rigorous mathematics to prove it. According to the second law of thermodynamics, Gibbs' free energy (G) tends to be minimum, once the thermodynamic equilibrium is restored in a system. To simplify further, let us consider the expansion series of Eq. (6.2) only up to the sixth order of the polynomial. Applying the concept of minimum state of free energy, we can take the liberty of setting $\frac{dG}{dP} \approx 0$ when our system reaches the thermodynamic equilibrium. Then Eq. (6.2) takes the form of Eq. (6.3).

$$E = \alpha P + \beta P^3 + \gamma P^5 \tag{6.3}$$

Equation (6.3) is nonlinear, and therefore, a plot between E and P gives rise to a hysteresis loop of a ferroelectric material. Let us now assume that in order to get a phase change in a ferroelectric material with respect to temperature, the coefficient α must pass through zero at some temperature. This we know from experiments is the case when $T \approx T_c$. This gives us the liberty of expressing α as in Eq. (6.4).

$$\alpha = \alpha_0(T - T_c) \tag{6.4}$$

Let us now express the electrical susceptibility, χ, as follows:

$$\chi = \frac{P}{E} = \frac{1}{a} \tag{6.5}$$

By combining Eqs. (6.4) and (6.5), we get Eq. (6.6).

$$\chi^{-1} = a_0(T - T_c) \tag{6.6}$$

We have come to the result that the inverse of the electrical susceptibly is linearly proportional to $(T - T_c)$, which is easily experimentally verifiable. We can also express it in a more familiar form as given in Eq. (6.7).

$$\chi = \frac{a_0^{-1}}{T - T_c} \tag{6.7}$$

Equation (6.7) is another formulation of the Curie–Weiss law given in Eq. (6.1). Substituting for α in Eq. (6.3), we come to Eq. (6.8).

$$E = \alpha_0(T - T_c)P + \beta P^3 + \gamma P^5 \tag{6.8}$$

Based on the LGD theory, we can now graphically represent the relationship between the Gibbs' free energy (*G*) and polarization (*P*) with respect to varying temperatures. The behavior is shown in Figure 6.10 and corresponds to the phase transition of the second order.

This is a very famous figure based on the LGD theory, and it has been reproduced hundreds of times ever since the theory came into existence approximately 80 years ago. Of course, over the decades, the original figure has gone through multiple modifications from time to time, but all cosmetic in nature. When $T > T_c$, the material finds itself in the paraelectric state, which is given by a parabola with a single minimum. At $T \approx T_c$, the phase change is in the process and yet not fully completed. That too is shown as a single minimum parabola. The most interesting thing happens when the process of phase transition has been completed, and the thermal equilibrium is established, which is the case for $T < T_c$. The material transforms from a paraelectric (for all practical purposes a linear dielectric material) to a ferroelectric material and thereby acquires the property of the spontaneous polarization, P_s.

Let us now examine the LGD theory with respect to the temperature dependence of the spontaneous polarization (P_s). We know that by definition, when $P \rightarrow P_s$, $E \rightarrow 0$. By substituting these parameters in Eq. (6.8), we can rewrite it as Eq. (6.9).

$$\alpha_0(T - T_c) + \beta P_s^2 + \gamma P_s^4 = 0 \tag{6.9}$$

For the second-order phase transition to be continuous, all the coefficients must be positive. Therefore, by just considering only the first two terms in the above equation, we can arrive at the relationship between the spontaneous polarization and temperature as expressed in Eq. (6.10).

$$P_s \approx \left\{ \frac{\alpha_0}{\beta}(T_c - T) \right\}^{\frac{1}{2}} \tag{6.10}$$

From this, we infer that the temperature-dependence of the spontaneous polarization should obey the $T^{1/2}$-relationship. In fact, experimentally, it is found to be true for most of the ferroelectric materials. The

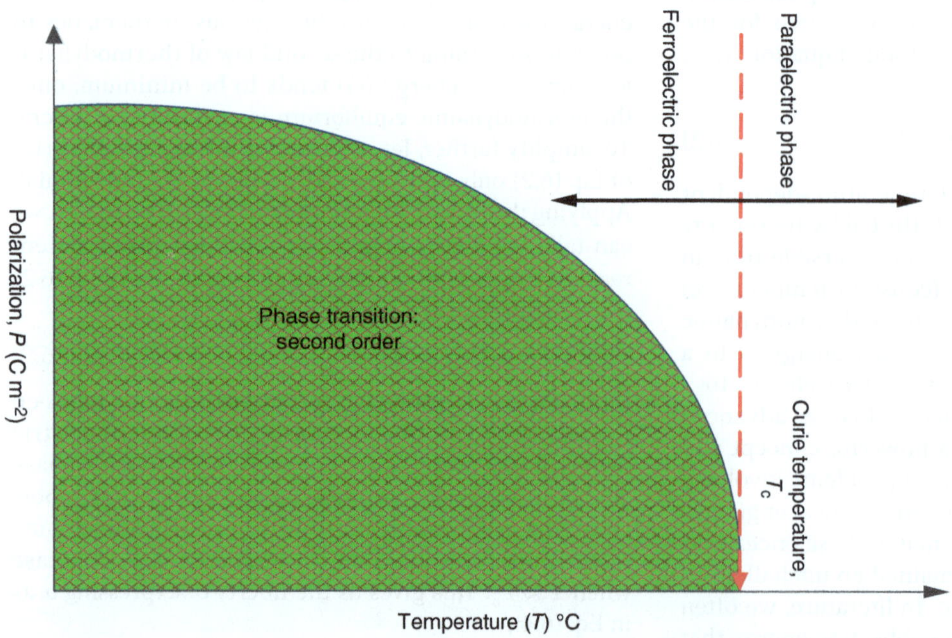

Figure 6.10 Temperature dependence of the spontaneous polarization for the second-order phase transition.

temperature-dependence of the spontaneous polarization for a ferroelectric material undergoing the second-order phase transition is shown in Figure 6.10.

The plot follows a smooth path, and at the Curie point, T_c, the spontaneous polarization abruptly becomes zero. Above this point, a true paraelectric phase comes into existence in which the polarization is only electric field induced and would follow the linear dependence with temperature. It will be there so long as the field is present and will cease to exist once the field is withdrawn. The switch from the ferroelectric state to the paraelectric state at the Curie point is continuous. The inverse of the relative dielectric constant, ϵ_r, as a function of varying temperature (T) for the second-order phase transition is given by Figure 6.11 from which we can infer that it follows accurately the Curie–Weiss law. We can also clearly see here that the switch from the ferroelectric phase to the paraelectric phase is smooth and continuous with absolutely no discontinuity at all.

We know that the relative dielectric constant (ϵ_r) and the electrical susceptibility (χ) are related as given by Eq. (6.11). It is common to substitute one for the other; the susceptibility is conventionally used in theoretical work, whereas the dielectric constant is commonly used in describing experimental results.

$$\epsilon_r \approx (1 + \chi) \tag{6.11}$$

Intuitively, we can guess that if the second-order phase transition occurs continuously, then the first-order phase transition must be represented by a discontinuous process. For this, we need to bring the electrical susceptibility (χ) in discussion. Simply by substituting for E/P with χ^{-1} in Eq. (6.8), we get Eq. (6.12) which would help us in examining the first-order phase change.

$$\chi^{-1} = \alpha_0(T - T_c) + \beta P^2 + \gamma P^4 \tag{6.12}$$

This enables us to consider the three possible scenarios that are when $T > T_c$, $T = T_c$, and $T < T_c$. Each of these will yield different values of the susceptibility, and let us label them as χ_1, χ_2, and χ_3, respectively. For example, when $T > T_c$ such that the thermal gradient is given by dT, then we can rewrite Eq. (6.12) as Eq. (6.13).

$$\chi_1^{-1} = \alpha_0 \, dT + \beta P^2 + \gamma P^4 \tag{6.13}$$

We can easily see that for each of the three conditions set above, the third and fourth terms of Eq. (6.13) will be common. To simply the results, let us set it equal to C. The three susceptibilities then can be expressed as Eq. (6.14).

$$\begin{aligned} \chi_1^{-1} &= \alpha_0 \; dT + C \text{ for } T > T_c \text{ (a)} \\ \chi_2^{-2} &= C \text{ for } T = T_c \qquad \text{(b)} \\ \chi_3^{-1} &= -\alpha_0 \; dT + C \text{ for } T < T_c \text{ (c)} \end{aligned} \tag{6.14}$$

We can therefore expect discontinuities to show up at the Curie point, T_c, in the plot representing the temperature-dependence of the relative dielectric constant for a ferroelectric material going through the first-order phase transition. Obviously the same trend will be found for the dielectric constant. This is shown in Figure 6.12.

The characteristic discontinuity associated with the first-order phase transition of a ferroelectric material can be seen in this figure. In the increasing temperature cycle as predicted ϵ_r^{-1} is decreasing following a linear relationship and at the Curie point, it drops almost in a straight-line and then reverses its course while entering the paraelectric phase in which once again it follows the linear dependence with increasing temperature. It is important to note that at $T = T_c$, it does not become zero. Its zero value can only be attained at $T = \theta$, which is the point where the extrapolation of the straight-line

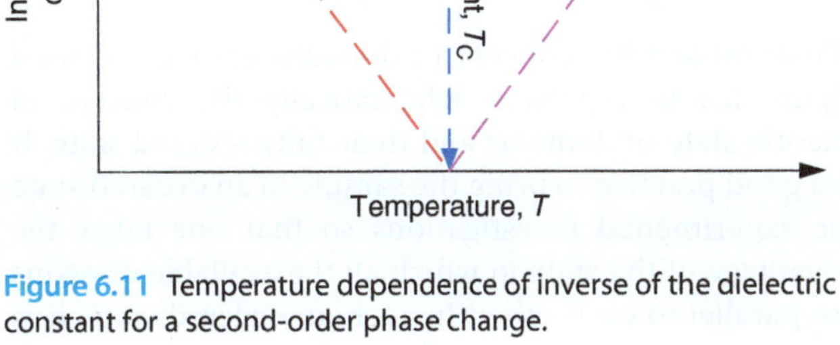

Figure 6.11 Temperature dependence of inverse of the dielectric constant for a second-order phase change.

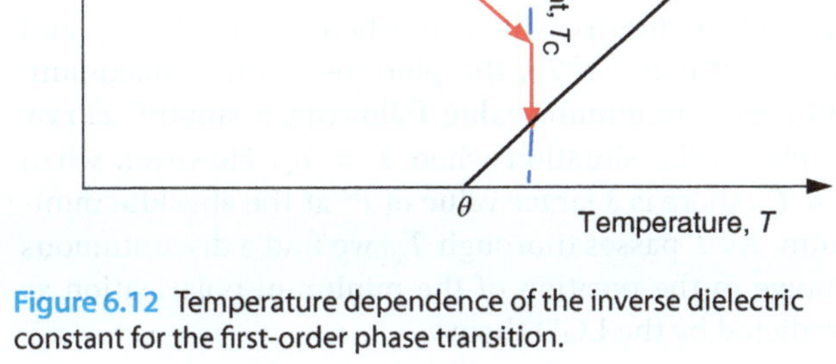

Figure 6.12 Temperature dependence of the inverse dielectric constant for the first-order phase transition.

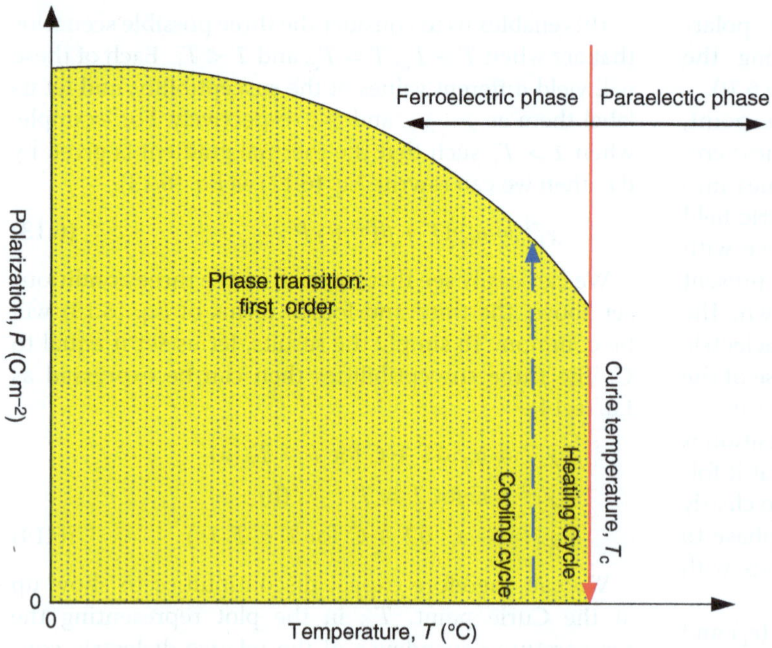

Figure 6.13 Temperature dependence of the spontaneous polarization for the first-order phase transition.

curve of the paraelectric region meets the temperature axis. Let us call θ as the asymptotic Curie point. For the first-order transition, we can express the temperature dependence of ϵ_r^{-1} by modifying the Curie–Weiss law slightly as expressed in Eq. (6.15). C is once again the Curie constant.

$$\epsilon_r^{-1} \approx \frac{C}{T - \theta} \tag{6.15}$$

The temperature dependence of the spontaneous polarization for the first-order transition is given by Figure 6.13. It also shows the discontinuous trend around the Curie point. Here we find that in the heating cycle, the spontaneous polarization reaches its minimum value that is greater than zero at the Curie temperature (T_c). In the cooling cycle, the P vs. T curve does not retrace its original path and merges with the original curve at the asymptotic Curie temperature (θ). During this process, the P–T curve acquires a hysteresis effect.

Now we can ask ourselves how does the polarization dependence on the Gibbs' free energy manifests itself for the first-order phase transition? To answer this question, we make use of Figure 6.14.

Here G vs. P^2 behavior has been plotted for three thermal conditions, which are when $T > T_c$, $T \approx T_c$, and $T < T_c$. When $T > T_c$, the plot goes from a maximum value to a minimum value following a smooth curve. Similar is the situation when $T = T_c$. However, when $T < T_c$, there is a larger value of P^2 at the absolute minimum. As T passes thorough T_c, we find a discontinuous change in the position of the minimum polarization as predicted by the LGD theory.

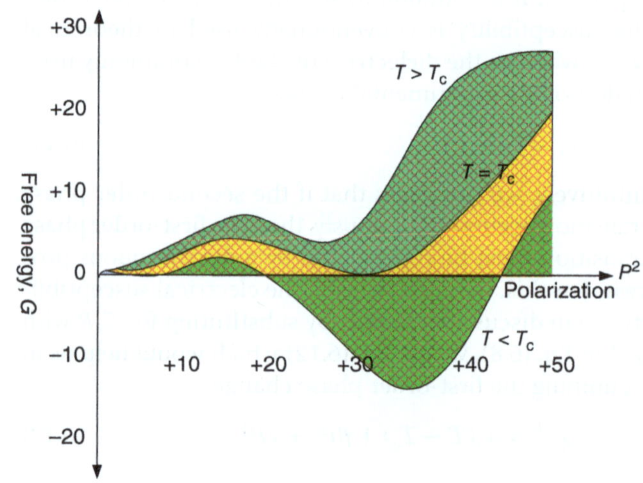

Figure 6.14 Relationship between Gibbs free energy and polarization for the first-order transition when $T > T_c$, $T \approx T_c$ and $T < T_c$ (arbitrary scale).

6.6 Experimental Determination of Ferroelectric Parameters

6.6.1 Poling of Samples for Experiments

We discussed the concept of domains earlier and used Figure 6.5 to represent schematically the concept of chaotic state of domains and their fully ordered state. It is a good practice to bring the sample in an ordered state for experimental investigations so that one takes the advantage of the state in which all the available domains are parallel to each other. In such an ordered state, it is easy to visualize that the polarization will be maximum and will point into one unique direction. How can we achieve such a state at room temperature when as we learned earlier that the thermal vibrations of the lattice destroy the internal order and makes the domains (or individual polarization vectors) disordered? Fortunately, we can overcome the effect of the thermally induced disorder rather simply by poling the sample. All we have to do is to heat up the sample above its Curie point in the presence of a strong electric field and leave it there for some time period and then slowly cool it to the room temperature while applying a strong electric field. The field, E, must be pure DC, and it should be obviously much greater than the coercivity, E_c. However, greater the field, the more successful the poling. What happens after the process is complete? The sample obviously y becomes polarized even at room temperature and now the thermal vibrations would not be able to destroy the order so easily. The domains will be frozen in the ordered state. The process of poling will go through these steps. First, with increasing temperature, the disordered state will keep on increasing reaching its maximum value when all the domains are totally chaotic. Second, in the cooling cycle, the domains will begin to nucleate as soon as the temperature is less than the Curie point. All these domains will experience the presence of the electric field and will automatically orient themselves in the direction of the electric field. With temperature reducing, the population of new domains will increase, and they would naturally follow the orientation of their neighbors. Once the samples comes back to the room temperature all available domains in the sample should be theoretically in their fully saturated state equivalent to the maximum polarization state that one can attain. The poled sample will display the maximum value of the polarization, the dielectric constant, the pyroelectric current, and, of course, the piezoelectric parameters.

Poling a ceramic sample is a simple and straightforward experiment because it is isotropic in nature, and any direction we chose for poling would be its principal direction for polarization. Obviously, we should be wise enough not to pole a single crystal sample arbitrarily. We know that there is a fixed polar axis in a single crystal because of the crystal symmetry. Therefore, one should pole it in the direction of the polar axis to maximize the advantage of poling. Similarly, wisdom demands that the films should be poled in the direction in which we wish to conduct experiments. Poling gives us the advantages that help us in developing good and practical devices.

6.6.2 Polarization vs. Electric Field

Sawyer and Tower were the first to configure a circuit to trace the hysteresis loop exhibited by a ferroelectric material [9]. Since its beginning in 1930, it has remained virtually the only circuit that is widely used for tracing the hysteresis loop of a ferroelectric material and experimentally determining the value of the spontaneous polarization. It has gone through many modifications over the years without any change in the principles of circuit theory behind the original circuit. The Sawyer–Tower circuit is basically a RC circuit, as can be seen from Figure 6.15, with the two resistances labeled $R(1)$ and $R(2)$ in series and the two capacitors identified as $C(X)$ and $C(R)$ also in series. The capacitor $C(X)$ is the ferroelectric sample whose polarization is needed to be measured. The capacitor $C(R)$ is the linear capacitor of known capacitance and serves here as the reference capacitor.

The series configuration of both the resistances and the capacitors act as voltage divider for the AC source

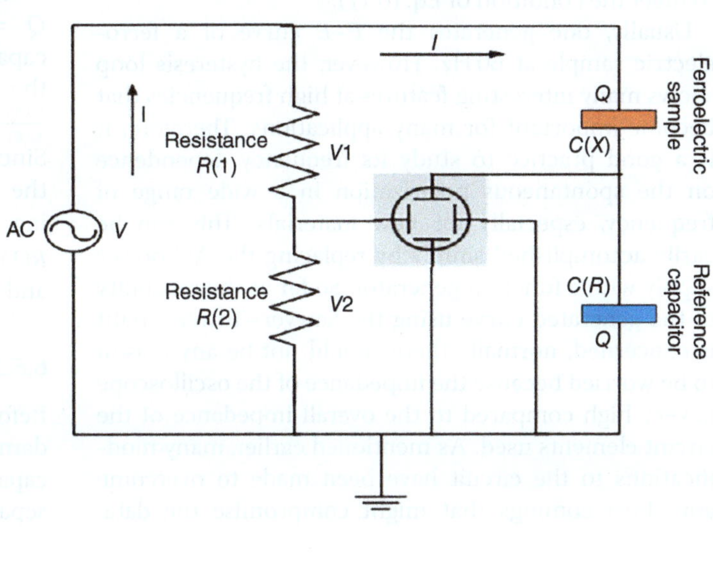

Figure 6.15 Sawyer–Tower circuit for determination of polarization of a dielectric material.

(V). An oscilloscope of high impedance is employed in the circuit to display the electric field-dependent polarization of the ferroelectric sample.

For the determination of polarization (P), as a function of the applied electric field (E) of the ferroelectric capacitor, $C(X)$, all we need to do is to monitor the voltage which is proportional to the polarization at the y-plates of the scope and monitor the voltage applied to the $C(X)$ capacitor at the x-plates. The purpose of the series resistances $R(1)$ and $R(2)$ is to split the source voltage, V, into the two components $V1$ and $V2$. They are used to polarize the two capacitors in a series.

Since we want to polarize the nonlinear capacitor $C(X)$ to its maximum level, we make $R(1) \gg R(2)$. That way we are assured that the sample capacitor $C(X)$ is polarized to its maximum capacity.

Before we begin, the experiment we should be familiar with two issues related the experiment. First, we know that by definition, the polarization, P, is related to the charge, Q, on the capacitor obeying the relationship: $P \approx Q \cdot A^{-1}$, where A is the electrode area on the ferroelectric capacitor. The charge Q (which is equal to $C(R) \cdot V2$) becomes the measure for the polarization of the ferroelectric crystal. Second, we need to fix the ratio between $C(X)$ and $C(R)$ so that voltage drop across $C(X) \gg C(R)$. Since the two capacitors are in a series, the charge, Q, is the same on both. Therefore,

$$Q = C(X)V1 = C(R)V2 \tag{6.16}$$

The ratio between the two capacitors will be governed by Eq. (6.17).

$$\frac{C(R)}{C(X)} = \frac{V1}{V2} > 1 \tag{6.17}$$

Since the spontaneous polarization of ferroelectrics, as can be seen from Table 6.3, varies between $0.26 < P_s < 60\,\mu C\,cm^{-2}$ it is easy to find a linear capacitor to meet the condition of Eq. (6.17).

Usually, one generates the P–E curve of a ferroelectric sample at 60 Hz. However, the hysteresis loop shows many interesting features at high frequencies that become important for many applications. Therefore, it is a good practice to study its frequency dependence on the spontaneous polarization in a wide range of frequency, especially for new materials. This can be easily accomplished simply by replacing the AC power supply with a function generator. So far as the reliability of the generated curve using the Sawyer–Tower circuit is concerned, normally there should not be any reason to be worried because the impedance of the oscilloscope is very high compared to the overall impedance of the circuit elements used. As mentioned earlier, many modifications to the circuit have been made to overcome any short comings that might compromise the data.

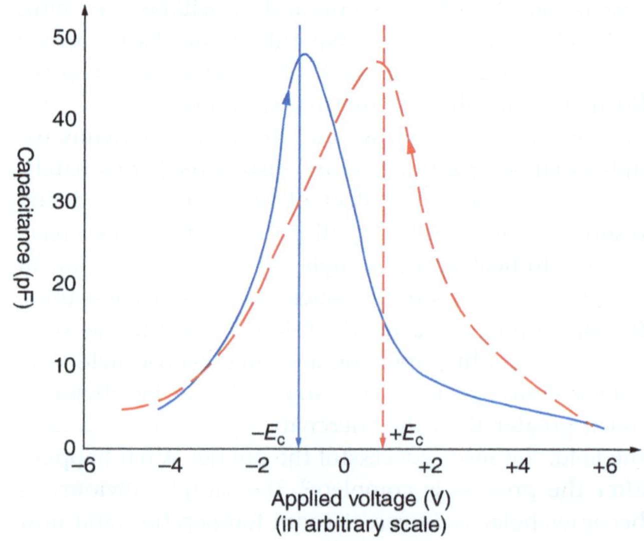

Figure 6.16 *C–V* hysteresis loop of a ferroelectric material.

Now high-class ferroelectric testers are commercially available (such as from Radiant Technologies testers) with advanced electronic circuits that eliminate possible sources of error. However, all of them are based on the fundamental principle of the original Sawyer–Tower circuit.

Exercise 6.2

Consider that the spontaneous polarization of a ferroelectric sample is $20\,\mu C\,cm^{-2}$ and 10% of the source voltage is resident on the reference capacitor $C(R)$. Determine the values of the components $C(R)$, $C(X)$, and $R(1)$ in the circuit of Figure 6.16. Assume that the circular electrodes on $C(X)$ capacitor are of 2 mm in diameter, resistance $R(2)$ is 1 kΩ, and the source voltage is 50 V.

Solution

Using the relationship $Q = P_sA = C(X)V(2)$, we find $Q = 20\ \mu C\ cm^{-2} \times (\pi r^2) \approx 6.28\ nC$. The reference capacitor $C(R)$ has the voltage on it of 5 V. That gives the value for $C(R)$ to be approximately 126 nF. Since $\frac{C(X)}{C(R)} = \frac{V(R)}{V(X)} = \frac{5}{45} = i = \frac{1}{9}$. That makes $C(X) = 0.14\,nF$. Since the two resistances $R(1)$ and $R(2)$ are in series the divided source voltage will bear the following relationship: $\frac{V(1)}{V(2)} = \frac{R(1)}{R(2)} = 9$. Since $R(2) = 1\,k\Omega$; $R(1) = 9\,k\Omega$. Summary: $C(X) \approx 0.14\,nF$, $C(R) \approx 126\,nF$, and $R(2) = 9\,k\Omega$.

6.6.3 Capacitance Measurement and C–V Plot

Before the emergence of digital electronics, the fundamental physical properties such as resistance (R), capacitance (C), and inductance (L) were measured separately and by multiple different methods. So far as capacitance is concerned the most frequently used methods adopted by scientists and engineers were the oscillator-based approach, charge-based approach, and compensated-bridge method. Each one has its merits and drawbacks. The data obtained were tainted by parasitic effects and instrumentation errors. In spite of these limitations, they produced fairly accurate results for bulk samples. But their limitations amplified when measuring the parameters of small magnitude such as found for thin film samples. Now the commercially available digital LCR meters with wide range of options are readily available, and all these three parameters can be determined with high precision and accuracy ranging in values form very small to very large.

A general-purposed LCR meter in the affordable range of price can measure all three parameters, *L*, *C*, and *R*, in a wide range of frequencies and voltages. The input signal can be of variety of shapes such as sine waves, square waves, rectangular waves, and so on. The biasing of samples by applying voltage between its two terminals can be accomplished rather effortlessly. The interfacing of the LCR meters with a computer facilitates the collection of data in dynamic modes in real time.

For a determination of capacitance of a ferroelectric material, the samples are prepared in a parallel plate configuration with the metallic electrodes of well-defined geometry. By knowing the electrode area (*A*), sample thickness (*d*) and sample capacitance (*C*), the dielectric constant can be readily determined $\left(\epsilon_r = \frac{Cd}{A\epsilon_0}\right)$. The accuracy with which the sample thickness is known is of vital importance. Any error in its measurement will make the values of the parameters measured inaccurate. This is particularly important if the sample happens to be thin film or of the nanometer scale. A sophisticated LCR meter provides the provision for eliminating the parasitic capacitance to enter into the value of the sample capacitance. This is obviously a big advantage.

A linear capacitor is independent of applied voltage, whereas a nonlinear capacitor shows strong voltage-dependence. Figure 6.16 shows the voltage-dependent capacitance of a ferroelectric material. The resulting C–V plot is obviously hysteric in nature with well-defined peaks separated by an electric field equivalent to the field separating the two coercivities in the ferroelectric hysteresis loop. The practical aspect of such a study is that the well-defined peak at the coercive points attests to the good quality of the sample.

Exercise 6.3

Why it is assumed that the capacitance, *C*, of a linear capacitor shows no dependence on applied electric field, whereas strong field dependence is expected from a nonlinear capacitor?

Solution

Let the sample have the capacitance = *C*, electrode area = *A*, thickness = *d*, and applied potential = *V*. The charge on the capacitor $Q = CV$. Therefore, the polarization, *P*, on the capacitor $= Q/A = (C/A)V$. That is, $P = (Cd/A)E$, where E = electric field. A plot between *P* and *E* will give a straight-line with the slope $= (d/A) \times C$. We see that *C* enters the equation as a coefficient and does not vary with the electric field. Now we consider the case of a nonlinear capacitor such as made of a ferroelectric material. Its *E* vs. *P* relationship is given by Eq. (6.3). It consists of three terms of *P* given by P, P^3, and P^5. Their coefficients $\alpha \neq \beta \neq \gamma$. Obviously, *P* vs. *E* relationship cannot be represented by straight-line plot. The plot will be nonlinear amounting to a changing value of *C* at each point.

6.6.4 Ferroelectric Domains (Experimental Determination)

Before piezoresponse force microscopes (PFM) became commercially available, a good polarization optical microscope with the magnification of 100× to 200× was the instrument of choice to investigate the domain structures of a ferroelectric material. Figures 6.17 and 6.18 are the domain configurations for barium titanate, and they were observed by using an optical polarization microscope. Figure 6.18 is the most cited ferroelectric domains of a barium titanate single crystal platelet [10]. It is the actual view seen along the polar *c*-axis of the tetragonal crystal. To enhance the patterns, the samples were first etched in concentrated HCl acid. We can

(a)

(b)

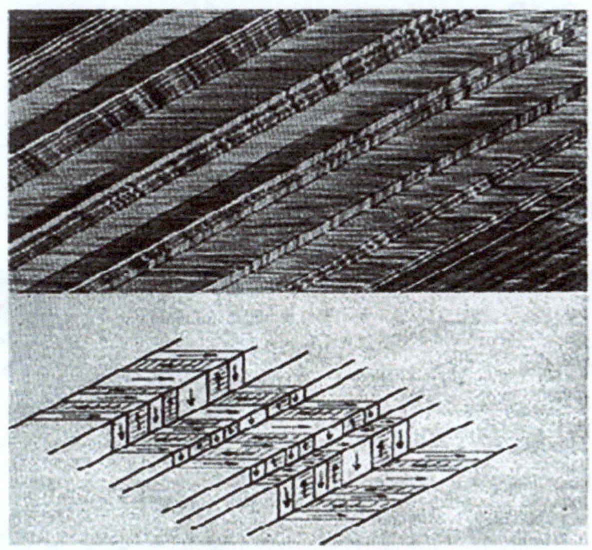

Figure 6.17 Actual view of ferroelectric domains of barium titanate single crystal oriented along the *a*-axis (a); and its simulated view (b). Source: Hooten and Merz 1955 [10].

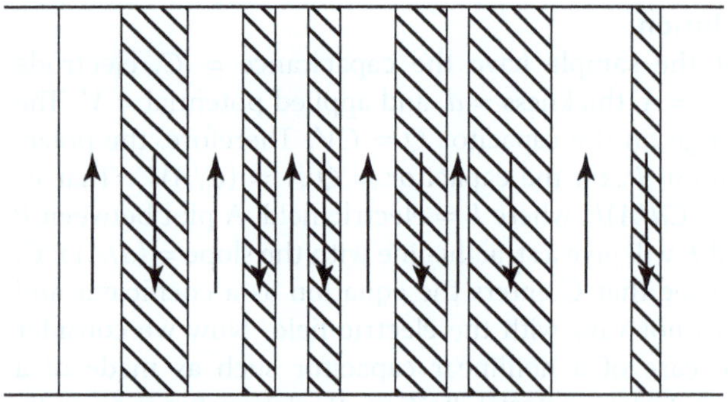

Figure 6.18 Antiparallel configuration of *c*-domains in a stressed barium titanate crystal platelet. Source: Hooten and Merz 1955 [10].

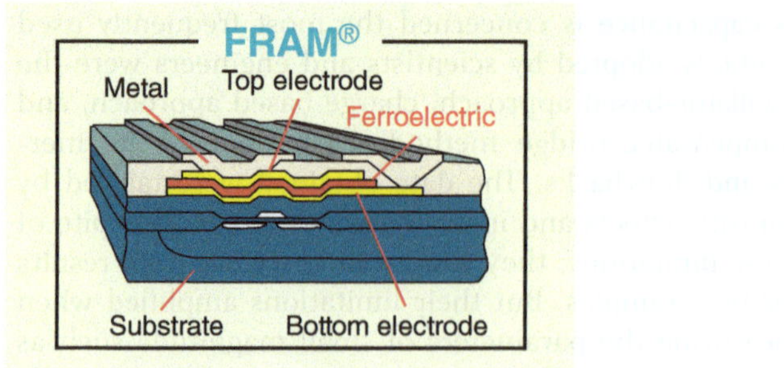

Figure 6.20 PZT based FeRAM structure consisting of one transistor and one capacitor. Source: From ROHM Co. Ltd. [12].

readily identify the fine features in the upper picture that are domain walls, 90° domains, and 180° domains. The bottom picture, which is simulated, shows clearly the 90° and antiparallel domains. Figure 6.18 is the representation of antiparallel domains when the platelets were subjected to a stress. The fact that the domains were influenced by stressing the platelets points to the piezoelectric nature of barium titanate.

The domain structures of barium titanate single crystals shown in Figure 6.19a,b were determined by using a PFM. Figure 6.20 shows 180° domains of barium titanate obtained by using a PFM [11]. The topograph of Figure 6.19a and the domains of Figure 6.19b were taken simultaneously by PFM. Here, we can clearly see the domains separated from each other by domain walls. The scale bar in the picture is 10 μm.

6.7 Recent Applications of Ferroelectric Materials

Ever since the field of ferroelectricity was discovered by Valasek in 1921 [1], it was identified as potentially a good field for novel applications. But for the first two decades, this dream had to wait for want of good material that could be processed as a stable material with reliable reproducibility. When in 1945 barium titanate appeared as the first ferroelectric material, it soon became the material of choice to explore its potential as a good device material with the possibility of being produced reliably and in large scale. Besides being rugged and atmospherically stable, it also did not contain any water molecules as the other sister ferroelectric materials. This obviously was an asset that made barium titanate the target for worldwide research and development efforts. It was first identified as an excellent capacitor material because of its high dielectric constant, and even today, it remains a heavily used material for commercial production of miniature capacitors.

For a long time, ferroelectrics were studied as potentially good substitute material to replace or at least partially displace magnetic ferrites as leading memory materials. Like in ferrites, its hysteresis loop is switchable, but unlike in ferrites, it is electrically controllable and can be manipulated with ease. Because of its remnant polarization that exists in two equal and opposite states, and the nonvolatile nature made barium titanate a very attractive material for fabrication of electrically controlled nonvolatile memory. We have discussed this point in good depth earlier while explaining the properties and parameters of the *P*–*E* hysteresis loop.

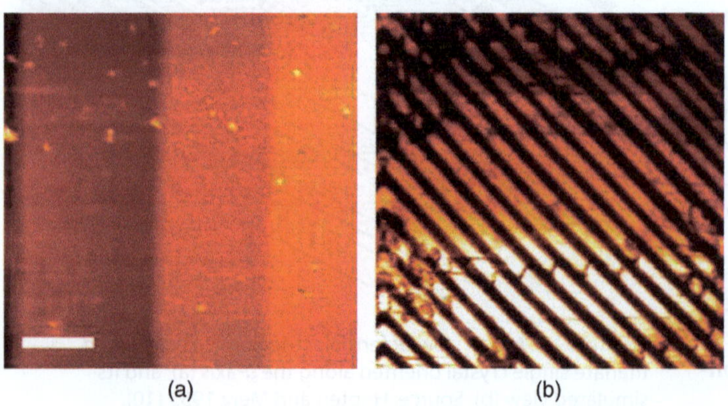

Figure 6.19 Barium titanate domains generated by a PFM: (a) topograph and (b) anti-parallel domains [11].

But one had to wait for a very long time before the emergence of reliable and fast nonvolatile ferroelectric memory. Once ferroelectric PZT became available in very high quality both as ceramic and epitaxial thin film, the rush for ferroelectric memory took off in high gear. Today, ferroelectric RAM (FeRAM) is produced in high volumes and is widely accepted by industry.

FeRAMs are fabricated as miniaturized film capacitors for data storage. Besides nonvolatility, the ferroelectric memory provides high-speed data access, low power consumption, high endurance in write mode, and excellent mechanical ruggedness making it durable and nonfragile. Today, it is considered an ideal memory for smart cards for which high security at low power consumption is the necessary condition. FeRAM has been widely adopted by technology associated with cellular phone and many other devices. FeRAMs are fully integrated today with the Si-technology, which has opened up many other sophisticated applications for thin-film ferroelectrics. With the availability of highly sophisticated methods for film growth, it is now possible to produce oxide ferroelectric films of various materials that are fully integrated with silicon in a vast array of complex structures and platforms.

The ferroelectric capacitor for the structure shown in Figure 6.20 is PZT (Pb-Zr-titanate) of unspecified composition [12]. Figure 6.21a shows the device structure consisting of p-type Si-substrate with heavily doped n-type source (N^+) and ferroelectric film capacitor of STN with the composition of $Sr(Ta_2Nb_2)O_7$ sandwiched between top and bottom metallic electrodes of platinum.

The gate oxide consists of layers of SiO_2, poly-Si, and IrO_2. This sophisticated MOSFET structure provides a reliable and stable output as shown in Figure 6.21b. The current–voltage characteristic is in the form of a hysteresis curve with fully saturated current in both the forward and reverse modes of the gate voltage. Notice the thickness of the insulating SiO_2 layer of only 13 nm, the active ferroelectric STN layer of 150 nm, and the memory cell itself being of just 0.6 μm. This type of structure fabrication on commercial scale naturally requires a highly sophisticated fabrication facility that only a microelectronic company can afford.

Some of the other recent developments related to the applications of ferroelectrics are (i) giant electroresistive switching (GERS) that is based on the piezoelectric and ferroelectric interface effects; (ii) ferroelectric tunnel junction (FTJ) in which electron tunneling is initiated using a configuration consisting of a nanometer thick ferroelectric film sandwiched between two metal electrodes; and (iii) multiferroic devices that are based on heterostructures consisting of ferroelectric and ferromagnetic materials.

In Table 6.4, we give the list of leading ferroelectric materials with their potential applications.

6.8 Antiferroelectricity

This is a very interesting group of materials, and like ferroelectrics, they also are fundamentally pyroelectric, and therefore, are governed by the same symmetry conditions. These materials exhibit paraelectric behavior in which the dielectric constant decreases with increasing temperature above the Curie point which is analogous to the behavior found for ferroelectrics. Both obey the Curie–Weiss law. However, unlike a ferroelectric material, antiferroelectric does not exhibit spontaneous polarization. This is the fundamental difference between these two classes of dielectrics.

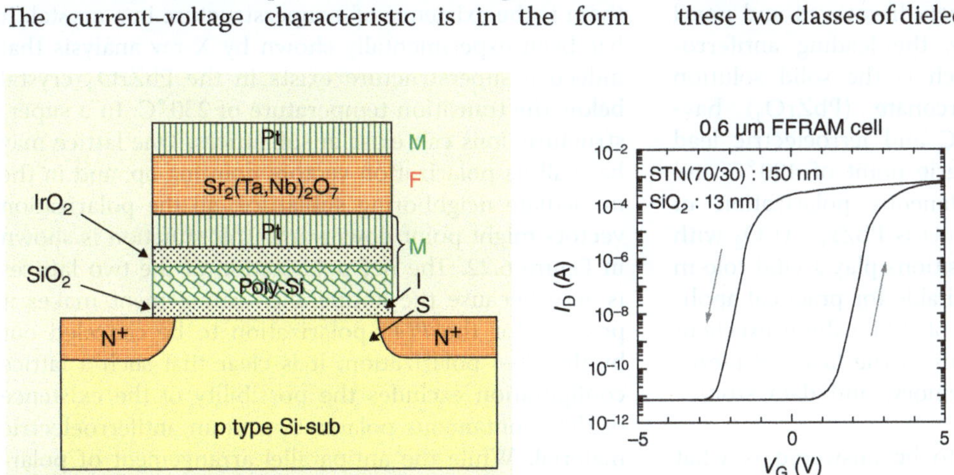

Figure 6.21 (a) STN based memory cell and (b) gate voltage, V_g, dependent output drain current, I_D, of the STN memory cell. Source: From ROHM Co. Ltd. [12].

Table 6.4 Selected ferroelectric materials and their applications.

Group	Name	Chemical formula	Abbreviation	Specialized applications
Perovskite, ABO_3	Barium titanate	$BaTiO_3$	BTO	Capacitors, electro-optics, piezoelectric devices
	Potassium niobate	$KNbO_3$	KN	Electro-optics
	Potassium tantalate	$KTaO_3$	KT	Electro-optics
	Potassium tantalate niobate	$KTa_{1-x}Nb_xO_3$	KTN	Electro-optics
	Lead zirconate titanate	$PbZr_xTi_{1-x}O_3$	PZT	Piezoelectric devices and memory
	Barium strontium titanate	$Ba_xSr_{1-x}TiO_3$	BST	Night vision
Ilmenite family, ABO_3	Lithium niobate	$LiNbO_3$	LN	Electro-optics
	Lithium tantalate	$LiTaO_3$	LT	Electro-optics and pyroelectric devices
Tungsten-bronze, AB_2O_6	Barium strontium niobate	$Ba_{5x}Sr_{5(1-x)}Nb_{10}O_{30}$	BSN	Electro-optics
	Barium sodium niobate	$Ba_{5x}Na_{5(1-x)}Nb_{10}O_{30}$	BNN	Electro-optics
	Potassium lithium niobate	$K_5Li_2Nb_{10}O_{15}$	KLN	Electro-optics
Miscellaneous	Potassium dihydrogen phosphate	KH_2PO_4	KDP	Infrared detector
	Tryglycine sulfate	$(NH_2CH_2COOH)_3{\cdot}H_2SO_4$	TGS	Infrared detector
	Bismuth titanate	$Bi_4Ti_3O_{12}$	BTO	Electro-optics
	Gadolinium molybdate	$Gd_2(MoO)_3$	GMO	Electro-optics
	Antimony sulfoiodide	SbSI	SbSI	Pyro-optic detector
	Rochelle salt	$NaKC_4H_4O_6{\cdot}4H_2O$	—	Ferroelectricity discovered

The typical example of an antiferroelectric material is the perovskite lead zirconate with the chemical formula of $PbZrO_3$. The concept of antiferroelectricity was first advanced by Kittel in 1951 [13]. Based on the theory of LGD, he developed a theory for antiferroelectricity that has gained almost universal acceptance.

Since the discovery of antiferroelectricity almost 65 years ago, it has come a long way. Many new antiferroelectric materials have been discovered and novel applications identified. Today, the leading antiferroelectric material is PZT which is the solid solution of antiferroelectric lead zirconate ($PbZrO_3$) having the Curie point of 230 °C and ferroelectric lead titanate ($PbTiO_3$) with the Curie point of 490 °C and the room-temperature spontaneous polarization of 50 μC cm^{-2}. Its chemical formula is $PbZr_{1-x}Ti_xO_3$ with $0 < x < 1$. As expected, compositions play a vital role in giving rise to its properties suitable for practical applications. Certain compositions of PZT exhibit excellent properties and find applications in the fields of piezoelectricity, ferroelectricity, memory, and data storage and electro optics.

The central question now to be answered is what causes antiferroelectricity to occur in a pyroelectric system of materials? The full theoretical explanation is beyond the scope of this book and therefore, we will not deal with it here. However, advanced readers may consider referring to the excellent book on dielectrics by Kao [4].

We recall that the central Ti^{4+} ions in the barium titanate unit cell is off centered giving rise to the non-centrosymmetry structure that makes barium titanate crystal a ferroelectric material. Displacement of an ion in a perovskite structure is not the only type of instability. It can have many different origins. One of them is the existence of a superstructure in a crystal. It has been experimentally shown by X-ray analysis that indeed a superstructure exists in the $PbZrO_3$ crystal below the transition temperature of 230 °C. In a superstructure, ions can exist in sublattices. One lattice may have all its polarization vectors pointing up, and in the immediate neighboring sublattice, all the polarization vectors might point downward. This situation is shown in Figure 6.22. The net polarization of the two lattices is zero because the antiparallel arrangement makes it possible for the $P(+)$ polarization to be canceled out by the $P(-)$ polarization. It is clear that such a lattice configuration excludes the possibility of the existence of the spontaneous polarization in an antiferroelectric material. While the antiparallel arrangement of polarization in sublattices of an antiferroelectric materials excludes the possibility of the spontaneous polarization to exist yet it does not eliminate anomalous behavior to occur in its other dielectric properties. Such changes

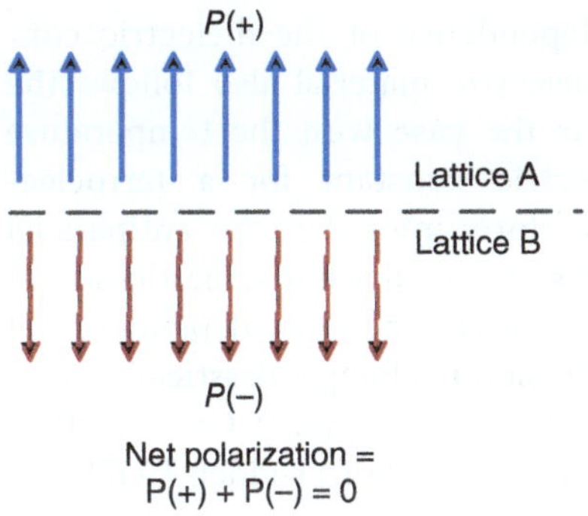

Figure 6.22 Polarization vectors in neighboring lattices A and B in a unit cell of an anti-ferroelectric crystal.

are most pronounced at the phase transition that can occur for a number of reasons. Temperature is definitely one of the prime reasons. The other agents for inducing a phase change could be electric field, pressure, materials processing, and so forth. Some of the readers might recall that there is a well-established phenomenon in the field of magnetism, which is called antiferromagnetism. However, the physics for the origin of antiferromagnetism lies at the core of quantum mechanics, whereas for antiferroelectrics, it is macroscopic in nature.

The phase change in an antiferromagnetic material takes place at a temperature which is called the Néel temperature (T_N) in honor of the French physicist, Louis Néel, who gave the theory of antiferromagnetism. It usually occurs at low temperatures. In contrast, the antiferroelectric Curie point occurs at relatively high temperature. They can cover a wide range of temperatures as can be seen from Table 6.5.

The table contains only the antiferroelectric materials of the perovskite group. They do come also in other structures, but perovskite is the leading group in which the most prominent antiferroelectric materials are found.

We see from the above table that antiferroelectric Curie temperatures of the solid solution series, PZT and BNT–BT are strongly dependent upon the composition. This provides an opening for tailor-making the materials with specific properties leading to certain novel applications. The Pb-based solid solution is currently the leading antiferroelectric material. The PZT materials are the materials of choice for ferroelectric and piezoelectric applications. However, they contain Pb that is environmentally damaging, which has led to vigorous research to replace it with nontoxic materials. BNT–BT is such a material, and it remains to be seen how successful it might prove itself to be valuable for applications and devices. We saw that PZT capacitors are widely used for FeRAM devices. Nontoxic

Table 6.5 Selected antiferroelectric materials and their Curie points.

Materials	Structure	Antiferroelectric Curie point, T_C (°C)
Lead zirconate, $PbZrO_3$	Perovskite	230
Bismuth ferrite, $BiFeO_3$	Perovskite	850
Cesium lead chloride, $CsPbCl_3$	Perovskite	47
Lead magnesium tungstate, $Pb(Mg_{1/2}W_{1/2})O_3$	Perovskite	38
Lead indium niobate, $Pb(In_{1/2}Nb_{1/2})O_3$	Perovskite	90
Lead zirconate titanate (PZT) solid solution, $PbZr_{1-x}Ti_xO_3$ with $0<x<1$	Perovskite	230-25 for $0<x<0.5$; x being concentration of $PbTiO_3$ in $PbZrO_3$
Bismuth sodium titanate barium titanate solid solution, (BNT–BT), $(1-x)$ $(Bi_{1/2}Na_{1/2})TiO_{3-x}BaTiO_3$ with $0<x<1$	Perovskite	Variable depending upon the value of x (see Figure 6.24)

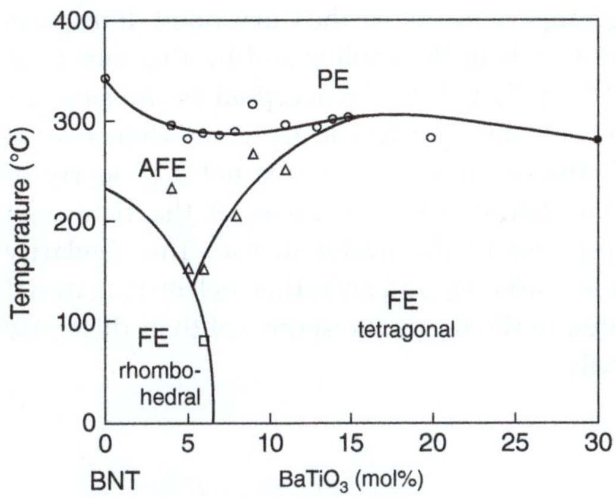

Figure 6.23 Phase diagram of bismuth sodium titanate and barium titanate (BNT–BT) solid solution. AFE refers to antiferroelectric phase, and FE to ferroelectric and PE to paraelectric phases, respectively. Source: From Tan et al. 2011 [14].

STN might replace it completely sometime in future in the field of FeRAM technology. Figure 6.23 depicts the phase diagram of BNT–BT solid solution system [14]. The antiferroelectric phase, AFE, exists for barium titanate concentration (x) range of $0<x<15$ mol%. The antiferroelectric Curie point varies with the change in concentration x. The AFE phase is stable approximately between 225 and 350 °C. The open circles represent the antiferroelectric Curie points for phase changes between the antiferroelectric state to paraelectric state and the solid triangles refer to the antiferroelectric Curie points at for transition occurring from the antiferroelectric state to ferroelectric state in tetragonal structure. This phase diagram indicates the coupled relationship between ferroelectricity, antiferroelectricity, and paraelectric phase. This is a classic example of structure–property relationship.

Now we must ask ourselves two pertinent questions. First, how does the temperature dependence of dielectric constant manifest anomaly; and second, how does the polarization residing in sublattices of antiferroelectric material behave when subjected to an electric field? The answer to the first question is simple, and for the second question, the short answer is that it behaves in an intriguing manner. We first discuss the anomalous behavior of dielectric constant with respect to temperature using Figure 6.24.

The above figure represents exactly the same trend what can be found in Figure 6.4 that gives the temperature dependence of dielectric constant of a ferroelectric material. The dielectric constant for $PbZrO_3$ antiferroelectric material reaches its maximum value at the Curie point of 230 °C by first increasing with temperature, and then follows the opposite trend of decreasing with increasing temperature above the Curie point. It displays a minor hysteresis in the cooling and heating curves at $T_C \approx 230$ °C. It should not be accepted as the singular behavior of an antiferroelectric material when cooled or heated. The thermal hysteresis is not a property of a material rather it occurs because of the nature of thermal response of the heater in use. The similarity between a ferroelectric and an antiferroelectric material is not limited to the thermal response of their dielectric constant only.

The temperature dependence of the dielectric constant for an antiferroelectric material also follows the Curie–Weiss law as is the case with the temperature dependence of dielectric constant for a ferroelectric material. We therefore, infer that the antiparallel arrangement of dipoles in the adjacent sublattices have no detectable impact on its temperature dependence of dielectric constant. The picture changes drastically, however, when a strong electric field is applied. Once the field reaches a certain critical value, all of a sudden *P*–*E* loops appear for both the forward and reverse mode of the applied field. The double-hysteresis loop, as seen from Figure 6.25, is the signature property of an antiferroelectric material. This along with anomalous thermal behavior of the dielectric constant define the antiferroelectric phenomenon. The emergence of a double-hysteresis loop with antiparallel configuration implies that the electric field induces ferroelectric phase in an antiferroelectric material, and this can occur when the field is large enough to switch the dipoles residing in the sublattices by 180°.

Notice the well-developed loops with the pronounced spontaneous polarizations labeled as $+P_s$ and $-P_s$ for each of the two directions of the potential applied. The critical fields are also identified in this figure as E_{crit}. In PZT, the double hysteresis loop occurs in the trigonal structure, which is its antiferroelectric state. The critical electric field, E_{crit}, is temperature-dependent; and it varies from its maximum value of 60 kV cm^{-1} at 200 °C to about 20 kV cm^{-1} at 230 °C. The double-hysteresis loops have been observed also for ferroelectric barium titanate. However, they nucleate only in the cubic phase (above T_c), where ferroelectricity ceases to exist. This point to the fact that the origin of its appearance in antiferroelectric phase of PZT is different than in the case of barium titanate.

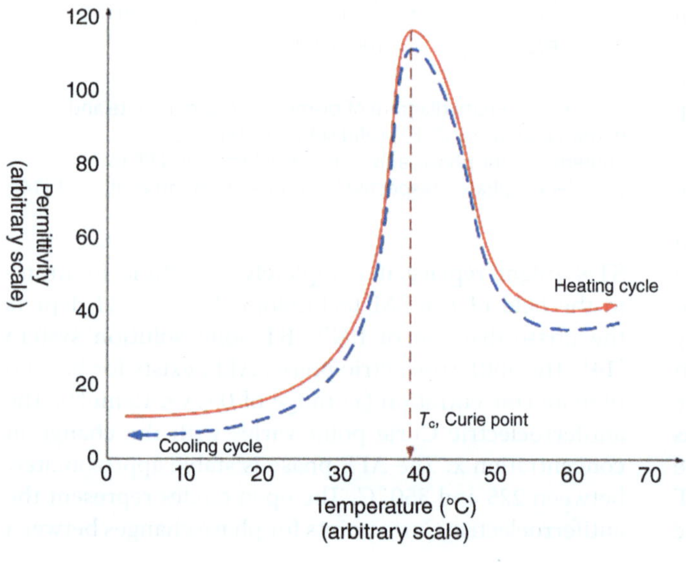

Figure 6.24 Temperature dependence of dielectric constant for lead zirconate antiferroelectric material.

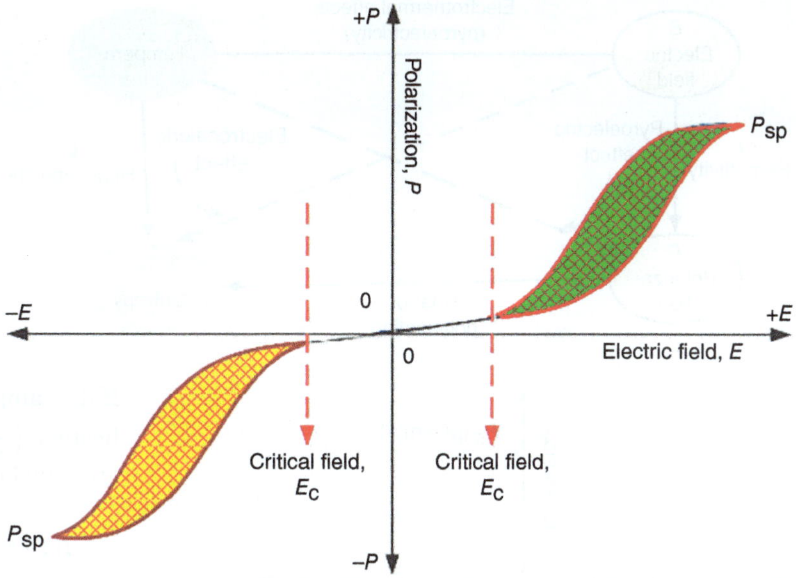

Figure 6.25 Double hysteresis loop of an antiferroelectric material.

6.9 Pyroelectricity

6.9.1 Historical Perspective

Pyroelectricity is one of the oldest branch of science; its roots going back to ancient times. According to one vision, it was known to Greeks already 24 centuries ago. Apparently, the Geek philosopher wrote that lyngourin had the property of attracting straws and bits of wood. What he referred to as lyngourin might have been the semiprecious mineral tourmaline, which is an established pyroelectric material. Another version of the history tells us that the pyroelectric effect was known to the natives of India and Ceylon (now Sri Lanka) long before the Europeans sailed to South Asia in search of spices. It is a recognized fact that tourmaline was abundantly found in South Asia, and the natives were amused to find that it attracted and repelled ash particles. Dutch merchants brought tourmaline to Europe around 1703. There it got the name of "Ceylon magnet." But in Tamil (language of the Tamil Nadu, India) and Sinhalese (language of Sri Lanka) it is called "Turmai" or "Thoramali." Obviously, its modern name of tourmaline bears strong resemblance with its original native name. Around 1747, Carl Linnaeus gave it the scientific sounding name of "Lapis Electicus." It was in 1756, that its pyroelectric nature was scientifically established when Aepinus observed that when heated tourmaline developed electricity with opposite polarity at the two ends. It was Brewster in 1824, who coined the term *pyroelectricity* based on the Greek word, "pyro" for fire. Whatever the exact history might be, pyroelectricity is now a well-established field of science and technology. Many applications have evolved over the years with infrared imaging and thermal detection leading the pack. Its impact on space missions is undeniable and pyroelectric sensors have greatly advanced the field of astronomy.

6.9.2 Pyroelectric Effect

In Figure 5.1, we discussed the principal and secondary interactions existing between the electric field (E), temperature (T), and mechanical stress (σ) in dielectric materials. These interactions, as we learnt, define the pertinent properties of polar dielectrics. Pyroelectricity is one of the products of such interactions. For dealing with this topic here in some depth, we need to examine once again, the interactions that exist between electric field (E), polarization (P), temperature (T), and entropy (S). For this, we reproduce once again Figure 5.3 here as Figure 6.26.

We know that the pyroelectric effect arises when temperature (T) and polarization (P) interact. Similarly, the electrocaloric effect, which is the converse of the pyroelectric effect, comes into existence when entropy and electric field couple. We know that ferroelectricity is the subset of pyroelectricity; both exhibit spontaneous polarization with the difference that in ferroelectrics, it is reversible, whereas it is nonreversible in the case of a pyroelectric materials. Figure 6.27 represents qualitatively the field-dependent polarization for a pyroelectric material.

Notice the difference between this plot and the hysteresis plot of Figure 6.3. A pyroelectric material does not display a hysteresis loop in its P–E behavior, which is the signature property of a ferroelectric material. The P–E plot for a pyroelectric material does not have crucial parameters such as remnant polarization and coercivity of the hysteresis loop of a ferroelectric material which

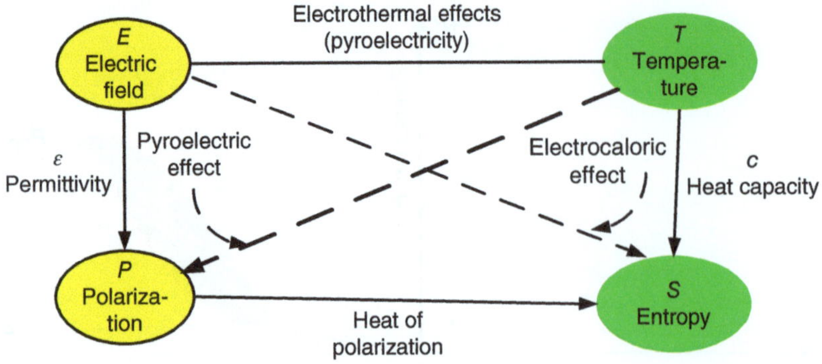

Figure 6.26 Electric field, polarization, temperature, and entropy diagram.

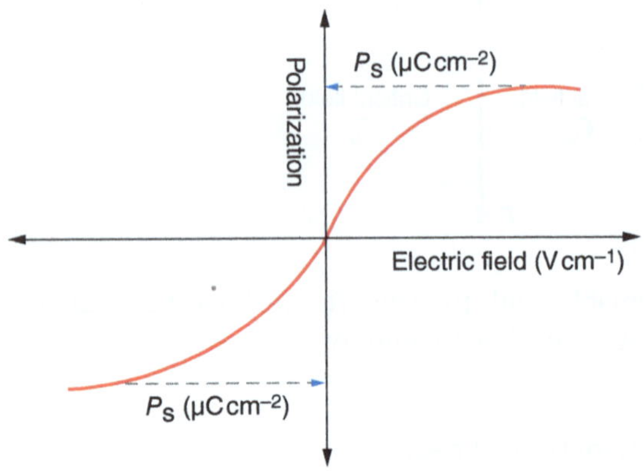

Figure 6.27 Electric field dependence of polarization of pyroelectric material.

makes the polarization reversible. The irreversibility of pyroelectric polarization makes a pyroelectric material less attractive for applications in data storage and memory.

For pyroelectrics, the change in temperature can induce a change in surface charge, which in turn can cause a change in its electrical polarization. The end result is the emergence of a thermal current, $I(T)$, which can be detected, in an external circuit. In order to understand the nature of this current, we need to consider the two components of polarization present when an electric field is applied to a pyroelectric material. One being the induced polarization, $P(I)$, and the other the spontaneous polarization, $P(T)$. Therefore, theoretically, it is possible that the measured thermal current, $I(T)$, will be the sum of the contributions from each of these two types of polarizations. However, the contribution made by the induced polarization is much smaller than the one made by the spontaneous polarization. Therefore, for all practical purposes, we stand on a solid ground by assuming that the measured thermal current, $I(T)$, will be equivalent to the current contributed by the spontaneous polarization only.

Now let us assume that heating gives rise to a charge Q in a pyroelectric capacitor having the electrode area of A. If the sample is heated uniformly at the constant rate of heating, $\left(\frac{dT}{dt}\right)$, the resulting thermal current, $I(T)$, will be given by Eq. (6.18).

$$I(T) = \frac{dQ}{dT} = A\left(\frac{dP_s}{dT}\right)\cdot\left(\frac{dT}{dt}\right) = AP(T)\cdot\left(\frac{dT}{dt}\right) \tag{6.18}$$

Rearranging the parameters, we get Eq. (6.21).

$$P(T) = \frac{I(T)}{A}\left(\frac{dT}{dt}\right)^{-1} \tag{6.19}$$

The parameter $P(T)$ is called the pyroelectric thermal coefficient and is an important parameter of a pyroelectric material. The higher its value, the greater is the thermal current generated. This parameter obviously is of importance for the design of an efficient thermal detector. We can compute its value by measuring the pyroelectric current, $I(T)$. However, we can also determine it directly by measuring charge Q_1 at temperature, T_1 and Q_2 at T_2. The magnitude of the pyroelectric coefficient, $P(T)$ is then given by Eq. (6.20).

$$P(T) = \frac{Q_2 - Q_1}{A(T_2 - T_1)} = \frac{\Delta Q}{A\Delta T} \tag{6.20}$$

However, the value so obtained might not be reliable since it might contain errors that we discuss in the next section.

Exercise 6.4

Consider a simple circuit consisting of a pyroelectric element with the electrode area $A = 0.25\ \text{cm}^2$ and a 1 MΩ resistor is series. When the pyroelectric element is heated at the rate of 1 °C min^{-1}, it generates a pyroelectric coefficient of 0.3 μC cm^{-2} °C^{-1}. Could this pyroelectric element be used as a source of green energy.

Solution

For a constant heating rate of 1 °C min^{-1}, Eq. (6.20) reduces to Eq. (6.21).

$$I(T) = P(T)\cdot A \tag{6.21}$$

Substituting the values for $P(T)$ and A, we get the pyroelectric current, $I(T) = 75$ nA. This current will pass through 1 MΩ giving a voltage drop, $V(T)$, of 75 mV across the resistor. This is sufficient to power many miniature devices. However, the equivalent electric power, $P = I(T) \times V(T) = 5.63$ nW. This level of power is very small to be of much practical use.

6.9.3 Experimental Determination of Pyroelectric Coefficient

There are many different ways in which we can determine both the thermal current and the pyroelectric coefficient. However, the reliability and accuracy of the results are not automatically given by each method. In the static method, the sample is heated that gives rise to a charge which is then monitored across a reference capacitor in parallel with the sample using an electrometer. In the dynamic method, the same circuit is used as for the static method, but the reference capacitor is replaced by a resistance of known value in series with the sample. The voltage generated across the resistance is monitored as a function of temperature, and from there, the value of thermal current computed. However, each method has its drawback. The values obtained may include the leakage current and the current associated with polarization switching. Leakage current correction can be made by determining the parameters under steady-state conditions, but elimination of the current because of the polarization switching is not trivial.

More reliable method was proposed in 1972 by Byer and Roundy [15]. The method is based on Eq. (6.20). Here instead of varying the temperature as is the case for the static and dynamic methods discussed, the rate of change of temperature ($\mathrm{d}T/\mathrm{d}t$) is maintained constant. Slower the value of ($\mathrm{d}T/\mathrm{d}t$), the more reliable the value of the parameters determined. The drawback of Byer–Roundy method is that it is difficult to separate the pyroelectric current from other nonpyroelectric currents that are, for example, thermally stimulated currents (TSC). TSC originates from the charged defects inside the crystal. Also the background noise from connecting cables and power lines, etc., can add errors in the values measured by the Byer–Roundy method. Nevertheless, it makes it possible to monitor the thermal current continuously as a function of temperature. A charge integration technique then is used, in accordance with Eq. (6.21), to obtain the temperature dependence of pyroelectric coefficient of a PNZT (Nb-doped PZT) film [16]. It was determined while keeping the rate of temperature at $1\,°\mathrm{C\,min^{-1}}$. The pyroelectric current increases nonlinearly with increase of temperature. It reaches its maximum value at the ferroelectric Curie point and then decreases with increasing temperatures

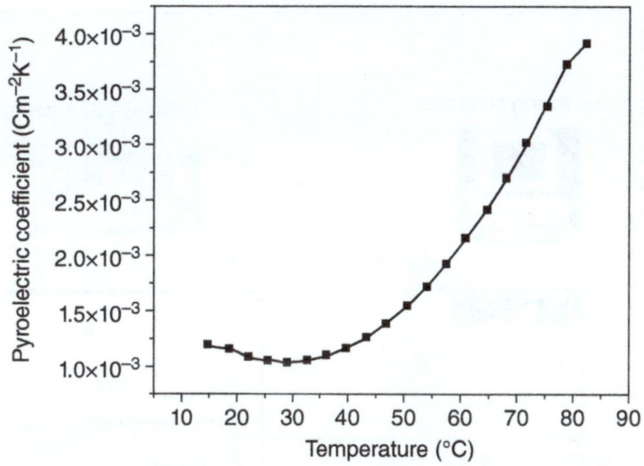

Figure 6.28 Temperature dependence of pyroelectric coefficient for a PNZT film. Source: Han et al. 2004 [16]. Reproduced with the permission of AIP Publishing.

(not shown in Figure 6.28). Figure 6.28 shows the temperature dependence of the pyroelectric coefficient, $P(T)$ for solgel grown ferroelectric PNZT film.

In 1982, Lynn and Edward proposed the low-frequency sinusoidal temperature (LFST) modulation technique to separate the pyroelectric current from the nonpyroelectric currents [17]. This is based on the assumption that Eq. (6.20) is valid only for the pyroelectric current, $I(P)$, and not for the nonpyroelectric currents that are proportional to the temperature, T, and not on the rate of change of temperature ($\mathrm{d}T/\mathrm{d}t$). In the LFST technique, the sample temperature is modulated in the sinusoidal form of $T = T_0 + \Delta T \cdot \sin(\omega t + \phi)$, where ω is the frequency, t being time, and ϕ the phase difference. This reduces Eq. (6.21) to Eq. (6.24).

$$I(T) = P(T)A\omega\Delta T \tag{6.22}$$

The phase difference between the temperature modulated curve and the pure pyroelectric current curve is 90°. Therefore, the true value of the pyroelectric coefficient, $P(T)$, can be derived using Eq. (6.24). Apart from producing the pure pyroelectric current, $P(T)$, free from errors, it also eliminates the DC offset current that might be present in the instrument. The experimental set up for current measurement using the LFST technique is given in Figure 6.29 [18]. The thermal stage in the figure is a thermoelectric element that is powered by a Keithley 228 A source meter which can provide currents with different wave forms. The sample is heated at the thermal stage by a current of desired wave characteristics. The resulting pyroelectric current, $I(T)$, is monitored by Keithley 6517 A electrometer. All current and temperature data are subsequently recorded by the controlling computer.

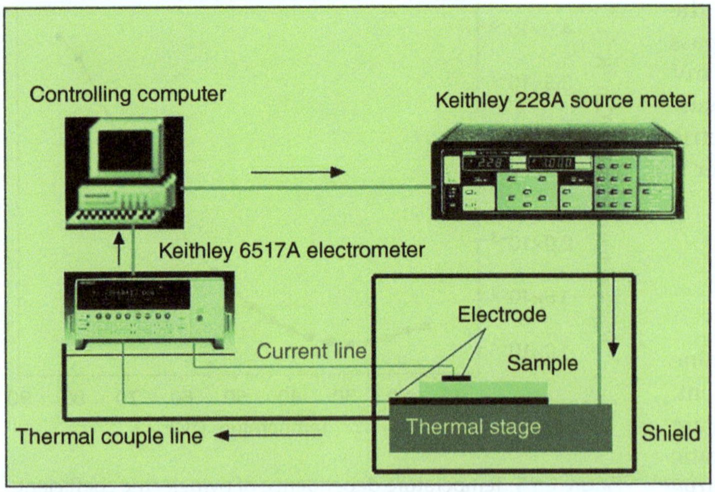

Figure 6.29 Experimental setup for measurement of pyroelectric current. Source: Han et al. 2008 [18]. Reproduced with the permission of Elsevier.

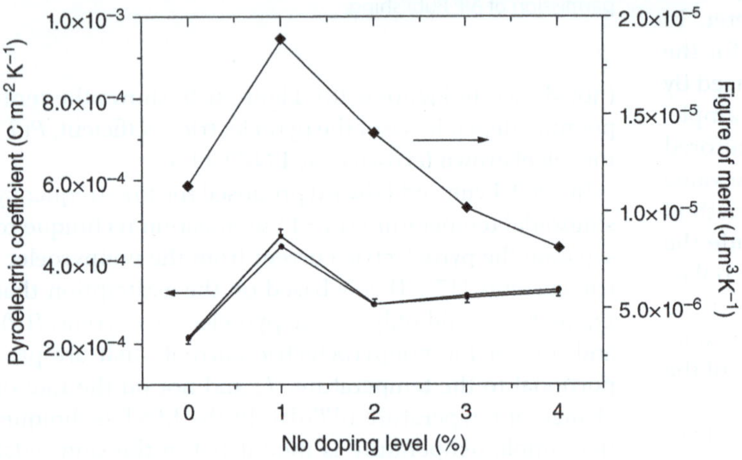

Figure 6.30 Room temperature pyroelectric coefficient and figure-of-merit determined by LFST technique for PNZT film as function of Nb-doping level. Source: Han et al. 2008 [18]. Reproduced with the permission of Elsevier.

Figure 6.30 shows the room temperature pyroelectric coefficient, $P(T)$, and figure-of-merit for a poled PNZT film as a function of doping level of Nb. Both these parameters show strong dependence on the Nb-doping level. We see that both peak when the doping level amounts to 1% of Nb. The figure-of-merit, η, is defined by Eq. (6.23).

$$\eta = \frac{P(T)}{c(\epsilon_0 \epsilon_r \tan \delta)^{\frac{1}{2}}} \tag{6.23}$$

In Eq. (6.23), ϵ_0 is the dielectric constant of free space, c is the volume specific heat, the relative dielectric constant, and $\tan \delta$ the loss tangent for the sample. Some examples of pyroelectric materials with their pyroelectric coefficients are presented in Table 6.6.

6.9.4 Applications of Pyroelectricity

The list of applications based on pyroelectricity is rather long. To discuss each of them in depth would require a chapter devoted to this topic alone. Instead, we are only providing here a very simple and brief background of some applications. The interested readers are encouraged to look at the excellent review article by Whatmore [19].

The pyroelectric detectors have a long and successful career since the effect was scientifically established. The heart and soul of all applications is the principles of thermal imaging. There are two types of atmospheric windows found in nature. One is in the wavelength range of 3–5 μm and the other between 8 and 14 μm. The thermal power emitted by the two wavelength ranges are different amounting to different temperatures. Sophisticated pyroelectric detectors are capable of responding to both these ranges. They can be made capable of meeting the requirements of different applications. Here is a list of applications and devices based on pyroelectric effect with a brief introduction for each.

1. *Infrared detection*: Based on the principles of thermal imaging, infrared detectors are developed that are now widely used in sophisticated cameras. They have

Table 6.6 Pyroelectric coefficients of selected materials.

Materials	Type	Pyroelectric coefficient, $P(T)$ ($\mu C\,m^{-2}\,K^{-1}$) at °C (temperature in parenthesis)
Triglycine sulfate, TGS	Single crystal	280 (35)
Deuterated TGS	Single crystal	550 (40)
Lithium tantalate, $LiTaO_3$	Single crystal	230 (25)
Strontium barium niobate (SBN) $Sr_{1-x}Ba_xNb_2O_6$	Single crystal	550 (23)
Lead zirconate titanate, PZT	Ceramic	400 (21)
$(Ba_{1-x}Sr_x)TiO_3$, (BST)	Ceramic	7000 (20)
$x = 0.33$		
Nb-doped PZT (Han)	Solgel grown film	1080 (21)

Source: Moulson and Herbert 2003 [22]. Reproduced with the permission of John Wiley & Sons.

both civilian and noncivilian uses. In particular, they are used by military for surveillance and by police for crime detection and fighting.

2. The principle of *thermal imaging* is applied also to develop night vision eye glasses that are valuable devices widely used by soldiers in night combats and by commercial pilots for navigation during nights.
3. *Intruder alarm*: When a building is unoccupied, its thermal scene is rather constant. As soon as an intruder enters, the building and moves around the thermal signature changes, which can be detected resulting in triggering an alarm.
4. *Smoke and fire detection*: Once the threshold of smoke and temperature exceeds in a building, the alarm sets in. This is a valuable tool for serving as a life-saving device and for protection of property from fire.
5. *Pollution monitoring and gas analysis*: This application is based on the physics of absorption spectra that is unique for each gas. The monitoring device consists of two pyroelectric detectors, each equipped with a filter for chosen wavelengths. For analysis, modulated broad band source is employed. The radiation falling on one detector is allowed to pass through the gas being analyzed, whereas the radiation falling on the other detector is allowed to pass through a reference cell. The comparison between the two outputs is a measure of the gas concentration in the atmosphere.
6. *Radiometry*: This application is a boon to the astronomers and for exploration of deep space. A radiometer based on pyroelectric effect is equipped with a lens that defines the field under observation. A chopper is employed to modulate the radiation. The magnitude of the output signal from the detector serves to measure the temperature of the target.
7. *Pyroelectric vidicon*: The infrared radiation are focused on a germanium lens and then allowed to fall on a pyroelectric target. The rear face of the target is scanned by an electron beam which reads the pyroelectric charges. The charges then produce a video signal on a linear array of pyroelectric detectors.
8. *Laser detectors:* Pyroelectric detectors are employed to monitor the pulses given by lasers. A variety of detectors are commercially available for this purpose.
9. *Energy conversion:* It is based on the principles of conversion of thermal energy into electrical energy. Pyroelectric materials when heated generate electric charges that can be harvested and then used as the source of electrical energy. The equivalent voltage developed across an external resistor can be used to power miniature devices.

6.10 Pyro-optic Effect

The pyro-optic effect can also be potentially exploited for detection of infrared radiation. The pyro-optic coefficient, $\theta(T)$, is defined as the change in refractive index with temperature. It is expressed as in Eq. (6.24).

$$\theta(T) = \frac{\eta_2 - \eta_1}{T_2 - T_1} = \frac{d\eta}{dT} \tag{6.24}$$

where $\theta(T)$ is the pyro-optic coefficient, η_2 the refractive index at temperature T_2, and η_1 the refractive index at temperature T_1.

Performance of a pyro-optic detector should be significantly superior to a pyroelectric or even photonic detector because no metallic contacts need to be used to connect pixel elements for the operation of a pyro-optic IR detector. The read out will be completely noncontact which is the major source of signal-to-noise ratio in a conventional detector. In addition, for a pyro-optic detector, the sample thickness need not be greater than the optimum thickness required for supporting the optical waves on reflection. That is, the pixel volume (or, mass) can be very small.

Originally, the concept of a pyro-optic detector was proposed in early 1990s [20, 21]. Antimony-sulfo-iodide, SbSI, was identified to be the most promising material for pyro-optic applications. Its pyro-optic coefficient is 6.5 m K^{-1} that is 25 times larger than the value for barium titanate ($\theta(T) = 0.31$ m K^{-1}). SbSI is a room temperature ferroelectric material with excellent electro-optic properties. But it is a very difficult material to produce either as a bulk single crystal or as an epitaxial film because of the high vapor pressure of antimony. Unless SbSI can be produced on a large scale as high-class single crystal, in bulk or film form, or if other equally promising materials are discovered this technology will remain unexploited.

Glossary

Antiferroelectrics The crystal superstructure is the origin of this effect. The parallel polarization vectors all pointing in the same direction occupy one sublattice, while in the very next neighboring sublattice, all of them point in the opposite direction. Taken as an aggregate, the antiparallel arrangement of polarization vectors cancel each other resulting in zero spontaneous polarization. Therefore, there is no hysteresis loop observed in the polarization vs. electric field plot of an antiferroelectric material. However, they appear all of a sudden when an electric field larger than a critical field is applied. They appear both in the positive and negative directions of the applied field giving rise to a double-hysteresis loop. Its dielectric constant increases with temperature reaching its maximum value at a critical temperature called the Curie point. At this temperature, a phase change occurs and the material either goes from its antiferroelectric state to a ferroelectric state or to the paraelectric state. These groups of materials are very fascinating for research, and they show potentials for energy storage and many other novel applications.

Curie point A well-defined temperature at which the phase transition can occur in some special class of materials is called the Curie point. It can be either of the first order or the second order and can be understood with the help of thermodynamics. A thermodynamic theory sets the rule to distinguish between the two types of phase changes. Curie point is the defining property of ferroelectrics, antiferroelectrics, and ferromagnetics. They lose their unique property of nonlinearity at this temperature and transform into a linear material above the Curie point. Its most used symbol is T_c. Its other names are Curie temperature and transition point. It is a material constant and its value varies from one material to another. This critical point is named in honor of the French Physicist Pierre Curie (1859–1906). He discovered it while studying the temperature dependence of ferromagnetic materials for his doctoral dissertation in Sorbonne (in Paris).

Electrets They too are dielectric materials in which the dipoles are frozen which gives rise to its polarization characteristics. It is counterpart of a permanent magnet and behaves like a battery. But it is not a ferroelectric material.

Energy storage A medium to store energy and be able to deliver it when needed. Capacitors are ideal examples of an energy storage system. Antiferroelectric materials are potentially good candidates for developing new class of energy storage medium.

Ferroelectrics They are very powerful group of nonlinear dielectric materials which are spontaneously polarized even when no electric field is present. They belong to the group of pyroelectrics and are considered a subset of pyroelectrics having in common the same crystal point groups as pyroelectrics. All ferroelectrics display piezoelectricity and pyroelectricity. Their defining properties are reversible spontaneous polarization, stable hysteresis loop between the polarization and electric field, high dielectric constant and nonvolatility. They are well suited materials for a host of applications including nonvolatile memory, miniaturized capacitors, and multiferroic devices.

Gibbs function It is a very important concept that touches many branches of Physics and Chemistry. It can be defined in many different ways. A very technical sounding definition is "In the transition by a system from a well-defined initial state to a well-defined final state during a reversible process, the work exchanged by the system with its surrounding equals Gibbs' free energy change." Alternatively, in simple language it can also be defined, "as equivalent to the chemical potential that is minimized when a system reaches the state of equilibrium at a constant temperature and pressure." This function is also known as Gibbs energy, Gibbs free energy and free enthalpy. This concept was given by Josiah Willard Gibbs, an American mathematician, in 1873.

Green energy A term referring to non-hydrocarbon derived source of energy. An example being the solar cells. They offer a non-polluting source of energy and an alternative to our petroleum and coal based energy. It is a very active field of research and ceramic capacitors are playing very important role already in this evolving technology.

Hysteresis loop The typical loop common to ferromagnetic and ferroelectric materials. In ferroelectrics, they are generated when a material is subjected to an alternating electric field while concurrently recording the resulting polarization. This is the most significant property of a ferroelectric material that is the basis for developing modern memory cells, called FeRAM (Ferroelectric Random Access Memory). In conjunction with a ferromagnetic loop a ferroelectric loop builds the foundation for novel multiferroic devices which is currently the most active field of world-wide research.

Paraelectric This is a state in which a ferroelectric material ceases to be nonlinear. The hysteresis loop disappears and is reduced to a single line. In this phase its polarization is linear with respect to the applied electric field. In paraelectric state ferroelectrics are equivalent to standard linear dielectrics.

Perovskite A special crystal unit cell with the chemical formula of ABO_3. It is the most prominent group in which most of the ferroelectric materials synthesize. It can be both centrosymmetric or non-centrosymmetric. $CaTiO_3$ is the classic example of a centrosymmetric perovskite cell (see Figure 6.6a); and $BaTiO_3$ that of the noncentrosymmetric unit cell (see Figure 6.6b). High temperature superconductor oxides and colossal magnetoresistive materials also come in perovskite structures.

Phase change Thermodynamically there can be only two types of phase changes. One being of the first order and the other of the second order. If the latent heat changes discontinuously then the phase change is said to be of the first order. Similarly, when the latent heat changes continuously the phase change is said to be of the second order. A simple way to recognize them is by observing experimentally how a particular property of a material changes with temperature. If it changes discontinuously, then the phase change is of the first order (see Figure 6.12) and if it is continuous, then it is of the second order (see Figure 6.11). The following two simple examples may give a clearer picture. *Example 1*: A solid material on heating can melt, and that is a phase change in the physical shape of the solid. When heated more, the liquid can go into the state of being a vapor. These changes occur at the specified temperature and pressure. *Example 2*: When a ferroelectric material goes from its nonlinear state to its paraelectric state at the Curie point is another example of a phase change. The material does not change its physical shape; it is still a solid. So is also the situation when a superconductor loses its superconductivity at the critical point and then becomes a simple normal conductor.

Phase transition A change in phase from one state to another is referred to as phase transition which can occur only at a specified temperature (and pressure). The Curie temperature is a good example at which the phase transition occurs. For example, barium titanate undergoes change in crystal structure from tetragonal to cubic at 120 °C, which is its Curie temperature.

Pyroelectrics The interaction between an electric field and temperature gives rise to pyroelectric effect in some dielectrics. When subjected to a temperature gradient a pyroelectric crystal such as tourmaline gets polarized. Pyroelectrics are spontaneously polarized but it is irreversible unlike that of a ferroelectric material. Pyroelectricity has its origin in the symmetry elements of crystals; for this effect to exist, the material must not have center of symmetry. Ten out of twenty point groups are specific to pyroelectricity, and these are also shared by ferroelectricity (see Table 6.1). That is why ferroelectricity is considered a subset of pyroelectricity.

Pyro-optic effect The origin of this effect lies in the change of refractive index of a material with respect to changing temperature. Pyro-optic effect is thought to be a very good effect that could be harnessed for developing extremely efficient thermal detectors. Unfortunately no suitable material has been discovered yet to meet the expectations of technology.

Relaxor ferroelectrics Some ferroelectric materials exhibit high electrostriction and dielectric dispersion which can spread over a wide range of frequencies. The transition from the ferroelectric state to paraelectric state is both temperature and frequency dependent. The temperature dependence of dielectric constant for relaxors exhibit double peaks unlike a typical ferroelectric material for which only one peak is found. The relaxor ferroelectric materials are found suitable for many applications and are therefore widely studied. PMN is a typical example of a relaxor ferroelectric. PMN is the abbreviation for Mg-doped lead niobate ferroelectric.

Spontaneous polarization The internal ionic configuration of certain class of nonlinear polar materials gives rise to polarization which is spontaneous and needs no electric field for its existence. Ferroelectrics and pyroelectrics are two nonlinear dielectrics which are spontaneously polarized.

Tungsten Bronze It is also a special class of crystal unit cell in which some prominent ferroelectric materials are found (see Figure 6.9). They are represented by the chemical formula of AB_2O_6. Some ferroelectric materials with excellent electro-optic properties are found in this group. SBN or Sr-Ba-Niobate is a good example of such a material.

Problems

6.1 Distinguish between antiferroelectricity, ferroelectricity, and pyroelectricity. Give some examples of their applications.

6.2 What are the two most important defining signature properties of a ferroelectric material? Describe each of them briefly.

6.3 Draw a hysteresis curve for a ferroelectric material and identify the salient features and define them.

6.4 Describe the crystallographic conditions necessary for ferroelectric phenomenon to be present in a material.

6.5 What is meant by the phase changes with respect to a ferroelectric material? Use temperature dependence of the spontaneous polarization and dielectric constant to amplify your answer.

6.6 Distinguish between a ferroelectric, a relaxor, and an antiferroelectric material. Name two typical materials exhibiting these phenomena and how they are used in developing devices and applications?

6.7 What is a perovskite structure? With the help of a figure represent it and discuss how does it differ from a normal fcc or bcc crystals?

6.8 Discuss how you would use a Sawyer–Tower circuit to display the hysteresis loop of a ferroelectric material.

6.9 Consider that in the above experiment, the spontaneous polarization of the sample is 10 μC cm^{-2}, and it has the electrodes of 1 mm in diameter on its opposite faces. The resistance $R(2)$ in the circuit of Figure 6.15 is 5 kΩ and the source voltage is 100 V. If only 20% of the source voltage is resident on the reference capacitor (R), find the values of $C(R)$, $C(X)$, and $R(1)$.

6.10 Explain why ferroelectric capacitors are considered to be nonlinear and how they can be used to develop a memory element.

6.11 Draw a simple diagram representing the configuration of a FeRAM and describe the essential features of this device.

6.12 Consider a simple circuit consisting of a pyroelectric element with the electrode area = 10 mm^2. The element is heated at a constant rate 5 °C min^{-1}. If a current of 10 μA is generated for a temperature change of 10 °C, calculate the accumulated charge and the pyroelectric coefficient.

6.13 Consider a pyroelectric element of PZT ceramic with the thickness of 100 μm, electrode diameter of 2 mm, dielectric constant of 300, and pyroelectric coefficient of 400 × 10^{-6} C m^{-2} K^{-1}. Calculate the voltage that develops across the element for the change in temperature of 10 K. Comment on the result.

References

1 Pandey, R.K. (2005). *Ferroelectric Materials*, Encyclopedia of RF and Microwave Engineering, vol. 2 (ed. K. Chang), 1505–1519. John Wiley and Sons. (Substantially reproduced with minor modifications).

2 Valasek, J. (1920). Piezoelectric and allied phenomena in Rochelle salt. *Phys. Rev.* 15: 537.

3 Fatuzo, E. and Merz, W.J. (1966). *Ferroelectricity*. Amsterdam: North-Holland.

4 Kao, K.C. (2004). *Dielectric Phenomena in Solids*. Elsevier Academic Press.

5 Merz, J. (1949). *Phys. Rev.* 76: 1221.

6 Jamieson, P.B., Abrahams, S.C., and Bernstein, J.L. (1968). *J. Chem. Phys.* 48: 5048.
7 Lines, M.E. and Glass, A.M. (1977). *Principles and Applications of Ferroelectricity and Related Materials.* Clarendon Press.
8 Fridkin, V. and Ducharme, S. (2014). *Ferroelectricity at the Nanoscale Technology*. Springer Verlag.
9 Sawyer, C.B. and Tower, C.H. (1930). *Phys. Rev.* 35 (3): 269.
10 Hooten, J.A. and Merz, W.J. (1955). Ferroelectric domains in barium titanate crystals. *Phys. Rev.* 85 (2): 409.
11 http://wikipedia.org/piezorespnse.
12 ROHM Co. Ltd., Bulletin No. 2011, June 21 (2011): source internet.
13 Kittel, C. (1951). Theory of antiferroelectric crystals. *Phys. Rev.* 82: 759.
14 Tan, X., Ma, C., Frederick, J. et al. (2011). The antiferroelectric–ferroelectric phase transition in lead containing and lead-free pervoskite ceramics. *J. Am. Ceram. Soc.* 94: 4091.
15 Byer, R.L. and Roundy, C.B. (1972). Pyroelectric coefficient direct measurement technique and application of a NSEC response time detector. *IEEE Trans. Sonics Ultrason.* 19: 333; see also Ferroelectrics, 3, 333, (1972).
16 Han, H., Song, X., Zhong, J. et al. (2004). Highly *a*-axis oriented Nb-doped $Pb(Ti_xZr_{1-x})O_3$ thin films grown by sol–gel technique for uncooled infrared detectors. *Appl. Phys. Lett.* 85 (12): 5310.
17 Lynn, E.G. and Edward, J.S. (1982). *J. Appl. Phys.* 53: 8974–8979.
18 Han, H., Kotru, S., Zhong, J., and Pandey, R.K. (2008). Effect of Nb doping on pyroelectric property of lead zirconate titanate films prepared by chemical solution deposition. *Infrared Phys. Technol.* 51: 216.
19 Whatmore, R.W. (1986). Pyroelectric devices and materials. *Rep. Prog. Phys.* 49: 1333–1386.
20 Cross, L.E., Bhalla, A.S., Ainger, F., and Demjanovic, D. (1991). Pyro-optic detector and imager. U.S. Patent No. 4,994,672.
21 Jie-Fang Li, D., Viehland, A.S.B., and Cross, L.E. (1992). *J. Appl. Phys.* 71 (5): 2106.
22 Moulson, J. and Herbert, J.M. (2003). *Electroceramics*, 2e. Wiley.

Further Reading

Jona, F. and Shirane, G. (1993). *Ferroelectric Crystals.* Dover Publications.

7

Elements of A Semiconductor

CHAPTER MENU

Science is not only a discipline of reason but, also, one of romance and passion.

Stephen Hawking

7.1 Introduction

Semiconductor materials dominate modern electronics and microelectronics technology and have greatly influenced our civilization. Their impact is second to none in each and every aspect of our lives. There is hardly any device that we use in our daily life that has remained immune from the impact of semiconductors be it a child's simple toy or a spaceship exploring the universe in quest of knowledge.

We all are familiar with the words diode, transistors, electronic switches, amplifiers, etc. They are ubiquitous in our lives, and it is hard to imagine how the modern society could even function without these wonder devices. We use a block diagram shown in Figure 7.1 to introduce our readers to the essential elements of semiconductor materials and devices. It is nevertheless just a brief representation of the field and, by no means, can the vastness of semiconductor technology be encompassed in one single figure. We have tried to do justice to this subject in this chapter particularly in describing the theoretical aspect of the semiconductor phenomenon relevant to the understanding of the nature and properties of electroceramic semiconductors that we will cover in the next chapter. It should be pointed out here that hardly any standard solid-state physics or electronics textbook devotes even one page to deal with electroceramic semiconductors in spite of the fact that ceramics, especially oxides and nitrides, play a crucial role in the success of conventional semiconductor technology.

7.2 Nature of Electrical Conduction in Semiconductors

As the name suggests semiconductors area special class of materials whose electrical conductivity is found to be between the two extremes of metal and insulators as already discussed briefly in Chapter 1. There we discussed the concept of electrical conductivity to distinguish between different classes of materials. In general, it holds true that the conductivity (σ) of a metal is greater than that of a semiconductor, which in turn is greater than the conductivity of an insulator. We can also express it in terms of resistivity as given in Eq. (7.1).

$$\rho_m \ll \rho_s \ll \rho_i \tag{7.1}$$

In these equations, the subscripts *m*, *s*, and *i* obviously refer to metals, semiconductors, and insulators, respectively. We give the values of electrical conductivity for some of the three categories of materials in Table 7.1. The conductivity of a metal is in the orders of magnitude greater than that of a semiconductor, which can be ascertained from the values tabulated in this table.

Besides electrical conductivity, we also can distinguish between solids on the basis of their respective bandgaps as also discussed in Chapter 1. In general, metals have

Fundamentals of Electroceramics: Materials, Devices, and Applications, First Edition. R. K. Pandey.

Companion website: www.wiley.com/go/Pandey/Fundamentals_Electroceramics

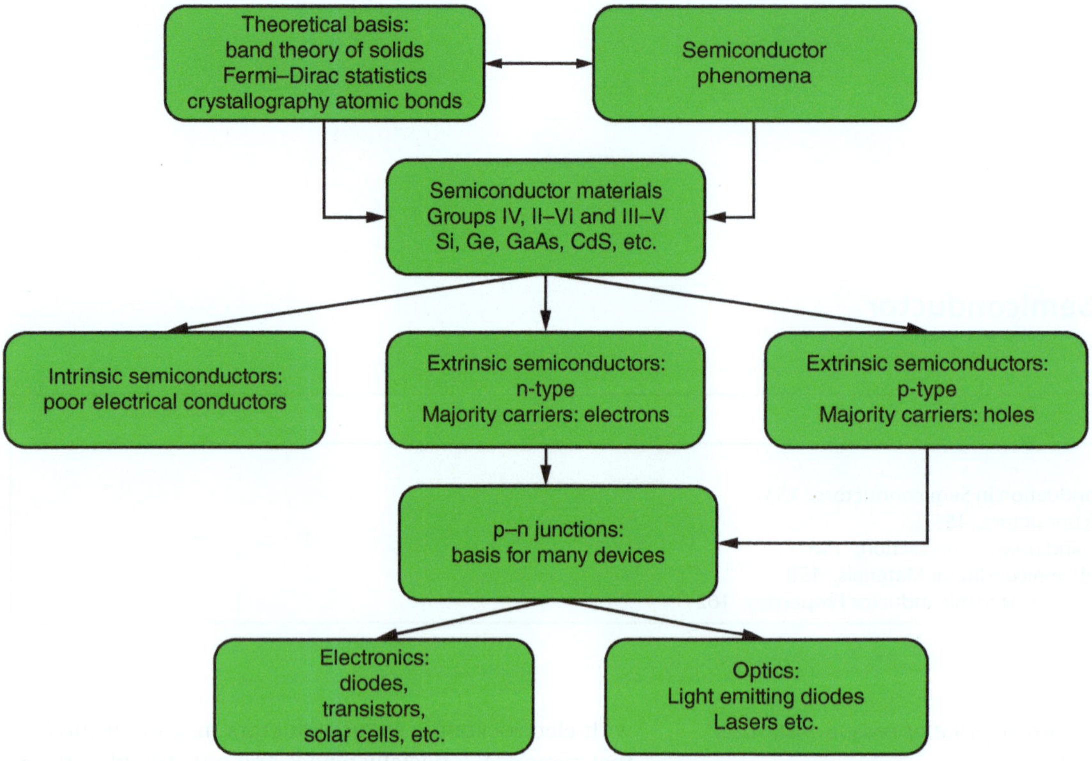

Figure 7.1 Block diagram showing a roadmap from theory to semiconductor materials to devices.

Table 7.1 Electrical resistivity for selected solids.

Room temperature resistivity	Metals	Semiconductors	Insulators
ρ (Ω m)	Ag: 1.59×10^{-8}	Si: 6.40×10^{2}	Al_2O_3: 1×10^{14}
	Au: 2.44×10^{-8}	Ge: 4.6×10^{-1}	Fused quartz:(SiO_2): 75×10^{16}
	Pt: 1.06×10^{-7}	GaAs: 1×10^{8}	SiO_2: 1×10^{13}
	Cu: 1.68×10^{-8}	Diamond (C): 1×10^{12}	Titania (TiO_2): 1×10^{12}

no bandgaps gaps, whereas insulators have very large bandgaps. The bandgap of a semiconductor lies between these two extremes. Figure 1.1 shows the distinction between metals, semiconductors, and insulator based on the bandgap criterion. Here, we find that the valence band and conduction band overlap for metals, whereas they are separated from each other by a bandgap for the other two cases. Relatively speaking, the insulators have large bandgaps; semiconductor bandgaps are much smaller than that of insulators, whereas there is no bandgap in metals.

The value of the bandgap plays a very important role in determining the technical importance of a semiconductor. In Table 1.2, values for the bandgap of some prominent semiconductors are given. On the basis of the bandgap, we can distinguish semiconductor materials in three main categories that are narrow bandgap, midlevel bandgap, and wide bandgap materials. Silicon with its 1.1 eV bandgap is considered to be a member of midlevel group. Materials with bandgaps greater than 2 eV are classified as wide bandgap materials, and practically, all electroceramic semiconductors such as ZnO and TiO_2 fall into this group. Electroceramic semiconductors are also called oxide semiconductors, and they are actively researched for new applications and technology. Indium antimonide (InSb) and lead telluride (PbTe) are members of the narrow bandgap group. Their bandgaps are 0.18 (InSb) and 0.30 eV (PbTe), respectively. Diamond (C) is a classic example of a wide bandgap material with the bandgap value of 5.47 eV. The other examples are aluminum nitride (AlN) and gallium nitride (GaN) with the bandgap values of 6.28 and 3.44 eV, respectively. In general, oxide semiconductors also have wide bandgaps. Some examples being: titanium oxide (rutile, TiO_2)

having the bandgap of 3.02 eV, copper(I) oxide (Cu_2O) of 2.17 eV, zinc oxide (ZnO) of 3.37 eV. It should be noted that before the advent of the silicon era, copper oxide was widely studied and used in devices especially in producing rectifying diodes.

7.3 Energy Bands in Semiconductors

The semiconductor models are primarily based on the band theory of solids according to which in every solid there are energy bands that are classified as either as allowed bands or forbidden bands. In a semiconductor, charge carriers are allowed in the two bands, whereas none in the forbidden gap. The allowed bands are also called the covalent band and valence band and the forbidden gap the bandgap.

The essential concepts of the band structure of solids are described by the Kronig–Penney model, which we have discussed in Chapter 1. We can picture a semiconductor material as the one in which the electrons can occupy energy states in the conduction band, whereas the holes are confined in the valence band and the bandgap is off limits to either electrons or holes. This is also the scenario for insulators with wide bandgaps.

In the case of a semiconductor material, some charge carriers can find their home in the forbidden gap as well when it is doped with suitable atoms to enhance its conductivity so that the material becomes technologically interesting. In its purest state with no defects, a single crystal of a semiconductor material does not contain any charge carriers in its bandgap region. Such a material is called an *intrinsic semiconductor* in which the number of electrons is equal to the number of holes such that $n_i = p_i$, where n_i is the number of electrons and p_i the number of holes, and "i" indicates an intrinsic semiconductor at a given temperature.

Another relevant theory for understanding of the occupancy states of the bands in a semiconductor is the Fermi–Dirac statistics that is one of the most elegant theories of physics and has led to the development of band theory of solids. We have seen in Chapter 1 that the Fermi–Dirac distribution correctly predicts the population and depopulation of energy states above the Fermi energy, E_F. The detailed analysis of the Fermi function, $F(E)$, in Eq. (1.56) led us to very important results summarized in Table 7.2.

We can infer from the above table that the Fermi energy can act as a gatekeeper between the occupied and unoccupied energy states in the valence and conduction bands. In an intrinsic semiconductor material at the absolute zero, the position of the Fermi energy (which is the Fermi level) lies in the midpoint of the bandgap. Its position can change as the temperature rises or when a material is heavily doped. It is important to realize, however, that the product of Boltzmann constant, k_B, and temperature, T (that is, k_BT) is approximately 0.025 eV ($1\,\text{eV} = 1.60\times 10^{-19}\,\text{J}$) at room temperature (293 K), which is very small in comparison to the width of the bandgap of commonly known semiconductor materials. Therefore, the temperature has to rise very high before the shift in its position from the midpoint would be of any consequence. Another important point to consider is the real nature of k_BT. The average thermal energy of a molecule that has three degrees of freedom is equivalent to $\left(\frac{3}{2}k_BT\right)$ and not just k_BT as is customary to use in statistical model. The reason being that when we speak of thermal energy in general terms, we actually mean the probability of finding a particle in the energy states between E and $E + \Delta E$, which is proportional to $\exp\left(-\frac{E}{k_BT}\right)$. If we were to consider an actual thermodynamic system, we would, of course, use the term $\left(\frac{3}{2}k_BT\right)$.

We find in Table 7.2 that at the temperature $T = 0\,\text{K}$ and when $E < E_F$, all electrons are confined below the Fermi energy and all states above it is totally empty. Consistent with the band theory of solids and the Fermi–Dirac statistics at absolute zero, all available electrons in an intrinsic semiconductor material must find their respective energy states in the valence band and occupy them. Since there are more energy states than the available number of electrons, nothing is left to go to the conduction band. Because of the full occupancy of the valence band, there is no vacant sites available at all for the electrons to roam around giving rise to even a minimal level of electrical conduction.

Table 7.2 Relationship between energy and Fermi function at a thermal equilibrium.

Energy	F(E)	Temperature (K)	Consequence
$E > E_F$	0	0	All energy states above the Fermi energy are totally empty
$E < E_F$	1	0	All energy states below the Fermi energy are 100% occupied
$E = E_F$	0.5	Any temperature	The probability of occupancy at the Fermi energy is always 50%
$(E - E_F) \gg k_BT$	None	High temperatures	Fermi–Dirac statistics merges into Maxwell–Boltzmann statistics

The energy band structure of an intrinsic semiconductor material is given in Figure 7.2. In this diagram, E_C is the energy level at the bottom edge of the conduction band, E_V is the energy level at the top of the valence band edge, and E_F is the Fermi energy. For an intrinsic semiconductor for which $n_i = p_i$, the Fermi level lies in the middle of the bandgap such that

$$E_F \approx \frac{E_g}{2} \approx \frac{E_C - E_V}{2} \tag{7.2}$$

From Figure 7.2, we can draw the conclusion that the Fermi level is below the bottom edge of the conduction band by an amount equal to $\frac{E_g}{2}$ and it is above the top edge of the valence band by an amount equal to $\frac{E_g}{2}$. The value of the bandgap plays a crucial role so far as applications of semiconductors are concerned. The conventional elemental semiconductors are found mostly in Group IV of the periodic table; whereas the compound semiconductors are combinations of Group II–VI and Group III–V elements. The list of semiconductor materials is too long to be reproduced in one single table, but we have put some selected examples of these three groups in Table 7.3. So far as the crystal structure is concerned, it is interesting to note that all elemental and compound semiconductors crystallize with a diamond-like unit cell. We have discussed the structure and properties of a diamond unit cell in Chapter 4.

7.4 Origin of Holes and n- and p-Type Conduction

In the previous section, we learnt that at absolute zero all electrons are frozen in the valence band with no mobility at all and therefore, unable to cause electrical conduction. Under such circumstances, we would

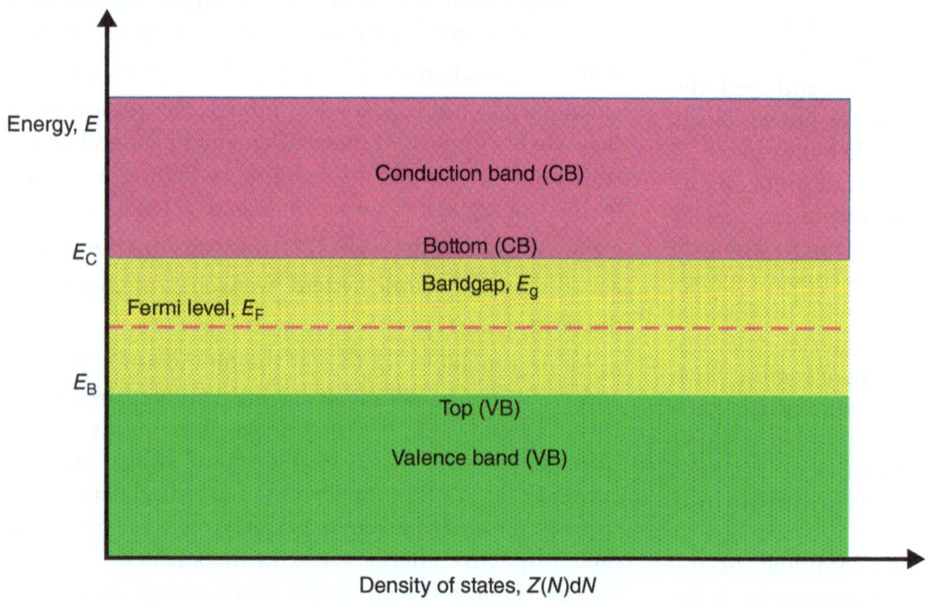

Figure 7.2 Idealized energy band diagram of an intrinsic semiconductor at 0 K.

Table 7.3 Bandgaps for some traditional semiconductors materials.

Materials	Bandgap (eV)	Group	Comments
Diamond, C	5.47	IV	Excellent thermal, conductivity, optical and mechanical properties
Silicon, Si	1.12	IV	Most used in microelectronics, solar cells, and for host of other applications
Germanium, Ge	0.67	IV	First material used for solid-state transistors. It has been replaced by silicon in microelectronics but is still used for some specialized applications
Gallium arsenide, GaAs	1.43	III–V	Widely used material for high-speed electronics, near infrared diodes (LED), and high-efficiency solar cells
Zinc sulfide, ZnS	3.54	II–VI	Commonly used for scintillators when doped

Source: Compiled from List of Semiconductor Materials, Wikipedia, the free encyclopedia. https://en.wikipedia.org/wiki/List_of_semiconductor_materials.

expect a semiconductor to have very large resistance, which indeed is the case. One can easily prove this experimentally by measuring the resistivity as a function of temperature. The plot of Figure 7.3 shows the experimental result found for practically all semiconductor materials.

We see here that as the temperature is lowered, the resistivity increases rapidly approaching the value of infinity at 0 K. In other words, all semiconductor materials tend to be insulating in nature at very low temperatures. Then the question arises under which conditions superconductors can conduct? The answer to this question can be found in the thermal energy that is absorbed by a material at a finite temperature. When the temperature rises, the crystal lattice can acquire sufficient thermal energy to set up phonon (another name for lattice) vibration allowing some electrons closest to the edge of the valence band to escape creating equivalent vacancies in the valence band. The absence of an electron at the allowed energy states is equivalent to having positively charged carriers that are called *holes*. At sufficiently elevated temperatures, the acquired kinetic energy can be sufficient for some electrons to overcome the forbidden gap and end up in the conduction band. Now we have a picture of a semiconductor with a few electrons in the conduction band and an equivalent number of holes in the valence band. Consequently, we should be able to measure the electrical conductivity, which would be the sum of the contributions made by each of the two types of carriers. This is expressed in Eq. (7.3).

$$\sigma_{net} = \sigma_{electron} + \sigma_{hole} \tag{7.3}$$

There are many possible ways in which holes can be created in the valence band and electrons pushed

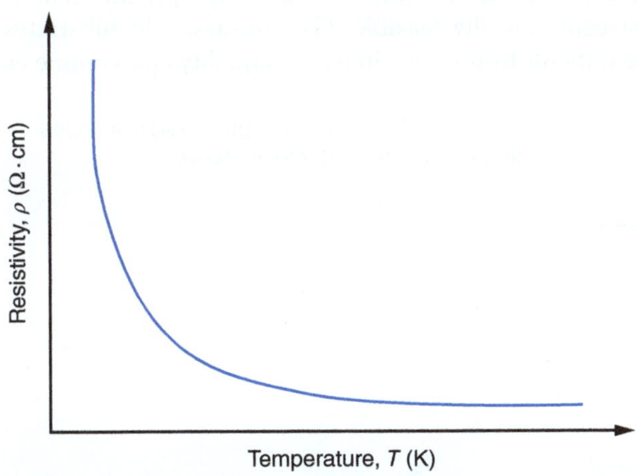

Figure 7.3 Temperature dependence of resistivity of a semiconductor material.

into the higher energy states of the conduction band. Temperature is one such agent; the other agents could be radiation of appropriate photon energy, ionizing radiation and by introducing electrons and holes by the process called *doping*. Doping is the most used method in practice. Once doped, the material is classified as extrinsic semiconductor material of which there are two types: one being the n-type and the other p-type.

To facilitate the understanding of these two types of semiconductors, let us consider the conductivity Eq. (7.3). There are three components of conductivity: the total conductivity (σ_{net}), the electron conductivity (σ_e), and the hole conductivity (σ_p). If $\sigma_e > \sigma_p$, the material is classified as an n-type semiconductor. Similarly a p-type material is the one for which $\sigma_p > \sigma_n$.

Whether it is an n-type or a p-type material, it plays a crucial role in determining their usefulness in technology.

Doping is the process in which we either introduce dopants in an intrinsic semiconductor either thermally or by ion implantation. When we add an electron, the host material accepts extra electrons in its lattice making it richer in electrons and its conductivity then is primarily because of the negatively charged electrons. Therefore, the converted material becomes an n-type semiconductor. Similarly, the dopants that rob electrons from the host material makes it a p-type material and its conductivity is primarily because of the positively charge holes. Now the question is how we accomplish doping mechanism in practice? Chemistry here comes to our help. For example, we know that Si is found in Group IV of the periodic table and therefore, it is called tetravalent atom meaning that it has four electrons in its outermost orbit (or valence electrons). It can be made n-type simply by doping it with an element from Group V. Group V elements are pentavalent, and in the process of doping, they can give up one electron to the Group IV elements to complete the task of converting it to an electron-rich atom. Similarly, elements of Group III can be used to convert a Group IV semiconductor to a p-type material. With respect to the host semiconductor of Group IV atoms, atoms of Group V are called *donors,* and those of Group III the *acceptors*. Now we can ask ourselves once again what is the mechanism through which the acceptors or donors can accomplish their assigned tasks? This is done by a process called *ionization*. When an element loses or gains an electron, we call the process ionization, and the energy involved in the process is called the *ionization energy*. For example, when we dope Si with antimony (Sb), the associated ionization energy is 39 meV. Similarly, ionization energy of 57 meV is involved when the dopant is Al. In Table 7.4, we give some examples of acceptors and donors for Group IV semiconductors.

Table 7.4 Examples of donors and acceptors for Si and Ge and their atomic numbers.

Donors	Group V elements	Atomic number
	Phosphorous, P	15
	Arsenic, As	33
	Antimony, Sb	51
Acceptors	Group III elements	
	Boron, B	5
	Aluminum, Al	13
	Gallium, Ga	31
	Indium, In	49

When an intrinsic semiconductor material is doped, we intentionally introduce impurity ions in the host lattice. In this process, for n-type dopants, most of the impurity atoms are ionized enabling the conduction band to be populated. The corresponding energy level that we call the donor level (E_d) lies below the edge of the conduction band. Similarly, for p-type dopants, the acceptor level (E_a) would reside right above the valence band edge of E_V. Both these ionization energy levels are of the order of 10 meV, which is just 40% of 25 meV of thermal energy at room temperature. The relative positions of the two ionization levels are shown in Figure 7.4.

7.5 Important Concepts of Semiconductor Materials

So far we have discussed the basic nature of the intrinsic and extrinsic semiconductors without going into any complicated aspects of this group of materials. Nevertheless, we do need to include the actual nature of the bandgap, concept of the mobility of charge carriers, and effective mass. Without this, our education in semiconductor will remain incomplete.

7.5.1 Mobility, μ

From Eq. (7.3), we learned that the conductivity of a semiconductor consists of the contributions made by both electrons and holes. We can deal with this equation now in some greater depth. By definition the conductivity of a solid is $\sigma_e = (ne)\mu$ as expressed in Eq. (1.11). The product (ne) has the special name of charge density. The mobility was defined in Eq. (1.10), which says that $\mu_e = \left(\frac{V_d}{E}\right) \approx \left(\frac{\Delta V_d}{\Delta E}\right)$ where V_d is the drift velocity.

When both electrons and holes are present, we can express the net conductivity more accurately as sum of contributions made by electrons and holes in Eq. (7.3). That gives us Eq. (7.4).

$$\sigma_{net} = e(n_e\mu_e + p_h\mu_h) \quad (7.4)$$

Obviously, here n_e, and p_h refer to electron and hole concentrations, respectively; and μ_e and μ_h their motilities.

At this point, we introduce a new term n_i and define it as the square root of the product np.

$$n_i = \sqrt{n_e p_h} \quad (7.5)$$

For an intrinsic semiconductor $n_e = p_h = n_i$, which amounts to calling n_i the intrinsic number of charge carriers. The product np is always n_i^2 irrespective of the fact if the material is intrinsic or extrinsic. This is because when n increases p decreases and vice versa making n_i^2 a constant.

The mobility, μ, is a very important parameter for semiconductors because its high value means a faster speed for the device. For example, the electron mobility for silicon is 15×10^2 cm^2 (V s)$^{-1}$, whereas for gallium arsenide, it is 85×10^2 cm^2 (V s)$^{-1}$. Therefore, the transistors built on gallium arsenide substrates are approximately six-times faster than those built on silicon. It is then obvious that everybody will like to have faster and faster devices especially computers and laptops, but that is not economically feasible. Gallium arsenide substrates are difficult to produce in as high quality and volume as

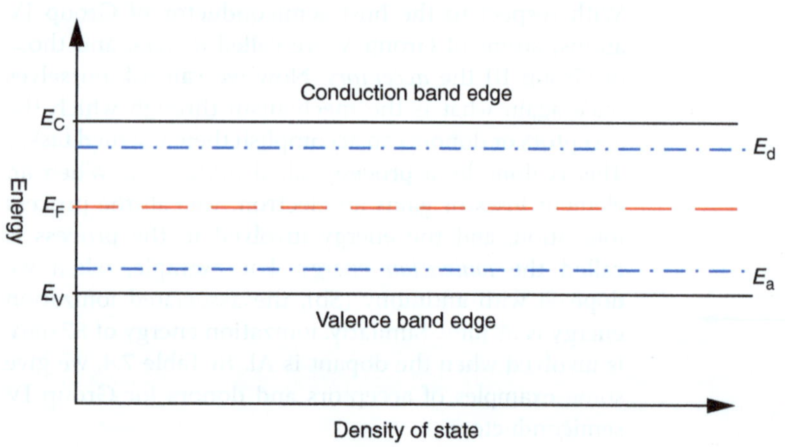

Figure 7.4 Donor and acceptor levels in a doped (impurity) semiconductor material.

silicon and it costlier than silicon. Therefore, the industry stays with silicon for consumer electronics and looks for other substrate materials such as gallium arsenide only when the device speed is a real issue. Another important point to notice is that the electron mobility is almost always much higher than the hole mobility. For example, for silicon, it is 600 cm^2 (V s)$^{-1}$, which is only 40% of the electron mobility. This tells us that n-type materials are preferable to p-type materials for practical applications. Exception is always there, but it is not abundant. Few devices must be built on p-type materials.

7.5.2 Direct and Indirect Bandgap, E_g

Semiconductors are also categorized on the basis of the nature of their bandgaps. There are two types of bandgaps: direct and indirect. How the transport of electrons takes place from the valence band to the conduction band defines the one type or the other. Simply stated, when during the transport, no loss of energy takes place, then the material is said to have direct bandgap. Otherwise, it is indirect. We offer now a more scientific explanation.

When a semiconductor surface is irradiated, it absorbs photon energy resulting in the transfer of this energy to the charge carriers that reside close to the valence band edge. When sufficient energy is absorbed some of the electrons might gain enough energy to overcome the bandgap and reach the closest available energy states near the conduction band edge. Let us say that the frequency of radiation is ν. According to the Planck's law of radiation, the photon energy associated with this radiation is given by

$$E_{ph} = h\nu = \frac{h}{2\pi} \cdot \left(\frac{2\pi}{\lambda_{ph}}\right) = \hbar k \tag{7.6}$$

where h is the Planck's constant, λ_{ph} the wavelength, $\hbar = \left(\frac{h}{2\pi}\right)$, and k the wave vector. The laws of physics demand that the conservation of energy and momentum must be conserved and therefore, the final radiative energy and the corresponding momentum can be expressed with the help of Eqs. (7.7) and (7.8).

$$E_{final} = E_{initial} + h\nu + E_{phonon} \tag{7.7}$$

$$k_{final} = k_{initial} + k_{photon} + k_{phonon} \tag{7.8}$$

The transition from valence band to conduction band takes place in one of the two possible ways that are direct and indirect. In a direct bandgap transition, there is no exchange of momentum, which means that $\Delta k = 0$, whereas in the indirect bandgap, transition exchange of momentum takes place and therefore, $\Delta k \neq 0$. Figure 7.5 shows the mechanisms enabling the two types of transitions to set in.

For direct bandgap transition, the maxima of the valence band must coincide with the minima of the conduction band, as seen in Figure 7.5a, in order for the momentum to remain constant. On the other hand, for an indirect bandgap transition, the condition that must be fulfilled is that the maxima of the valence band must not coincide with the minima of the conduction band (Figure 7.5b). This requires a change in moment during the process of transition. When a direct bandgap semiconductor absorbs light of frequency ν, the associated photon energy will be equal or greater than the bandgap. That is,

$$E_{ph} \geq E_g \geq h\nu \geq \frac{ch}{\lambda} \tag{7.9}$$

where c is the velocity of light in free space, h the Planck constant, and λ the wavelength of light. From Eq. (7.9), we conclude that the minimum photon energy that is absorbed during the direct bandgap transition equals

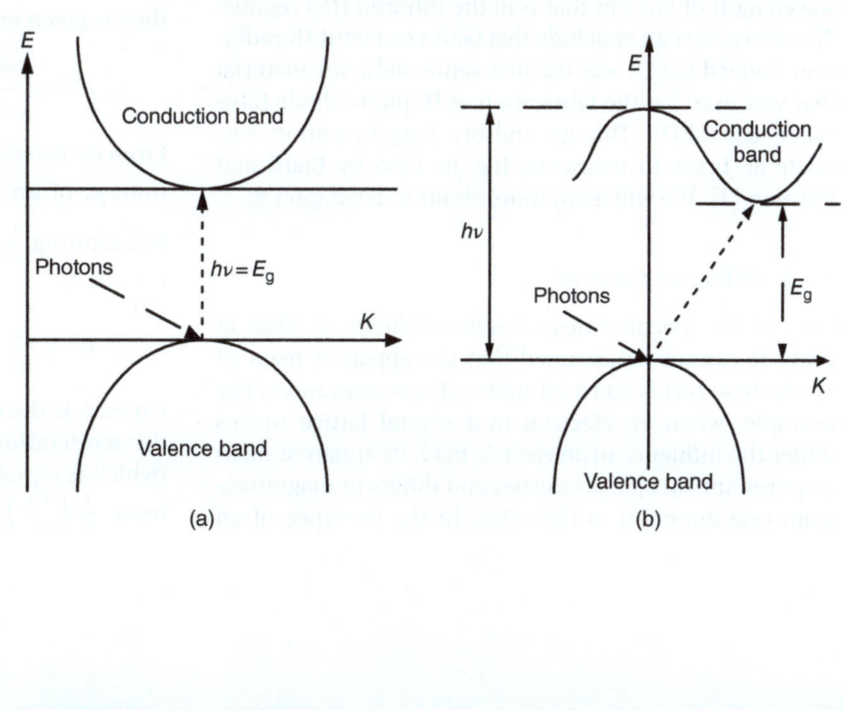

Figure 7.5 (a) Direct band-to-band transition, and (b) indirect band-to-band transition. Source: From Srivastava and Srinivasan 1987 [1].

to the bandgap, E_g. This is a handy tool to find out the photon energy necessary for the completion of electron transition from the valence band to the conduction band in a direct bandgap material such as GaAs having the bandgap of 1.43 eV.

The knowledge of these two types of bandgaps plays a crucial role in the design and fabrication of light-emitting diodes (LEDs). For LEDs, direct bandgap materials are preferred because the emission takes place without loss of intensity as would be the case if indirect bandgap materials were considered.

LEDs are rapidly becoming the technology of choice to replace the conventional incandescent bulbs where the loss of energy is substantial. We will discuss this issue in greater depth in the next chapter, when we would consider direct bandgap oxide semiconductors.

Exercise 7.1

Find the minimum wavelength of light absorbed for an electron to go from the valence band edge to the conduction band edge in GaAs. Discuss the result with respect to emitted radiation for an electron to transit from the conduction band edge to the valence band edge. Can you attach any technical importance to your observations?

Solution

Using (Eq. (7.9)) and the values of $c = 3\times10^8\ \mathrm{m\,s^{-1}}$, $h = 6.62\times10^{-34}$ J s, and 1 eV = 1.60×10^{-19} J we get:

$$\lambda_{\text{absorbed}} = \frac{ch}{E_g} = \frac{3\times6.62}{1.43\times1.60}\times10^{-7}\ \mathrm{m} = 868\ \mathrm{nm}$$

It is obvious that when the transition takes place from the conduction band edge to the valence band edge, the emitted photon energy would be equal to the absorbed energy corresponding to the 868 nm wavelength of photon if we can neglect any loss during the transition. That would mean that the emitted radiation would have a wavelength of 868 nm that is in the infrared (IR) regime. Therefore, we can conclude that GaAs can emit IR radiation. Indeed GaAS was the first semiconductor material that was used for the fabrication of IR photo diode (also called the LED). This ground-breaking invention was made at Texas Instruments, Inc. in 1966 by Biard and Pittman [2]. We will learn more about it in Chapter 8.

7.5.3 Effective Mass, m^*

One of the peculiarities of semiconductor is that at times it cannot be assumed that the apparent mass of an electron (m) is constant under all circumstances. For example, when an electron in a crystal lattice moves under the influence of an electric field, its apparent mass acquires anisotropic properties and differs in magnitude from one direction to the other. In the presence of an electric field, the electron inside the crystal lattice experiences a force and moves in the direction determined by the field with an average drift velocity giving rise to electrical conduction.

The force that the electron experiences because of the applied electric field has two components. One is the classical Newtonian force and the other due to the interaction with the lattice. The classical force is easy to explain on the basis of the second law of motion (i.e. $F = ma$, where F is the force, m the mass, and a the acceleration). But the force due to the interaction with the lattice can be explained only on the basis of the band theory of solids. According to this theory, as we already discussed in Chapter 1, the movement of an electron in a periodic potential over distances larger than lattice spacing's can be very different from its motion in vacuum. The concept of effective mass (written as m^*) was conceived to explain the band structures of solids. In certain situations, electron effective mass cannot be ignored in dealing with the theory of semiconductors. Let us now examine this statement in some detail.

When an electron propagates through space such as in a periodic lattice as a wave its velocity is not the same as the drift velocity. The characteristic velocity of an electron when treated as a wave is called the *group velocity*, V_g. It is defined as the ratio between the wave frequency, ω, and wave vector, k ($V_g = \omega/k$). We can also write it more elegantly as in Eq. (7.10).

$$V_g = \frac{d\omega}{dk} \approx \frac{d}{dk}(2\pi\nu) \tag{7.10}$$

Since $E = h\nu$ and $\hbar = \left(\frac{h}{2\pi}\right)$, we can also express Eq. (7.10) as Eq. (7.11).

$$V_g \approx \frac{1}{\hbar}\left(\frac{dE}{dk}\right) \tag{7.11}$$

The acceleration, a^*, associated with the group velocity then is given by Eq. (7.12).

$$a^* = \left(\frac{dv_g}{dt}\right) \approx \frac{1}{\hbar}\left(\frac{d^2E}{dk^2}\right)\cdot\left(\frac{dk}{dt}\right) \tag{7.12}$$

From de Broglie relationship, we know that the momentum, p, of an electron is ($\hbar k$) such that $\frac{dp}{dt} = \hbar\left(\frac{dk}{dt}\right)$. Substituting $\frac{1}{\hbar}\left(\frac{dp}{dt}\right)$ for $\left(\frac{dk}{dt}\right)$ in Eq. (7.12), we arrive at Eq. (7.13).

$$a^* = \frac{1}{\hbar^2}\left(\frac{d^2E}{dk^2}\right)\cdot\left(\frac{dp}{dt}\right) \approx \frac{1}{\hbar^2}\left(\frac{d^2E}{dk^2}\right)\cdot F \tag{7.13}$$

where F is the classical force of the electron. Comparing the acceleration, a^*, with the Newtonian acceleration, a (which is equal to force/mass) we can conclude that the term $\frac{1}{\hbar^2}\left(\frac{d^2E}{dk^2}\right)$ must be equal to $\left(\frac{1}{m^*}\right)$. From this, we learn that the effective mass, m^* is inversely proportional to the curvature of the E–k diagram of the band theory discussed in Chapter 1. It is also inversely proportional to mobility such that the product $\mu m^* \approx \rho\tau$, where ρ is the resistivity and τ the average time between two collisions of electrons in a lattice. It is also called the relaxation time and mean free time.

The concept of the effective mass is not just theoretical. Its existence can be verified. It has been determined experimentally by the cyclotron resonance experiment for many metals and semiconductors including Si and Ge. For example, for Si, the ratio between the effective mass (m^*) and the rest mass (m_0) is 0.49 for electrons and 0.19 for holes. Without going into the details of the theory of cyclotron resonance, let us simply accept the fact that for the resonance to occur it must satisfy the condition of $m^* \approx \left(\frac{eB}{\omega_c}\right)$, where B is the magnetic field and ω_c the critical frequency. We would expect number of resonance in the absorption versus frequency plot and each of them corresponding to the effective mass of either the electron or the hole in a semiconductor sample. We find it to be true as can be seen from Figure 7.6.

Notice the strong resonances that occur when electrons and holes are encountered by the magnetic field. The peaks occur at ω_{c1} for electrons and for holes, they are at ω_{c2} and ω_{c3}.

7.5.4 Density of States and Fermi Energy

The density of states is a very important concept for computing the available energy states in solids but is particularly important for understanding some of the very important characteristics of semiconductor materials. An insight in this concept helps us in determining the carrier concentrations and energy distributions of carriers in semiconductors.

The carriers in semiconductors can acquire zero, one or two spatial dimensions. For example, quantum dots are zero dimensional, whereas the quantum wires and quantum wells are one dimensional (1D) and two dimensional (2D), respectively. Their density of states must be known to understand these very specialized semiconductor phenomena and devices.

The calculation of density of states is based on the quantum mechanical concept of infinite potential well of certain width with rigid walls so that electrons can't escape. Its energy, E, is given by

$$E = \left(\frac{h^2}{8m_e}\right)\left(\frac{1}{W^2}\right)n^2 \tag{7.14}$$

where n is the principal quantum number, and W is the width of the potential well.

Let us extend this model to the case of a cubic crystal with the lattice constant of 1 nm. Using the values of the physical constants, we get the value of $E \approx 3.75 \times 10^{-2}$ eV from Eq. (7.14) for the crystal lattice.

In the case of a 3D structure that a crystal is, n can be treated as a vector to a point in space defined by (n_x, n_y, n_z). Then according to the rule of quantum mechanics n must satisfy the following condition.

$$n^2 = (n_x^2 + n_y^2 + n_z^2) \tag{7.15}$$

For the cubic structure, then $n^2 = 3n_x^2$. Therefore, we can write the energy, E_{nx}, corresponding to $n^2 = 3n_x^2$.

$$E_{nx} \approx 3.75 \times 10^{-2}(3n_x^2) \tag{7.16}$$

If, $E_{nx} = 1$ eV, we compute the value for $n_x \approx 3$ from Eq. (7.16). Substituting this in Eq. (7.15), we get $n \approx 5$, which is also an integer. Therefore, by replacing a model of the potential well with a cubic crystal, we have not violated any rules of quantum mechanics.

Let us now find the value for the energy that resides right below the energy E_{nx}. This we can do simply by

Figure 7.6 Absorption vs. frequency plot for a semiconductor sample in cyclotron experiment.

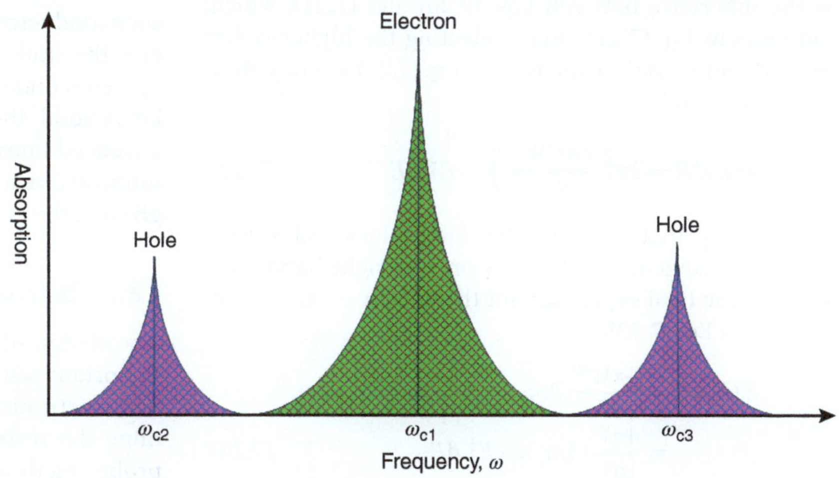

substituting $(n_x - 1)^2$ for one of the three n_x^2 terms in Eq. (7.16).

$$E_{nx-1} \approx 3.75 \times 10^{-2}\{(n_{x-1})^2 + 2n_x^2\} \tag{7.17}$$

The difference of energy between these two neighbors is given by Eq. (7.18).

$$\Delta E \approx (-2n_x + 1) \approx -5 \text{ eV} \tag{7.18}$$

This level of energy is approximately five-times larger than the bandgap of silicon. With this level of energy an electron can easily reach the upper level of the conduction band from the valence band by completely skipping the bandgap during its migration. There are plenty of available energy states waiting to be populated. We now venture to find an expression for the density of state, $Z(E)\mathrm{d}E$. To facilitate this calculation, we need to know that in a quantum number space, each integer represents one energy state. This then translates to a theorem, which states that

$$\text{Number of energy states in any volume} = \text{Numerical value of the space} \tag{7.19}$$

From this, we can conclude that in a sphere of radius n, the number of states is equal to $\left(\frac{4\pi}{3}n^3\right)$. This leads us to Eq. (7.20).

$$\frac{4\pi}{3}n^3 = \frac{4\pi}{3}\left(\frac{8m_e W^2}{h^2}\right)^{\frac{3}{2}} \cdot E^{\frac{3}{2}} = \text{Number of states having energy less than } E \tag{7.20}$$

Similarly,

$$\text{Number of states having energies less than } E + \mathrm{d}E = \frac{4\pi}{3}\left(\frac{8m_e W^2}{h^2}\right)^{\frac{3}{2}} \cdot (E + \mathrm{d}E)^{\frac{3}{2}} \tag{7.21}$$

Therefore, the density of states, $Z(E)\mathrm{d}E$ for E and $E + \mathrm{d}E$ is the difference between Eqs. (7.20) and (7.21), which amounts to Eq. (7.22) after neglecting the higher-order negative terms is the expansion of Eq. (7.21) because they are not allowed.

$$Z(E)\mathrm{d}E = 2\pi\left(\frac{8m_e W^2}{h^2}\right)^{\frac{3}{2}} \cdot E^{\frac{1}{2}}\,\mathrm{d}E \tag{7.22}$$

Considering that each spin has two values and a little more rearrangement of terms on the right-hand side gives us the final expression for the density of states that is given in Eq. (7.23).

$$Z(E)\mathrm{d}E = \frac{4\pi W^3}{h^3}(2m_e)^{\frac{3}{2}} \cdot E^{\frac{1}{2}} \cdot \mathrm{d}E = \frac{4\pi V}{h^3}(2m_e)^{\frac{3}{2}} \cdot E^{\frac{1}{2}}\,\mathrm{d}E \tag{7.23}$$

where V is the volume of the cube structure. Substituting A for $\left\{\frac{4\pi V}{h^3}(2m_e)^{\frac{3}{2}}\right\}$, the above equation becomes identical to Eq. (1.59).

The density of states gives us the number of available states, but not the number of states that are populated. This we can also get by integrating $Z(E)\mathrm{d}E$ between $0 < E < E_F$ and setting it equal to the product NV, where N = number of electrons per unit volume and V the volume. In other words,

$$\int_0^{E_F} Z(E)\mathrm{d}E = NV \tag{7.24}$$

And by integrating Eq. (7.23), we are setting it to get Eq. (7.25).

$$\int_0^{E_F} Z(E)\mathrm{d}E = \frac{4\pi V}{h^3}(2m)^{\frac{3}{2}}\int_0^{E_F} E^{\frac{1}{2}}\,\mathrm{d}E \tag{7.25}$$

Comparing these two equations and doing some rearrangements, we can get an expression for the Fermi energy that is given by Eq. (7.26).

$$E_F \approx \left(\frac{h^2}{8m_e}\right) \cdot \left(\frac{3N}{\pi V}\right)^{\frac{2}{3}} \approx \frac{h^2}{8m_e}\left(\frac{3n}{\pi}\right)^{\frac{2}{3}} \tag{7.26}$$

This is the same equation as Eq. (1.60). Values for Fermi energy for some metals are given in Table 1.18.

7.6 Experimental Determination of Semiconductor Properties

In this section, we should discuss experimental methods to determine at least the properties that are relevant to the goal of this chapter. They are the resistivity, band gap, and how to distinguish between a direct bandgap material and an indirect bandgap material. The concept of mobility also warrants our attention as it plays a crucial role in designing semiconductor devices. Of course, there are many more important properties of a semiconductor material that must also be addressed to give the students a thorough background in the physics of semiconductor materials, which obviously would be beyond the scope of this book. Nevertheless, for advanced knowledge in semiconductor phenomena, the interested readers are advised to consult the list of books given at the end of this chapter.

7.6.1 Determination of Resistivity, ρ

Knowledge of the resistivity of a sample is extremely important when it comes to semiconductor materials and electronic devices. The standard method to determine the resistivity of a bulk sample is the four-point probe method, which was originally developed by

L.J. van der Pauw in 1958 at Philips Labs in the Netherlands. The greatest benefits of this method are that samples of any shape can be used as well as it can completely eliminate the contributions originating from the contact resistance in the circuit. So the result obtained is the true representative of the material and nothing else.

What is the contact resistance and what is its source? The contact resistance is the resistance of the metallic contacts and necessary to connect the sample to the measuring equipment. The current–voltage (I–V) characteristics of the contact resistance can be either linear or nonlinear. The nonlinear behavior depends on many factors and plays a crucial role in the design and configuration of a rectifying device such as a diode. The contacts that can produce linear current–voltage (I–V) characteristic are called the ohmic contact, and the ones that produce nonlinear I–V behavior are called the nonohmic contacts. Its another name is rectifying contacts. Both types of contacts are necessary for semiconductor technology, and we need to study them. In the next chapter, we will discuss the roles they play in device design and its efficiency.

7.6.2 Four-Point Probe (van der Pauw) Method

The principle of the van der Pauw method is pretty simple. The configuration shown in Figure 7.7 is the simplified version of the actual experimental arrangement used for measuring the resistivity of a bulk sample. This configuration is, however, universally used for the determination of sheet resistance (or surface resistance) of a semiconductor sample resistance (Figure 7.7).

Four metallic contacts are placed evenly spaced on the polished surface of the sample, and they are usually a millimeter or less apart and of small dimensions. Current is applied between the two outermost contacts and voltage measured between the two innermost contacts using a precision high impedance voltmeter. Circular or square shapes for the contacts are the most used configurations because the well-defined geometry makes it more precise to determine the current density (current/area, I/A). By definition the relationship between resistance (R) and the resistivity (ρ) for a bulk sample with well-defined geometry is given by Eq. (7.27).

$$\rho = R\left(\frac{A}{L}\right) \tag{7.27}$$

where A is the cross-section area of the metal electrode and L the sample length. Assuming that the width of the sample is W and t its thickness, we can rewrite Eq. (7.28) as follows:

$$\rho = R\left(\frac{Wt}{L}\right) = R\left(\frac{W}{L}\right) \times t \tag{7.28}$$

If we know the film thickness, we can get its resistivity simply by multiplying it with the sheet resistance.

7.6.3 Two-Point Probe Method

We can also determine the resistivity of a material using the two-point probe method in which the measured resistance is the sum of sample resistance and the contact resistance. The experimental configuration that we can use is shown in Figure 7.8.

A row of small collinear dot electrodes are put on the surface of the sample. The interspatial distance between any two dots need not be the same. We apply a potential V, say between any two dots, and measure the resulting current to be I. The current I will see three series resistances in the circuit, namely two contact resistances (R_c) and one sample resistance (R_s). A simple analysis of the circuit leads us to Eq. (7.29). It states that

$$\frac{V}{I} = R_{net} = 2R_c + R_s \tag{7.29}$$

Now by some simple manipulation, we can determine the value of the contact resistance. By plotting a series

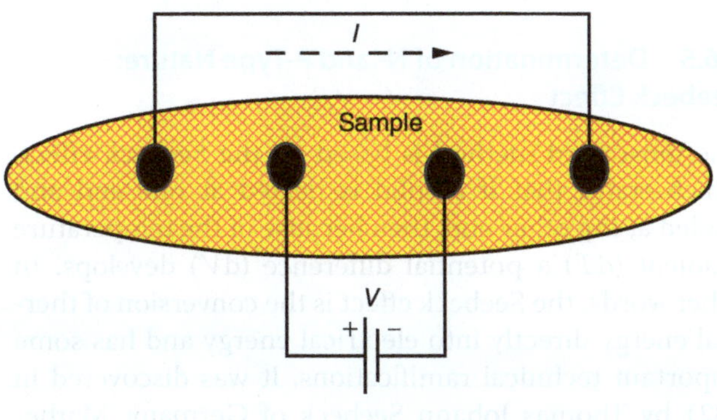

Figure 7.7 Schematic of four-point probe method for determination of current–voltage characteristics of a sample.

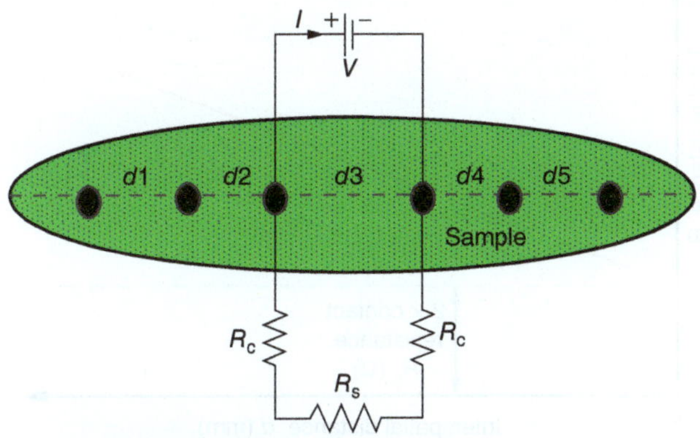

Figure 7.8 Configuration for measurement of sample resistance using two-point probe method.

of R_{net} corresponding to different segments between the first and the last dot, we would get a straight-line as seen in Figure 7.9.

This line will intercept the y-axis at $R_{net} = 2R_c$. The contact resistance remains constant and so we can find the value of sample resistance R_s simply by subtracting it from R_{net}. By knowing the sample geometry, we can easily get the value of the resistivity for both bulk and film samples.

Exercise 7.2

Silicon at room temperature has the following parameters: $n_i = 1.72 \times 10^{18}$ per m^3, $\mu_n = 0.15\,m^2\,(V\,s)^{-1}$ and $\mu_p = 0.45\,m^2\,(V\,s)^{-1}$. Find the sample resistance considering it is 1 cm long × 1 mm wide × 0.1 mm thick.

Solution

From Eq. (7.5) we get $n_e = p_h = 1.78 \times 10^{18}$ per m^3. And from Eq. (7.4) we get the conductivity, $\sigma = 173\,mS\,m^{-1}$ using $e = 1.602 \times 10^{-19}$ C. That is, the sample resistivity, $\rho = 5.78\,\Omega\,m$. Then using this value in Eq. (7.27), we get sample resistance, $R = 578\,k\Omega$.

7.6.4 Determination of Bandgap, E_g

The temperature dependence of the resistivity is a well-established method to determine the bandgap of semiconductors. The optical absorption method is another widely used method. We will discuss both these methods.

According to the theory of semiconductor, the temperature dependence of the conductivity is given by Eq. (7.30).

$$\sigma = CT^{\frac{3}{2}} \exp\left(-\frac{E_g}{2k_B T}\right) \tag{7.30}$$

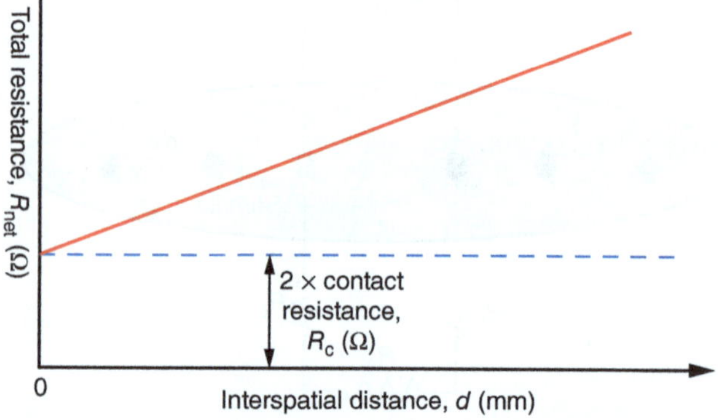

Figure 7.9 Total resistance as a function of interspatial distances in two-point probe method.

The constant of proportionality C consists of many parameters like the effective mass of electrons and holes, Boltzmann constant, k_B, and Planck constant, h. The full derivation of this equation is based on complex mathematical formulation. We give this in Appendix II for the benefit of interested readers. The exponential term in this equation dominates over the $T^{3/2}$ term, and this can be neglected for all practical purposes. The plot of $\ln \sigma$ vs. T^{-1} will be a straight-line with the slope equal to $-\frac{E_g}{2k_B}$. The value of the bandgap so obtained is found to be consistent with values determined from other experiments. The general nature of the plot between $\ln \alpha$ vs. T^{-1} for a typical semiconductor material is given in Figure 7.10.

From Figure 7.10, we learn that the intrinsic nature dominates at the high temperature regime and extrinsic behavior dominates at low temperature. They are separated by a region called the ionization region. Now we can ask ourselves a question whether a material with only one type of carrier (n- or p-type) will exhibit both intrinsic and extrinsic behaviors. The short answer is yes. The reason being that, like electrical conductivity, the Fermi level is also temperature-dependent making it possible for the intrinsic and extrinsic behaviors to be found in the same material but in different temperature regimes. From the slope of the two straight-line plots seen in Figure 7.10, we can easily evaluate the respective values of the bandgaps. At low temperatures, the impurities present in the material are responsible for the origin of the charge carriers by the process of ionization of impurity atoms. This process takes place over a certain temperature range that lies, as we can see, between the intrinsic and extrinsic regimes. Once all the impurity atoms are completely ionized, the intrinsic behavior sets in and prevails at high temperatures. From the temperature dependence of conductivity, we learn that all doped semiconductors are intrinsic or pseudo-intrinsic at high temperatures and extrinsic at low temperatures.

7.6.5 Determination of N- and P-Type Nature: Seebeck Effect

The most used method is based on the Seebeck effect, which states that if a solid is heated at one end and cooled at the other end, then because of the temperature gradient (dT) a potential difference (dV) develops. In other words, the Seebeck effect is the conversion of thermal energy directly into electrical energy and has some important technical ramifications. It was discovered in 1821 by Thomas Johann Seebeck of Germany. Mathematically, it can be expressed as the ratio (S) between the thermal gradient (dT) and the resulting potential

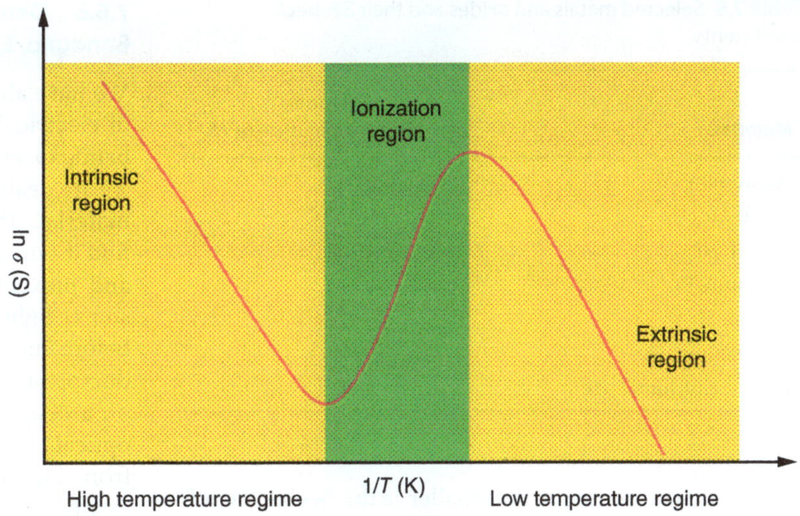

Figure 7.10 Log conductivity vs. reciprocal temperature plot for an extrinsic semiconductor material.

difference (dV) as in Eq. (7.31).

$$S = \left(\frac{dV}{dT}\right) = \frac{dV}{T_2 - T_1} \tag{7.31}$$

Here $T_2 > T_1$ and S is called the Seebeck coefficient. It is also referred to as thermoelectric power in scientific literature. The cold junction, T_1, in practice is kept at 0 °C. Nowadays, it is achieved electronically by compensating for potential drop at T_1 if it is greater than the ice point.

The sign of S can be either positive or negative, depending upon the nature of the material. If electrons diffuse from the hot end to the cold end, the coefficient S is negative, and it is positive when the electron diffusion occurs from the cold end to the hot end. It is negative for an n-type semiconductor and positive for a p-type material. The plot of the Seebeck coefficient as a function of T^{-1} yields a straight-line with the slope equal to the thermoelectric potential, dV. The advantage of such a plot is that we can get the value of the thermoelectric potential without trying to determine it using a voltmeter, which would not give us the correct value because the measured voltage using a voltmeter will always include the contributions made by the connecting leads.

But if we are interested in knowing only whether a semiconductor sample is of n-type or p-type we can easily find it out by using as simple experimental arrangement shown in Figure 7.11. By introducing a thermal gradient between two metallic contacts placed on the surface of the sample, we would enable the migration of charge carriers between the two contacts. By monitoring the deflection of the voltmeter connected between the two junctions, we can find out the nature of the sample.

If the deflection is negative, the material is n-type and when it is positive, it is p-type. By doing this simple experiment, we can find out whether the semiconductor material is of n-type or p-type, but not the exact value of the Seebeck coefficient. It is customary to express the value of the Seebeck coefficient with respect to the value of platinum, which is set at zero. We tabulate the values of Seebeck coefficient for some common metals and semiconductors in Table 7.5.

A simple inspection of the above table tells us that in general, the value for the Seebeck coefficient for semiconductors is 1–2 orders of magnitude higher than for metals. This is because the Seebeck coefficient is related to the electrical conductivity. The rule of thumb is that

Figure 7.11 Schematic for the sign determination of the Seebeck coefficient.

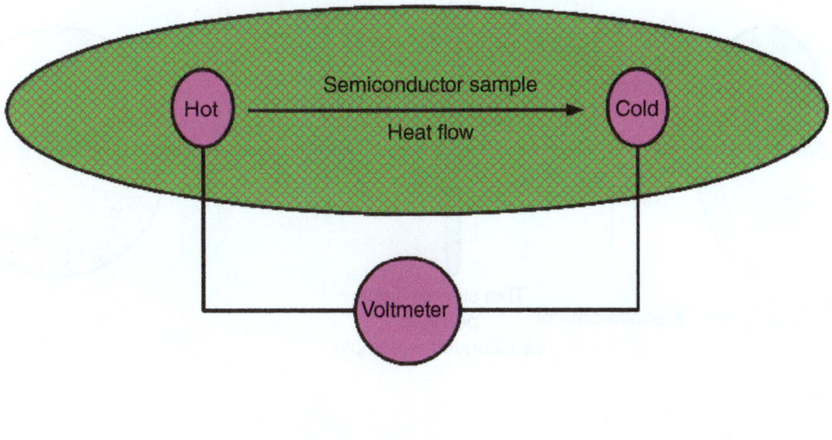

Table 7.5 Selected metals and oxides and their Seebeck coefficients.

Materials	S at room temperature (μV K^{-1})
Aluminum, Al	3.5
Gold, Au	6.5
Nickel, Ni	−15
Platinum, Pt	0
Silicon, Si	440
Germanium, Ge	300
Lead–Tellurium, PbTe	−180

higher the conductivity, the smaller is the Seebeck coefficient. Insulators have even higher values of the Seebeck coefficient than semiconductors.

This Seebeck effect is the basis of thermocouples, which are the classical temperature-detecting devices and have been in active use for centuries. It is also the basis for thermoelectric generators, which operate as a heat engine. These devices are used in power plants for energy recycling and in automobiles as automotive thermoelectric generators for increasing fuel efficiency. They have also found their applications in space probes.

Exercise 7.3

The Seebeck coefficient of Si is 440 μV K^{-1}. Calculate the potential difference developed when the sample temperature changes from room temperature of 20–220 °C. Comment whether Si could be used as practical thermal generator.

Solution

Using Eq. (7.31), we get $dV = 88$ mV. In comparison, all other materials listed in Table 7.5 including Ge would produce smaller potential difference than Si. The value of 88 mV produced by silicon is significant. For energy recycling in power plants and automobiles, silicon thermoelectric generators could have some applications.

7.6.6 Determination of Direct and Indirect Bandgap, E_g

We have already discussed the temperature dependence of electrical conductivity to determine the value of the bandgap. However, this method does not tell us whether we are dealing with a direct bandgap or indirect bandgap material. The optical absorption method enables us to find it out. The experimental set up used is rather simple and inexpensive. In this experiment (Figure 7.12), the source light is allowed to pass through a monochromator before entering a highly polished thin slice of a semiconductor sample. The light exiting the sample is recorded on a detector. The typical frequency (ν) dependence of optical absorption (α) is given by Figure 7.13a,b. We see from Figure 7.13a, that the absorption increases very slowly as the frequency increases and as soon as the frequency corresponding to the bandgap is reached, the absorption coefficient increases rapidly at a constant frequency. This happens when the bandgap is direct. At the conclusion of the transition absorption, coefficient once again increases very slowly. On the other hand, for the indirect bandgap, the absorption keeps on increasing with increasing frequency and its rate of increase slows down once the transition is complete (Figure 7.13b). The value of the bandgap is determined by interpolating the absorption edge to the frequency axis as shown by the dashed lines. In Table 7.6, we give a few examples of direct and indirect bandgap semiconductors.

7.6.7 Determination of Mobility, μ

The mobility is an important parameter, which plays a vital role in the microelectronic technology. Whether a transistor is a high-speed device or a low-speed device depends on the value of mobility of the semiconductor material used as we have already discussed before. We describe here two experimental methods to determine the mobility. They are Haynes–Shockley method and Hall effect method.

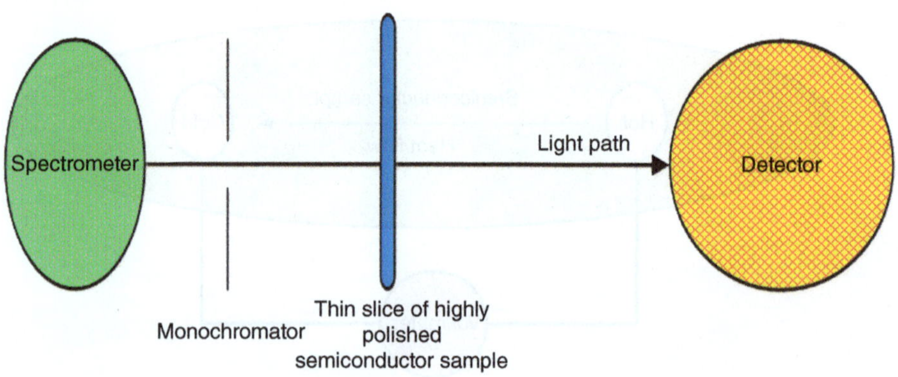

Figure 7.12 Experimental setup for determining the bandgap type of a semiconductor material using the optical absorption method.

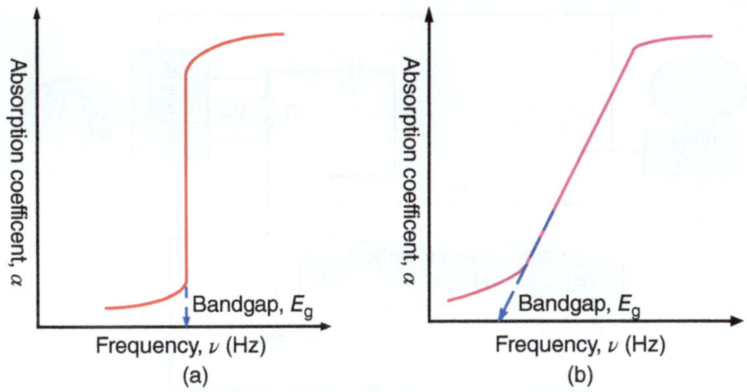

Figure 7.13 Frequency dependence of optical absorption coefficient for (a) direct bandgap and (b) for an indirect bandgap semiconductor material.

Table 7.6 Some examples of direct and indirect bandgap materials and their technical importance.

Material	Bandgap (eV)	Type of bandgap	Comments
Diamond, C	5.47	Indirect	Excellent optical, thermal and mechanical properties
Silicon, Si	1.12	Indirect	King of microelectronics technology
Silicon carbide (6H), SiC	3.0	Indirect	Used initially for blue LED, potentially useful for high temperature electronics
Gallium arsenide, GaAs	1.43	Direct	Excellent material for high-speed electronics, solar cells, near infrared LED
Indium antimonide, InSb	0.17	Direct	High quantum efficiency material, used for infrared detectors and thermal imaging
Cadmium sulfide, CdS	2.42	Direct	Potentially a good material for optoelectronics and high efficiency solar cells
Lead selenide, PbSe	0.27	Direct	Infrared detector and for thermoelectric devices

Source: Compiled from List of Semiconductor Materials, Wikipedia, the free encyclopedia. https://en.wikipedia.org/wiki/List_of_semiconductor_materials.

7.6.7.1 Haynes–Shockley Method

In a crystal lattice, an electron wanders randomly at the Fermi energy giving rise to lattice scattering. In a doped semiconductor material, there are two types of scattering: one is the lattice scattering and the other the ionized impurity scattering. Both types influence the conductivity of a semiconductor material. We encountered the contribution made by the ionized impurity while discussing the temperature dependence of conductivity (Figure 7.10). When an electric field is applied, the electron begins to move in the direction opposite to the applied filed (the direction of the field is defined by the direction of a positive test charge) with a uniform drift velocity, V_d, which by definition is equal to the product of mobility, μ, and electric field, E (see Eq. (1.10)).

We can use Haynes–Shockley experiment to determine the drift velocity directly and then get the values of both types of mobility: the n-type and p-type. The experimental configuration of this method is shown in Figure 7.14 [1]. The substrate is an n-type material. The two metallic contact labeled emitter (E) and collector (C) are a distant, d, apart from each other. Once the sample sees a potential between the two contacts, a rectifying current begins to flow that is the characteristics of a diode. We will discuss this concept in greater detail in the next chapter.

Some holes are injected at the emitter contact (E) by a pulse generator. These holes begin to move toward the collector terminal in the presence of the electric field $E_{EC}(=\frac{V_{EC}}{d})$, where V_{EC} is the applied voltage between the two contacts. The current begins to flow from one electrode to the other. If the switch in the circuit is open, a small parasitic signal is received at the oscilloscope. Once the switch is closed, the circuit is complete, and the real current flows through the resistance R resulting in a potential drop of V_R across it. The oscilloscope records it as a pulse. The time dependence of V_R is qualitatively given in Figure 7.15.

The holes cover a distance of d in time $\Delta T = (T_2 - T_1)$ the presence of the electric field E with the uniform drift velocity of $V_d\left(=\frac{d}{\Delta T}\right)$. From this, then the value for the hole mobility is obtained using the relationship $\mu_p\left(=\frac{V_d}{E}\right)$. The charge density, (ne) can then be obtained simply from the ratio between the conductivity, $(\frac{\sigma_p}{\mu_p})$ (see Eq. (1.11)).

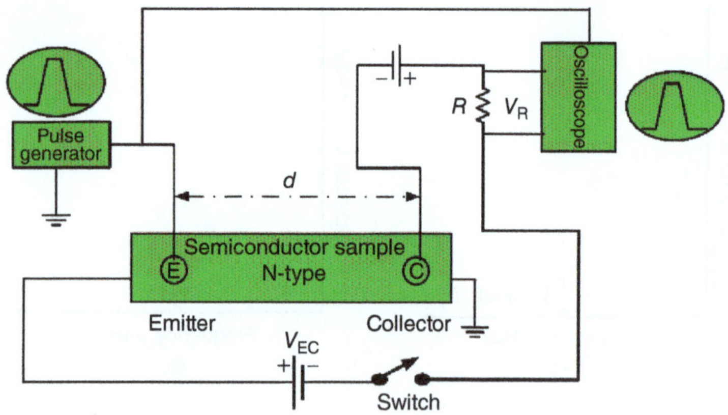

Figure 7.14 Configuration for Haynes–Shockley experiment for the determination of mobility. Source: From Srivastava and Srinivasan 1987 [1].

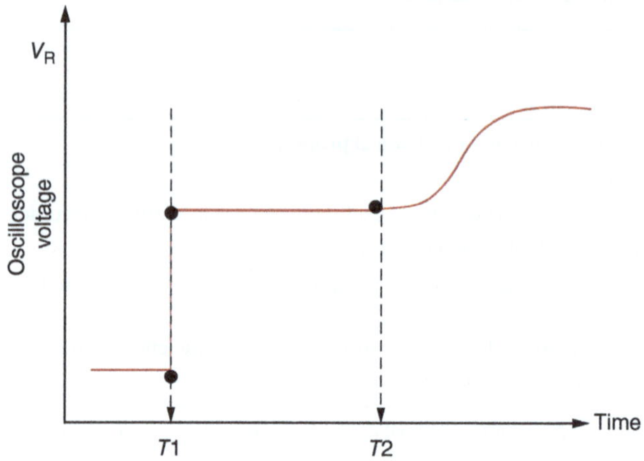

Figure 7.15 Plot for time-dependent voltage output for Haynes–Shockley experiment to determine the mobility of a semiconductor sample.

Similarly the electron mobility, μ_e, can be found by replacing the substrate with a p-type material and injecting electrons at the emitter site. Of course, one will have to be careful about the polarity of the two power sources used. The unit for mobility is $m^2\ V^{-1}\ s^{-1}$. It is also expressed in $cm^2/(V\,s)$. More modern version of the Haynes–Shockley experiment uses a laser source to inject carriers in the semiconductor samples. We give the values of mobility for some materials in Table 7.7.

Here we note that the electron mobility is much larger than the hole mobility. The reason being that the effective mass of hole (m_0^*) is much greater than the corresponding effective mass of electron (m_e^*). We know that the mobility (μ) is inversely proportional to the effective mass.

7.6.7.2 Hall Effect

We discussed in the previous section a direct method to determine the mobility. It can also be determined by the Hall effect method. The effect was discovered by Edwin Hall of the USA in 1879. Today, it is the most used method for the determination of mobility, μ, and carrier concentration (ne) of semiconductors.

Table 7.7 Electron and hole motilities of some semiconductor materials [3].

Materials	Group	Electron mobility, μ_n $(cm^2\ (V\,s)^{-1})$	Hole mobility, μ_p $(cm^2\ (V\,s)^{-1})$
Si	IV	1 450	500
Ge	IV	3 800	1 820
GaAs	III–V	8 800	400
InSb	III–V	78 000	750
GaN	III–V	1 350	13
CdS	II–VI	400	50

The experimental set up is rather simple as can be seen from Figure 7.16 [4]. The current, I, between the two metallic contacts on the surface of a semiconductor sample flows in the x-direction from left to right with a constant drift velocity. Let us assume that the substrate is an n-type semiconductor with electron as its primary charge carriers. When a magnetic field, B, is applied to the sample along the z-direction a force, F_H, develops along the y-axis because of the Lenz's law of magnetic force. This force is perpendicular to both the magnetic field and the electric current.

According to the Lenz's law, the force due to the magnetic field acting on the electrons is described by the following equation.

$$F_H = e(V_d \times B) \approx eV_dB\ \sin\theta \qquad (7.32)$$

where V_d is the drift velocity of the electrons and θ is the angle between the drift velocity and the magnetic field. Since F_H is perpendicular to both V_d and B, the magnetic force will be given by Eq. (7.33).

$$F_H \approx eV_dB \qquad (7.33)$$

In honor of Hall, this force is called the Hall force that a sample experiences because of the magnetic field in the experimental setup. In the presence of this force, the electrons are deflected upwards and accumulate

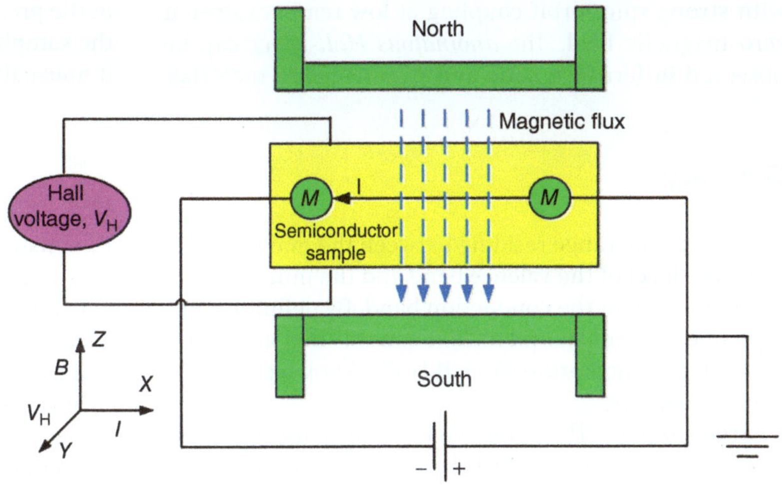

Figure 7.16 Experimental set up for Hall effect measurements. Source: Pandey et al. 2012 [4]. Licensed under a Creative Commons Attributes, 3.0 Unported License.

on the upper surface of the sample along the positive y-direction. This will cause positive charges to accumulate on the opposite face to restore the charge neutrality of the sample. Consequently, an electric field, E_H, is set up along the y-axis.

Once the equilibrium is restored, the force, eE_H, must be equal to the Hall force such that

$$eE_H = eV_dB \tag{7.34}$$

Combining Eqs. (1.10) and (1.11), we get $V_d = \frac{\sigma_e E}{ne} \approx \left(\frac{J}{ne}\right)$ where J is the current density. Substituting it in Eq. (7.34), we get Eq. (7.35).

$$E_H = \left(\frac{1}{ne}\right) JB \approx R_H JB \tag{7.35}$$

Here $R_H = \left(\frac{1}{ne}\right)$, and it is called the Hall constant. E_H is called the transverse electric field.

Assuming the sample to be of width, W, and thickness, and substituting them in Eq. (7.35) we get

$$R_H^{-1} = (n_e e) = \frac{IB}{V_H t} \tag{7.36}$$

where V_H is the Hall voltage with a negative sign for the electrons.

Similarly, we can find an expression for the mobility in terms of I, B, V_H, and t using the definition of mobility as given by Eq. (1.11) as $\mu_e = \frac{\sigma_e}{ne}$. That is,

$$\mu_e = \left(\frac{\sigma_e}{ne}\right) = \frac{\sigma_e V_H t}{IB} \tag{7.37}$$

Both the carrier concentration and mobility can be evaluated from Hall effect experiment for an n-type material. However, these parameters can also be evaluated for a p-type semiconductor simply by replacing the n-type substrate with a p-type substrate. The sign of Hall voltage would naturally be positive.

It becomes more involved when both n- and p-type carriers are present in the same sample. For such a case, the Hall constant is given by Eq. (7.38).

$$R_H = \frac{p_h \mu_h^2 - n_e \mu_e^2}{e(p_h \mu_h + n_e \mu_e)^2} \tag{7.38}$$

The experiment described here is the classical experiment for observing the Hall effect. Besides being very important to characterize semiconductor materials, the Hall effect is also the basis for very practical devices. As Hall probes, they are widely used to measure the strength of magnetic fields accurately. As sensors, they are used for speed detection, current sensing, for navigation, positioning, and proximity switching.

Exercise 7.4

Consider a semiconductor sample having the drift velocity of 1 mm s^{-1} is subjected to a magnetic field that is perpendicular to current flowing in the sample. If this field of 2 V m^{-1}, compute the value of the magnetic field applied.

Solution

From Eq. (7.34), we have $E_H = V_dB$. Substituting for $V_d = 10^{-3}$ m s^{-1} and $E_H = 2$ V m^{-1}, we get for $B = 2000$ G.

There are different types of Hall effects that are classified as quantum Hall effect, spin Hall effect, and anomalous Hall effect. Very brief description is provided here for each of them so that our readers become familiar with these associated effects.

The *quantum Hall effect* is observed in 2D electron systems at low temperatures and at very large magnetic fields. The result is the quantization of the Hall conductance. The *spin Hall effect* is the accumulations of spin on the boundaries of a current carrying samples with no magnetic field present. This effect can be observed in metals and semiconductors at room temperature as well as at cryogenic temperatures. It has been observed in 2D quantum wells (such as of mercury telluride) with strong spin–orbit coupling at low temperatures at zero magnetic field. The *anomalous Hall effect* can be observed in ferromagnetic and paramagnetic materials in the presence of a magnetic field. The magnetization of the sample contributes to the Hall conductivity making it unusually large and therefore, the name *anomalous.*

Glossary

Band gap The range residing between the maximum energy level of the valence band and the minimum energy level of the conduction band. By definition no charge carrier can find its home in this region at absolute temperature (0 K). It is also known as the forbidden gap.

Charge carriers There are two types of charge carriers in semiconductors – electrons and holes. Both can contribute to the overall conductivity of a semiconductor. When the contribution is primarily because of electrons, the material is classified as an n-type semiconductor. When the main contribution to the conductivity comes from the holes, the material is called a p-type semiconductor. Holes are assigned a positive charge in value equal to the electron charge that is negative in sign.

Conduction band The uppermost energy band in a semiconductor material that is completely empty at absolute temperature. No charge carrier can find its home there because all are confined in the valence band.

Doping A chemical process through which it becomes possible to convert an intrinsic semiconductor material to an extrinsic semiconductor. During the process of doping, electron sharing takes place between the host element and the doping element.

Effective mass The mass of an electron in a periodic lattice (like a single crystal) is not constant in all directions. According to the band theory of solids, it is different in different directions (meaning that it is anisotropic) under the influence of an electric or magnetic field. This can be confirmed by cyclotron resonance experiment. The concept of effective mass plays an important role in understanding the theory of semiconductor. In general, the effective mass of an electron is much larger than the effective mass of holes.

Extrinsic semiconductor At temperatures greater than absolute zero, vacancies (void) arise in the valence band because of the migration of electrons to the conduction band. For this to happen, the electrons must acquire an energy greater than the bandgap. This process when complete can put sufficient electrons in the conduction band to influence the overall conductivity. Also when a semiconductor material is doped with impurity ions, an intrinsic material gets converted into an extrinsic material.

Fermi energy see Chapter 1.

Fermi level see Chapter 1.

Group velocity When an electron propagates through space as a wave it acquires a velocity called the *group velocity*. It is different than the drift velocity that is the characteristic associated with a particle.

Intrinsic semiconductor When a semiconductor material is pure and free of defects and impurities it is called intrinsic material. It is a poor conductor of electricity. As stated above, all available electrons are confined in the valence band at absolute zero.

Mobility It is defined by Eq.(1.10). It is a very important parameter of a semiconductor material. If its value is high, the material is supposed to be a good candidate for high-speed electronics. Its unit is m^2 $(V\ s)^{-1}$.

Semiconductor This classifies solids that are electrically (and also thermally) less conductive than metals but more conductive than insulators. Semiconductors are the heart and soul of microelectronic technology.

Valence band The energy band in a semiconductor material found at the bottom of the energy band diagram. All charge carriers can be found only in this energy band at absolute zero. At this temperature, all allowed energy states are fully occupied resulting in no electrical conductivity.

Problems

7.1 Consider the case of a GaAs sample with $\mu_n \approx 0.850\ m^2\ (V\ s)^{-1}$ and $\mu_p \approx 0.040\ m^2\ (V\ s)^{-1}$. Its intrinsic concentration is $1.8 \times 1012\ m^{-3}$. Assume that electrons to hole ratio is 4. Find the intrinsic conductivity of the sample.

7.2 Find the atomic density of Si assuming its atomic mass to be 28.09, density = $2.33\ g\ cm^{-3}$. If the ratio between the intrinsic carrier concentration and atomic density is 3×10^{-11}, find the value for the intrinsic carrier concentration.

7.3 The conductivity of a semiconductor sample increases by 10-fold from 100 Sm^{-1} at room temperature to 1000 Sm^{-1} at 250 °C. Calculate the Fermi energy of the material. Comment on your result.

7.4 In a p-type Ge semiconductor, the Fermi level is 0.05 eV above the intrinsic level ($=E_g/2$). The energy gap of Ge is 0.66 eV. What is the probability of finding an electron in the conduction band at 300 K?

7.5 Consider the case of an n-type doped Ge sample. The dopant is P. The sample conductivity is 100 Sm^{-1} at 300 K and its $\mu_n \approx 0.35\ m^2\ (V\ s)^{-1}$. Calculate the concentration of the dopants in parts per million by weight assuming that all dopants are ionized.

7.6 Discuss the difference between the four-point and two-point probe methods for the determination of resistivity of a semiconductor sample.

7.7 What is the Seebeck effect and what is meant by the Seebeck coefficient? What is the reason that it can have either positive or negative value?

7.8 The Seebeck coefficient of Ge is 300 $\mu V\ K^{-1}$ and of Au it is 6.5 $\mu V\ K^{-1}$. Calculate the potential difference developed by each of these materials for the temperature gradient of 500 °C. Comment on the results whether the results can have any practical significance.

7.9 Describe Haynes–Shockley method for the determination of mobility, μ, of a semiconductor sample. Discuss the advantages of using this method compared to others.

7.10 Describe the Hall effect experiment and explain why it is an indispensable tool for the characterization of a semiconductor sample.

7.11 (1) Find the magnitude of the magnetic field that would be required to produce the drift velocity of 5 $\mu m\ s^{-1}$ in a semiconductor sample. Assume that the magnetic field is perpendicular to the current in the sample and the sample produces a transverse electric field of 5 $mV\ m^{-1}$.
(2) Find also the carrier concentration of the sample assuming that the current density to be 2 $\mu A\ m^{-2}$.

References

1 Srivastava, C.M. and Srinivasan, N. (1987). *Science and Engineering Materials*. New Delhi: Wiley Eastern Limited.

2 Biard, R. and Pittman, G.E. (1966). Semiconductor radiant diode. US Patent Number 3,293,512.

3 Solymar, L. and Walsh, D. (1999). *Electrical Properties of Materials*, 6e. Oxford Science Publications, Oxford University Press. ISBN: 0-19-856272-1.

4 Pandey, R.K., Stapleton, W., Padmini, P. et al. (2012). Magnetically tuned varistor-transistor hybrid device. *AIP Adv.* 2: 042188.

Further Reading

Streetman, B.G. and Banerjee, S. (2000). *Solid State Electronic Devices*, 5e. Prentice Hall. ISBN: 0-13-025536-6.

8

Electroceramic Semiconductor Devices

CHAPTER MENU

Research is what I'm doing when I don't know what I'm doing.
We can lick gravity, but sometimes the paperwork is overwhelming.
I have learned to use the word impossible with the greatest caution.

Wernher von Braun

8.1 Introduction

Electroceramic semiconductor materials, especially oxide semiconductors, have gained prominence in the past few years because of their many unique properties not easily found in conventional semiconductors. Being ceramic in nature, they are rugged, capable of withstanding high temperatures, and can also be processed as transparent substrates in large volumes. In general, oxides are wide bandgap semiconductors and show remarkable immunity to high doses of radiation such as neutrons, protons, and heavy ions. These desirable material properties make oxide semiconductors suitable for transparent electronics, high temperature electronics, and space electronics. The devices built on such substrates can be useful also for a wide range of general-purpose applications. By now, oxide semiconductors have become one of the most active fields of fundamental and applied research. Because of the demand for flat panel displays, research in transparent electronics has gained prominence. Of special interest are binary semiconductor oxides that exhibit interesting physical properties. Prominent members of this group include zinc oxide (ZnO), tin oxide (SnO_2), indium oxide (In_2O_3), indium tin oxide (In_2O_3–SnO_2 alloy), gallium oxide (Ga_2O_3), tungsten oxide (WO_3), and titanium oxide (TiO_2). The lesser-known oxide semiconductors are members of the iron titanate group, and they have been recently found to be good materials for many devices with applications in space electronics, high temperature electronics, and many other conventional applications. In the nonoxide group, silicon carbide (SiC) stands out as an established material for power electronics. In this chapter, we will cover some of the devices built on electroceramic semiconductors and their potential applications. Before we can do this, we need to gain some background in the fundamentals of electronic devices. With this objective in focus, we will also study in this chapter, the mechanics of forming contacts between two metals and between a metal and semiconductor, and examine how the energy band diagram gets modified and what are its ramifications. The knowledge so gained will help in understanding the operation of metal–semiconductor and other devices that are indispensable for microelectronic technology and a vast array of applications. The block diagram of Figure 8.1 outlines the semiconducting nature of electroceramics and their wide range of applications.

Fundamentals of Electroceramics: Materials, Devices, and Applications, First Edition. R. K. Pandey.

Companion website: www.wiley.com/go/Pandey/Fundamentals_Electroceramics

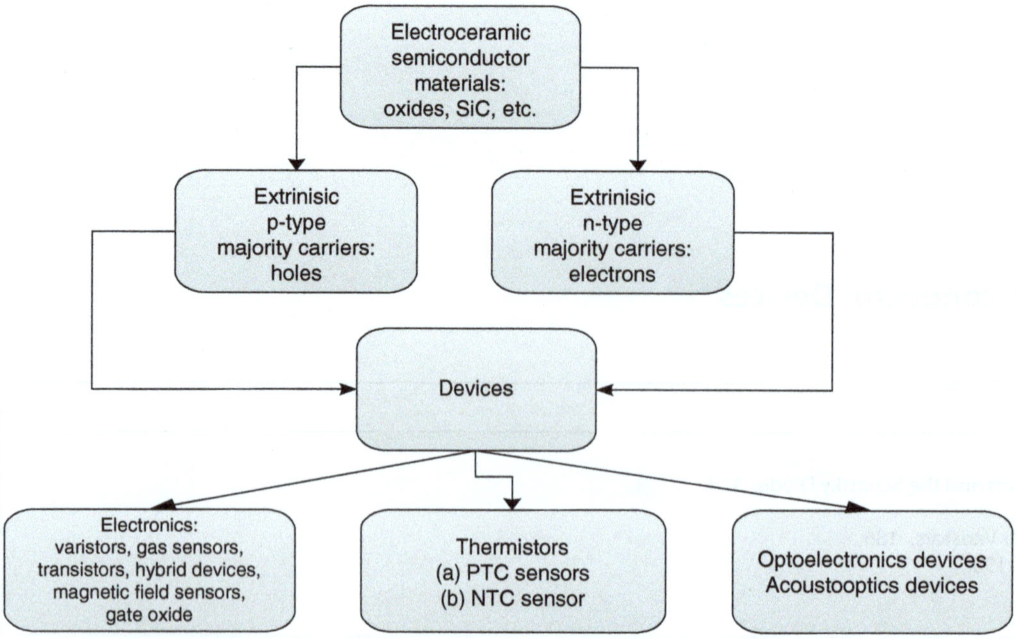

Figure 8.1 Roadmap from materials to devices of electroceramic semiconductors.

8.2 Metal–Semiconductor Contacts and the Schottky Diode

Here we will learn what happens when a metal is brought into contact with another metal, and then when a metal and a semiconductor are brought into contact. That will lead us to the better understanding of Schottky diode, which is a very useful device.

8.2.1 Metal–Metal Contact

For all electrical measurements, we need metal contacts on the sample, and this is especially crucial in the case of semiconductors and insulators. Whether they produce linear current–voltage ($I-V$) characteristics or nonlinear characteristics decides whether the device is ohmic or nonohmic. It is important to remember that while the metallic contacts themselves may be ohmic, the devices they might produce may be nonohmic. We briefly discussed this point in the previous chapter. Obviously, nonohmic devices are of great interest to the electronic technology because of the large number of applications one can find for such devices. Diodes and transistors are the leading semiconductor devices that dominate the field of microelectronics and both are good examples of nonlinear devices. In the following sections, we will study the physical principles involved behind the nonlinear $I-V$ behavior of diode and transistors. First, let us understand what happens when two metals are brought in contact?

The result is the formation of an interface with a potential barrier between the two metals that can be created only when the work functions of the two metals are different. We encountered the term *work function* while discussing the photoelectric effect in Chapter 2. We defined it as the minimum energy required for an electron to escape from the surface of a solid to reach the vacuum level. By convention, the energy of the vacuum level is assigned the value of infinity. This is a fundamental property of all materials and can also be explained by the concept a chemical potential.

Let us consider two metals M1 and M2 with work function W1 for M1 and W2 for M2 are in contact as shown in Figure 8.2. Let us also assume that W1 > W2.

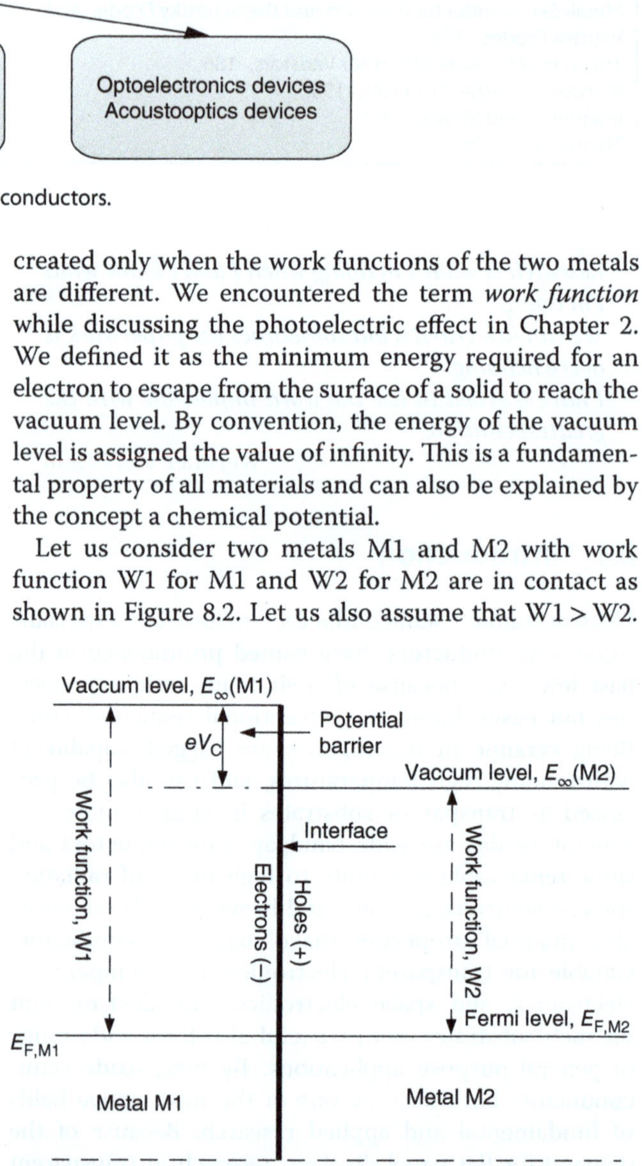

Figure 8.2 Origin of potential barrier when two metals are in contact.

An interface is formed with accumulation of negatively charged electrons on one side and the positively charged holes on the opposite side. Consequently, an electric field is set up with a defined polarity at the interface, which inhibits the flow of electrons from one side to the other. A dynamic equilibrium is set up when equal number of electrons cross from one side to the other. It is puzzling to imagine an equilibrium state to be dynamic. But it is indeed allowed by thermodynamics and is handled by the tough mathematics of statistical thermodynamics. For all practical purposes, it is wiser for us to just accept the truth that when two dissimilar metals are brought in contact a potential barrier (V_C) results.

During the process of interface building, the two Fermi levels, $E_{F,M1}$ for metal M1 and $E_{F,M2}$ for metal M2 merge to become a single level when the system reaches a thermal equilibrium. This is one of the most important concepts of solid-state physics and is an important tool in building models to explain many phenomena.

We now know that when an equilibrium is established, the difference in work functions of the two metals gives rise to a contact potential, V_C. Alternatively, we can also say that the formation of an energy potential barrier with energy equal to eV_C is the final result that has to be equal to W1 – W2. We can generalize this result by stating that the potential barrier is also equal to the difference, ΔE_F, between the two initial Fermi levels. This is a general law valid for all systems at a thermodynamic equilibrium and worth remembering while treating the issues related to potential barriers in semiconductors.

The contact potential (V_C) that develops when two dissimilar metals are brought into contact is real and can be measured. We already mentioned in Chapter 1 that the work function is a fundamental materials property and can be experimentally determined by contact potential measurements, photoelectric effect, thermionic emission, and field emission. The values obtained by these methods can vary somewhat for the same material, as well as whether the sample is single crystalline or polycrystalline or even amorphous. Also, it varies when measured for different crystal orientations of the same single crystal. In spite of the fact that contact potential is real, we cannot use it as a battery. The reason being that it will be in violation of the second law of thermodynamics that forbids the extraction of power from an equilibrium state. We have already given in Table 1.6 a list of materials with the values of their respective work functions.

8.2.2 Metal Semiconductor Contact

Let us consider a metal with work function of W_M and an n-type semiconductor with the work function of W_S, such that $W_M > W_S$. When they are apart, they are represented by Figure 8.3. We also assume that the width of the conduction band (CB) is W_{CB}, which is by definition smaller than W_S. Nothing happens so far either to the metal or the semiconductor until the distance (d) separating them begins to diminish. As soon as it vanishes, a contact is made between the metal and the semiconductor, and a new picture emerges as represented by Figure 8.4. First of all, as expected, the two Fermi levels merge to form a single level shown by a dashed line on the semiconductor side. The other important observation we can make is that there are two potential barriers developed; one right above the top of the conduction band with the barrier height equal to the difference between the two work functions which is $W_M - W_S$. The other barrier is found right above the Fermi level on the metal side with the designation of W_B, which is called the Schottky barrier height.

Migration of electrons from the metal side to the semiconductor side can take place only by overcoming the

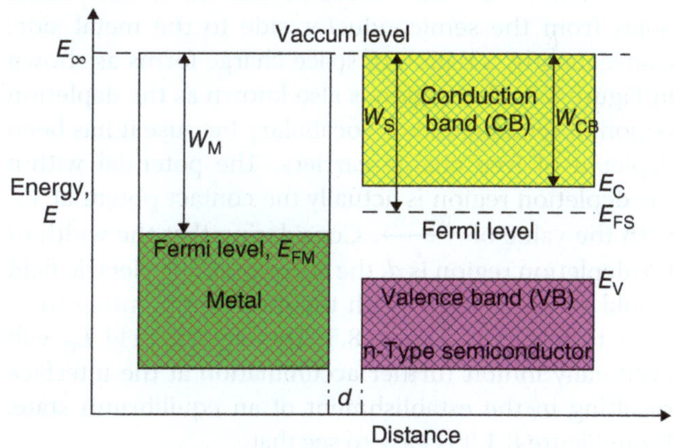

Figure 8.3 A metal and a semiconductor not in contact.

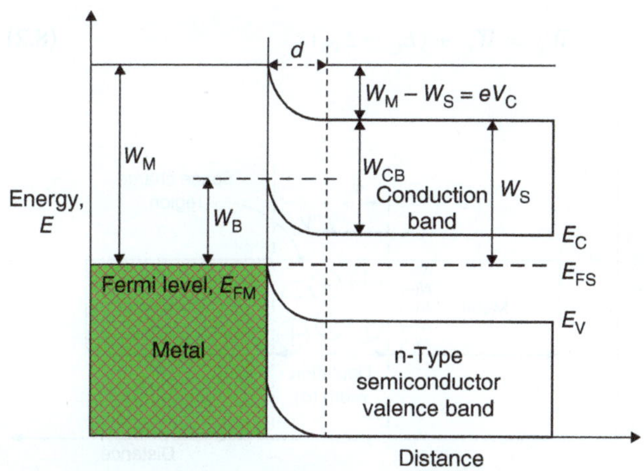

Figure 8.4 Energy band diagram when a metal and an n-type semiconductor are in contact.

Schottky barrier, W_B. On the other hand, the potential barrier ($W_M - W_S$) must be overcome for the electrons to go from the semiconductor side to the metal side. The two potential barriers are related, which we will find out after we examine what might cause bending of the bands near the interface with a downward curvature.

As the contact is made and the thermal equilibrium established, a new picture emerges that is given by Figure 8.5. The accumulation of electrons, which are indicated by $N(-)$ in this figure, happens on the metal side of the interface.

Here the depletion width, d, is shown purposely grossly exaggerated because we need it for explaining what exactly happens within the interface when contact is made between a metal and a semiconductor. In reality, the interface is very narrow and all is happening at atomic scale.

Because of the electrostatic mirror imagining principle, positively charged holes develops on the semiconductor side of the interface. Some of the electrons from metal side will diffuse to semiconductor side and similarly some holes from the semiconductor side to the metal side. Consequently, a region of space charge forms as shown in Figure 8.5. This region is also known as the depletion region in semiconductor vocabulary because it has been depleted of free charge carriers. The potential within the depletion region is actually the contact potential V_C with the value of $\frac{W_M - W_S}{e}$. Considering that the width of the depletion region is d, the corresponding electric field would be E_C ($=V_C/d$) with the direction pointing from right to left as in Figure 8.5. The electric field E_C will eventually inhibit further accumulation at the interface resulting in the establishment of an equilibrium state. From Figure 8.4 it is easy to see that

$$W_B = (W_M - W_{CB}) = eV_C + (E_C - E_{FS}) \tag{8.1}$$

In other words,

$$W_B = W_C + (E_C - E_{FS}) \tag{8.2}$$

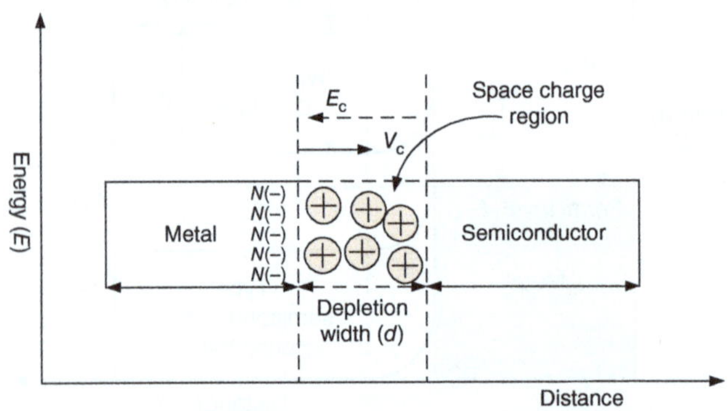

Figure 8.5 Formation of depletion layer between a metal and a semiconductor in contact.

When no external current flows through the structure, the thermal current from metal to semiconductor side is equal to the thermal current flowing from the semiconductor side to the metal side but of opposite polarity. They cancel each other leaving the structure intact. But when a potential is applied to the structure such as by connecting it to a battery, an important modification takes place and the picture changes resulting in the onset of a rectifying current under certain conditions, which we will study in the next section. However, when the work function of metal is smaller than that of the n-type semiconductor, two significant changes take place in the band diagram. First, the band in the depletion region bends with an upward curvature, and second, no Schottky barrier is formed. The contact potential still remains the same with $W_S - W_M = eV_C$.

Exercise 8.1

Using the values for work function given in Table 1.6, calculate the potential barrier that will arise if Al is brought in contact with Au, Ag, and Pt. Comment on the results.

Solution

The potential barrier, eV_C, is equal to (a) for (Cu/Al) = 0.66 eV; (b) for Au/Al = 1.13 eV; (c) for Ag/Al = 0.34 eV; and for Pt/Al = 1.37 eV. Notice that these values are comparable to the bandgap of many semiconductors. For example, the contact potential between Au and Al is nearly the same as the bandgap of Si. Similarly, the contact potential between Pt and Al is 1.37 eV which is comparable to the bandgap of GaAs. Since no semiconductor is involved, we cannot produce any semiconductor device using just two metal contacts.

8.2.3 Schottky Diode

While discussing the band diagram given in Figure 8.4, we identified the formation of the potential barrier W_B, which we called Schottky potential barrier or simply Schottky barrier. This barrier plays a crucial role in giving rise to nonlinear current–voltage characteristics (also called rectifying I–V), of a Schottky diode. It was invented by Walter H. Schottky of Germany in early 1900, and ever since it has been in use in electronics. Over the years, it has gone through many modifications and is produced today in a variety of configurations to meet the requirements of a wide range of applications. The counterpart of the Schottky diode in electroceramics is a varistor, which also is based on the band structure similar to that shown in Figure 8.4. We can represent a Schottky diode using a very simple configuration consisting of a metal and an n-type semiconductor as in Figure 8.6. The band diagram of this device would be equivalent to the structure of Figure 8.4 if the metal side

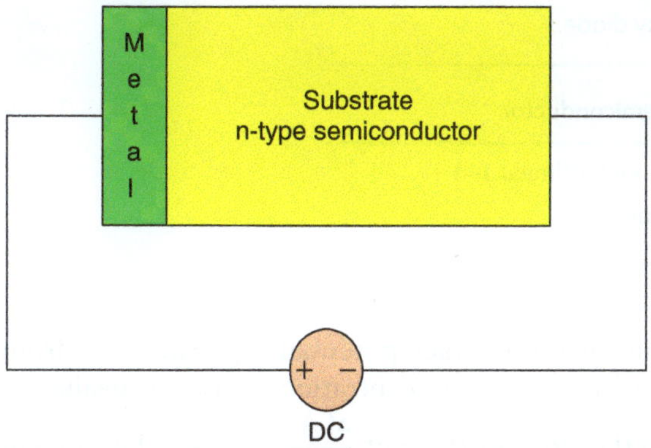

Figure 8.6 Schematic of a forward biased Schottky diode.

is connected to the positive terminal of a battery and the semiconductor side to the negative terminal. The current begins to flow causing some significant changes in the band diagram. Since the resistance in the depletion region (equivalent to an insulator) is much larger than the combined resistance of the metal and the semiconductor, most of the potential drop will occur across the depletion width. Since the contact potential, V_C, and the applied potential, V, are in opposite directions, the new contact potential would be equal to $(V_C - V)$ and the resulting potential barrier height equal to $e(V_C - V)$. But there will be no change in the Schottky barrier height, W_B. Under the changed scenario, now more current would flow from the semiconductor side to the metal side than from the metal to the semiconductor side making the net current present in the structure nonzero.

Let us assume that I_1 is the current from the semiconductor side to the metal side and I_2 the current from the metal side to the semiconductor side. We can also express them as follows:

$$I_1 \propto \exp\left[-\frac{e(V_C - V)}{k_B T}\right] \tag{8.3}$$

$$I_2 \propto \exp\left(-\frac{W_B}{k_B T}\right) \tag{8.4}$$

Before we connected the structure in the forward bias configuration, the thermal currents I_1 and I_2 were identical. Since no change has taken place in the Schottky barrier height W_B with the application of potential V, we can express Eq. (8.4) as follows:

$$I_2 \propto \exp\left(-\frac{eV_C}{k_B T}\right) \tag{8.5}$$

Therefore, the net current I_d flowing in the forward biased diode would be the difference of I_1 and I_2 which is given by Eq. (8.6).

$$I_d \approx I_0\left[\exp\left(\frac{eV}{k_B T}\right)\right] - 1 \tag{8.6}$$

I_0 in Eq. (8.6) is constant, and it can vary from material to material, and the surface characteristics of the junction. Obviously, the current described by Eq. (8.6) is not a linear equation, rather it is the classic representation of a rectifying current that makes diodes a special class of nonlinear device. Theoretically speaking, the configuration of Figure 8.6 would allow currents to flow only in one direction and not in the other. This is the hall mark of a diode. But in practice, the current when the diode is reverse biased is not zero rather it has a finite value. But in comparison to the current in the forward biased mode, it is very small and therefore not much of a consequence. The current–voltage plot of a typical Schottky diode is shown in Figure 8.7. The current, I_0, in the reverse biased mode of the device is usually of the order a few microampere, whereas the current, I, in the forward biased mode can be as large as few milliampere. I_0 is also called the leakage current of the device. The smaller it is, the superior is the diode in its performance.

One can also build a Schottky diode using a p-type semiconductor substrate, but the structure must fulfill the condition of $W_M < W_S$. From the discussions so far, we can conclude that the relationship between W_M and W_S is critically important for the fabrication of a good rectifying Schottky diode. Polarity of the semiconductor substrate also plays a crucial role in the fabrication of a good Schottky diode. The golden rules, as summarized in Table 8.1, are worth remembering while venturing to build a good Schottky diode.

It is interesting to note that the conditions are exactly opposite to each other for the two types of substrates. The most commonly used Schottky diodes are built using n-Si substrates. For high voltage applications, SiC-based Schottky diodes are preferred. Ag, Al, Au, Cr, Ni, Pt, and W are the most used metal for contacts for Schottky diodes based on Si, Ge, and GaAs.

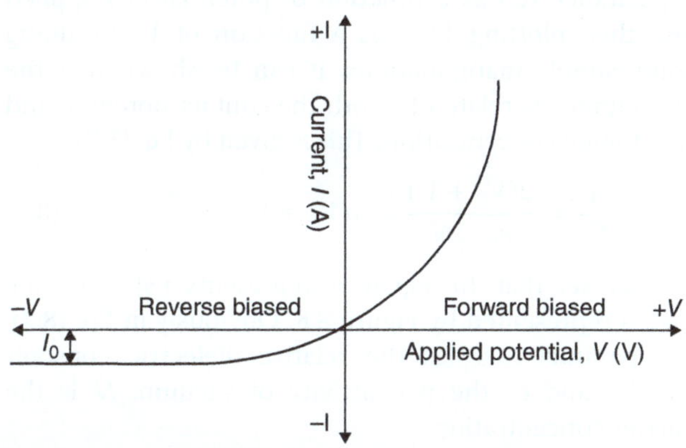

Figure 8.7 Characteristic current–voltage plot of a typical Schottky diode.

Table 8.1 Conditions for ohmic and nonohmic (rectifying) contacts for a Schottky diode.

Condition	n-Type semiconductor	p-Type semiconductor
1. When $W_M > W_S$	Nonohmic contact (nonlinear $I–V$)	Ohmic contact (linear $I–V$)
2. When $W_M < W_S$	Ohmic	Nonohmic

It is amazing that though this diode was invented more than 115 years ago, it is even today in constant demand because it has many advantages over the p–n junction diode. In p–n junction diode, the depletion layer is formed by the diffusion of charge carriers in the p–n junction interface, which gives rise to a potential barrier resulting in the rectifying characteristic of current–voltage relationship. Compared to a p–n junction diode, the Schottky diodes are far easier and less costly to fabricate especially because it does not require any specialized lithographic steps for fabrication which is essential for making p–n junction diode and requires that the semiconductor can be doped both n-type and p-type. The switching voltage (also called turn on voltage) of a silicon p–n junction diode is of the order of 600–700 mV, whereas for a Schottky diode, it is only between 200 and 300 mV. Additionally, this has fast recovery time because of the small amount of stored charge that makes the Schottky diodes suitable for high speed applications. Another advantage of the Schottky diode is its very small junction capacitance because the metal contact can be made very small using even a thin wire for a point contact. In comparison, the capacitance of a p–n junction diode is significantly higher.

8.2.4 Determination of Contact Potential and Depletion Width

The standard experimental method is to measure the capacitance (C) as a function of potential (V) applied and then plotting $1/C^2$ as a function of V. By doing some simple manipulations, it can be shown that the capacitance is related to both the contact potential and the doping concentration. This is given by Eq. (8.7).

$$\frac{1}{C^2} = \frac{2(V_C + V)}{e\epsilon_r\epsilon_0 N} = a(V_C + V) \tag{8.7}$$

We can see that this equation represents a straight-line and is represented by Figure 8.8. Obviously in Eq. (8.7), the parameters ϵ_r is the relative dielectric constant ($=\epsilon/\epsilon_0$) and ϵ_0 the permittivity of vacuum. N is the carrier concentration.

Both the contact potential V_C and the carrier concentration N can be evaluated by analyzing the plot of Figure 8.8. The intercept of the x-axis when $\frac{1}{C^2} = 0$ gives the value of the contact potential V_C. Similarly, from the slope, the carrier concentration, N, can be evaluated because the slope of the plot $\frac{\Delta\left(\frac{1}{C^2}\right)}{\Delta V} = \frac{2}{e\epsilon_r\epsilon N}$. The value of the contact potential, V_C, obtained from the $C–V$ plot is usually smaller than the real value. For accuracy sake, it should be corrected by adding a correction factor of $\frac{2k_BT}{e}$ where k_B is the Boltzmann constant, T the room temperature, and e the electronic charge. The correction factor amounts to approximately 50 mV by using 293 K for room temperature. However, if the contact potential is large such that $V_C \geq 500$ mV, the correction factor need not be considered.

Exercise 8.2

Consider that the work function for n-type Si is 4.01 eV. Suppose that Al, Cu, Au, Ag, and Pt are used to make contacts with Si. Using the values given in Table 1.6 for work function for these metals, find out the Schottky barrier height that would form between these metals and Si. Then interpret the results.

Solution

By definition, the potential barrier $W_C = (W_M - W_{Si}) = eV_C$. Then W_C (a) for Al/Si contact is = 0.15 eV; (b) for Ag/Si contact is = 0.49 eV; (c) for Cu/Si contact is = 0.81 eV; (d) for Au/Si contact is = 1.28 eV; and (e) for Pt/Si contact is = 1.52 eV. We see that the largest contact potential W_C would be formed between Pt/Si and the lowest between Al/Si. All contacts calculated with the

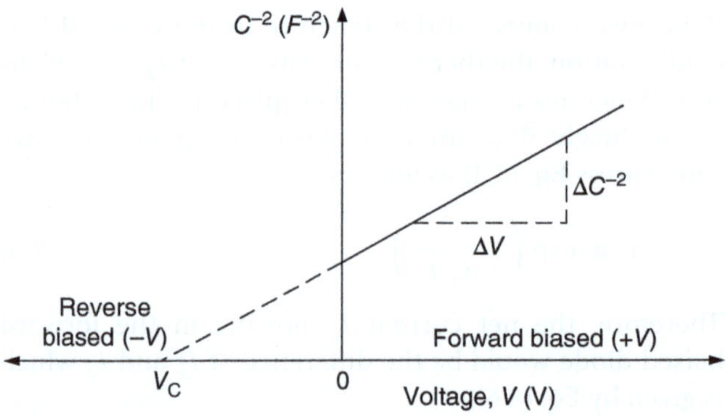

Figure 8.8 Capacitance vs. applied voltage plot for a Schottky diode.

given metals would form a Schottky barrier with n-type Si substrate. According to Eq. (8.2), the Schottky barrier W_B is equal to $[W_C + (E_C - E_{F,S})]$.The lowest Schottky barrier height will be between Al/Si and the largest between Pt/Si. The maximum number of electrons will migrate from the metal side to the semiconductor side at the Al/Si contact giving rise to a large current. On the other, hand the least number of electrons can switch sides from the metal to semiconductor side at the Pt/Si contact. Consequently, the resulting current will be the smallest. Therefore, for an efficient device, we would choose Al as the contact metal with Si.

8.2.5 Oxide Semiconductor Materials and Their Properties

Oxides are members of a fascinating group of electroceramics. Because of their multifunctional nature and projections of their impact on a variety of technology ranging from information technology to flexible electronics to spintronics, they have become a subject of intensive research and development throughout the world. The number of publications produced dealing with one or the other aspect of oxides has surged exponentially during the last decade. Oxides are being processed and studied as bulk and nanosized ceramic, single crystal, thin and thick film for fundamental studies and device development. Data storage, high speed computation, multiferroics, transparent electronics, flexible electronics, flat panel display, and optoelectronics are some prominent examples of technology that is supposed to benefit greatly from the new knowledge gained by research in oxide semiconductors. Obviously, thin-film transistors (TFT) are of particular importance because they will play crucial role in the success of at least three emerging technologies which are transparent electronics, flexible electronics and flat panel displays. The list of emerging technologies where oxide semiconductors have crucial roles to play is already big as discussed. Still to this large list we can also add such recent developments as varistor-transistor hybrid devices and magnetic sensors where wide bandgap ceramic oxide semiconductors have proven to be very useful materials. We will include later these topics also in our discussion in this chapter.

Novel and exotic oxide materials showing interesting multifunctional properties are obviously in demand. Binary metal oxides are prominent members of the semiconductor groups and most of them can be processed as transparent substrates. Therefore, they are of particular interest for transparent electrodes, flexible electronics, flat panel display, and opto-electronics. Almost all of them are wide bandgap materials, and some among them are even direct bandgap materials. That makes them potentially viable materials for opto-electronics, light-emitting diodes (LEDs), and lasers besides being of interest for conventional semiconductor technologies. Some even show piezoelectric properties making them attractive for acousto-optic devices. Some members of this group are suitable materials for gas-sensing applications.

Examples of transparent binary metal-oxide semiconductor group are tin oxide (SnO_2), indium oxide (In_2O_3), zinc oxide (ZnO), cadmium oxide (CdO), and their alloys such as ZnO–SnO_2, ZnO–In_2O_3, and In_2O_3–SnO_2. Other lesser-known examples of oxide semiconductors are members of the iron titanate group; many members of this group are also room temperature ferromagnetic apart from being also wide bandgap semiconductors. These are established radhard materials and suitable for the development of devices for sensor applications, space electronics, and high temperature electronics. Also because iron titanates consist of iron oxide (FeO) and titanium oxide (TiO_2), both of which are proven biocompatible, it is conceivable that electronic devices developed using these as substrate materials might find their applications in bioelectronics. But first their biocompatibility must be ascertained before we can even talk about their applications in bioelectronics. In Table 8.2, we tabulate the properties of some oxide semiconductors with potentials in emerging electronic technology.

There are many other metal oxide semiconductors, but they rarely come as transparent materials. Some examples of these oxides are MnO, FeO, CoO, NiO, and Cu_2O with the bandgap values (in eV) of 4, 3.2, 3.2, 2.86, and 2.04, respectively. Cuprous oxide comes both as n-type and p-type and form p–n junction diodes with rectifying properties. In Table 8.3 we give the frequency, wavelength, and associated photon energy for visible radiation, which should make it easier to appreciate the importance of transparent direct bandgap materials.

The electromagnetic spectrum covers a wide range of wavelengths varying from very short γ rays to very long radio waves. Visible radiation is only a very small part of the electromagnetic spectrum with the range of wavelengths between 410 and 710 nm as shown in Table 8.3. This range is called visible because the human eye can detect it. We are incapable of detecting radiations with wavelengths that fall outside this range. In Table 8.3, we have given the average wavelength for each color. In reality, each color covers a range of wavelengths such as it is 390–430 nm for violet and 650–770 nm is the range for red. Radiation with wavelength smaller than 390 nm is classified as near ultraviolet (UV) and the one with wavelength greater than 770 nm is called near infrared (IR). Both UV and IR consist of a wide spread of wavelength that are appropriately classified as near, mid, and far UV or IR. According to Eq. (7.9), there is a direct relationship

Table 8.2 Properties of some oxide semiconductors [1–4].

Material	E_g (eV)	E_g type	ρ (Ω cm)	n (cm^{-3})	μ (cm^2 (V s)$^{-1}$)	Polarity n- or p-type
SnO_2	3.8–4.0	Direct	7.5×10^{-5}–7.5×10^{-4}	2.7×10^{20}–1.2×10^{21}	18–31	n
TiO_2 (rutile)	3.0–3.5	Direct	3×10^{-3}–2×10^{-1}	10^{18}–10^{21}	0.05–0.2	n
ZnO	3.3–3.6	Direct	1.9×10^{-4}–5.1×10^{-4}	1.1×10^{20}–1.5×10^{21}	28–120	n
In_2O_3	3.5–4.0	Direct	4.3×10^{-5}	1.40×10^{21}	103	n
CCTO	3.40	Indirect	≈1000 at 200 °C	?	?	n
IHC 33	2.28–2.30	Indirect	2.28	10^{19}–10^{21}	0.31	n and p
IHC 45						
Fe_2TiO_5	2.77	Indirect	364.25	10^{15}	6.30	n

Here E_g is bandgap; ρ the resistivity; n the carrier concentration, μ the mobility; CCTO is $CaCu_3Ti_4O_{12}$; IHC 33 and IHC 45 are ilmenite–hematite with 33 at.% hematite and 45 at.% hematite, respectively; and Fe_2TiO_{12} is pseudobrookite (PsB). *Note:* Resistivity of CCTO, IHC 33/45, and Fe_2TiO_5 are orders of magnitude higher than those of the typical binary metal-oxide semiconductors; also they could not be measured precisely at room temperature because of instrumental limitation. Based on resistivity alone, they can be classified as insulators; yet they are good oxide semiconductor materials. Values of all parameters are at room temperature unless otherwise stated.

Table 8.3 Parameters defining visible radiation.

Radiation	Range of wavelength, λ (nm)	Average wavelength, λ (nm)	Average frequency, $v\times10^{14}$ (Hz)	Average photon energy, E_{photon} (eV)
Violet	390–430	410	7.30	3.02
Indigo	430–460	445	6.74	2.78
Blue	460–500	480	6.38	2.64
Green	500–570	535	5.75	2.38
Yellow	570–590	580	5.17	2.14
Orange	590–650	620	5.00	2.07
Red	650–770	710	4.60	1.90

between the bandgap and the wavelength of emitted radiation. Substituting for the product (ch) with 1240 eV m, we get it as in Eq. (8.8).

$$\lambda = \frac{ch}{E_g} \approx \frac{1240}{E_g}\text{nm} \tag{8.8}$$

We find from Table 8.2 that the bandgap of the four binary metal oxides vary between 3 and 4 eV, which means that they could emit radiation of wavelength between 282 and 376 nm which would be obviously in the UV range. Another important observation, we can make is that oxides in general have much lower values of mobility compared to the mobility of conventional semiconductors. It is a big drawback that works against oxide semiconductors. ZnO, In_2O_3, and SnO_2 show the most potential for wide range of applications and therefore have been studied the most. The ease and relatively low cost of processing high-quality binary oxide materials are factors in their favor. TFT is one area of technology where transparent oxide semiconductors have made the most contribution. Other areas are optical filters, transparent electrodes, and coating where oxide semiconductor performs exceedingly well.

The TFT based on ZnO, In_2O_3, and SnO_2 exhibit proven properties that would facilitate their adoption in real-life electronics on a wider scale. To validate this point, we present in Table 8.4 parameters comparing oxide-based TFT with the competing technologies based on amorphous Si, poly Si, and organic semiconductor technologies.

From Table 8.4, we can conclude that oxide-based TFT devices exhibit superior properties compared to those based on the other three classes of materials. The history of TFT is long and goes back to 1960s. Most of the TFTs are field effect devices. So far p–n junction devices have not been produced with either good performance or using a simple configurations to make them acceptable for commercial applications. But intense research is going on to produce good p-type ZnO that shows the most promise for producing p–n junction

Table 8.4 Comparison between oxide semiconductor thin-film transistor (TFT) and the other available technologies.

TFT properties	Oxide semiconductors	Amorphous Si	Poly Si	Organic semiconductors
Carrier mobility (cm^2 $(V\ s)^{-1}$)	1–100	1 (maximum)	50–100	0.1–10
Leakage current (A)	10^{-13}	$\approx 10^{-12}$	$\approx 10^{-12}$	$\approx 10^{-12}$
Manufacturing cost	Low	Low	High	Low
Long term reliability	High (forecast)	Low	High	Low in air
Yield	High	High	Medium	High
Processing temperature (°C)	≈ 350	≈ 250	<500	Room temperature

Source: Fortunato et al. 2012 [5]. Reproduced with permission of John Wiley & Sons.

devices such as a diode or a transistor. Some success has been reported, and it is hoped that it is just the matter of time when ZnO both in n- and p-polarity will become available opening the flood gate of new electronic and optical technologies. Now let us discuss two devices, namely TFT and LED, in some depth in which oxides are already used or are likely to be used in not too distant future.

8.2.6 In Search of UV-blue LED

We claimed earlier that metal-oxide semiconductors are capable of emitting UV radiation and could be used for UV LED. But in reality much more is involved than just the transition of electrons across the direct bandgap satisfying the conditions of conservation of energy and momentum. For a LED to evolve two conditions must be met; one is the direct bandgap nature of the semiconductor material and the other that p–n junction can be formed. This entails that the material must come both as n-type and p-type. Because of the transparency to visible radiation and direct bandgap nature of metal-oxide semiconductors, they are materials of choice for fabrication of LEDs emitting in near UV and blue. Because high-quality ZnO single crystals are widely available commercially, it is actively researched with this goal in focus but so far, the progress has been slow. The reason being that no reliable source of p-type ZnO exists so far. Consequently, p–n junction in ZnO cannot be produced which is the necessary condition for the emission of desired radiation. To illustrate the role that a p–n junction plays in successful fabrication of a LED, we go back to the time when the first LED appeared. It was the one that emitted in near IR and was invented at Texas Instruments, Inc., in 1966 by Biard and Pittman [6]. This was the landmark discovery that set in motion the intense race throughout the world in search of LEDs that could emit in green, blue, and other colors. The race culminated in Nobel Prize in Physics in 2014 to Isamu Akasaki and Hiroshi Amano (Japan) and Shuji Nakamura (Japan, USA) for the discovery of blue LEDs. But the material for this LED is not ZnO or any other transparent oxide semiconductors with direct bandgap. Gallium nitride, GaN, and complex configuration consisting of AlGaInP/GaAs were used for blue emission. Indium tin oxide (ITO) was used as transparent contact material. To understand the physical mechanism involved in a LED device, we examine Figure 8.9 that represents the modified original configuration used by the inventors at Texas Instruments, Inc. in 1966. In fact, this configuration is still very much used in practice for fabrication of efficient LED. Here a p–n junction is sandwiched between n-GaAs and heavily doped p-GaAs. Ohmic metallic contacts are evaporated on the two opposite faces of the structure leaving part of the upper surface exposed for the radiation to pass through.

A battery is connected to these electrodes to achieve forward biasing of the structure. This puts the structure under the influence of an electric field enabling the injection of charge carriers in the p–n junction region, leading to recombination of electrons and holes. This sets the right condition for the initiation of the electroluminescence effect in the device resulting in

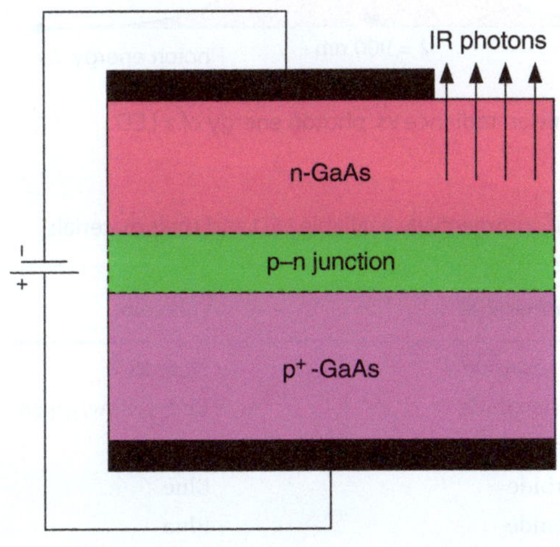

Figure 8.9 Experimental configuration for LED for red emission.

emission of photons with energy equivalent to the direct bandgap of GaAs which is 1.43 eV. The wavelength associated with this photon energy is equivalent to 790 nm which is in near-IR region. A LED is a classic example of a device that is capable of converting electrical energy into light energy.

The near IR emitting LED is widely used today for remote communication such as for accessing our TV screen and programming it remotely. Figure 8.10 is representation of the intensity vs. wavelength plot of the GaAs-based LED discussed here. The relative light output reaches its peak value around the wavelength of 900 nm ($E_g \approx 1.26$ eV). This value is about 0.17 eV lower than the value cited above. Such small discrepancy can be attributed to the quality of single crystal and the experimental method used for the determination of the bandgap. It is interesting to note that the inventor in 1966 named their device, "Semiconductor Radiant Diode." Today, it is called light-emitting diode or LED. By now LED devices are commercially available for wavelengths covering all visible colors, near IR and UV. In Table 8.5, we list the commercially available LED for each color and the semiconducting material used for their fabrication.

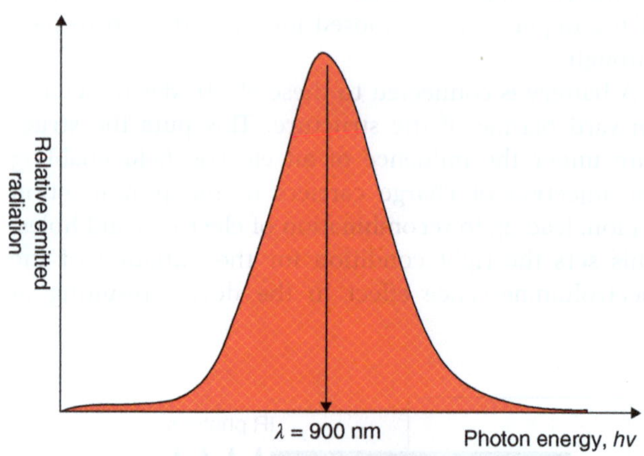

Figure 8.10 Emitted radiance vs. photon energy of a LED.

Table 8.5 List of commercially available LED and their materials.

Semiconductor material	LED color
GaAs, gallium arsenide	Near IR
GaP, gallium phosphide	Red, yellow, green
GaN, gallium nitride	Green
SiC, silicon carbide	Blue
ZnSe, zinc selenide	Blue
AlGaN, aluminum gallium nitride	UV

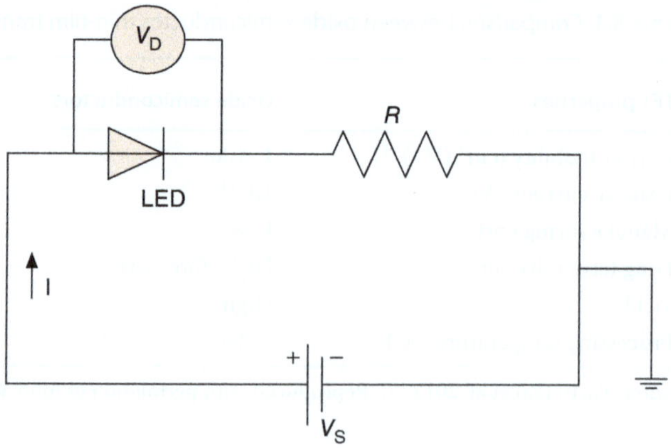

Figure 8.11 Circuit for determining the current–voltage characteristics of a LED.

8.2.7 Determination of *I–V* Characteristics of a LED

The rectifying nature of the LED diode can be easily obtained by making use of the circuit shown in Figure 8.11 and by putting the LED device electrically in the forward bias mode. Here V_s is the source potential capable of generating the circuit current, I, which depends upon the LED and resistance R connected in series. The purpose of using the resistance, R, in series with the LED is to keep the current, I, within the prescribed limit of the LED; otherwise, it might exceed the uppermost current limitation resulting in damaging the device for good. Because of the resistance in series, the voltage drop across the LED device is V_D, which by definition has to be less than the supply potential of V_s. By plotting I as a function of V_D, we should get a plot analogous to the plot seen in Figure 8.12. Obviously, it is a nonlinear $I–V$ that is the defining characteristic of a diode. It will obey the same relationship between I and V as given by Eq. (8.6). The voltage, V_F, is a characteristic

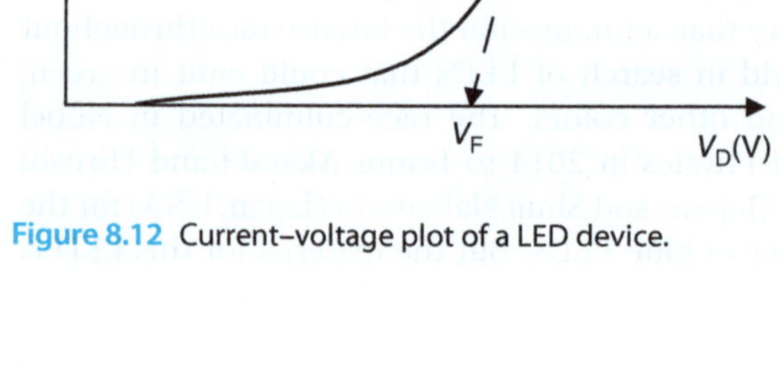

Figure 8.12 Current–voltage plot of a LED device.

parameter of a LED and is called *forward voltage drop*. This is the threshold voltage that must be overcome before the device enters the region where conduction begins and radiation is produced. In other words, we can also define it as the point where the LED switches into the active mode and begins to emit radiation. Its value depends upon the wavelength of the emitted radiation by a LED and the substrate material used to fabricate the device. Usually, it is of the order of a few volts and is defined with respect to the current of 20 mA. This is the industry standard. In a LED device, now the forward current can reach values of a few amperes; a value of 2–3.5 A being rather common.

Exercise 8.3

Consider that the source potential (V_s) is 10 V, and it is 3 V at the LED. If the circuit current is 7 mA, find the value of the series resistance R.

Solution

According to Ohm's law, the voltage drop across R is $V_R = V_s - V_{LED} = 7\,\text{V}$ and therefore $R = \frac{V_R}{I} = \frac{7}{7\times10^{-3}}\left(\frac{\text{V}}{\text{A}}\right) = 10^3\ \Omega$.

8.2.8 Thin-film Transistor (TFT)

The literature is full of reports about fabrication and characterization of TFT based on a variety of semiconductor substrates. A TFT has obviously some advantages over thick film or more widely used bulk transistors for which high-quality bulk single crystal substrates are used. They are ease and cost of fabrication, suitable for flexible electronics, and integration with flat panel displays. Already they are used widely in liquid crystal displays that are the preferred mode of display for laptops and many handheld devices with displays. As we already discussed in Section 8.3, oxide semiconductors, particularly of the binary metal-oxide group, have some unique mechanical, electrical, and optical properties that make them materials of choice for fabrication of TFT with potential for large-scale applications. Their transparency to visible light is a property that is of great advantage not only for opto-electronic devices rather also for transistor technology. Most of the TFT based on binary oxide semiconductor are based on ZnO substrates because they exhibiting superior performance compared to those based on amorphous Si, poly Si, or organic semiconductors (see Table 8.5). To date, almost all oxide TFT are field effect devices similar to the metal-oxide-semiconductor field effect transistor (MOSFET) transistors. Because of relatively high carrier mobility, transparent ZnO is also the preferred material in search for a p–n junction-based TFT. But a reliable and reproducible method has not been discovered as yet to produce p-type ZnO in large scale with the consequence that ZnO-based homogeneous p–n junction devices have not been produced. But this has not impeded the advancement of TFT technology using alternative material and alternative approach for transistor fabrication.

The TFT technology is not new. Already in the 1960s, TFT based on SnO_2 were produced. But the first truly transparent TFT was reported in 2004, and it was based on SnO_2 [7]. We discuss this device here. Instead of a classical field effect transistor (FET), it is of the enhancement-and-depletion mode. These transistors show excellent properties and are suitable for many applications where transparency is a requirement. Naturally, these devices are also suitable for all applications where a TFT is preferred. The transparent SnO_2 TFT is also thought to be a good candidate for gas sensing. We will discuss this point in the next section.

The original device configuration of the transparent SnO_2 is reproduced in Figure 8.13. It consists of clear glass as substrate over which thin films of ITO and ATO (super lattice of aluminum oxide, Al_2O_3, and titanium oxide, TiO_2) are deposited using techniques such as sputtering for ITO and atomic layer for ATO. Both these materials are also transparent. ATO acts as an insulator and ITO as a gate contact. Then a SnO_2 thin film is grown on the ATO layer by RF magneton sputtering followed by rapid thermal annealing (RTA) at 600 °C in the presence of oxygen. Finally, two ITO contacts are made on the top surface of SnO_2, and they serve as source and drain terminals. The potential between gate (G) and source (S) is shown as V_{GS} and between source and drain (D) as V_{DS}. This completes the device configuration. In Figure 8.13 we also find two plots for transmittance as a function of wavelength of light. The plot labeled (b) is the raw data as measured and the one labeled (a) is the plot after correction for reflectance. Both plots exhibit

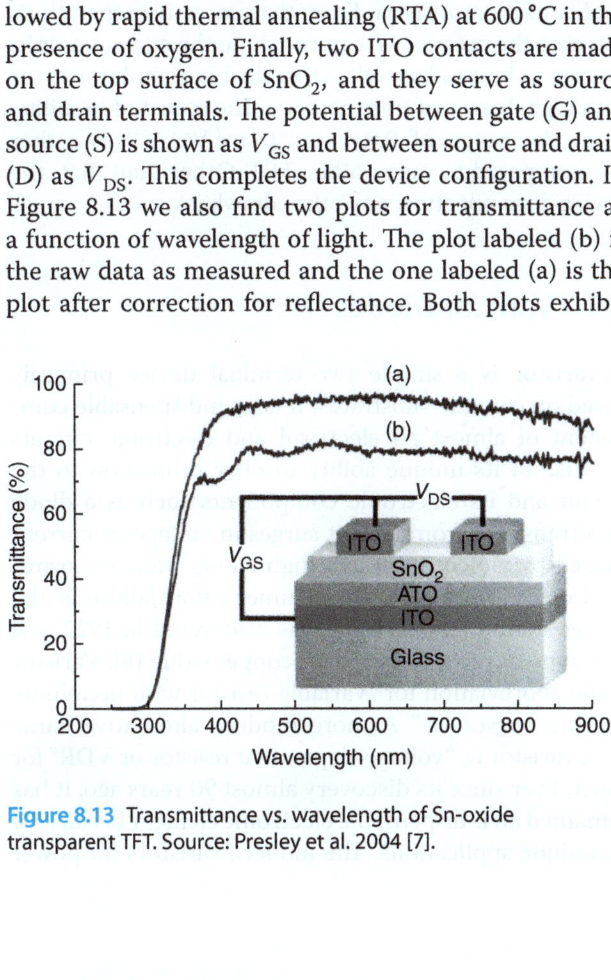

Figure 8.13 Transmittance vs. wavelength of Sn-oxide transparent TFT. Source: Presley et al. 2004 [7].

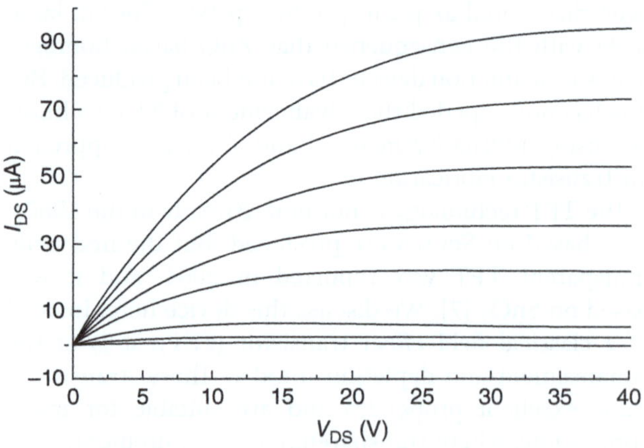

Figure 8.14 *I–V* characteristics of a SnO_2 transparent thin film transistor (TFT). Source: Presley et al. 2004 [7].

high degree of transparency in the wavelength range of $400 < \lambda < 900$ nm, which is impressive.

The transistor $I-V$ characteristics of the device in steps of 5 V between the successive curves are displayed in Figure 8.14. Here we notice that the driving voltage V_{DS} varies between $0 < V_{DS} < 40$ V that produces the transistor current, I_{DS}, in the range of $0 < I_{DS} < 90\,\mu A$ indicating a high resistance of the device structure.

Each curve shows saturation suggesting the good quality of the transistor. From the successive separations between the curves, we can conclude the current amplification to be good as well. All in all, it is a good transistor with high degree of transparency. The reported mobility, μ, in the range of $0.8 < \mu < 2.0\,cm^2\,V^{-1}\,s^{-1}$ is rather low compared to many other TFT. Other than that, the device does not show any other drawback.

8.3 Varistor Diodes

A varistor is a simple two-terminal device primarily based on ceramic substrates. It is an indispensable component of almost all electrical and electronic circuits because of its unique ability to offer protection to the circuit and its electronic components such as a diode or a transistor from abrupt surges in voltage or current caused by violent weathers (lightening, thunderstorms, etc.) or by some sort of instrumentation failure at the power station. The varistor was discovered in 1927 as a new type of rectifier based on copper oxide [8]. Varistor is the abbreviation for "variable resistor with nonohmic varying resistance." A more modern alternative name for a varistor is, "voltage-dependent resistor or VDR" for short. Ever since its discovery almost 90 years ago, it has remained an indispensable electronic element because of its unique applications. The modern varistors for power electronics are based exclusively on metal-oxide semiconductor ceramic substrates of highly modified zinc oxide (ZnO) with the exception of SiC substrates. However, SiC varistors are far less wide spread in use than the ZnO varistors. In spite of its practical importance and long productive life, a varistor remains until today an unsung hero of electronics especially when compared to more glamorous silicon-based transistors and diodes. First, we will discuss here the important properties of a varistor and later in this chapter, we will examine it as a source for the development of novel electronic devices such a transistor and a magnetic sensorbased on ceramic substrates.

It is customary to represent a varistor, when it is working as a diode, with its current–voltage ($I-V$) characteristics. The rectifying nature of current is the signature property of a varistor.

In Figure 8.15 the generic $I-V$ characteristic of a varistor is shown. It consists of three distinct regions, namely, a region of "high impedance" that is sandwiched between two regions of "low impedance." The two dashed parallel lines identify the position of the two switching voltages, $\pm V_s$. Initially, the device remains dormant until the voltage applied exceeds the switching voltage, $+V_s$, while the device is in forward biased mode. Once this point is reached, the varistor begins to conduct and the current increases rapidly. The same pattern repeats itself when the device is in the reverse biased mode. The varistor becomes an active device above $\pm V_s$. The switching voltage controls the flow of current and therefore is also called the threshold voltage. Both names are widely used while dealing with varistors and their applications. This unique property of the varistor diode is exploited in protecting an electronic component from abrupt surges in power supply. To achieve this, the varistor is connected in parallel to the circuit (or load), as shown in Figure 8.16. Varistors are modeled as two Schottky diodes in back-to-back configuration for which

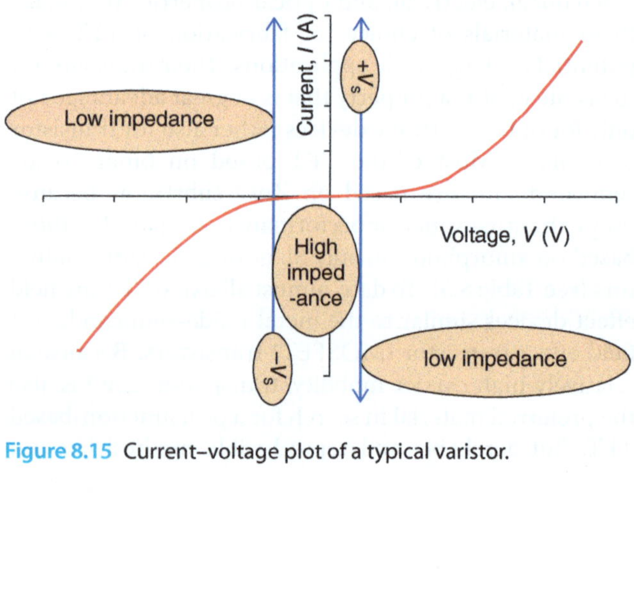

Figure 8.15 Current–voltage plot of a typical varistor.

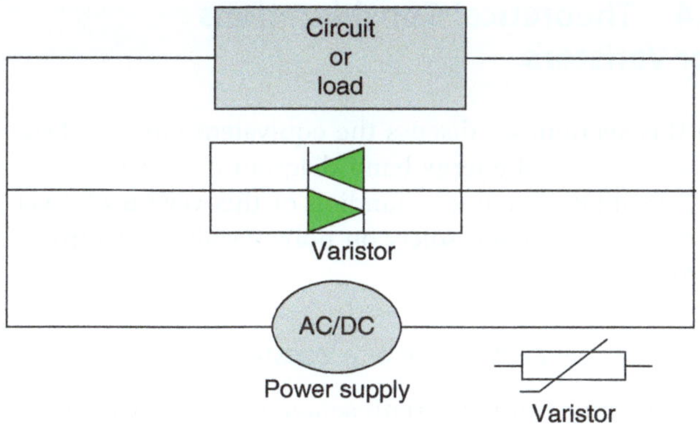

Figure 8.16 Configuration for circuit protection using a varistor.

the symbol of a diac is used (shown parallel to load). Another electronic symbol that is also used is shown at the bottom right of the same figure.

As seen from this circuit, we find that the varistor acts like a shunt and when the voltage from the power supply exceeds the tolerance limits of the components, the varistor switches and absorbs the excess voltage thus preventing it to reach the components and destroy them. As soon as the surge ends, the varistor goes back to its dormant state, and the circuit or load can resume its activity uninterrupted. The varistor works both for DC and AC supply and is therefore a superior circuit protector than a fuse or other classes of diodes.

In order to find the relationship between the current, I, and voltage, V, let us assume that the initial resistance of the device changes from R to $R + \mathrm{d}R$ when the input voltage changes from V to $V + \mathrm{d}V$. If we define the ratio between the initial resistance, R, and its change, $\mathrm{d}R$, as α such that

$$\alpha = \frac{R}{\mathrm{d}R} = \left(\frac{V}{I}\right) \cdot \left(\frac{\mathrm{d}I}{\mathrm{d}V}\right) \tag{8.9}$$

By rearranging the parameters, we get

$$\int \frac{\mathrm{d}I}{I} = \alpha \int \left(\frac{\mathrm{d}V}{V}\right) \tag{8.10}$$

That is,

$$\ln(I_2 - I_1) = \alpha \ \ln(V_2 - V_1) \tag{8.11}$$

We can rewrite Eq. (8.11) in another form as in Eq. (8.12).

$$\alpha = \frac{\ln (I_2 - I_1)}{\ln(V_2 - V_1)} \tag{8.12}$$

Therefore,

$$I = AV^{\alpha} \tag{8.13}$$

Equation (8.13) represents the general equation that describes the current–voltage relationship obeyed by a varistor diode. Here A is obviously a constant whose value would depend upon the material. The exponent α is called the nonlinear coefficient (NLC) of a varistor and is its figure-of-merit. Higher the value of NLC, the better the varistor is supposed to be.

Varistors can be broadly subdivided into two categories: one for high-voltage applications and the other for low-voltage applications. For practical applications, a varistor is preferred to have the following characteristics: (i) it should have the capacity to operate smoothly in normal and short-term voltage fluctuations; (ii) it should be able to absorb the excess electrical energy over and above the uppermost prescribed limit for the load; and (iii) it should provide protection from unexpected voltage and current fluctuations on a long-term basis; and (iv) the device must not show fatigue and electrical failure originating from the physical nature of the material on which it is based. Varistor by now is an age-old proven device with a large number of applications covering the wide spectrum ranging from high voltage circuits, power electronics, consumer electronic and electrical circuits, microelectronics, automobile electronics, displays, sensors, and many others.

8.3.1 Metal Oxide Varistors

Zinc oxide (ZnO) varistors are the prime examples of metal oxide varistors. It is so-called because the material for varistors is never pure ZnO, rather it has many metal oxide additives some of which are cobalt oxide, manganese oxide, and bismuth oxide. The additives are dispersed uniformly in small percentage throughout the ZnO matrix. The additives promote grain growth with well-defined grain boundaries sandwiched between the adjacent grains of ZnO that lends the varistor device the desirable electrical properties that are essential for reliable applications of the device under field conditions. The most essential property that is expected from a good varistor is its well-defined I–V characteristics with a high value of NLC (α), as well as the well-defined switching voltage. Another parameter of importance is the dielectric breakdown voltage of the ceramic material itself. Another example of metal oxide varistor is a SnO_2 varistor. It too has many desirable properties but is inferior to ZnO varistors for high-voltage applications.

Today, the ZnO varistors are the most used varistors in industry, and it has almost completely replaced silicon carbide varistors that were widely used before the appearance of ZnO varistors on the scene. The I–V characteristic of ZnO varistor in comparison with the I–V of SiC varistor is depicted in Figure 8.17. The NLC, α, for ZnO varistors can range from 20 to 30 and sometimes can reach even higher values. This can be achieved by adding proper additives during the ceramic processing stage. Furthermore, it can be fabricated with the voltage

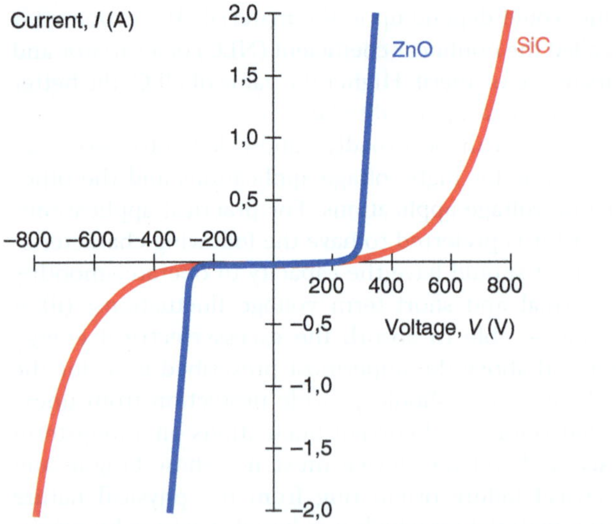

Figure 8.17 Comparative current–voltage characteristics of ZnO and SiC varistors. Source: File: Typische Varistorkennlinien.gif-Wikipedia Commons.

rating to cover the range from 10 to 10^6 V. These properties are conducive to the needs of power electronics and high-voltage power engineering. The configuration of Figure 8.16 is commonly used for circuit protection from over voltage. It can be easily modified for protection from over current in supply by including an impedance in series with the varistor that is connected in parallel to the circuit (or load) which needs to be protected.

From the I–V characteristic of Figure 8.17, we can easily identify its switching voltage to be ≥200 V, which makes it particularly suitable for high-power applications where it remains irreplaceable. On the other hand, because of its high threshold voltage, ZnO varistors are obviously not suitable for low-voltage applications. Varistors with low-switching voltage and high value of the NLC, α, are being actively pursued that would meet the requirements of microelectronic circuit components. Some success has been achieved in this pursuit with the development of double-layered low-voltage ZnO varistors. They are produced by dry extrusion molding process. The ZnO matrix is doped by small percentage of large number of binary metal oxides some of which are TiO_2, La_2O_3, Cr_2O_3, Cu_2O_3, MnO_2, and SiO_2. Such layered-structure varistors exhibit threshold voltage of approximately 10 V mm^{-1}, leakage current of 9 μA, and NLC of 23. SnO_2 varistors also show promise for low-voltage applications. When SnO_2 is doped with Bi_2O_3, CoO_2, Nb_2O_5, and Cr_2O_3, the breakdown voltage is found to be in the range of 28–60 V mm^{-1}, and the NLC between 11 and 20. These parameters are highly dependent on the sintering temperatures used for processing doped SnO_2 ceramic.

8.4 Theoretical Considerations for Varistors

In this section, we discuss the equivalent circuit, idealized model, and energy band diagram of varistor. They will facilitate our understanding of this very useful yet simple device based solely on polycrystalline electroceramics.

8.4.1 Equivalent Circuit of a Varistor

A varistor is commonly represented by an equivalent circuit consisting of an intergranular resistance R_{iG} and an intergranular capacitor C_{iG} and grain resistance, R_G. The actual configuration is given by Figure 8.19. The intergranular capacitance and resistance are in parallel, while the grain resistance R_G is in a series through which the current I exits the device. The intergranular resistance is equivalent to the varistor's VDR in which the current is almost zero at low voltage and high for the potential greater than the switching voltage, V_s. It is assumed that $R_G \ll R_{iG}$. This assumption has found its validity in the experimental determination of these parameters for ZnO varistors.

8.4.2 Idealized Model of Varistor Microstructure

We know that a polycrystalline ceramic substrate consists of large number of irregularly shaped single crystals commonly called the grains that are separated from each other by a very thin layer of highly insulating material referred to as grain boundary. Both grains and grain boundaries are made visible by a high-power scanning electron microscope (SEM). The current transport in a varistor is mediated from grain to grain by tunneling through the grain boundary. This process is unlike the process of current transport in a metal or even a p–n junction diode. For a better understanding of the current transport process involved in a varistor, we now consider an idealized picture representing a typical microstructure as shown in Figure 8.19 [9]. Here, as expected, the grain boundaries are sandwiched between the neighboring metal oxide grains (MOG), which are identified as square boxes.

In actual ZnO varistors, the width of a grain (labeled as d in the figure above) is usually of the order of a 10–12 μm and the thickness of the grain boundary (labeled as t) of the order of 100–110 nm. That is, $\left(\frac{d}{t}\right) \approx 10^2$. When a voltage is applied between the two metal contacts separated by a distance D, a current begins to flow in the circuit by overcoming the Schottky barrier which, as we know, is the consequence of the depletion layers formed.

We can visualize a varistor as consisting of large number of diodes that are connected parallel to each other.

This facilitates the transport of a bipolar rectifying current in a varistor, resulting in $I-V$ characteristics when the device is biased either in forward or reverse mode. If the voltage applied is very small, a reverse leakage current of very small magnitude is generated across each diode junction. Since the leakage current ($\approx\mu$A) is orders of magnitude smaller than the varistor current ($\geq$mA), it remains buried, and does not diminish the efficiency of the varistor under field conditions. The ratio between the leakage current and the varistor current is even greater than 10^3 in the case of a ZnO varistor, and the leakage current has no discernible role to play in the device performance.

Now the question arises how we could evaluate the values of the two device parameters, C_{iG} and R_{iG}, present in the equivalent circuit of Figure 8.18? For this let us make use of the configuration given in Figure 8.19 for finding appropriate equations that we can use.

Let us assume that in the structure shown in Figure 8.19 A is the area of the metal electrode, and there are n number of grains of width d and equal number of grain boundaries of width t, also the relative dielectric constant of the varistor material between the upper and lower electrodes covering the distance D is ϵ_r. Considering the configuration to be equivalent to that of a parallel plate capacitor, we can express the capacitance C_1 which is primarily due to the insulating grains as

$$C_1 = \frac{\epsilon_r A}{D} = \frac{\epsilon_r A}{nd} \tag{8.14}$$

That is not all. The contribution made by the grain boundaries must also be considered. Let C_2 be the capacitance because of the grain boundaries present. Then

$$C_2 = \frac{\epsilon_r A}{nt} \tag{8.15}$$

Since the grains and the grain boundaries are in parallel configuration, the sum of these two capacitances will give us the capacitance, C_{iG}. That is,

$$C_{iG} = C_1 + C_2 = \frac{\epsilon_r A}{n}\left(\frac{1}{d} + \frac{1}{t}\right) \tag{8.16}$$

Substituting n with (D/d) and rearranging a bit Eq. (8.16) becomes

$$C_{iG} = \frac{\epsilon_r A}{D}\left(1 + \frac{d}{t}\right) \approx \frac{\epsilon_r A}{D} \cdot \frac{d}{t} \tag{8.17}$$

In Eq. (8.17), we could ignore 1 within the bracket because $\frac{d}{t} \gg 10^2$. We can draw two conclusions from Eq. (8.17). First, that the effective dielectric constant is increased by a factor of $\left(\frac{d}{t}\right)$. Second, the true thickness is not D, but it increases by an amount equivalent to $\frac{Dt}{d} \approx nt$.

Now let us figure out the value of the resistance in parallel in the equivalent circuit of Figure 8.18. We know from the theory of a capacitor that the parallel impedance R_{iG}, is related to the frequency, ω, of the AC source, the loss tangent, $\tan\delta$, and the capacitance C. That is,

$$R_{iG} = \frac{1}{\omega C \tan\delta} \approx \frac{1}{\omega\epsilon_r \tan\delta}\left(\frac{D}{A}\right) \tag{8.18}$$

Alternatively, the intergranular resistivity, ρ_{iG}, can be written as follows:

$$\rho_{iG} = R_{iG}\left(\frac{A}{D}\right) = (\omega\epsilon_r \tan\delta)^{-1} \tag{8.19}$$

Since both the dielectric constant and the loss tangent are material properties and remain constant at room temperature, we can conclude that $\rho_{iG} \propto \omega^{-1}$. It is indeed found experimentally that the varistor impedance drops as the inverse of frequency.

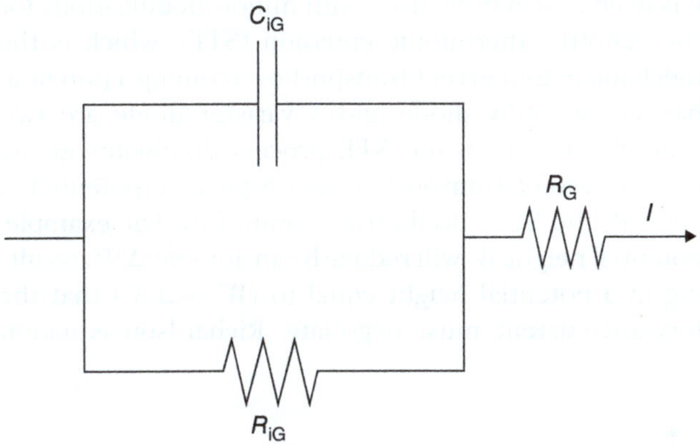

Figure 8.18 Equivalent circuit of a varistor.

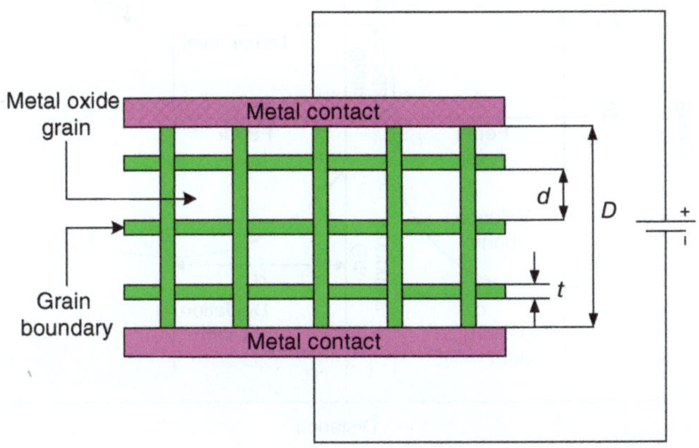

Figure 8.19 Schematic of idealized structure of a metal oxide varistor. Source: Modified from Levinson and Phillip 1986 [9].

Exercise 8.4

Find an expression for the potential across a single grain for the block model of Figure 8.19.

Solution

Let us apply a voltage V between the two electrodes with distance D. Assume that there are n numbers of single grains in the structure. Then the voltage experienced by each grain, V_G, would be equal to V/n. Also $D = nd$. Therefore,

$$V_G \approx \frac{V}{n} \approx \frac{Vd}{D}$$

Note that we are here ignoring the contributions of the grain boundaries because $R_{GB} \gg R_G$. Therefore, the voltage at the grain boundaries will be much smaller compared to V_G.

If $V = 10$ V, $d \approx 50$ μm, and $D = 2$ mm, then $V_G \approx V\left(\frac{d}{D}\right) \approx \frac{10\times50\times10^{-6}}{2\times10^{-3}} \approx 250$ mV. Also $n = \left(\frac{D}{d}\right) \approx \frac{2\times10^{-3}}{50\times10^{-6}} \approx 40$.

8.4.3 Energy Band Diagram: Grain–Grain Boundary–Grain (G–GB–G) Structure

We have already discussed in this chapter the origin of the formation of Schottky barrier, while discussing when a metal and a n-type semiconductor are in contact. We can expand that knowledge now to construct an energy band diagram when a grain boundary is sandwiched between two grains which is the situation that we encounter in the case of a varistor. This grain–grain boundary–grain, or simply G–GB–G, configuration forms as a consequence of double Schottky barriers that are generated when majority carriers are trapped at the interface states at the grain boundaries. The position of the Fermi level can be determined considering the density of states of acceptors (N_c) and donors (N_d) at a thermal equilibrium at temperature, T. This is expressed in Eq. (8.20).

$$E_c - E_v = k_B T \ln\left(\frac{N_c}{N_d}\right) \quad (8.20)$$

It is a famous equation of semiconductor physics and its full derivation can be found in any standard textbook. The energy band diagram of a G–GB–G configuration is depicted in Figure 8.20. For simplicity, we are considering only two grains, A and B, and one grain boundary (GB) that is sandwiched between the two grains. The grain boundary is to be envisioned as a very thin or infinitely thin narrow interfacial layer; usually of the order of only a few nanometers.

The potential barrier height that exists for this model with zero potential applied ($V = 0$) is shown as Φ_0 in the conduction band of grain A. When the structure is forward biased, the barrier height Φ_0 decreases to the height of Φ as a consequence of the Schottky effect. It is shown in grain B. The difference between these two heights is $\Delta\Phi \approx (\Phi_0 - \Phi)$. To be consistent with the block model of Figure 8.19, the grain width in Figure 8.20 is labeled as d and the width of the GB as t. Once the equilibrium is established, a current begins to flow across the G–GB–G structure.

We can now try to find an expression for the varistor current, I, as a consequence of the applied potential. This can be best achieved by making use of the famous Richardson law originally derived to explain the thermionic emission from hot filament. Owen William Richardson, a British physicist, was the first one to explain the thermionic emission phenomenon and was awarded the Nobel Prize in Physics in 1928 for his groundbreaking theory. According to this theory, the thermionic current is related to the work function of a metal and is given as follows:

$$I = A^* \exp\left(-\frac{W}{k_B T}\right) \quad (8.21)$$

The constant A^* is called Richardson constant, and it varies from material to material. All other parameters in this equation have their standard meaning that we have already used many times before. This famous equation has been successfully used with minor modifications for the Schottky thermionic emission (STE), which is the mechanism for current transport overcoming a potential barrier. Schottky diode and a varistor diode are two examples in which the STE process dominates giving rise to current transport. When a positive potential is applied, the Schottky barrier diminishes. For example, potential height W will reduce by an amount ΔW, resulting in a potential height equal to $(W - \Delta W)$ that the forward current must negotiate. Richardson equation

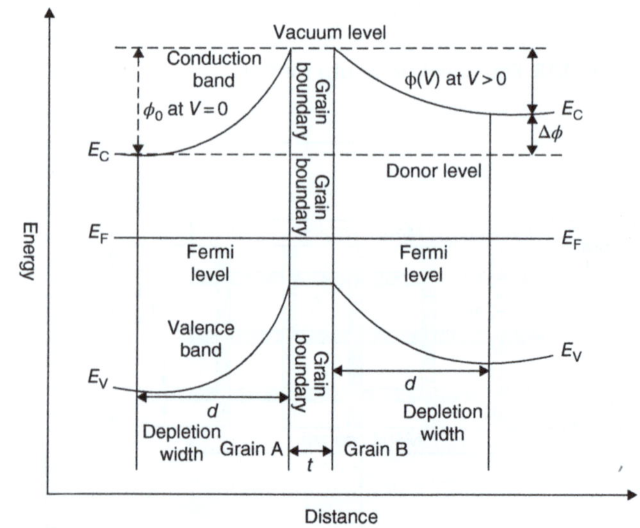

Figure 8.20 Energy band diagram of a grain-grain boundary-grain (G–GB–G) structure.

when applied for such a situation takes the form of

$$I = A^* T^2 \exp\left(-\frac{W - \Delta W}{k_B T}\right) \tag{8.22}$$

Alternatively, we can also rewrite the above equation with respect to the potential barriers given in Figure 8.20 and considering that a potential V is applied to the structure. This transforms Eq. (8.22) into Eq. (8.23).

$$I = A^* T^2 \exp -\left\{\frac{\phi(V) + E_F}{k_B T}\right\} \cdot \left\{1 - \exp\left(-\frac{eV}{k_B T}\right)\right\} \tag{8.23}$$

This equation correctly predicts the value of the current flowing through a varistor diode and represents its rectifying nature with respect to applied potential V. But finding a practical solution of this equation is not trivial. It requires the talents of a skilled theoretician to come forward with a simpler and practical formulation of Eq. (8.23). Even then let us now try to find a practical solution to Eq. (8.23) using the parameters given in Figure 8.18. Life would become much less stressful if we could neglect the effect of charge carriers and doping densities while solving STE equations. In fact, this is a very common practice in semiconductor physics, and we are not doing anything that is not allowed.

To reach this goal, we make use of the Poisson equation, which relates the potential barrier height (ϕ) to the space charge distribution $\{\rho(x)\}$ as expressed mathematically in Eq. (8.24). As a side remark, we should mention that the Poisson equation is a powerful tool in the hand of theoretical physicists and mathematicians in solving complex problems of science and engineering.

$$\frac{\delta^2 \phi(x)}{\delta x^2} = -\frac{\rho(x)}{\epsilon \epsilon_0} \tag{8.24}$$

Here ϵ is the permittivity of the medium, and ϵ_0 is the permittivity of free space. After double integration, the Poisson equation yields a double parabolic potential energy barriers resembling qualitatively the energy band diagram seen from Figure 8.20. On completion of the integration process, we can get Eq. (8.25) with respect to the potential barriers, ϕ (at V) and ϕ_0 (at $V = 0$).

$$\phi(V) \approx \phi_0 \left(1 - \frac{eV}{4\phi_0}\right) \quad \text{and} \quad eV \leq \phi_0 \tag{8.25}$$

After some rearrangement, we can rewrite Eq. (8.23) as:

$$I = A^* T^2 \exp -\left(\frac{\phi(V) + E_F}{k_B T}\right) \cdot \left(1 - \exp\left(\frac{eV}{k_B T}\right)\right) \tag{8.26}$$

Furthermore, let us consider that the voltage experienced by each grain is $V_{G.}$ Once again, taking into consideration the parameters of Figure 8.19, we can express V_G (see also Exercise 8.4) as follows:

$$V_G \left(\approx \frac{V}{n} \approx \frac{Vd}{D}\right)$$

Now we can readily rewrite Eq. (8.26) as Eq. (8.27).

$$I = A^* T^2 \exp -\left\{\frac{\phi\left(\frac{Vd}{D}\right) + E_F}{k_B T}\right\} \cdot \left\{1 - \exp\left(-\frac{eVd}{Dk_B T}\right)\right\} \tag{8.27}$$

Making use of the solution of Poisson equation (see Eq. (8.25)), we can express $\phi\left(\frac{Vd}{D}\right) \approx \phi_0$, and $\exp\left(-\frac{eVd}{Dk_B T}\right) \approx 1 - \frac{eVd}{Dk_B T}$; Eq. (8.27) takes the form of Eq. (8.28).

$$I = A^* T^2 \exp -\left\{\frac{\phi_0 - E_F}{k_B T}\right\} \cdot \left(\frac{eVd}{Dk_B T}\right) \tag{8.28}$$

Alternatively, Eq. (8.28) can also be written as Eq. (8.29).

$$I = \left[A^* T^2 \exp -\left\{\frac{\phi_0 - E_F}{k_B T}\right\} \cdot \left(\frac{ed}{Dk_B T}\right)\right] \cdot V \approx G(T) \cdot V \tag{8.29}$$

In the above equation, for simplicity let us substitute the identity within the square bracket with $G(T)$. Equation (8.29) represents another form of Ohm's law with $G(T)$ being equivalent to conductance which is temperature-dependent. Equation (8.29) is obviously a nonlinear equation resulting in the rectifying nature of the I–V characteristics of a varistor diode. Furthermore, if V is small, then the current is small, and when it is large, the current is large and rises rapidly even with small increment of applied voltage. Therefore, we can consider Eq. (8.29) as the true representation of varistor I–V behavior.

Exercise 8.5
Find the breakdown voltage for the ZnO grain from the information given in Eqs. (8.30) and (8.31).

Solution
For a ZnO varistor the critical voltage, V_C, is defined as the breakdown voltage of a single GBG junction. It is also called the critical value. To a first approximation, V_C can be estimated by using the second part of Eq. (8.25) after substituting V with V_C. This gives us Eq. (8.30).

$$V_C \approx \frac{4\phi_0}{e} \tag{8.30}$$

And the NLC, α, is related to the potential barrier height ϕ_0 as Eq. (8.31).

$$\alpha \approx \frac{\phi_0}{k_B T} \tag{8.31}$$

Assuming that $\alpha = 30$ in ZnO varistor, and $T = 300$ K the critical value of the potential for the breakdown of the GBG junction.

$$V_C \approx \frac{4\alpha k_B T}{e}$$
$$\approx \frac{1.2 \times 10^2 \times 3 \times 10^2 \times 1.38 \times 10^{-23}}{1.60 \times 10^{-19}} \approx 3.11 \text{ V}$$

This level of voltage on GBG junction that is very thin amounts to a large electric field.

8.5 Varistor-Embedded Devices

In the previous section, we discussed extensively ZnO varistor and established that it remains unchallenged for the last many years as a circuit-protecting device. But only recently, it has been reported that the nonlinear I–V characteristics of a varistor can be tuned with the help of some external agents leading to the discovery of embedded transistors and magnetic field sensors. In the process, the tuned varistor acquires the unique properties of transconductance and signal amplifications that have been so far associated with only transistors. Furthermore, under certain conditions, its response to frequency gives rise to low pass filters with bandwidths appropriate for many applications.

It is interesting to know that varistor precedes the famous bipolar junction transistor (BJT), which was discovered in Bell Labs. in 1947 by 20 years. These two devices in their fabrication, processing, and applications are so different that normally one would not suspect a link between a varistor and a transistor. Varistor is a ceramic-based device, whereas transistors are built on highly sophisticated platforms of single crystals and epitaxial films of semiconductor materials found in Group IV, Groups III–V, and Groups II–VI of the periodic table. The discovery of transistor was recognized as a giant step forward in the field of solid-state physics and consequently (and well deserved) Walter Brattain, John Bardeen, and William Shockley, the three giants who discovered this wonder device, received the Nobel Prize in Physics in 1956. The transistor is the king of microelectronics today while the varistor, which has been around for 90 years, still remains grossly understated in spite of its very productive life. It is interesting that when the first solid-state device, now universally known as the transistor, was discovered a name for it needed to be found. A search was launched at its birthplace (Bell Labs.) to find a generic name for the new device. Multiple suggestions followed, and out of six proposals, the name, "transistor" emerged as the finalist. This name was proposed by John R. Pierce [10]. He suspected an inherent relationship between varistors and transistors. But this was perhaps never conclusively demonstrated. The word *transistor* is the contraction of "transconductance varistor." Recent reports in scientific literature confirm the coupled relationship between a varistor and a transistor. This concept is the thrust of our discussion in this section.

The block diagram of Figure 8.21 leads us from a varistor to a transistor and a magnetic sensor. We encounter in this schematic the existence of three different types of embedded transistors when the I–V characteristics of a varistor is modified (or tuned) with the help of a biasing potential, an electric field, and a magnetic field. Also, we find here that the magnetically tuned varistor leads us to a magnetic field sensor. We also find from the same block diagram that the varistor itself acquires additional properties after being tuned by an external agent resulting in expanding its field applications beyond its traditional role of circuit protection. These varistor–transistor hybrid devices can coexist simultaneously on a single chip and so adding to the possibility of integrating them with integrated circuits and accessing them either as a varistor, or a signal amplifier or a transistor. These devices were discovered using varistors based on iron-titanate electroceramics. They are listed in Table 8.6.

It should be appreciated that this is just the beginning and many more new materials will be found with time that could be used to develop these hybrid devices with superior properties suitable for a wide range of practical applications.

Determination of radiation hardness of IHC 45 and pseudobrookite (PsB) was done using primarily three types of radiations. Details are summarized in Table 8.7. Samples of Mn-PsB were not evaluated for their radhard properties, but we assume it also to be a good radhard material.

8.5.1 Voltage Biased Varistor and Embedded Voltage Biased Transistor (VBT)

For studying the effect of a bias voltage on the I–V characteristic of a two terminal varistor, a circuit similar to the one shown in Figure 8.22 is used in conjunction with a semiconductor parametric analyzer [11]. It represents the typical circuit used for the determination of I–V characteristic of an n–p–n BJT. The series resistance R between source (S) and bias (B) is introduced for controlling the ratio between the drain current, I_d, and

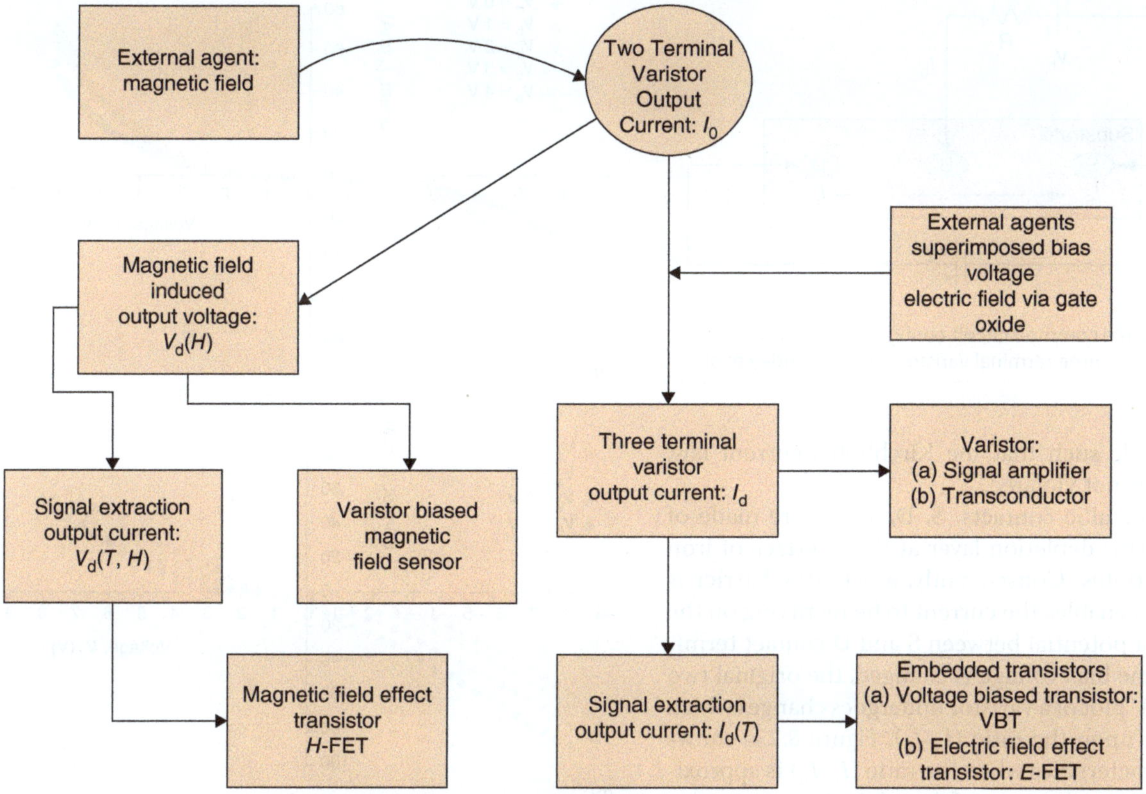

Figure 8.21 Roadmap from varistor to transistor and magnetic sensor.

Table 8.6 Materials for hybrid devices and their physical properties.

	IHC 45 ceramic (55% ilmenite, $FeTiO_3$ + 45% hematite, Fe_2O_3)	PsB single crystal (pseudobrookite, Fe_2TiO_5)	Mn-PsB ceramic (Mn-doped pseudobrookite) (0.40% Mn_2O_3 + 0.60 Fe_2TiO_5)
Magnetic Curie point, T_c (K)	Ferromagnetic, $T_c \approx 610$	Feebly magnetic to nonmagnetic	Strongly magnetic; T_c = ?
Resistivity, ρ (Ω cm)	2.38	365	23.22
Seebeck coefficient, S (μV K^{-1})	160	−132	−135
Polarity/mobility, μ (cm^2 $(V\ s)^{-1}$)	n/p-Type; 0.31[a)]	n-Type; 6.30	n-Type; 2.3
Bandgap, E_g (eV)	2.28; indirect	2.77; indirect	2.36
Carrier concentration, n (cm^{-3})	10^{19-21}	10^{15}	10^{16-17}
Radiation hardness	Excellent	Excellent	Not determined

a) The polarity of IHC 45 is strongly dependent upon temperature and environment of annealing.

Table 8.7 Radiation sources, their energy and fluence/doses, and labs where this determination was done.

Source	Energy	Fluence/dose	Labs
Neutron	73 MeV[a)]	2.1×10^9 neutrons cm^2	Los Alamos National Labs (US DoE)
Proton	10 and 40 MeV	5×10^{10} protons cm^{-2}	Cyclotron Institute at Texas A&M University
Heavy ions (Fe)	3.07 MeV/nucleon	3 krad	Brookhaven National Labs (US DoE)

a) This represents a weighted average of a broad neutron spectrum that ranges from 1 to 800 MeV.

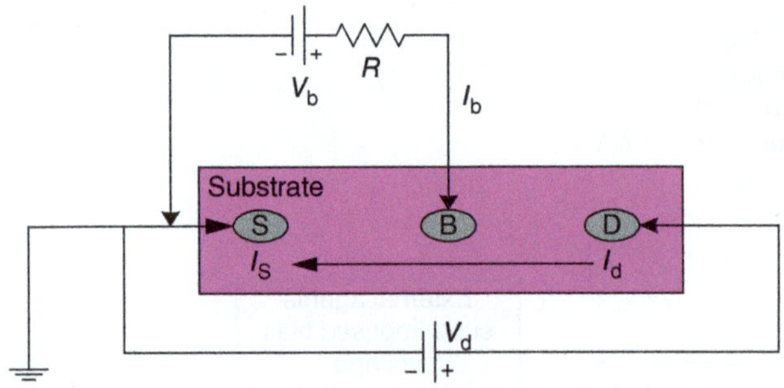

Figure 8.22 Source common circuit configuration for I–V determination of a three terminal varistor. Source: Pandey et al. 2015 [11].

bias current, I_b such that the Kirchhoff's current law, $I_d = (I_b + I_s)$ is not violated.

All three metallic contacts, S, D, and B are made of silver that forms depletion layer at the interface of iron titanate substrates. Consequently, a Schottky barrier is formed which enables the current to be rectifying on the application of potential between S and D contact terminals. When the bias voltage is changed, the original two terminals $I-V$ plot of a varistor undergoes changes which is dependent upon the ratio (I_d/I_b). Figure 8.23a shows the $I-V$ characteristics when the ratio (I_d/I_b) is approximately 10^3. The nonlinearity of the original curve remains basically unchanged as the biasing voltage increases from 1 to 4 V.

On the other hand, major changes take place in the nature of the $I-V$ plots when $I_d \leq 10^2 I$. This is shown in Figure 8.23b. Here we find that the successive curves do not pass through the origin but shifts along the x-axis and the output signal is slightly larger for the same bias potential applied. Moreover, each successive curve is separated from each other in larger units making them easier to identify. It is important to note that the shape of the $I-V$ plots suffer significant asymmetry compared to what we had in Figure 8.23a. When the ratio between the drain current and the bias current is very large, such as in Figure 8.23a, the NLC of the varistor decreases with increasing bias, but it is never greater than 15%. In this mode, varistor loses its efficiency to a certain degree as a protective device yet never fails completely in protecting a load. However, when $(I_d \leq 10^2 I_b)$, the NLC decreases rapidly making the varistor unfit for circuit protection. But the varistor in this mode gains in signal amplification potential. Another property acquired by the biased varistor is the *transconductance* that is usually found in transistors. It is also known as *transfer function or mutual conductance.* By biasing the varistor with a potential we can introduce in it both the signal amplification and transconductance potentials which are normally ascribed to a transistor as its signature property.

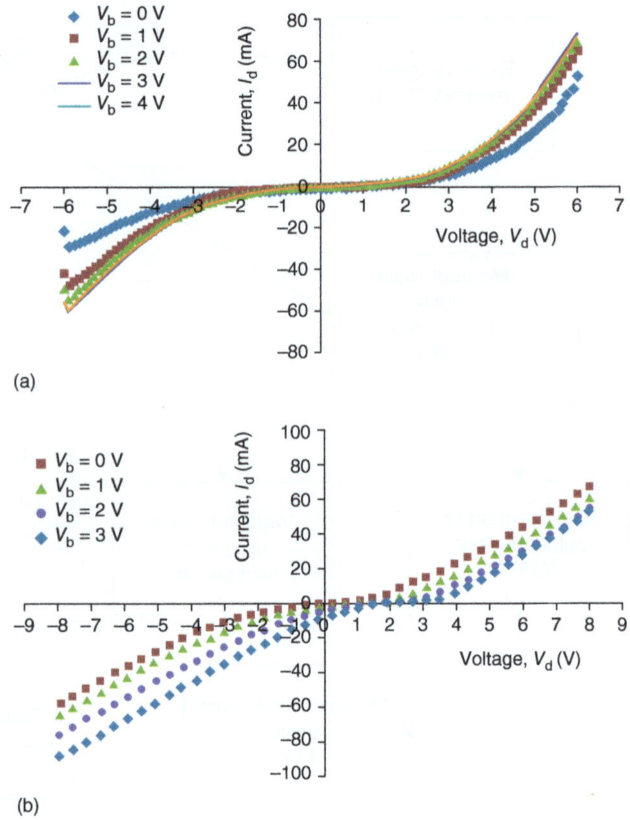

Figure 8.23 Drain current (I_d) vs. drain voltage (V_d) plots of an IHC 45 varistor when (a) $I_d \geq 10^3 I_b$. (b) $I_d \leq 10^2 I_b$. Source: Pandey et al. 2015 [11].

Now let us estimate the contribution made by the bias potential to the drain current, I_d, of the three terminal device.

Taking into consideration that the applied bias potential (V_b) will either add to or subtract from the original I_0 depending upon how it interacts with the potential barrier. We know that a positive potential diminishes the Schottky height whereas a negative potential increases it. Assume that the drain current (I_d) of the three terminal device changes by an amount, $I_d(\mathrm{T})$ on the application of a bias potential (V_b). Then we can express the relationship between three currents, I_d, I_0, and $I_d(\mathrm{T})$ as in Eq. (8.32).

$$I_d(\mathrm{T}) = I_d \pm I_0 \tag{8.32}$$

To be consistent with the convention used in practice that a positive voltage gives rise to a positive current and similarly a negative voltage results in a negative current, we can rewrite the above equation as a set of two equations as in Eqs. (8.33).

$$\begin{aligned} +I_d(\mathrm{T}) &= (I_0 - I_d) \quad \text{for} + V_d \text{ (a)} \\ -I_d(\mathrm{T}) &= (I_d - I_0) \quad \text{for} - V_d \text{ (b)} \end{aligned} \tag{8.33}$$

The two current, I_d and I_0, are experimentally determined whereas the third current, I_d(T) can be obtained with the help of Eqs. (8.33).

After determining the contributed current for each value of the applied bias potential we can generate the I–V characteristics of the resulting new three terminal device as shown in Figures 8.24 and 8.25.

Figure 8.24 may not be easily associated with the I–V characteristic of a typical transistor especially because none of the plots therein show any sign of saturation, which is the hall mark of a transistor. But for the condition of $I_d \leq 10^2 I_b$, we can see from Figure 8.25 that each plot corresponding to individual bias potential is fully saturated making it easier to accept them as the I–V characteristics of a transistor. This new device is also bipolar in nature, which is consistent with its origin in the very nature of its parent varistor. This embedded device is called the *voltage biased transistor* (VBT). It is a new discovery that was made in 2014 by the author of this book [5]. Unlike the traditional semiconductor transistor, the VBT device is based on an electroceramic ceramic substrate. It is important to point out here that the classical solid-state BJT is so named because its structure consists of the bipolar nature of its substrate, that is, n-type and p-type semiconductor materials. The I–V plots for a typical BJT are not necessarily bipolar as in the VBT transistor. The BJT devices operate only in the forward mode of the biasing potential, which is in contrast to the VBT that operates irrespective of the fact if the applied bias is positive or negative. In Figure 8.26, we show the I–V plots of a VBT transistor buried in a varistor based on another electroceramic substrate, which is Mn-doped PsB (R.K. Pandey, unpublished result). Here we can see that the plots are saturated for the forward bias as well as reverse bias mode of the device. In the forward bias mode though the device is capable of boosting currents to a greater percentage than in the reverse mode. The internal impedance of the MnPsB device seems to be greater than for the IHC 45 device, resulting in smaller output current. This is supported by the fact that according to Table 8.6, the resistivity of MnPsB is approximately larger by a factor of 10 compared to the resistivity of IHC 45.

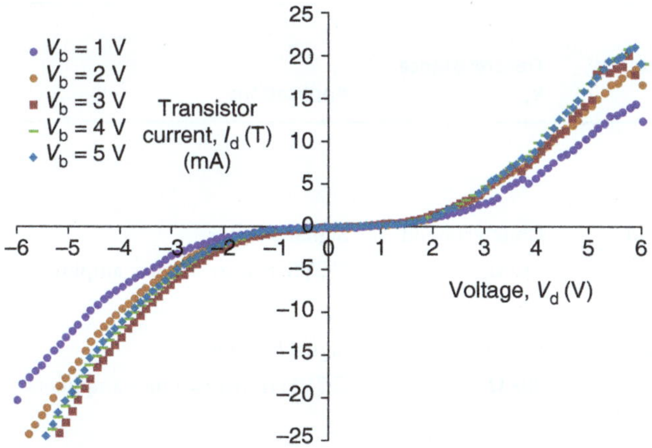

Figure 8.24 I_d(T)–V plots for IHC 45 voltage biased transistors (VBT) with $I_d \geq 10^3 I_b$. Source: Pandey et al. 2015 [11].

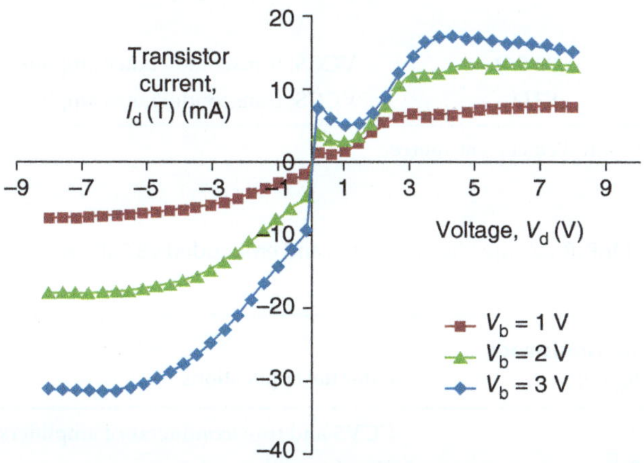

Figure 8.25 I_d(T)–V plots for IHC 45 voltage biased transistors (VBT) with $I_d \leq 10^2 I_b$. Source: Pandey et al. 2015 [11].

We would recall that earlier we said that signal amplification and transconductance are the two signature properties that define the efficiency of a transistor. Now we need to determine these parameters for the electroceramic transistors described here. We can define the signal amplification (S_a) and transconductance (G_m) as expressed by Eqs. (8.34) and (8.35).

$$S_a = \frac{\text{final current at a constant drain voltage}}{\text{intial current at the same drain voltage}} = \frac{(I_{d,f})}{(I_{d,i})} \times 100 \qquad (8.34)$$

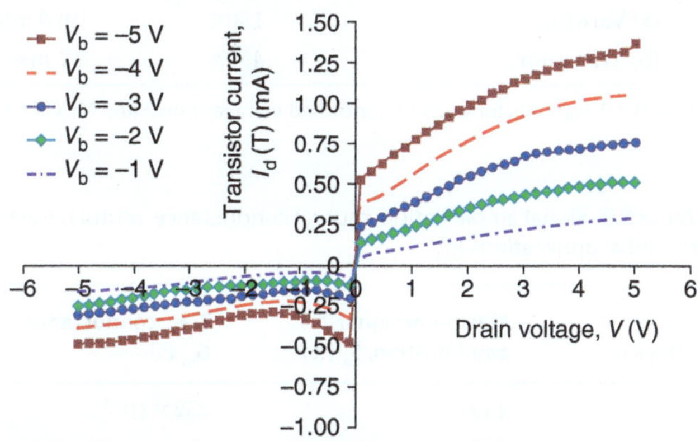

Figure 8.26 I_d(T)–V plots for Mn-PsB ceramic voltage biased transistor (VBT) with $I_d \leq I_b$. Source: Pandey et al. 2015 [4].

$$G_m = \frac{\text{current at a constant drain voltage}}{\text{bias volage}} = \frac{I_d(T)_{V_d}}{V_b} \approx \frac{\Delta I_d(T)}{\Delta V_b} \quad (8.35)$$

The reciprocal of G_m is called transresistance or simply mutual resistance, R_m. Both these terms are widely used to identify the potential applications of a device. In Table 8.8 the values of these parameters for an IHC 45 based VBT transistor and its parent varistor are tabulated. Similarly, pertinent properties for a MnPsB based transistor are tabulated in Table 8.9.

8.5.1.1 Frequency Dependence of IHC 45 VBT Device

The knowledge of frequency dependence of an electronic device, especially of a transistor, has important practical ramifications in determining its usefulness in alternating current circuits. Frequency dependence of the varistor and its embedded VBT transistor was determined using the same circuit as depicted in Figure 8.22 simply by replacing the DC power source by an AC power source. The drain current was kept constant at 2 V_{pp}, while the frequency was varied from 20 Hz to 4 MHz. The output current was determined as a function of frequency for the varistor. The transistor current for the VBT device was then could be found by the extraction process using Eq. (8.32). The frequency dependence of the two devices is plotted in Figure 8.27 [4]. We can conclude from this figure that currents of the two devices show a remarkable constant nature with respect to frequency for the entire range of 20 Hz to 4 MHz. The varistor current is naturally larger than the transistor current. The bandwidth of the

Table 8.8 Maximum current amplification (S_a), transconductance (G_m), transresistance (R_m), and potential applications of IHC 45 voltage biased varistor and its embedded transistor (VBT) [11].

Devices	Maximum signal amplification	Transconductance, G_m	Transresistance, R_m	Applications
Case I. Devices with $I_d \geq 10^3 \times I_b$				
Drain voltage kept constant at +5 V				
(a) Varistor	163%	Undetermined	Undetermined	Signal booster
(b) Transistor	160%	1.62 mS	617 Ω	CCVS; transresistance amplifiers
Drain voltage kept constant at −5 V				
(a) Varistor	196%	Undetermined	Undetermined	Signal booster
(b) Transistor	158%	3.24 mS	308 Ω	CCVS; transresistance amplifiers
Case II. Devices with $I_d \leq 10^2 \times I_b$				
Drain voltage kept constant at +6 V				
(a) Varistor	160%	7.17 mS	139 Ω	VCCS; transconductance amplifiers
(b) Transistor	220%	4.48 mS	223 Ω	VCCS; transconductance amplifiers
Drain voltage kept constant at −6 V				
(a) Varistor	150%	10.8 mS	92.5 Ω	VCCS; transconductance amplifiers
(b) Transistor	450%	12 mS	83 Ω	VCCS; transconductance amplifiers

Note: CCVS stands for current controlled voltage source and VCCS for voltage controlled current source.

Table 8.9 Signal amplification, mutual conductance, mutual resistance for MnPsB voltage biased varistor, and embedded VBT along with potential applications [4].

Devices	Maximum signal amplification, S_a (%)	Transconductance, G_m (S)	Transresistance, R_m (Ω)	Potential applications
VBT	465	252×10^{-3}	4	CCVS and transconductance amplifiers
Varistor	240	??	??	Signal amplifier

Drain voltage kept constant at +5 V.

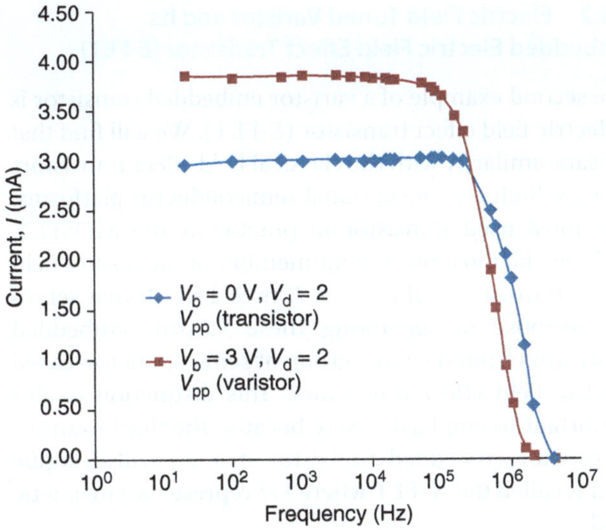

Figure 8.27 Frequency dependence of IHC 45 varistor and its VBT embedded transistor. Source: Pandey et al. 2014 [3].

varistor is determined to be 780 kHz, and it is 310 kHz for the embedded transistor. This means that both these devices should perform well as low pass filters that are important electronic components and widely used. Normally, electronic filters are made using combination of various resistors, capacitors, and inductors. They have to be carefully selected to meet the necessary mathematical conditions for reliable operation in field. Compared to the commercially available filters, the filter based on IHC 45 varistor and its embedded transistor are built on very simple platforms without the need of any other electronic components. They are rugged and not at all prone to be easily damaged in field conditions.

8.5.1.2 Comparison Between a VBT, BJT, and Schottky Transistor

It is rather natural to confuse the VBT transistors with the well-established BJT transistor because of their similarity in device configuration and mode of current transport. But these are only superficial similarities. In reality, they are distinctly different devices because of the physical principles involved in their very origin and the way the current transport takes place in each of them. Let us consider the three devices shown schematically in Figure 8.28 (R.K. Pandey, unpublished result).

In Figure 8.28a, we find three metallic contacts labeled as source (S), bias (B), and drain (D). This configuration makes a VBT device equivalent to three different varistors, namely S–B, B–D, and D–S, which are interconnected on a single substrate. This configuration is crucial for the origin of a signal amplifying varistor and

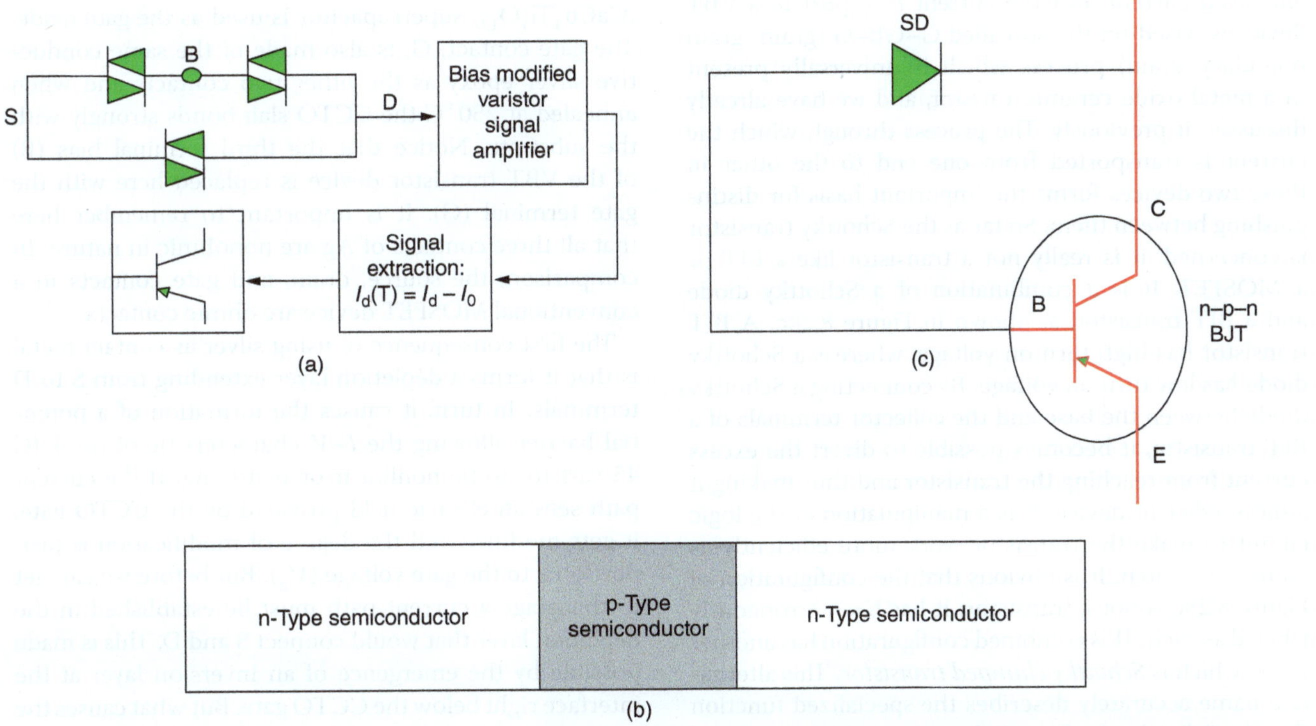

Figure 8.28 (a) Voltage biased transistor (VBT); (b) bipolar junction transistor (BJT), and (c) Schottky transistor.

an embedded VBT transistor. In comparison, a BJT transistor, as can be seen from Figure 8.28b, consists of a n–p–n configuration (alternatively, it can be p–n–p configuration too, though this structure is not usually preferred in practice). A typical BJT transistor consists of three terminals, namely, emitter (E), collector (C), and base (B). The emitter (E) and collector (C) contacts are made on the two n-arms and base (B) on the p-arm. In a VBT device, the bias terminal (B) is used to control the current between source (S) and drain (D). Similarly, the base terminal (B) controls the current between emitter (E) and collector (C) in a BJT transistor. It is important to remember that in a varistor, the source (S) and drain (D) are nonohmic in character so that a depletion layer can be formed at the interface leading to the formation of a potential barrier. This facilitates the nonlinear I–V characteristic of a varistor. The third terminal bias (B) may or may not be nonohmic. In a BJT device, on the other hand, all three metal contacts are ohmic and as such no potential barrier forms; and in fact they are not even required. The current transport between the emitter (E) and the collector (C) takes place by diffusion of charge carriers from n-side to the p-junction. The thermionic emission is the underlying process responsible for current transport in both these devices. In a BJT transistor, the charge carriers are thermally generated, and the current transport is because of the injection of charge carriers at the "base" terminal which are subsequently collected at the "collector" terminal resulting in enhanced current. But the current transport in a VBT device is based on the so-called G–GB–G (grain–grain boundary–grain) process which is universally present in a metal oxide ceramic varistor, and we have already discussed it previously. The process through which the current is transported from one end to the other in these two devices forms the important basis for distinguishing between them. So far as the Schottky transistor is concerned, it is really not a transistor like a BJT or a MOSFET. It is a combination of a Schottky diode and a BJT transistor, as shown in Figure 8.28c. A BJT transistor has high turn on voltage, whereas a Schottky diode has low turn on voltage. By connecting a Schottky diode between the base and the collector terminals of a BJT transistor, it becomes possible to divert the excess current from reaching the transistor and thus making it a more efficient device. It is a manipulation of the logic circuit to make the transistor work more efficiently as an on–off switch. It is obvious that the configuration of Figure 8.28c is not a transistor; it has been erroneously labeled as such. This combined configuration has another name which is *Schottky clamped transistor*. This alternative name accurately describes the specialized function that the Schottky diode provides in conjunction with the BJT transistor.

8.5.2 Electric Field Tuned Varistor and Its Embedded Electric Field Effect Transistor (*E*-FET)

The second example of a varistor embedded transistor is a electric field effect transistor (*E*-FET). We will find that it bears similarity with the classical field effect transistors that are built on conventional semiconductor platforms. The most used transistor in practice is the MOSFET, and it is the most prominent member of the host of field effect transistors. The letter *E* in *E*-FET device serves the purpose of separating these varistor embedded transistors from the large group of semiconductor-based electric field effect transistors. This distinction is also important to emphasize here because the third example of varistor embedded transistor that we will describe next is called the *H*-FET where *FH* represents a magnetic field.

Let us discuss now how the varistor I–V characteristics can be modified by the application of an electric field to produce another example of a transistor-based on a varistor. A careful consideration of the schematic of Figure 8.29 with its number of components should facilitate this goal [4]. Students should recognize its similarities with the MOSFET structure that is used to explain the basic principles behind the operation of a MOSFET transistor. Here too the metallic contacts source (S) and drain (D) are made of conductive silver epoxy that diffuses in substrate of p-type IHC 45 when annealed at about 250 °C. A thin slab of CCTO ($CaCu_3Ti_4O_{12}$) supercapacitor is used as the gate oxide. The gate contact, G, is also made of the same conductive silver epoxy as the other two contacts and when annealed at 250 °C the CCTO slab bonds strongly with the substrate. Notice that the third terminal bias (B) of the VBT transistor device is replaced here with the gate terminal (G). It is important to remember here that all three contacts of Ag are nonohmic in nature. In comparison, the source, drain, and gate contacts in a conventional MOSFET device are ohmic contacts.

The first consequence of using silver as contact metal is that it forms a depletion layer extending from S to D terminals. In turn, it causes the formation of a potential barrier allowing the I–V characteristic of the IHC 45 varistor to be nonlinear or rectifying. If the current path sees an electric field provided by the CCTO gate, it gets modified and the degree of modification is proportional to the gate voltage (V_g). But before we can get to this stage a current path must be established in the depletion layer that would connect S and D. This is made possible by the emergence of an inversion layer at the interface right below the CCTO gate. But what causes the formation of the inversion layer? When the gate voltage exceeds the critical voltage of the structure, the inversion

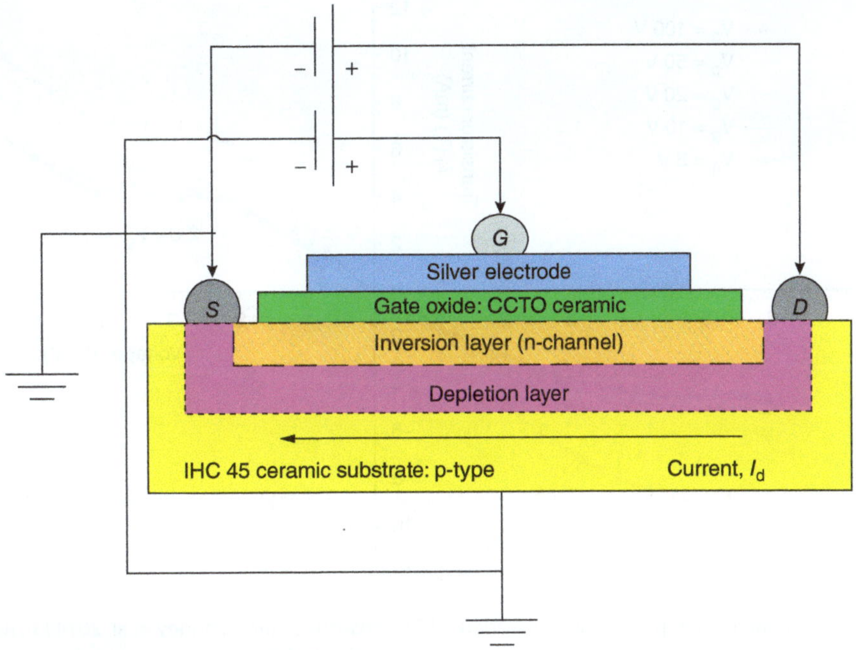

Figure 8.29 Modified schematic for producing a varistor embedded electric field effect transistor (*E*-FET). Source: Pandey et al. 2014 [3].

layer forms and it is occupied by electrons. The accumulated electrons in the inversion layer forms n-channel, which is conductive in nature. Now the current transporting between the S and D contact will experience the presence of the gate voltage, and in the process, undergo some changes altering the original varistor current from I_0 to I_d.

Let us pause for a minute here and try to understand from where the electrons come to populate in inversion layer in a p-type substrate? First of all, they can come from the p-substrate itself where they are the minority carriers. But there is another source of electrons also. We know that currents are thermally generated in a diode and a transistor. That would mean that electron–hole pairs can form in the depletion layer, when it is subjected to an electric field which is provided by the CCTO gate oxide. These are the two main sources of elections in the inversion layer. Another question that we must also answer now is what is the physical picture of the inversion layer? When in a p-medium the number of electrons is greater than the number of holes, then a conducting n-channel can form within the depletion region resulting in the onset of an *inversion layer* to form.

Having resolved some of the inherent issues in understanding the nature of the structure of Figure 8.29, we can now move to the understanding of the modified current, I_d. Once again, applying Eqs. (8.33), we can find the value of the *E*-field contributed current. Let us call this I_d(T) again, the same as we did for the VBT current. By plotting the I_d(T)–V as in Figure 8.30, we get the characteristics of the embedded electric field effect transistor (*E*-FET) which was buried in the varistor until we extracted the transistor current, I_d(T) from the field modified varistor current, I_d [4].

The curves of Figure 8.30 resemble those of a conventional field effect transistor except that in the forward biased mode they are very different in shape. They consist of three parts: (i) ohmic region from $0 < V_d < 2.5$ V: (ii) nonohmic region between $2.6 < V_d < 4.2$ V, here each curve is in saturated state; and (iii) again ohmic range for $V_d > 4.2$ V. What causes the saturated region that is sandwiched between the two ohmic regions to develop is not fully understood because a detailed band structure of the device is not known. On the reverse mode of the device, however, we find that each I–V plot is what we would expect in a typical transistor. Here for each plot, the state of saturation appears when $V_d > -4.5$ V approximately. As in the case of the VBT transistor, we can now find out the values of transconductance and current amplification parameters for the *E*-FET transistor. In order to do it, we can define first the transconductance for the *E*-FET transistor simply by replacing bias voltage, V_b, with gate voltage, V_g in Eq. (8.35). This gives us Eq. (8.36).

$$G_m(E) = \frac{\text{transistor drain current at a constant drain voltage}}{\text{gate volage}} = \frac{I_d(\text{T})_{V_d}}{V_g} \approx \frac{\Delta I_d(\text{T})}{\Delta V_g} \tag{8.36}$$

By plotting I_d(T) as a function of gate voltage, V_g, at a constant drain voltage, we should get a straight-line with the slope equal the transconductance, $G_m(E)$. Now using Eq. (8.34) we can easily calculate the values for signal

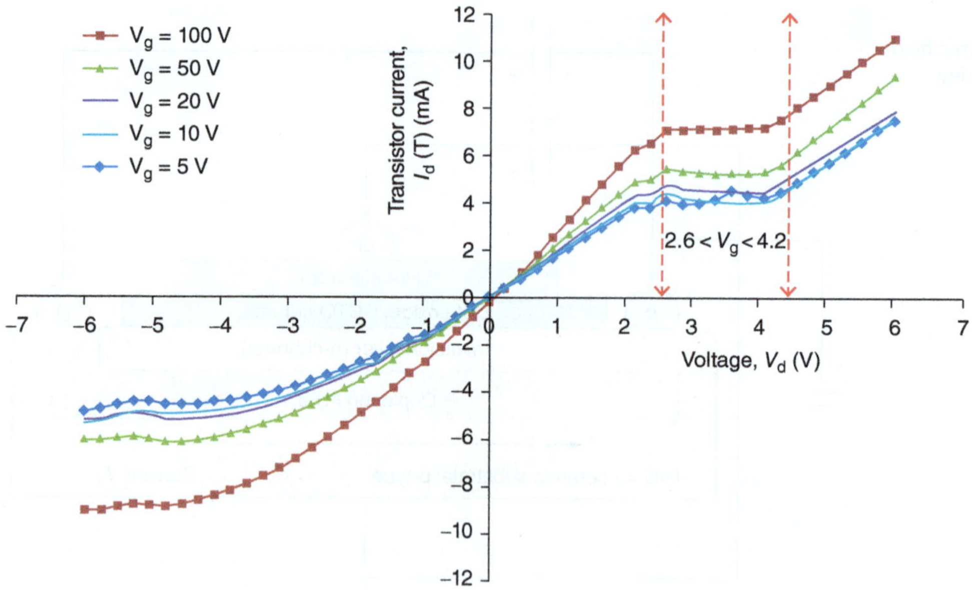

Figure 8.30 I_d(T)–V plots of an embedded E-FET transistor. Source: Pandey et al. 2014 [3]. Reproduced with permission of Springer.

Table 8.10 Pertinent properties of IHC 45 E-FET transistor.

Constant drain voltage	Transconductance, $G_m(E)$ (μS)	Transresistance, $R_m(E)$ (kΩ)	Maximum current amplification, S_a (%)	Potential applications
+4 V	33.8	28.6	180	Constant current voltage supply (CCVS); and transresistance amplifier
−5 V	40.1	24.94	190	Constant current voltage supply (CCVS); and transresistance amplifier

Source: Pandey et al. 2014 [3]. Reproduced with permission of Springer.

amplification, S_a. The values for transconductance, associated transresistance, maximum signal amplification, and potential applications are summarized in Table 8.10 for the E-FET transistor.

8.5.2.1 Frequency Dependence of IHC 45 *E*-FET Device

The frequency dependence of the output currents for the parent IIHC 45varistor and its embedded E-FET transistor are depicted in Figure 8.31 [4]. The varistor drain current peaks at 1.25 MHz with the bandwidth of 2.4 MHz; whereas for the transistor it peaks at 325 kHz with a bandwidth of 1.65 MHz. Also the two currents show strong frequency dependence. Both exhibit resonances, whereas it was not the case with the VBT device. We can assume that here the resonances appear because of the CCTO capacitor in the circuit. It is known that the frequency dependence of the dielectric constant and loss tangent of CCTO exhibit strong resonance [4]. So far as IHC 45 itself is concerned it shows no resonance in its dielectric behavior with respect to frequency. We can now conclude that the frequency response of the varistor as well as the E-FET transistor could be utilized to build a good low pass filter as we determined also for the VBT devices.

8.5.3 Magnetically Tuned Varistor and Embedded Magnetic Field Effect Transistor (*H*-FET)

The third class of varistor embedded transistor is a *magnetic field effect transistor* with the acronym of H-FET. The letter H is the symbol for a magnetic field just like E is for an electric field. We will see later that H-FET and E-FET transistors are very different from each other and have some distinguishing characteristics.

The history behind the development of magnetic transistor is interesting. In 1959, the first metal-oxide-semiconductor was invented in Bell Labs. [12]. It was almost immediately recognized as a major breakthrough in the field of solid-state devices. Its signal amplification was impressive, and therefore, it was readily adopted by engineers and scientists. Soon after that the search began in all earnest to find a magnetic counterpart to

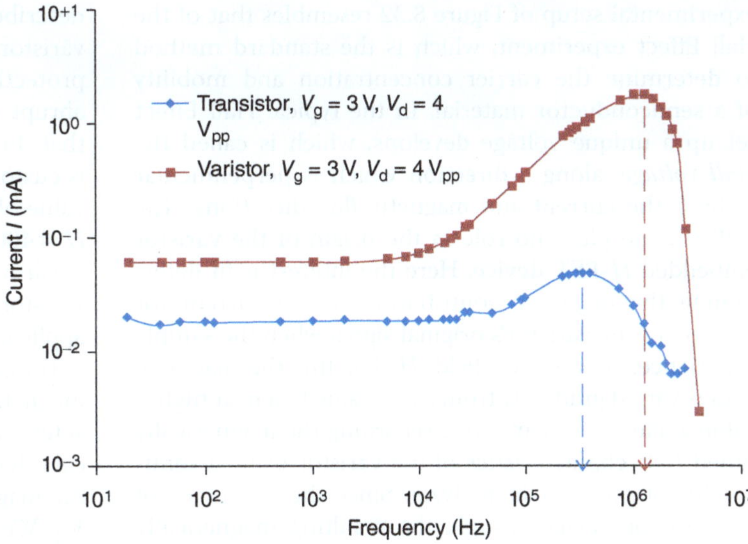

Figure 8.31 Frequency dependence of IHC 45 *E*-FET transistor and biased voltage biased varistor. Source: Pandey et al. 2014 [3].

the semiconductor MOS transistor. In 1966, a truly magnetic field effect transistor (MAGFET) was discovered in Westinghouse Electric Corporation [13]. It had all the attributes that the semiconductor MOS transistor had but was slow. Therefore, it was never accepted by industry, and so it remained unrecognized and basically forgotten. Then in 1971, a split drain MOS structure was invented and was named MAGFET [14]. However, it did not perform as a transistor in a conventional sense but could measure the differential Hall voltages. It would take another 20 years before a new magnetic transistor appeared on the scene. In 1990, a new class of transistor was developed which is now recognized as a *spintronic transistor*. It is also known as *spin mediated transistor* [15]. It is based on the configuration similar to what is used for the conventional semiconductor MOSFET transistors with certain subtle distinctions. As the name suggested, this device is based on the quantum mechanical property of an electron called *spin* which, as we might recall, can have only two values ($\pm\frac{1}{2}$or ↑ and ↓ polarized states). These two polarized states of the spins are manipulated to detect the magnetic state of the device. Then in 1995, another spintronic transistor was discovered based on the concept of spin valve effect that can be observed in the multilayers of giant magneto-resistive materials. Its configuration is similar to that used for the semiconductor BJT transistors [16]. Here the transmission characteristics of the spin-dependent transport exhibit fully saturated characteristics similar to the I–V characteristics of a conventional transistor. These sophisticated devices and the *H*-FET device differ in many subtle ways.

First of all an *H*-FET transistor is an extremely simple device based on equally simple physics and does not require any specialized skills to produce. This is in contrast to the other magnetic transistors that require highly sophisticated skills to produce and are based on advanced physics. They are built on single crystal or thin film platforms in contrast to the *H*-FET transistor for which the substrate is electroceramic IHC 45. Having acquired the background knowledge we are ready to discuss the fabrication, properties, and applications of an *H*-FET device.

The experimental set up of Figure 8.32 is used to tune magnetically the current of a varistor [17]. Here the sample is so oriented that the current (I) flowing from D (drain) to S (source) contacts of an IHC 45 varistor is perpendicular to the magnetic flux lines (H) produced between the two poles of a variable gap magnet. The

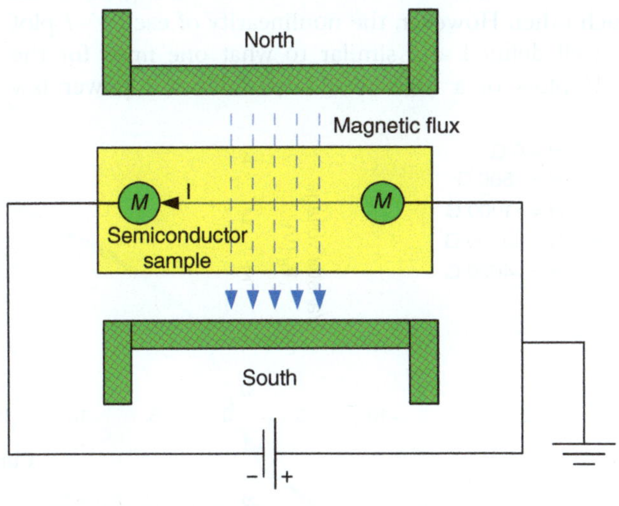

Figure 8.32 Experimental set up for determining *I–V* of an IHC 45 varistor with varying magnetic field, *H*. The symbols S and D represent source and drain, respectively. Source: Pandey et al. 2012 [17].

experimental setup of Figure 8.32 resembles that of the Hall Effect experiment, which is the standard method to determine the carrier concentration and mobility of a semiconductor material. In the typical Hall Effect set up a unique voltage develops, which is called the *Hall voltage*, along a direction which is perpendicular to both the current and magnetic flux directions. The Hall voltage plays no role in the origin of the varistor embedded *H*-FET device. Here the interest is in determining the magnetic contribution to the current (or voltage) of the varistor's original signal when the sample experiences a magnetic field, H. For this the magnetic field is varied gradually from a low value to a next higher value while at the same time recording the magnetically tuned $I-V$ characteristics of the varistor using a semiconductor parametric analyzer. Since the resistance of the varistor device is high, the resulting magnetically induced current is very small of being of any practical value. For example, it changes merely by 2.4 mA for a change of magnetic field by 3500 G. Therefore, no efforts are made to present here the $I-V$ characteristics of the magnetically induced current. They do show, however, well-defined nonlinearity, similar to that can be seen from Figure 8.24, but no saturated regions at all. But the picture changes drastically when instead of the $I-V$ characteristics the voltage–current ($V-I$) characteristics of a varistor are considered for the determination of the effect of a magnetic field on the output signal (V). For this, a current is injected in small units, and the resulting voltage recorded. In Figure 8.33, we present the magnetic field dependent $V-I$ characteristics of an IHC 45 varistor [18].

The plots hardly show any effect of the applied magnetic field, and they are difficult to distinguish from each other. However, the nonlinearity of each $V-I$ plot is well defined and similar to what one finds for the $I-V$ plots of a varistor. It also follows a power law described by $V \propto I^{\gamma}$, where γ is the NLC of the $V-I$ varistor. This mode of the varistor can also be used for protecting electronic components and circuits from abrupt current fluctuations. It is worth remembering that the NLC, γ, should not be confused with α which is customarily used for the NLC of an $I-V$ varistor. The value of NLC, γ, changes from 0.53 at $H = 0$ to 0.49 at $H = 4000$ G amounting to a reduction of just by 7%. This small amount of loss in efficiency of the $V-I$ mode of the varistor does not make the device unsuitable for practical applications.

The question now arises if there is significant amount of magnetically induced output signal so that the $V-I$ characteristics of the resulting device would be of any practical value? To determine this, we need to extract first of all the magnetic contribution to the varistor drain voltage, V_d. We are here dealing with three types of drain voltages. First, being V_0 corresponding to $H = 0$, the second being V_d when $H > 0$ and the third being $V_d(T)$ which is equivalent to the magnetic contribution to the unbiased potential V_0. The two voltages, V_d and V_0 are measured, but the magnetically induced component, $V_d(T)$, which remains buried in V_d must be extracted using Eq. (8.37).

$$V_d(T) = V_d \pm V_0 \tag{8.37}$$

The polarity of $V_d(T)$ will depend upon whether V_d is greater or smaller than V_0. Therefore, we can rewrite Eq. (8.37) as set of two dependent equations as expressed in Eqs. (8.38)(a and b).

$$\begin{aligned} +V_d(T) &= (V_d - V_0) \text{ when } V_d > V_0 \text{ (a)} \\ -V_d(T) &= (V_0 - V_d) \text{ when } V_d < V_0 \text{ (b)} \end{aligned} \tag{8.38}$$

Using these two equations, both $\pm V_d(T)$ can be extracted for each value of V_d. Then we get the resulting $V_d(T)-I$ characteristic of the embedded transistor. This is shown in Figure 8.34 [16].

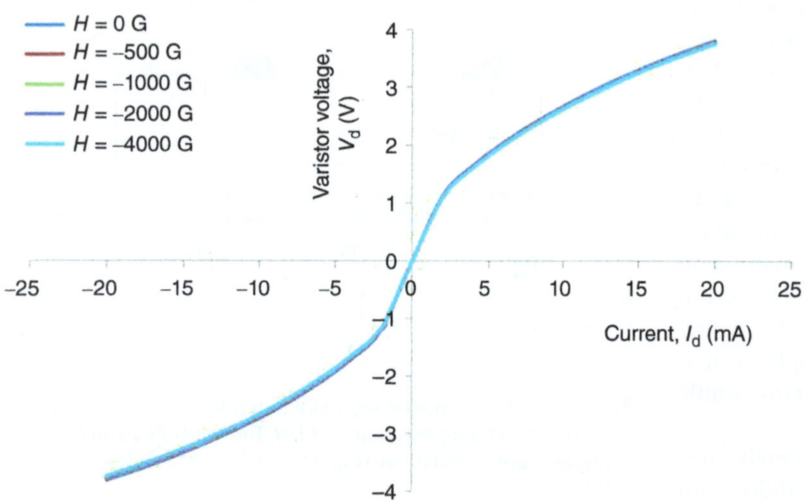

Figure 8.33 $V-I$ plots of an IHC 45 varistor with varying magnetic field. Source: Pandey et al. 2016 [18].

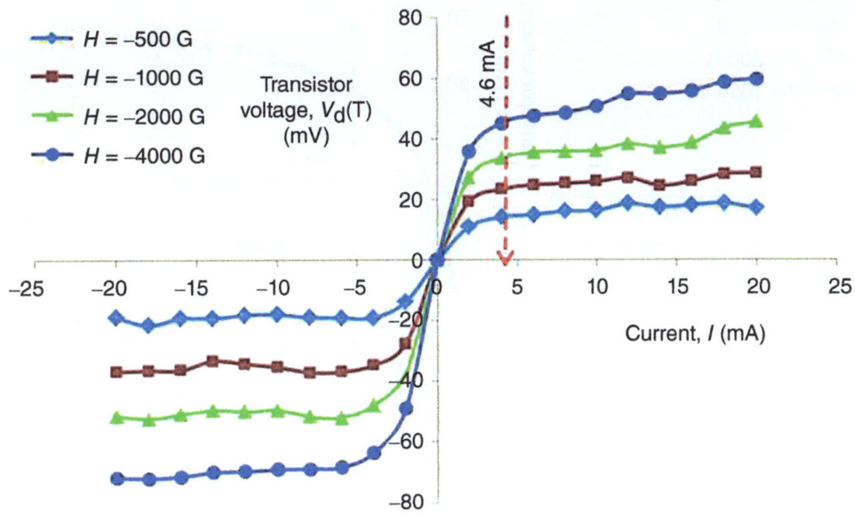

Figure 8.34 V_d(T)–I of IHC 45 H-FET transistor with varying magnetic field, H at room temperature. Source: Pandey et al. 2016 [18].

The data used for this figure and Figure 8.33 were collected at room temperature. The same experiment was also done at 100 K to determine the effect of low temperature on the performance of the H-FET transistor. In this figure, we can notice that the V_d(T) is in mV range that is 3 orders of magnitude smaller than the drain voltage, V_d, of varistor when $H > 0$. Furthermore, we find that each plot in the figure is well defined, and separation between them makes it easier to identify them individually. Initially, the output voltage increases linearly until the current = 4.6 mA, and from there, they tend to saturate. If this device is biased by a current of 4.6 mA, the H-FET transistor can be used as an "off" and "on" switch. The saturated output signal and electronic switching are its two important properties. Now let us examine its capacity for signal amplification and voltage detectivity. We also need to find out its magneto-resistive coefficient (MRC). The MRC parameter is equivalent to transconductance of conventional transistor. Let us use the symbol $\eta(H)$ for MRC and S_a for maximum signal amplification. We can evaluate these two parameters using Eqs. (8.39) and (8.40).

$$S_a = \left[\frac{\{V_d(\text{T})_{\text{at max } H}\}}{\{V_d(\text{T})_{\text{at min } H}\}} \times 100\right] \text{ with } I_d \text{ constant} \tag{8.39}$$

$$\eta(H) = \frac{V_d(\text{T}) \cdot I^{-1}}{H} \approx \frac{R(T)}{H} \approx \frac{\Delta R(T)}{\Delta H} \tag{8.40}$$

We can also find the voltage detectivity, V^*, of the device simply by defining it as $\frac{\Delta V_d(\text{T})}{\Delta H}$ while keeping the drain current constant. These parameters are tabulated in Table 8.11 for this device at room temperature. The table also gives the potential applications of the device based on these parameters. We find from this table that the maximum signal amplification with $-H$ and constant current at −20 mA is 380. It increases to 1260 when the applied field is $+H$ with current remaining at −20 mA is applied. In this table, also the parameters for the H-FET transistor are included when the applied field is $+H$ though no corresponding figures have been provided.

Table 8.11 Room temperature properties and potential applications of magnetically embedded H-FET transistors.

Room temperature	Constant current, I_d (mA)	Voltage detectivity, V^* (μV G^{-1})	MRC, $\eta(H)$ (μΩ G^{-1})	Maximum amplification, S_a (%)	Potential applications
$+H$	+20	50.9	206	830	Good voltage amplifier
$+H$	−20	76.9	227	1260	Good voltage amplifier
$-H$	+20	12.2	610	350	Voltage amplifier; excellent magnetic field detector and voltage sensor
$-H$	−20	15.1	12.2	380	Voltage amplifier, voltage and resistance sensor

Source: Pandey et al. 2016 [18]. MRC stands for magneto-resistive coefficient. Its symbol is $\eta(H)$ [18].

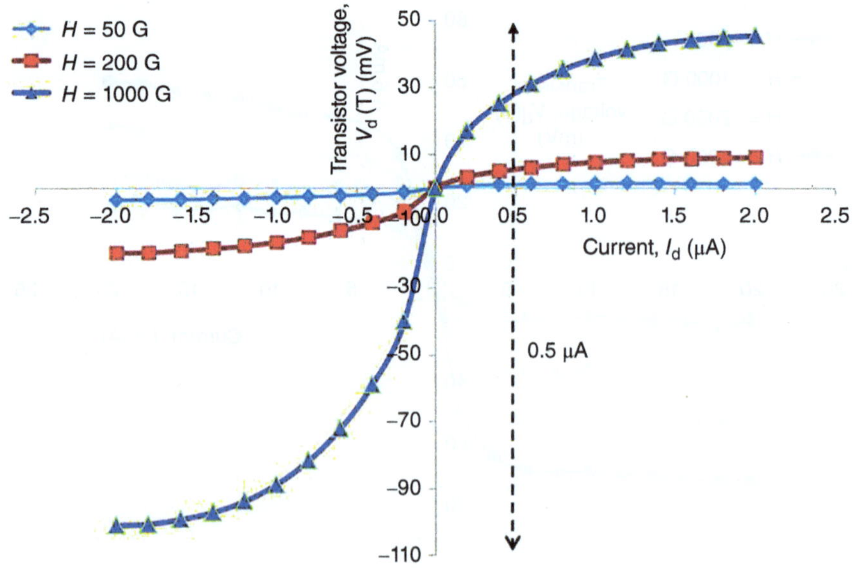

Figure 8.35 V_d(T) vs. I plots for an IHC 45 *H*-FET transistor at 100 K. Source: Pandey et al. 2016 [18].

The reason being that $V_d(T)–I$ characteristics are identical in shape and nonlinearity with the exception of the magnitude of the output signal.

Now let us examine what happens to the device when it is cooled. The magnetic moment of a ferromagnetic material and the resistivity of a semiconductor increase when cooled. IHC 45 is both ferromagnetic as well as semiconductor. Its devices should naturally exhibit the effect of low temperatures. To ascertain that the experiment was done by cooling the device to 100 K, the resulting $V_d(T)–I$ plots are shown in Figure 8.35 for $H = 50$, 200, and 1000 G. Higher values of H could not be applied because the sample was getting saturated. Properties of the HFET transistor are given in Table 8.12.

We see some interesting features in Figure 8.35. First, the current applied is in μA range that is about 3 orders of magnitude smaller than the currents applied for the same device at room temperature. Yet the voltage output is in mV range; the same as for the room temperature device. The maximum signal amplification for the device at 100 K is 3500, which is much larger than for the room temperature device. The switching current for this device is estimated to be approximately 0.5 μA. In comparison, it was 4.6 mA for its room temperature counterpart (Table 8.12).

Exercise 8.6

Suppose that the NLC (α) of a varistor in $I–V$ mode is 3. What would be the value of the NLC (γ) for the same device in its $V–I$ mode?

Solution

The $I–V$ varistor obeys the power law of $I \propto V^{\alpha}$, and the $V–I$ varistor obeys $V \propto I^{\gamma}$. These two relationships can be expressed as: $\alpha = \frac{\ln I}{\ln V}$ and $\gamma = \frac{\ln V}{\ln I}$. That leads us to the relationship: $\alpha \cdot \gamma = 1$. Since $\alpha = 3$, $\gamma \approx 0.33$.

8.6 Magnetic Field Sensor

Here we will examine a nonclassical way of producing a magnetic field sensor. Magnetic field sensor technology has a very long history and a productive life. The very first magnetic compass, which is the classic example of a

Table 8.12 Properties and potential applications of magnetically embedded *H*-FET transistors at 100 K [18].

	Constant current, I_d (μA)	V^* detectivity (μV G^{-1})	MRC, $\eta(H)$ (μΩ G^{-1})	Maximum amplification (%)	Potential applications
+*H*	+2	46.5	1.57	3500	Excellent resistance sensor and excellent voltage amplifier
+*H*	−2	102	0.56	2800	Excellent resistance sensor and excellent voltage amplifier

MRC stands for magneto-resistive coefficient.

magnetic sensor, appeared in China around 1088. Lodestone that is the naturally magnetized piece of mineral magnetite (Fe_3O_4) was used to construct the oldest form of magnetic compass. It is interesting that, before the compass appeared in China, already lodestone was found to be magnetic by Greek philosophers Thales of Miletus and Socrates. Yet this knowledge was not transferred by Greeks to construct a magnetic compass. The simplest form of a magnetic compass is still around and in use. It has played a vital role in the development of our civilization and globalization. This navigational instrument has advanced international trade, science, and engineering, as well as international travel. Magnetic sensing technology has advanced greatly from its origin in China, and this technology is still expanding rapidly with the new discoveries made in the field of magnetic materials and phenomena. Table 8.13 gives a few examples of magnetic sensors and includes the varistor-based sensor that we are going to study subsequently.

Hall effect sensor, usually called a gauss-meter, is the most used device for measuring the actual magnitude of a magnetic field and it can cover a range of $10^0 < H < 10^5$ G. It is based on the interaction between moving electric carriers (electrons and holes that give rise to an electric current) and an external magnetic field. Our earth itself is a magnet whose field strength fluctuates between $0.25 < H <$ to 0.65 G. Only a very sensitive sensor can detect this level of field. The magnetic field sensor that we will study here is a novel device only recently discovered [19]. The magnetic field response of a varistor in its resistive mode (VDR) is the basis on which this discovery is based. For this, the substrate used is a single crystal of PsB that is feebly magnetic at room temperature. The varistor and its corresponding VDR characteristics of PsB is shown in Figure 8.36. The PsB VDR exhibits interesting magnetic field dependence that has been utilized for developing a simple, rugged, and practical magnetic field sensor that covers a range of $0 < H < 4500$ G.

The R–V characteristic is obtained directly from the I–V characteristic of a varistor simply by substituting resistance for current. Figure 8.36 represents the resulting R–V plot for a PsB varistor. In the same figure, the associated I–V characteristic of the varistor is plotted for comparison. The unique R–V characteristic makes a

Table 8.13 Some examples of magnetic sensors and their operating range.

Magnetic sensors	Range of operation (approximate) in Gauss (1 G = 10^{-5} T)
SQUID (superconducting quantum interference device)	10^{-7}–10^4
Search coil	10^{-7}–10^9
Magnetotransistor	10^{-2}–10^4
Giant magnetoresistive	10^{-2}–10^7
Hall effect Gaussmeter	10^1–10^5
Varistor based sensor	10^2–0.5×10^4

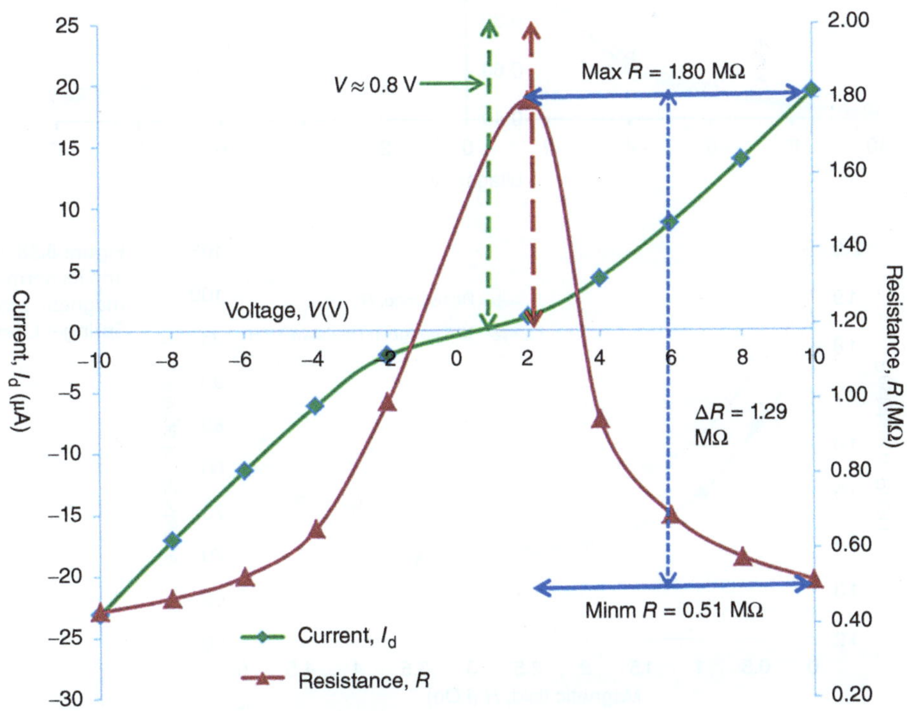

Figure 8.36 I–V plot of a varistor and its R–V counterpart (VDR characteristic) of single crystal PsB. Source: Pandey et al., 2015 [19].

VDR device a very useful electronic component and it is used widely as such.

Notice that the peak of the resistance does not coincide with zero of the I–V curve for the varistor. This shift can be attributed to the incremental units chosen for sweeping voltage between its maximum and minimum range in the experimental set up of the parametric analyzer that was used to collect the I–V data. The R–V plot of a VDR also follows a simple power law given by $R \propto V^{-\beta}$ where β is a negative quantity and is the NLC of the VDR device. This is related to the NLC (α) of the varistor. At this point one, might wonder how a single crystal of PsB can exhibit varistor like I–V characteristics where no grain boundaries are supposed to be present. We established on the basis of Figure 8.20 that the current transport in a varistor is because of the G–GB–G (grain–grain boundary–grain) mechanism. A similar mechanism according to which the varistor effect in a single crystal oxide can be explained is called "crystal plane (*hkl*)-metal oxide (mo)-crystal plane (*hkl*)" model first proposed by Schwing and Hoffmann [20]. The model assumes that during the course of crystal growth of an oxide material a layer of metal oxide (for which the oxide flux such as PbO and V_2O_5 could also be the source) gets trapped between two crystal planes of *hkl*-orientation. In other words, a metal oxide layer is sandwiched between two (*hkl*) planes of a single crystal giving rise to a (*hkl*)-mo-(*hkl*) structure which can facilitate the varistor like I–V characteristic also in a single crystal. By using the configuration of Figure 8.32, it is easy to obtain magnetic field induced R–V characteristics. In Figure 8.37, the V–R plots are depicted corresponding $0 < H < 4500$ G [19].

At $V = +2$ V, the resistance peaks around 1.81 MΩ for $H = 0$ G and drops to approximately 1.40 MΩ with $H = 4500$ G; amounting to a difference of about 410 kΩ. The largest drop, as seen from Figure 8.38, occurs as soon as a field of 1000 Oe is applied [19]. The separation becomes much smaller between the successive fields. For example, it is only 80 kΩ between $1000 < H < 4500$ G. These features could be exploited for developing a simple yet practical magnetic fields sensor. For some applications, it is conceivable that this could be considered an

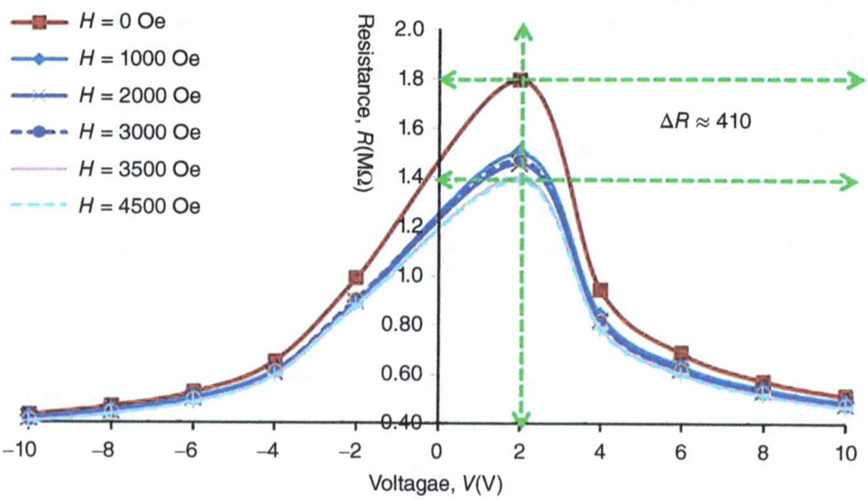

Figure 8.37 Resistance vs. voltage plot of a PsB based VDR with varying magnetic fields. Source: Pandey et al. 2015 [19].

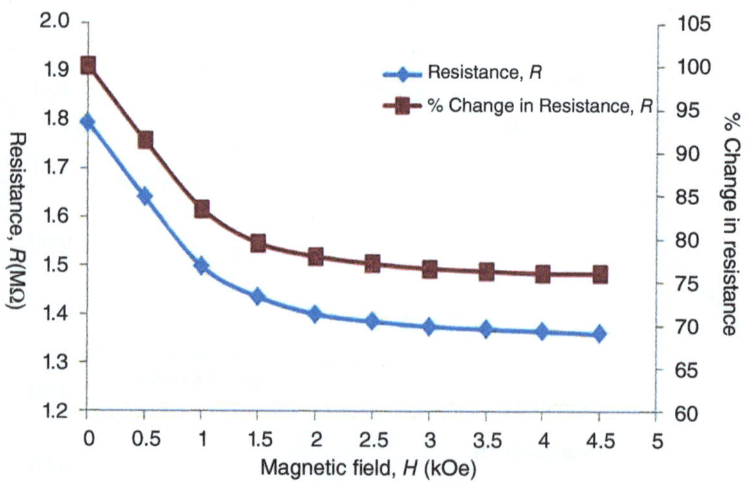

Figure 8.38 Magnetic field dependence of resistance and its normalized counterpart for a PsB based VDR magnetic sensor. Source: Pandey et al. 2015 [19]. Springer License.

alternative to the well-established Gaussmeter based on Hall Effect. Being built on a ceramic substrate, it should be rugged and strong and should be able to withstand high temperatures, hostile atmospheres, and abuses encountered in field operations. In Figure 8.38, we find the magnetic field dependence of resistance and its normalized component (NR) which is simply the ratio expressed as percentage between the resistance at field H, $R(H)$, to the initial resistance, $R(0)$, at $H = 0$. The values for NR can be readily computed using Eq. (8.41).

$$NR = \frac{R(H)}{R(0)} \times 100 \tag{8.41}$$

We can identify two distinct regions in Figure 8.38, first, for $0 < H < 500$ Oe and the second, for $500 < H < 4500$ Oe. The slope for the first region is found to be $\approx 300\,\Omega\,Oe^{-1}$, which would make the device very sensitive for fields up to 1000 Oe. When the entire range of $0 < H < 4500$ Oe is considered the relationship between R and H appears to follow the trend given by $R \propto H^{-0.83}$. This would amount to the device sensitivity, $S(R)$, of approximately $60\,\Omega\,Oe^{-1}$ which is more than sufficient for designing a good magnetic sensor to cover the range of $0 < H < 4500$ Oe. The relationship given in Eq. (8.42) defines the sensitivity, $S(R)$, of the sensor.

$$S(R) = \frac{\Delta(\ln R)}{\Delta(\ln H)} \tag{8.42}$$

The magnetic field also affects the NLC of the two devices. Varistor's NLC (α) decreases with the increasing magnetic field from its maximum value of 2.30 at $H = 0$ to its minimum value of 2.05 at $H = 4500$ G which amounts to a loss of about 10% in its efficiency. By the same token, the NLC (β) of the VDR device changes from its initial value of −1.30 at $H = 0$ to −1.05 at $H = 4500$ meaning thereby that it changes by approximately 19%, which is almost the double the value of change in NLC of varistor. While the varistor might remain useful for some practical applications in the presence of magnetic fields, the VDR suffers a loss in its efficiency, which might not be acceptable for some applications.

Another approach for building a magnetic sensor would be to exploit the magneto-resistive property of the VDR device. The resistance of all materials experience changes when subjected to an external magnetic field and the resulting parameter is called *magneto-resistance (MR)*. It is a universal phenomenon whether the material is magnetic or nonmagnetic. MR can have both negative and positive values. Ferromagnetic materials exhibit large values of MR compared to the nonmagnetic materials and therefore, many of the sensors are based on ferromagnetic materials. Let us now examine the response of the magneto-resistive property of PsB VDR to find out its potential for magnetic sensor applications. MR is defined as the difference in resistance between its initial value, $R(0)$ when $H = 0$ and the value, $R(H)$, when $H > 0$. It can be expressed as in Eq. (8.43).

$$MR = \pm\{R(H) - R(0)\} \tag{8.43}$$

This equation implies that the MR is zero at $H = 0$ (Figure 8.39). For each curve, the maximum MR peaks at $V = +2$ V just as we found it to be the case for the R–V plots of Figure 8.37. Here too the peak values increases as the applied magnetic field increases. The difference between the peak MR at $H = 1000$ Oe and $H = 4500$ Oe is 120 kΩ. In comparison, it was just 80 kΩ for the same two fields for the resistive mode device (Figure 8.37). From this, we can conclude that the MR-based sensor could offer a better resolution than the resistive mode sensor. Having established the superiority of the MR sensor, let us now examine how exactly the peak MR

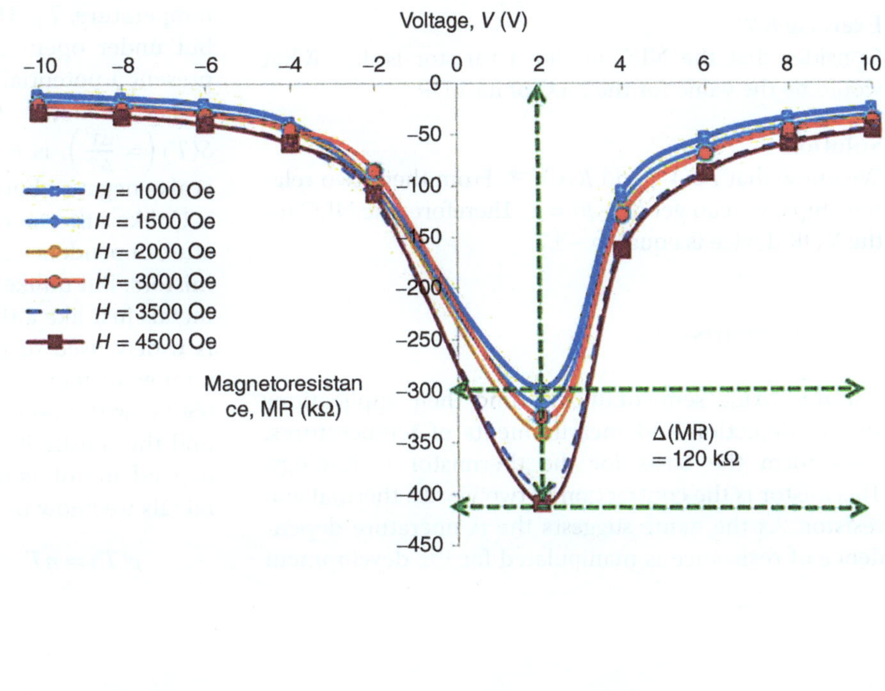

Figure 8.39 Magneto-resistance (MR) vs. driving voltage with varying magnetic field for a PsB based VDR magnetic sensor. Source: Pandey et al. 2015 [19].

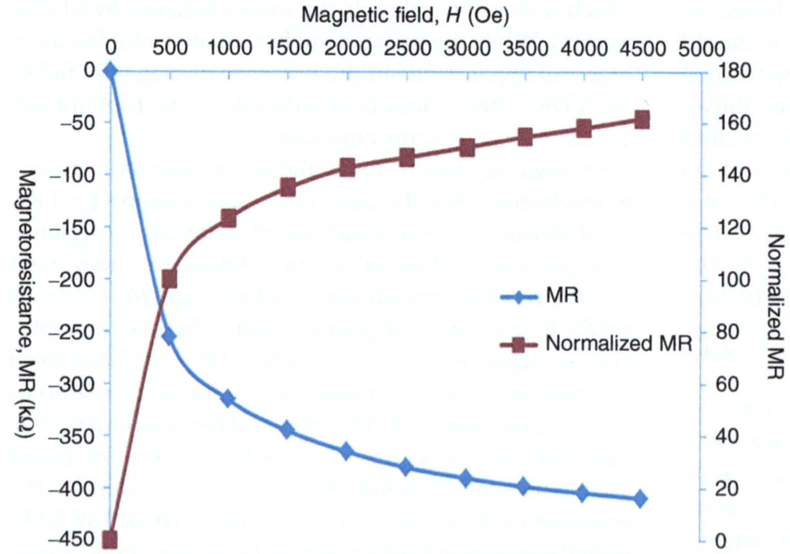

Figure 8.40 Magnetic field dependence of peak MR (at $V = +2$ V) (blue) and its normalized counterpart (red). Source: Pandey et al. 2015 [19].

varies with the magnetic field. Figure 8.40 shows its trend [19].

We find from this figure that MR increases by 60% for the field range of $500 < H < 4500$ G, and for the same range, it follows the relationship $\mathrm{MR} \propto \ln H$. This helps us in finding an expression for the sensitivity $S(\mathrm{MR})$ for this device using the same logic as used for deriving Eq. (8.42). This leads us to Eq. (8.44).

$$S(\mathrm{MR}) \approx \frac{\Delta(\mathrm{MR})}{\Delta(\ln H)} \tag{8.44}$$

For such a sensor the sensitivity, $S(\mathrm{MR})$ would be approximately equal to $70\,\Omega\mathrm{G}^{-1}$ which is approximately 17% more than the sensitivity of the resistive mode sensor. That is, the sensors both in resistive mode and magneto-resistive mode could be useful for general purpose magnetic field sensor.

Exercise 8.7

Consider that the NLC (α) for a varistor is 4.7. What would be the value for the NLC of its VDR?

Solution

We know that $I \propto V^{\alpha}$ and $R \propto V^{-\beta}$. From these two relationships, we can get $(\alpha - \beta) = 1$. Therefore, the NLC for the VDR device is equal to -3.7.

8.7 Thermistors

Ceramic oxide semiconductors find their applications also in detection and measurements of temperatures. They form the basis for the thermistor technology. Thermistor is the contraction of two words: thermal and resistor. As the name suggests the temperature dependence of resistance is manipulated for the development of a thermistor device. The electronic symbol for a thermistor is the same as for varistor and is given in in the insert of Figure 8.16.

Three types of solid-state devices are used for the detection of temperature and its measurement. They are thermocouples, thermistors, and RTDs which is the abbreviation for *resistance temperature detectors* (RTDs).

Thermocouple is a very simple device consisting of two dissimilar metallic wires that are brought in contact by forming a junction by fusing two wires at one end and leaving the remaining two ends unfused. The physical principle of the thermocouple is based on the Seebeck effect which states that a temperature gradient, $\mathrm{d}T$, gives rise to an electromotive force, $\mathrm{d}V$, at the open ends of a thermocouple which is kept at a lower temperature, T_2, with respect to the junction which is kept at a higher temperature, T_1. This emf is capable of performing work but under open-circuit condition when no current is present a potential difference, $\mathrm{d}V$, develops at the open ends. For metals, the sign for the Seebeck coefficient, $S(T)\left(=\frac{\Delta V}{\Delta T}\right)$, is negative. For n-type semiconductor, it is also negative but is positive for a p-type material.

Unlike a thermocouple, a RTD is based on the temperature dependence of resistance as is also the case for thermistors. But unlike a thermistor, RTDs are also based on metals just like a thermocouple, and thin-film platinum is widely used to produce RTD elements. It can cover a range of temperature from $-240\,°\mathrm{C}$ to approximately $650\,°\mathrm{C}$ with good accuracy and linearity. We can easily find the sensitivity of a RTD from the temperature (T) dependence of its resistivity, $\rho(T)$, of the metal used. For metals we know that

$$\rho(T) = \eta T \tag{8.45}$$

Here the parameter, η, stands for the temperature coefficient of resistance. Obviously, this is the figure-of-merit of the RTD device. Larger its value, the greater is the sensitivity of the device. Replacing $\rho(T)$ with resistance, R, we can also define the temperature coefficient of resistance as in Eq. (8.46).

$$\eta = \left(\frac{A}{L}\right) \times \left[\frac{\Delta R(T)}{\Delta T}\right] \tag{8.46}$$

In the above equation, A is the area and L the length of the device.

In contrast to a thermocouple or a RTD, thermistors are based on ceramic or more precisely on semiconducting oxide ceramic materials. Commonly used materials for thermistors are magnetite doped (Fe_3O_4), nickel oxide (NiO), and ferroelectric barium titanate ($BaTiO_3$). It is interesting to note that the foundation of thermistor was laid around 1833 by Michael Faraday of England who is considered as one of the greatest physicists of all times. While studying the properties of silver sulfide, he found that unlike for a metal, the resistance of silver sulfide dropped dramatically with increasing temperature. This was the beginning of the semiconductor science which dominates today's microelectronics. The first commercially viable thermistor was produced around 1930. Thermistors can measure temperatures between −100 and +500 °C with accuracy and linearity comparable to that of RTD but with better sensitivity, especially between −80 and 130 °C.

We know that for metals, resistance increases linearly with temperature, but for semiconductors, resistance decreases with temperature obeying an exponential relationship as shown in Eq. (8.47).

$$R = R_\infty \exp\left(\frac{c}{T}\right) \tag{8.47}$$

Here R_∞ is the temperature-independent resistance and c a constant. Equation (8.47) is the simplified form of Eq. (7.30) whose interpretation we have discussed in detail. Alternatively,

$$\ln R = \frac{c}{T} + \ln R_\infty \tag{8.48}$$

The parameter c can be determined from the slope of $\ln R$ vs. $\left(\frac{1}{T}\right)$ plot. It is called the figure-of-merit of the thermistor device. If it is negative, the device is named "negative temperature coefficient thermistor of simply NTC" and when it is positive the device is called "positive temperature coefficient thermistor or PTC." Notice that Eq. (8.48) does not include any reference temperature. In practice though the absolute temperature, T, is measured with reference to a temperature which we might call T_{ref}. Including this parameter gives us the modified form of Eq. (8.47) which is used in practice by industry.

$$R = R_\infty \exp\left\{c\left(\frac{1}{T} - \frac{1}{T_{ref}}\right)\right\} \tag{8.49}$$

The value for the parameter c varies from material to material but is usually in the range of $3000 < \kappa < 5000$ K.

The conventionally used materials for NTC are doped Mn_3O_4, $Fe_3O_4–ZnCr_2O_4$, and $Fe_3O_4–MgCr_2O_4$, whereas doped ferroelectric barium titanate, $BaTIO_3$ (BT) is the preferred material for the fabrication of PTC devices.

When doped with La^{3+}, Nb^{5+}, and Ta^{5+} BT becomes a good n-type semiconductor. These thermistors are found to have good performance in the temperature range of approximately $140 < T < 240$ °C. The reason being that in this temperature regime the resistance vs. temperature dependence of doped BT exhibits a linear dependence as can be seen from Figure 8.41. The slope of such a plot will give directly the value for c which can also be considered as the sensitivity of the device and in that respect the BT thermistor will be more akin to a RTD than a typical thermistor. We may recall that BT undergoes structural phase change from tetrahedral structure to cubic phase at the ferroelectric Curie temperature, T_c at 120 °C. Above this temperature, BT is no more a ferroelectric material and enters a phase in which there is no spontaneous polarization present.

Because of the structural change at the Curie point, the resistance vs. temperature exhibits a linear dependence instead of the exponential dependence which is the normal trend for a semiconductor phase. The widely accepted theoretical model of the PTC effect in ferroelectric doped-BT is based on the existence of double Schottky barriers at the grain boundary. We discussed this topic previously to explain the formation of double Schottky barriers with respect to the G–GB–G transport mechanism (see Figure 8.20). The resistance vs. temperature behavior of doped-BT is also consistent with the temperature dependence of dielectric constant (ϵ) of doped BT as we can see also from Figure 8.41. The maximum of ϵ at the ferroelectric Curie temperature corresponds with the minimum of the resistance, R. Above this critical point, dielectric constant decreases in the cubic phase, whereas the resistance increases linearly with temperature with a positive slope until the temperature reaches approximately 250 °C. Then also, it keeps increasing but with significantly smaller slope meaning that the value of the temperature coefficient, κ, of the PTC thermistor decreases vastly making the thermistor less efficient.

8.7.1 Heating Effects in Thermistors

When a voltage (V) is applied to a thermistor, it gives rise to a current (I) which in turn can induce Joule

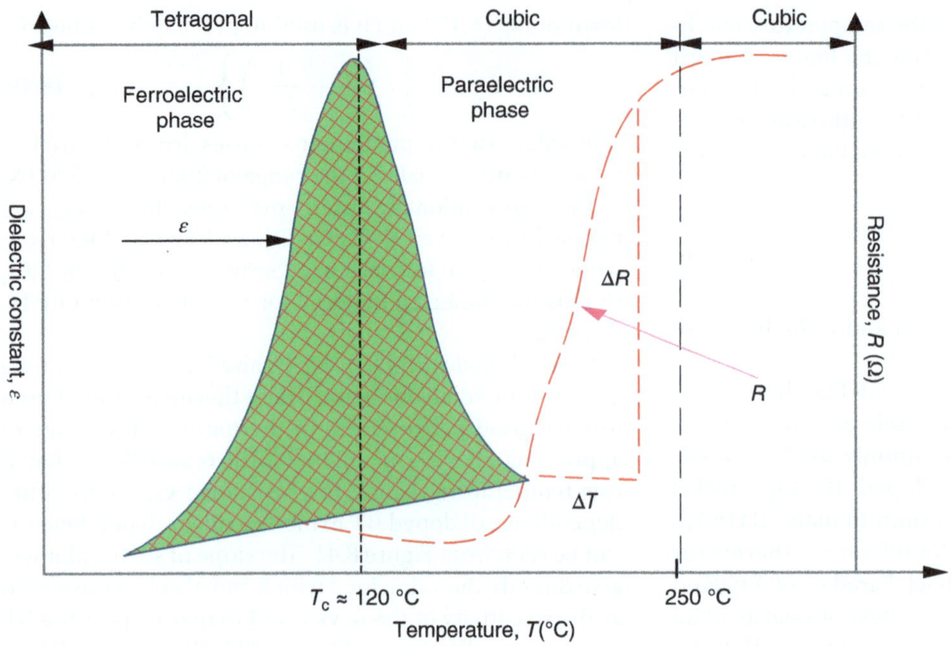

Figure 8.41 Trend showing temperature dependence of resistance (R) red curve and dielectric constant (ϵ) blue curve for doped barium titanate (BT) semiconducting ceramic.

heating and if its magnitude is large the thermistor would fail to monitor temperature correctly. Unless this error is corrected, thermistor becomes an unreliable temperature-monitoring device. However, the effect of Joule heating can be exploited for good uses resulting in many of its practical applications. The electric power, P_E, experienced by the thermistor is simply the product between its conduction, G ($=R^{-1}$), and V^2.

$$P_E = I \cdot V = GV^2 \tag{8.50}$$

Historically, it is interesting to note that already around 1841, it was discovered by James Prescott Joule that a current passing through a metallic conductor produces heat, Q, in time, t, such that it is equivalent to ($I^2R \cdot t$). This is known as Joule's first law. This effect was also discovered independently by Heinrich Lenz in 1842; and therefore, it is also called Joule–Lenz law in German speaking countries. This discovery has impacted the high-temperature technology in a very significant way. Joule's heating is known also as *ohmic heating* or *resistive heating* because of its relationship to Ohm's law.

Now the question arises what happens to the heat accumulated in the thermistor device? The answer is simple. Because of the principle of conservation of energy, the electric power (which is equivalent to electrical energy) is converted into heat energy. It will dissipate to the environment and would be governed according to the Newton's law (1701) of cooling. This law states that the rate of heat transfer of an object is proportional to the thermal gradient, ΔT, existing between its own temperature and the ambient temperature. Assuming that the rate of heat transfer per unit time is, Q_T, between the thermistor at temperature, T_R, and its surrounding which is at temperature, T_0, we can formulate their relationship as in Eq. (8.51).

$$Q_T \propto \Delta T = H^*(T_R - T_0) \tag{8.51}$$

Here H^* is the heat transfer coefficient which is constant for a material. Its other name is *heat dissipation factor*. Once the thermal equilibrium is established, P_E must be equal Q_T. By rearranging the parameters in the two Eqs. (8.50) and (8.51), we should arrive at Eq. (8.52).

$$(T_R - T_0) = \frac{1}{H^*} GV^2 \tag{8.52}$$

Alternatively,

$$H^* = \frac{GV^2}{\Delta T} \tag{8.53}$$

By knowing the thermal gradient, ΔT, we can determine the heat dissipation factor, H^*. Thermistors can be used as a flow rate sensor of a fluid that passes through the thermistor because the dissipation factor of a liquid increases with its flow rate. Other applications are outlined in Table 8.14.

Metal oxide electroceramic materials are widely investigated as potential sensors for detecting variety of gases. Of special interest are sensors for detection of hazardous and life-threating gases in automobile, in chemical industry, semiconductor industry, and environment in general. Because of concerns about climate

Table 8.14 Applications of PTC and NTC thermistors.

Positive temperature coefficient (PTC) thermistors	Negative temperature coefficient (NTC) thermistors
Used as a current-limiting device for circuit protection. Joule heating leads to an increase in thermistor resistance which in turn can reduce the current in the circuit to the safe level	Used as resistance thermometers in low-temperature measurements of the order of 10 K
For temperature compensation of voltage controlled oscillators	Used as inrush-current limiting devices in power supply circuits
For circuit protection of lithium batteries	For monitoring and control of temperature of coolant liquid and oil in automobiles
Used in automobile electronics to provide additional heat in cold climate inside the cabin	For maintaining temperature within the prescribed level of incubators, 3D printers and possibly of large data storage units
Used as timers in the degaussing coil circuits. If the heat generated in the circuit exceeds the prescribed tolerance limit of the degaussing coil, it will shut off. Particularly important for CRT displays	For monitoring and control of temperature of variety of consumer products such as toasters, coffee makers, refrigerators, freezers, hair dryers
For control of current of large and complex circuitry commonly found in defense and aviation electronics	In Food Processing and Handling industry for maintaining proper temperature to prevent waste of products and food borne diseases

Source: https://en.wikipedia.org/wiki/Thermistor. Licensed under CC BY 3.0.

warming and protection of environment, gas sensing technology has experienced a huge growth as a result of which new materials and novel sensing approaches are constantly evolving. The semiconductor-based chemical sensors have the advantages of being small in size; they have high sensitivity and are based on simple operating principles requiring simple electronics. Nevertheless, they have poor shelf life and not good stability at higher temperatures.

There is a large array of ceramic materials found suitable for gas sensing technology. Many of them are of complex chemical formulations, but some are simple metal-oxides. Some examples of complex ceramic configurations are: Yattria stabilized zirconia (YSZ), Ca-Sr-Co-oxide ($Ca_{0.4}Sr_{0.6}CoO_3$), Ca-Zr-In-oxide ($CaZr_{0.9}In_{0.1}O_{3-x}$), Sr-Ce-Yb-oxide ($SrCe_{0.95}Yb_{0.05}O_3$), and Li-conducting Li-alumintirano-phosphate. The simpler group consists of members of binary metal oxides semiconductors such as ZnO and SnO_2 and their alloys. Since the emergence of nanomaterials electroceramics are now routinely synthesized with nanometer sized grains for a variety of applications including for detection and monitoring of gas species encountered in our atmosphere and in workplace that have detrimental consequences on health, environment and food supply. The range of gases detected and monitored by ceramic sensors can vary widely; some can be harmless while others can be highly toxic to very dangerous. Typically, they can detect hydrogen (H_2), oxygen (O_2), nitrogen (NO_2), carbon mono-oxide (CO), carbon dioxide (CO_2), and gases with complex chemical formula. Some are used also to detect and measure humidity and atmospheric conditions in general.

The physical principles involved in gas sensing are as varied as the gases themselves. To develop models based on sound scientific basis to explain the interaction of chemical processes with the physical properties of materials is not a trivial issue. In this respect, the sensor technology is more advanced than its science. Obviously, it is beyond the scope of this book to provide a sound scientific argument to explain why and how each sensor works. Nevertheless, there are excellent review papers written on this topic, and it may be worth-while for the interested readers to refer to them. We recommend to our readers particularly two publications given here in Refs. [21, 22]. In short, how certain gases react with the surface of substrates may involve proton conduction mechanism, current limiting process, and conduction through carbonate ions.

There are basically three categories of gas sensors available in the market place. They are solid electrolyte sensors, catalytic combustion sensors, and semiconductor sensors which are of our interest in this chapter. The prominent members of binary metal-oxide semiconductor family used in sensor technology are TiO_2, ZnO, SnO_2, WO_3, La_2O_3, and their alloys. These sensors can operate at high temperature between 100 and 500 °C and therefore are particularly well suited for applications in high temperature environments as encountered, for example, in automobiles, engines, and airplanes. Of special interest to the transportation industry are sensors that can detect and monitor air to fuel ratio (A/F).

The automobile industry is the largest market for solid-state sensors based on metal-oxides. TiO_2 sensors are widely used for the determination of this parameter because of its sensitivity and simplicity. Its resistance (R) shows a strong dependence on the partial pressure of

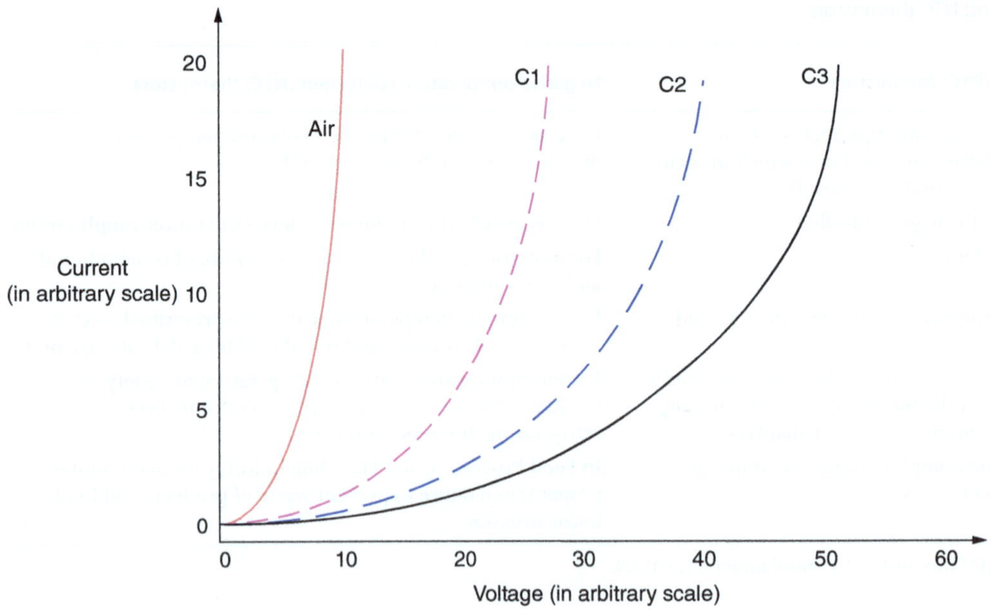

Figure 8.42 *I–V* characteristics of a typical gas sensor based on varistors. C1, C2, and C3 are concentration of toxic gas in air–gas mixture.

oxygen (P_{O_2}). The relationship between resistance and oxygen partial pressures can be expressed by modified Arrhenius equation (1889). It is

$$R = R_\infty P_{O_2} \exp\left(-\frac{E_{ac}}{k_B T}\right) \tag{8.54}$$

Here R_∞ is a temperature-independent constant and E_{ac} the activation energy, $k_{B,}$ and T have their usual standard meaning. This equation is routinely used whenever a chemical process is involved. Notice its similarity with Eq. (8.47) or more precisely with Eq. (7.19), which is the standard equation to describe the temperature dependence of electrical conductivity of a semiconductor. If the temperature is kept constant, $R \propto P_{O_2}$. The activation energy, E_{ac}, can be found from the slope of the plot $\left(\ln R \text{ vs. } \frac{1}{T}\right)$ and from its intercept ($\ln R_\infty + \ln P_{O_2}$). The sensors built on substrates consisting of two metal oxides have been found to exhibit significantly higher sensitivity than those based on a single metal oxide [23].

Binary metal oxides such as ZnO, SnO_2, and WO_3 have been used since long for detection and monitoring of gases at elevated temperatures. However, these devices were operated at low fields when their $I–V$ characteristics were ohmic in character. Yet they showed significant changes in the presence of gases. Building gas sensors on metal oxide varistors with its typical rectifying $I–V$ characteristics is of relatively recent origin. It was around 1998 that the ZnO and SnO_2 varistors built on porous ceramics were recognized as potential candidates for building practical gas sensors [24].

However, only recently, WO_3 varistors with large breakdown voltages were identified as potentially good candidates for fabrication of gas sensors [25]. Tungsten oxide, WO_3, is a prominent member of the binary metal oxide series, and it is now used as a preferred material for gas sensing technology. Figure 8.42 depicts the $I–V$ characteristics of a gas sensor based on a varistor. The exposure to toxic gases induces chemical changes on the surface layers causing modifications in the original $I–V$ characteristics of the device. The $I–V$ curves shift toward higher and higher potential as the concentration (C1, C2, and C3) of the gas mixture increases. It is interesting, however, that the basic shape of the typical nonlinear $I–V$ characteristics of the device remains unchanged inspite of the fact that the surface might suffer chemical changes.

Glossary

Bandwidth The range of frequencies allowed by an electronic filter to pass through while blocking all frequencies outside the allowed domain. The upper most allowed frequency is called cutoff frequency. It corresponds to the frequency at the 3 dB (decibel) point on the signal vs. log frequency plot where the signal reduces to 50% of its maximum value. In signal processing, bandwidth refers to the highest reliable transmission rate of data. In general, the bandwidth is the difference of frequencies between the maximum and minimum frequencies in a range (band) of an electronic signal.

Depletion layer In a semiconductor field, the depletion layer is referred to that region in a device which has been robbed of active charge carriers and is populated only by ionized acceptors and donors impurities. Its other names are depletion range, depletion zone, junction region, space charge region, or space charge layer. Since the depletion zone cannot conduct because of absence of electrons and holes it is insulating in nature.

Diode A two terminal semiconductor device that allows the flow of current in only one direction, usually when biased with a positive potential, and not in the other direction. That is, it will block the passage of current when biased with negative potentials. The nature of the unidirectional current transmission in a diode is nonlinear which is also called rectifying current. There are two types of diodes: one is called the p–n junction diode and other Schottky diode.

Gate oxide An electric field is generated by placing a thin capacitor layer on top of the field effect transistor structure (such as a MOSFET) to control the output current. The material for the capacitor is usually an oxide insulator such as SiO_2. Many different oxides can be used in conjunction with a field effect transistor (FET). In transistor vocabulary, it is common to call this capacitor layer as *gate oxide*.

Grain boundary A polycrystalline material consists of large number of single crystalline grains, each with different crystal orientation. They are isolated from each other by a very thin region called grain-boundary, which is insulating in nature. Ceramic is another name for a polycrystalline material. Unlike a ceramic a single crystal does not have any grain boundaries. Oxides are prominent members of electroceramic group. Some of them can be semiconductors, some magnetic and yet others might possess some unique electrical properties not commonly found in normal ceramic materials. They are used for many applications in electronics and microelectronics. Nitrides and carbides are also members electroceramic group but not as prominent in electronics as oxides.

Integrated circuit An integrated circuit, also called IC, consists of a large number of tiny transistors, diodes, capacitors, resistors, and other electronic components in millions on a semiconductor wafer to perform multiple functions such as an amplifier, oscillator, time counter, computer memory elements, or microprocessor. This is the very spine of the modern microelectronics.

Magnetic field As an electric field is associated to an electrical potential so is the magnetic field associated with a magnet. The unit to express a magnetic field, also referred to as H-field, is called Tesla (T), or Gauss (G), or Oersted (Oe). In principle, G and Oe are synonymous, but Tesla is different. The three units are interrelated: $1\ \text{T} = 10^4\ \text{G} = 10^4\ \text{Oe}$.

Partial pressure It is a common term used to identify a mixture of two or more gases. According to the Dalton's law of partial pressure, which states that in a mixture of nonreacting gases the total pressure exerted is the sum of partial pressures of each individual gas present in the mixture. For example, air mainly consists of N_2 and O_2 such that N_2 is 78% by volume and O_2 by 21%. Argon and carbon dioxide are the two other gases but they are much less than 1% by volume. Partial pressures are commonly referred to as ppm (parts per million).

Potential barrier It is a very important concept in semiconductor device technology. It refers to a potential height that is encountered by electrons during their transport from one point to another that must be surmounted before a current can flow. It forms when two dissimilar materials are brought in physical contact such as two dissimilar metals or a metal and a semiconductor. The potential barrier that forms as a result of a metal and a semiconductor brought into contact is commonly referred to as Schottky potential or Schottky barrier height.

Schottky diode It is a metal semiconductor diode with the same general nature of the current–voltage characteristics as found for a p–n junction diode. It is important to remember that the physical principles involved in the operation of the two types of diodes are based on entirely different physics.

Space change It is electrically a neutral region that is insulating and not capable of current transmission. See also the description for depletion layer.

Thermistor A solid-state device to detect and monitor temperature. As the name suggests the word thermistor stands for *thermal resistance*. It is built on semiconducting electroceramic substrates of materials such as ZnO, SnO_2, and WO_3. Unlike thermocouples and RTDs (*resistance temperature detectors*) thermistors can operate satisfactorily at high temperatures. There are two categories of thermistors based on the sign of the slope of resistance vs. temperature characteristics of the substrate material. If its slope is positive the device is called *positive temperature coefficient (PTC) thermistor* and if it is negative the device becomes a *negative temperature coefficient (NTC) thermistor*.

Transconductance It is a device property exclusive to a transistor. It is defined as the ratio of the incremental change in drain output current (ΔI_d) corresponding to the incremental change in bias voltage (ΔV_b) or the gate voltage (ΔV_g). Its another name is *mutual resistance.* Its magnitude identifies the device as a CCVS (current controlled voltage source), or a VCCS (voltage controlled current source). Its unit is S (Siemen) just like Ω is for resistance. The inverse of transconductance is obviously called *transresistance or mutual resistance.*

Transistor A three terminal semiconductor device capable of producing current amplification and transconductance. It is a very important device and is the heart and soul of microelectronics. There are two principle categories of transistors; one is the BJT mostly of n–p–n configuration (commonly called BJT), and field effect transistors (FET). It is the most powerful component present in an IC. The word transistor is the abbreviation of "*transconducting variable resistor, or transconducting varistor.*"

Varistor It is a special class of diode based on metal–semiconductor platforms. Like a p–n junction diode, its current–voltage characteristics are rectifying (nonlinear), but unlike a p–n-junction diode, it can conduct both in forward (+voltage) and reverse (−voltage) mode and is uniquely bipolar. It is exclusively built on oxide electroceramic substrates and can carrying large electric fields without breakdown happening. It is present in practically all electrical circuits and used for protecting electronic components especially diodes and transistors from sudden voltage and current surges in power supply. Its signature properties are "switching voltage" (also called "threshold voltage" or breakdown voltage) and NLC (α).

Varistor embedded devices Only recently, it has been discovered that a varistor can be the origin of embedded devices when its current–voltage characteristics are tuned by an external agent. Four types of embedded devices have been identified: VBT; electric field effect transistor (E-FET), MAGFET (H-FET), and magnetic sensors. All these devices are based on electroceramic oxides just like a varistor is. They are potentially good devices for high-temperature operations, they show high immunity to radiations and being built on ceramics, they are rugged and can tolerate abuses encountered under field conditions.

VDR The voltage-dependent-resistor (VDR) is a solid-state device based on the voltage-dependent response of resistance of a varistor diode. Depending on the electroceramic material used for fabrication of the parent varistor, a VDR device can exhibit magnetic field-dependent resistive change or magneto-resistive change in the order of many kiloohms or more enabling the VDR to be used as a sensitive magnetic sensor. VDRs can have obviously many electronic applications. Its signature property is also called nonlinear coefficient, NLC (β).

Work function We have already defined it also in Chapter 1.

Problems

8.1 Discuss the formation of a potential barrier (a) when two metals are in contact and (b) when a metal and an n-type semiconductor are in contact. What are the key differences between the two mechanisms? What is the importance of a potential barrier?

8.2 What is a Schottky barrier and under what condition this can form? Can the barrier height be changed and if so what would be its effect on the $I-V$ characteristics of a diode?

8.3 Describe the operational mechanisms of a p–n junction diode and a Schottky diode. Under what conditions a contact between a metal and semiconductor will become rectifying?

8.4 The workfunction of n-type Si is 4.75 eV. It forms contacts with Au, Al, and Pt. Considering that the workfunction of Au is 5.25 eV, it is 4.15 eV for Al and 5.75 eV for Pt. Find the values of the potential barriers formed once the contacts are made. Comment on your results.

8.5 Can any of the contacts formed in Problem 8.4 be used for fabricating a Schottky diode? Justify your answer. Calculate the contact potentials of the three contacts given in Problem 8.4.

8.6 Find the current in a Schottky diode having a potential barrier of 0.05 eV at 300 K. Assume the modified Richardson constant, A, to be 3.44 AK^{-1}

8.7 Describe with the help of a figure the configuration of a typical LED based on AlGaN which emits in UV of 300 nm. Calculate the bandgap of the material. Using this value compute the value of the Planck constant, h. Compare it with its standard value.

8.8 (1) How would you determine experimentally the $I-V$ characteristics of a LED device? Discuss the meaning of forward voltage drop.
(2) If the series resistance is R_s, the source potential is 5 V, the potential drop at the LED is 2 V and the circuit current is 5 mA, find the value of R_s.

8.9 What is the fundamental difference between a conventional transistor and a transparent thin film transistor? Discuss the essential requirements for fabricating a transparent transistor.

8.10 What are varistors? How do they differ from a typical Schottky diode? Discuss the $I-V$ characteristics of a varistor emphasizing the unique properties of the device.

8.11 What is a VDR? How could you use it for building a sensor and what would be the benefit of using such a sensor?

8.12 Varistors and transistors are supposed to be coupled devices. Describe one type of such an embedded transistor and how would you separate its $I-V$ characteristics from that of the varistor. Discuss its attributes.

8.13 Describe the transport mechanism in a varistor on the basis of energy diagram of a G–GB–G structure.

8.14 Calculate the capacitance of a varistor device considering that its relative dielectric constant is 10^3, dot contacts of diameter 1 mm, its thickness is 1 mm, the separation between two adjacent grains is 1 μm and grain thickness is 0.01 μm and there are 1000 grains present. What is the contribution of grain boundaries in the total capacitance of the device?

8.15 What is a thermistor? Consider a thermistor having the resistance of 1 kΩ, which is subjected to a voltage of 10 V. Calculate its dissipation factor of the device when the temperature gradient of 200 °C. Calculate also the power and the amount of heat that would be generated.

References

1 Fukumura, T., Toyosaki, H., and Yamada, Y. (2005). Magnetic oxide semiconductors. *Semicond. Sci. Technol.* 20: S103.
2 Pandey, R.K., Stapleton, W.A., Tate, J. et al. (2013). *AIP Adv.* 3: 062126.
3 Pandey, R.K., Stapleton, W.A., Sutanto, I. et al. (2014). *J. Electron. Mater.* 43 (5): 1307.
4 Pandey, R.K., Stapleton, W.A., Sutanto, I. et al. (2014). *Ceram. Trans.* 249: 175.
5 Fortunato, E., Barquinha, P., and Martins, R. (2012). *Adv. Mater.* 24: 2945–2986.
6 Biard, J.E. and Pittman, G.E. (1966). Semiconductor radiant diode. Texas Instruments, Inc. Richardson, Dallas, TX. US Patent 3,293,513 of 20 December 1966.
7 Presley, R.F., Munsee, C.L., Park, C.-H. et al. (2004). *J. Phys. D: Appl. Phys.* 37: 2818.
8 Grondahl, L.O. and Geiger, P.H. (1927). *Trans. AIEE* 46: 357.
9 Levinson, L.L. and Phillip, H.R. (1986). *Ceramic Materials for Electronics* (ed. R.C. Buchanan). Merkel Dekker. ISBN: 0-8247-7501-5.
10 Bell Telephone Laboratories – Technical Memorandum (1948). May 28, 1948.
11 Pandey, R.K., Stapleton, W.A., and Sutanto, I. (2015). *IEEE J. Electron Devices Soc.* 3 (3): 276. doi: 10.1109/JEDS 2409023.
12 (John) Atalia, M.M. and Kahing, D. (1963). Electric field controlled semiconductor device. US Patent # 3,102, 27 August 1963.
13 Gallagher, R.C. and Corak, W.S. (1966). A metal-oxide semiconductor (MOS) Hall element. *Solid State Electron.* 9: 571–580.
14 Mohan, G.R. and Carr, W.N. (1971). Magnetic sensitivity of a MAGFET of uniform channel current density. *Solid State Electron.* 14: 995–1001.
15 Datta, S. and Das, B. (1990). Electronic analog of the electro-optic modulator. *Appl. Phys. Lett.* 56: 665–667. doi: 10.1063/1.102730.
16 https://en.wikipedia.org/wiki/Thermistor.
17 Pandey, R.K., Stapleton, W.A., Padmini, P. et al. (2012). *AIP Adv.* 2: 042188. doi: 10.1063/1.4773328.
18 Pandey, R.K., Stapleton, W.A., and Schad, R. (2016). *IEEE J. Electron Devices Soc.* 4 (6): 473–479.
19 Pandey, R.K., Stapleton, W.A., Sutanto, I., and Shamsuzzoha, M. (2015). *J. Electron. Mater.* 44 (4): 1100.
20 Schwing, U. and Hoffmann, B. (1980). *J. Appl. Phys.* 51: 4558.

21 Azad, A.M., Akbar, S.A., Mhaisalker, S.G. et al. (1992). *J. Electrochem. Soc.* 139 (12): 3690.
22 Wang, C., Yin, L., Zhang, L. et al. (2010). *Sensors* 10: 2088. ISSN 1424-8220.
23 De Lacy Costello, B.P.J., Ewen, R.J., Ratcliff, N.M., and Sivanand, P.S. (2003). *Sens. Actuators, B* 2 (92): 159.
24 Shimizu, Y., Lin, F.C., Takao, Y., and Egashira, M. (1998). *J. Am. Ceram. Soc.* 81: 1633.
25 Hua, Z., Wang, Y., Want, H., and Dong, L. (2010). *Sens. Actuators, B* 150: 588.

Further Reading

Moulson, A.J. and Herbert, J.M. (2008). *Electroceramics: Materials, Properties and Applications*, 2e. Wiley. ISBN: 978-0-471-49747 (HB).

Solymar, L. and Walsh, D. (1999). *Electrical Properties of Materials*, 6e. Oxford Science Publications, Oxford University Press. ISBN: 0-19-856272-1.

9

Electroceramics and Green Energy

Science of today is the technology of tomorrow.
Edward Teller

9.1 Introduction

Today the world depends primarily upon energy derived from coal, petroleum, and natural gas. The ever increasing demand of our industrial society for a reliable and ample supply of energy available at will has contributed to our over-dependence on energy produced by fossil fuels, which as we know, emit hydrocarbons and many other pollutants causing serious damage to environments. Chemical pollutants are the major source of many health issues that we face today. It takes millions of years for Nature to produce fossil fuels. The rate at which they are depleted today they may be exhausted in just few more decades. We access them by mining or by digging deep into the earth that causes serious environmental damage to the landscape and adversely impacts the sensitive ecosystem. Furthermore, as more and more countries are becoming industrialized the environmentally induced health problems are also becoming acute and dangerous. This has made it necessary to develop alternative sources of energy that must be clean and nonpolluting. This so-called *green energy* is the focus of this chapter. We are going to examine the role electroceramics can play in the advancement of the *green energy*.

9.2 What is Green Energy?

Green energy is supposed to be clean because it does not emit polluting chemicals as do the conventional energy sources of today do. Not only that, it is supposed to be renewable because its origin is embedded in our environment and therefore of unlimited supply. These green energy resources are there around us waiting to be used for harvesting green energy without damaging the earth and environment by mining and digging as for coal and petroleum. We recognize five natural resources for harvesting green energy as outlined in Figure 9.1.

These five natural energy resources are now actively pursued for producing energy that is reliable and cost-effective. However, the field of renewable energy technology is vast as we can see from Figure 9.2.

Currently, approximately 80% of energy comes from the conventional sources of fossil fuels, and the market share of alternative energy is just about 20%. However, the picture is slowly changing in favor of renewable energy. Solar energy and wind energy are big players on this team followed by vibration, heat, and water.

As with any other technology, renewable energy is also materials-dependent to be competitive in the marketplace. History tells us how materials have played a role over the centuries in advancing the growth of technology. Demand for new materials grows in tandem with the growth of a new technology. How the partnership between energy-driven technology and materials has evolved from 1700 until today is summarized in Figure 9.3 [1].

As the technology evolved from windmills to solar panels, the need for raw materials increased rapidly. In the last 300 years, we have used metals at an alarming rate resulting in acute shortage of them. Like fossil fuels, nature produces metals, and once they are gone, they cannot be easily replenished. It is expected that man-made materials such as plastics and ceramics will

Fundamentals of Electroceramics: Materials, Devices, and Applications, First Edition. R. K. Pandey.

Companion website: www.wiley.com/go/Pandey/Fundamentals_Electroceramics

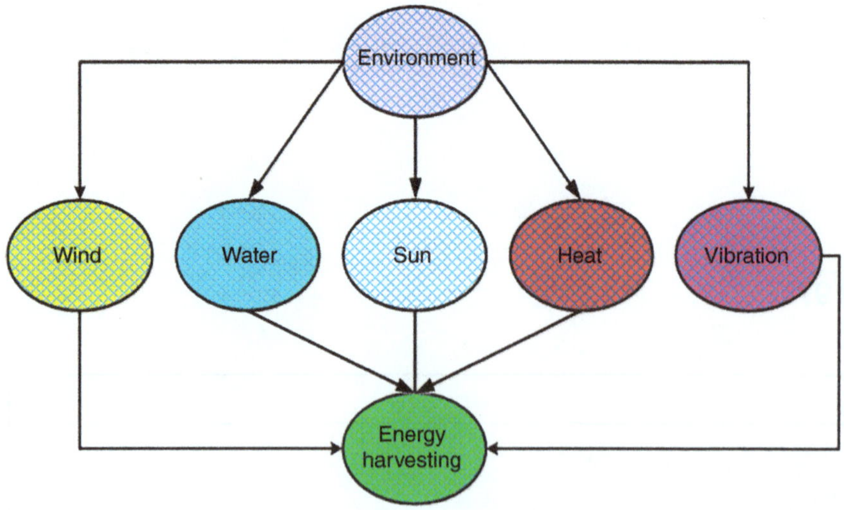

Figure 9.1 A block diagram showing environmental sources of renewable energy pertaining to green energy harvesting.

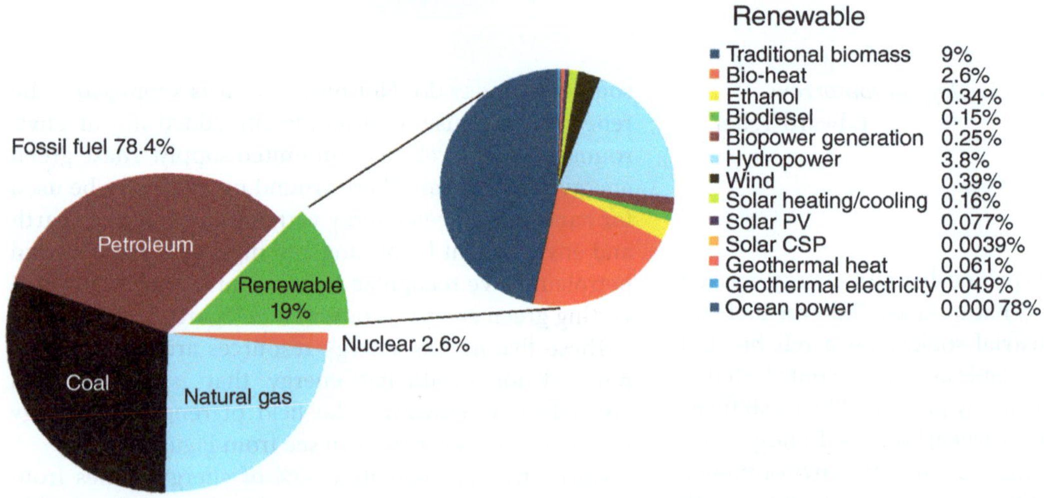

Figure 9.2 Comparison between conventional energy and renewable energy as they are used today. Source: https://upload.wikimedia.org/wikipedia/commons/b/b6/Total_World_Energy_Consumption_by_Source_2013.png.

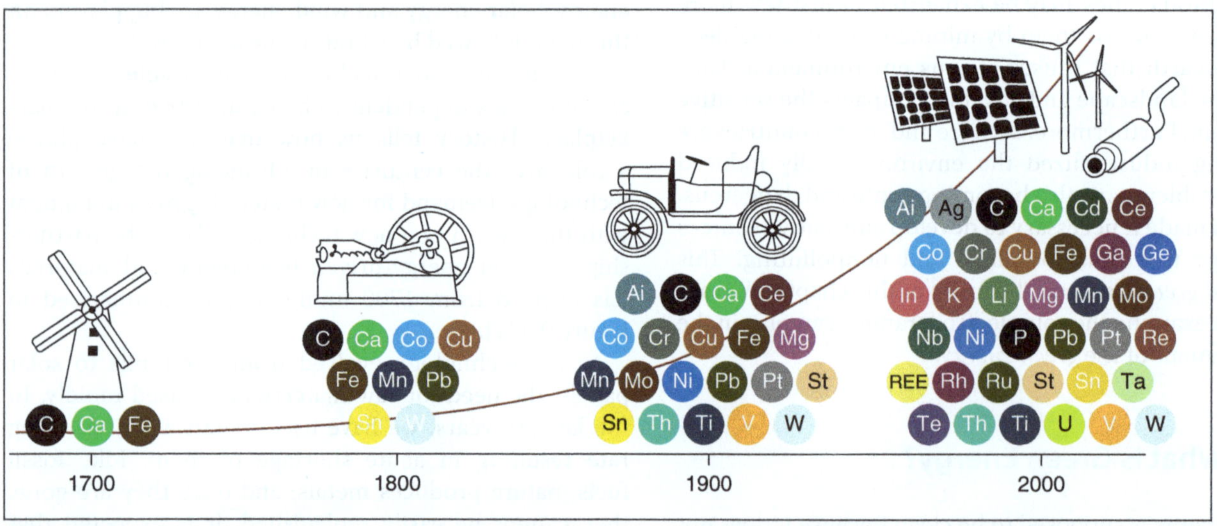

Figure 9.3 Evolution of technology from windmills to satellites and demand on raw materials from 1700 until today. Source: Zepf et al. 2014 [1]. https://www.bp.com/energysustainabilitychallenge.

have to compensate for loss of natural materials, but they cannot be substituted for some of the critical applications, where metals are the only option Role of ceramics in energy harvesting.

To examine the impact of ceramics in the green energy arena, we need to understand the concept of *energy harvesting*. This is a newcomer in the class of renewable energy. Energy harvesting refers to the process by which energy can be derived from sources such as the Sun, water, wind, heat, and vibration. Once harvested, the energy must also be stored for subsequent uses. Energy harvested from the environment is specifically tailored to meet the power requirements for small devices such as our mobile phones, mobile computers, and hand-held radios. The power delivered would not be sufficient for large power-hungry machines.

Energy harvesting has many other names such as power harvesting, energy scavenging, or ambient power. Even though energy harvesting can provide only power for small devices, it has become important for civilian and military applications because of its flexibility and availability under field conditions where conventional electricity may not be available.

Electroceramics are already playing an important role in this arena. For example, the piezoelectric effect can be used for conversion of mechanical energy into electricity. We have seen previously that a mechanical stress from wind and vibrations can induce electric polarization by converse piezoelectric effect. Similarly, some ferroelectrics can be a suitable medium for producing electricity from the Sun via the photovoltaic (PV) effect. We have learnt about piezoelectricity and ferroelectricity in Chapters 5 and 6, respectively. We know that these effects are most prominent in perovskites that are found to be the ideal platforms for energy harvesting. Electroceramic cells are essentially nonlinear capacitors and therefore, by their very nature, they are suitable for energy storage. Ceramics can withstand high temperatures and therefore they are well suited for high temperature fuel cell research and development. In the subsequent sections, we will discuss some ceramic energy storage devices and show how these materials are being used in advancing the field of *green energy*.

9.3 Energy Storage and Its Defining Parameters

Energy is stored whenever an electric or magnetic field is present. Without going into the physics of it we can express the relationships with the help of two standard equations such as Eqs. (9.1) and (9.2). Both these equations show the quadratic relationship between the energy and the field. Equation (9.1) represents the case of an electric field, E where ϵ_0 is the permittivity of vacuum and Ω_E the electrical energy.

$$\Omega_E = \frac{1}{2}\epsilon_0 E^2 \tag{9.1}$$

Similarly, Eq. (9.2) represents the equation for magnetic energy of Ω_H when a magnetic field H is present. Here μ_0 represents the vacuum permeability.

$$\Omega_H = \frac{1}{2\mu_0}H^2 \tag{9.2}$$

As a capacitor can store electrical energy, an inductor can store the magnetic energy. A solenoid is a good example of a magnetic storage device. Superconductors can give us highly efficient storage devices, but as we know they are not practical because superconductivity occurs only at cryogenic temperatures. So far as *green energy* is concerned, magnetic energy does not play an important role.

Exercise 9.1

Calculate the magnetic energy stored in vacuum corresponding to the earth's magnetic field of 5×10^{-5} T. Find the value of the electrical field corresponding to the same amount of energy.

Solution: Using Eq. (9.2), we get Ω_m is $10^{-3}/\text{m}^3$. Substituting this in Eq. (9.1) and using $\varepsilon\epsilon_0 \approx 8.854\times10^{-12}$ F m^{-1}, we get for $E = 15\,\text{kV m}^{-1}$. *Comment:* We see from this simple result that a large electric field is required to produce an energy equivalent to the energy corresponding to the earth's magnetic field, which is just equal to 0.5 G and that is a very weak magnetic field.

Once the energy is produced, no matter from which sources, it must be stored such that it becomes readily available when a need arises. Both the energy harvesting and energy storage are very important components of *green energy*. There are different types of electrical energy storage devices. Each of them has their place in the overall picture of green energy for daily use.

For applications, energy is classified on the basis of two inter-dependent parameters that are energy density and power density. Combined together they define an energy storage device and its scope for practical applications.

The electrical power is defined as the rate per unit time (t) at which electrical energy is transferred from one point to another in an electric circuit. Numerically, it is given by the product of current (I) generated when a voltage (V) is applied to the circuit. When it is normalized with respect to volume (or mass), we get the power density (η_P) as expressed in Eq. (9.3).

$$\eta_\text{P} \approx \frac{\text{Electrical Power}}{\text{Volume}} \approx \frac{V\times I}{m^3}\ (\text{J s}^{-1}\ \text{m}^{-3}) \tag{9.3}$$

The commonly used unit for the power density is $W\,m^{-3}$ because $1\,W = 1\,J\,s^{-1}$. In industrial circles the unit often used for power density is $W\,kg^{-1}$. The energy density (η_E) is defined as in Eq. (9.4)

$$\eta_E \approx \frac{\text{Electrical energy}}{\text{Volume}} \approx \frac{V \cdot I \cdot t}{m^3} (J\,m^{-3}) \qquad (9.4)$$

Again, as in the previous case, this parameter is expressed in the unit of $Wh\,kg^{-1}$ by industry. The symbol "h" is here for hour. The ratio between power and energy is simply equivalent to the duration for a one-time energy flow as shown in Eq. (9.5).

$$\frac{P}{E} \approx \frac{V \cdot I}{V \cdot I \cdot t} \approx t^{-1} \qquad (9.5)$$

The standard energy storage devices are capacitors, batteries, and fuel cells. Among batteries, we are familiar with lead-acid batteries, NiCd batteries, and lithium-ion batteries, each of which varies in their capacity for storing energy and delivering power. We have different types of capacitors, some of which are double-layer capacitors, ultra-capacitors, supercapacitors, and electrolytic capacitors. Then there are fuel cells and hybrid storage devices. They differ vastly among themselves with respect to energy density and power density as illustrated in Figure 9.4. This figure is based on the original Ragone chart which was developed around 1968 to compare batteries with respect to their capacity for storing energy and delivering it [2]. Notice in this figure that the units common in industrial work have been used for describing the relationship between the two parameters.

The energy density and power density are plotted in log–log scale in the Ragone chart. On this plot, any point on the Y-axis represents not only the energy but also the time that would be required to deliver the power on the X-axis. The Ragone plot is a valuable tool in the hands of power engineers who can use it to determine the steady delivery of power to operate a miniature device or a large power-hungry system.

Development of efficient and cost-effective energy storage devices for *green energy* is of high priority not only because these devices are environmentally friendly but also because they can advance many key technologies such as electric vehicle technology, smart electric grids, miniaturized electronic and optical devices, high speed-high density computer memories, mass transportation, and transit systems. These are only a very few examples. We can easily visualize that an efficient system of storing green energy and delivering it at high rates is the ultimate objective of this emerging technology as well as any other energy technology. Its impact on environment, mankind, and our planet cannot be underestimated.

9.3.1 Capacitor as an Energy Storage Device

Capacitors are one of the oldest and most reliable work horses of energy storage technology. In this arena, electroceramics play a vital role. Both linear and nonlinear dielectrics are used to fabricate all types of capacitors on a commercial scale. In fact, barium titanate, which is the most prominent member of the ferroelectric family, is

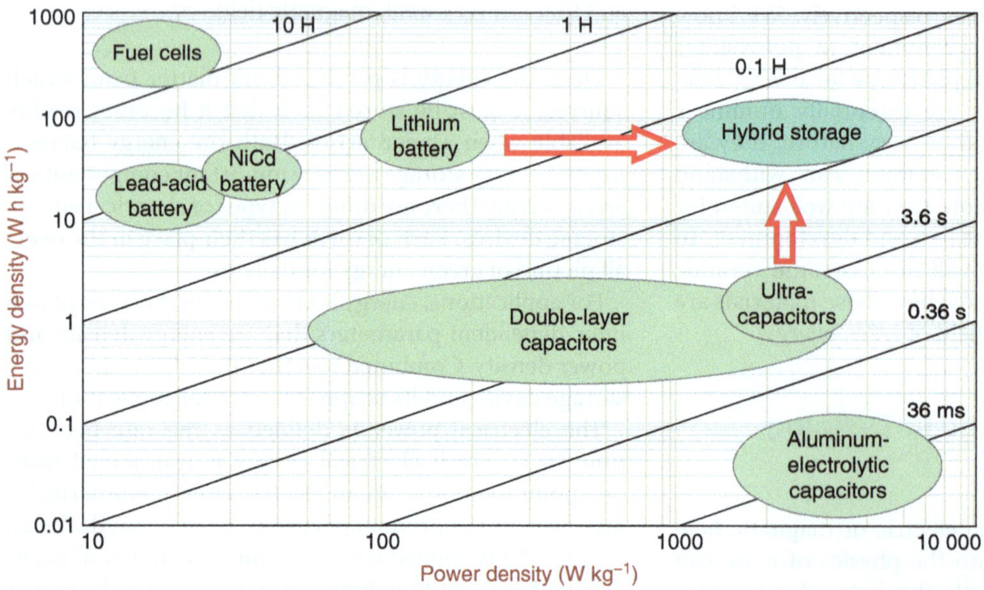

Figure 9.4 Energy density vs. power density of commonly used electrical energy storage devices. Source: http://www.generalcapacitor.com/. Note: Published as Ragone Plot under General Capacitor title, published by US Defence Logistics Agency.

used because of its high dielectric constant by industry to produce commercial-grade capacitors in large volumes throughout the world. They can be produced in various shapes and sizes, but the parallel plate capacitor is the most preferred configuration. Capacitors are present in almost all electrical circuits as a vital component because they can perform multiple operations ranging from simple energy storage to highly sophisticated electronic filters.

The two most important properties of a capacitor device are (i) how much energy it can store and (ii) how long it can hold the stored energy. The first condition is dependent on the permittivity of the dielectric material used; and the second is related to its *dissipation factor*, which is also called the loss tangent or simply $\tan\delta$. The loss tangent is the defining property of a capacitor because it is the measure of the loss rate of energy of the device. In fact, the inverse of loss tangent ($\frac{1}{\tan\delta}$) is the figure-of-merit of a capacitor. It is also called the *quality factor, Q*.

Based on these two parameters, capacitors are divided into two classes. The first class of capacitors is fabricated out of dielectric materials with permittivity ranging between 10 and 500 and loss tangent ($\tan\delta$) around 3×10^{-3}. The second category of capacitors are of materials with permittivity ranging between 1000 and 20×10^{3} and having $\tan\delta < 3\times10^{-3}$. It is obvious that so far as the energy storage is concerned the capacitors of second category are preferred. This is where ferroelectric materials become important because they usually have very high dielectric constant with very small loss tangent.

From basic physics, we know that the energy (E) of a parallel plate capacitor is simply one half the product of its charge (Q) that develops when a potential (V) is applied. Mathematically, expressed it is given by Eq. (9.6).

$$E = \frac{1}{2}QV = \frac{1}{2}CV^2 \tag{9.6}$$

where C is the capacitance of the device that is related to the permittivity of the dielectric and to the area (A) of the metal electrode and thickness (d) of the capacitor. We express this relationship as Eq. (9.7).

$$C = \epsilon_r\left(\frac{A}{d}\right) \tag{9.7}$$

where ϵ_r is the relative permittivity which is simply the ratio between the permittivity of the medium and that of free space. Combining the above two equations, we get capacitor energy as shown in Eq. (9.8).

$$E = \frac{1}{2}\left(\frac{\epsilon_r A}{d}\right)V^2 \tag{9.8}$$

From Eq. (9.8), we can conclude that because the energy is proportional to V^2, we can greatly increase energy simply by applying more and more voltage. But we

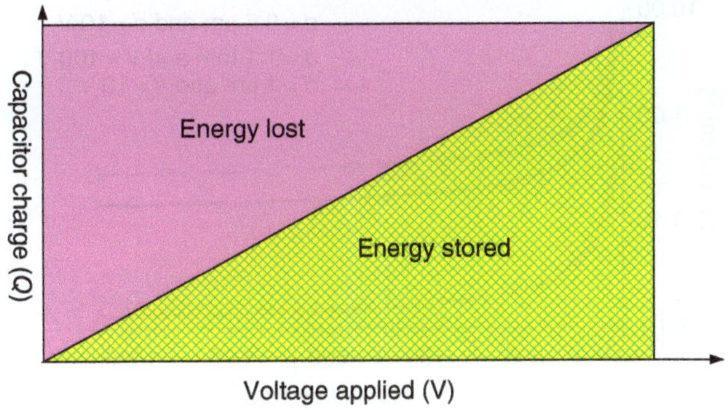

Figure 9.5 Energy density stored in a parallel plate capacitor as function of potential applied.

cannot keep on increasing voltage at will because we might reach the break-down point of the dielectric and second, because it is an impractical solution. But we can achieve this goal by selecting a material of high permittivity for the device with the ratio $\left(\frac{A}{d}\right)$ large as well. The latter condition we can achieve by using a thin film substrate with point electrodes.

The energy density (η_E) of the capacitor can be expressed as in Eq. (9.9).

$$\eta_E = \frac{\text{energy}}{\text{volume}} = \frac{1}{2}\epsilon_r E^2\ (\text{J m}^{-3}) \tag{9.9}$$

Figure 9.5 shows the energy stored by a parallel plate capacitor. We see from this figure that only half of the available energy can be stored by a capacitor. What happens to the other half? It is simply dissipated as heat through the internal resistance of the capacitor.

Exercise 9.2

Calculate the energy density of a capacitor considering the following data: Potential applied = 100 V, relative permittivity = 1000, and its thickness = 1 µm.

Solution: From Eq. (9.9) we have the energy density = $\frac{1}{2}\epsilon_r E^2$. It is given that $\epsilon_r = 10^3$, applied potential = 10^2 V and dielectric thickness = 1×10^{-6} m. Therefore, $E = 10^8\ \text{V m}^{-1}$. Substituting for E and permittivity value in Eq. (9.9), we get energy density = 5×10^{18} J. *Comment*: Notice the very large value of the energy density, which may not be achievable by the very thin capacitor. The very large value of the electric field will most certainly cause the breakdown of the dielectric material and thus make the capacitor fail completely.

The power density, (η_P) of the capacitor is given by Eq. (9.10).

$$\eta_P = \frac{\eta_E}{t} \approx \frac{1}{2}\left(\frac{\epsilon_r}{t}\right)E^2 \tag{9.10}$$

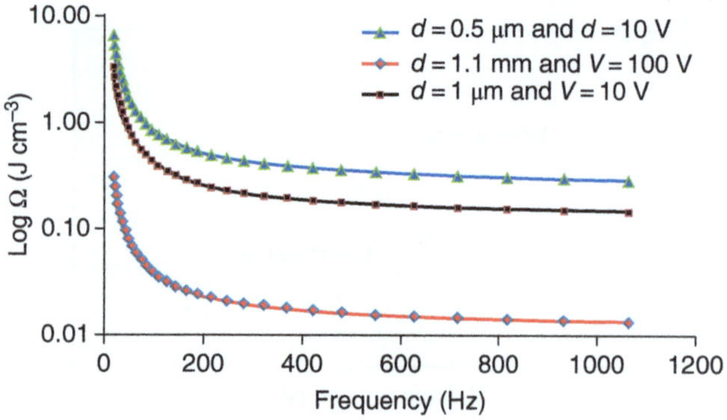

Figure 9.6 Frequency dependence of energy density of CCTO with different thickness. Source: Pandey et al. 2013 [3]

Since half of the energy is lost in a capacitor, one must look for ways to overcome this unavoidable device limitation that is inherent to the nature of a capacitor. We have two obvious options, increasing both the permittivity and the $\left(\frac{A}{d}\right)$ ratio. To these two options, now we can add the third option of using a *supercapacitor* that exhibits giant permittivity such as CCTO. CCTO stands for Ca-Cu-titanate, with the chemical formula of $CaCu_3Ti_4O_{12}$. CCTO is also a member of the perovskite family, but since it is cubic, it is neither piezoelectric nor ferroelectric. It is a linear dielectric. Its relative permittivity is 5×10^4, which can be improved by a factor of 10 by process optimization [3]. Figure 9.6 shows the frequency dependence of the energy density of a CCTO capacitor (with $\varepsilon_r \approx 53 \times 10^4$) with varying thickness. In general, the energy density decreases with increasing thickness and frequency. The sample with the thickness of 0.5 μm meets the industrial benchmark for energy density which is $5\,J\,cm^{-3}$.

9.3.2 Battery-Supercapacitor Hybrid (BSH) Devices

Battery-supercapacitor hybrid (BSH) devices are of great interest. They also advance the frontiers of *green energy*. Here the word "supercapacitor" refers to a device configuration and not CCTO supercapacitor material. An inspection of Figure 9.4 tells us that the hybrid storage devices hold great promise for the future because they have both high energy density and high power density. These devices are constructed by combining a high-capacity battery-type electrode and a high-rate capacitive electrode. They are essentially electrochemical energy storage systems capable of delivering high power density with a long cycling life. Some examples of these devices are Li-Na-ion BSH, acidic-alkaline BSH, and redox electrolytes BSH.

9.3.3 Piezoelectric Energy Harvester

Ever since its discovery by the Curie brothers in 1880, it has been known that a piezoelectric material can convert mechanical energy into electric energy and vice versa. The relationship between the two types of energy is defined by the electromechanical coupling coefficient (k) of the material. We recall that we have studied piezoelectricity in Chapter 5, where we also learned about transducers, actuators, and electromechanical coupling.

It is only in about the last 20–25 years that energy harvesting using piezoelectric effects has appeared on the horizon. The reasons are twofold: first the emergence of renewable energy, and second, the vastly changing face of electronic technology. The power need changes as the new technology evolve. For example, a desktop computer can consume 100 W for its operation, while a thermoelectric device may not need more than 100 nW of energy. These two simple examples show us that the range of needed power from small to large systems can vary by an order of 10^9. We can therefore conclude that there is a room for the piezoelectric energy harvester in the vast spectrum of power requirements. They can certainly play an important role in providing power to a host of small devices.

A piezoelectric power generator consisting of a micromachined silicon substrate with an integrated thin-film structure (as in Figure 9.9) can deliver power equivalent of $35\,mJ\,cm^{-3}$. This level of energy density is more than sufficient to power a host of small devices like electronic watches, digital cameras, smart phones, tablets, miniature radios, and two-way communication devices. Even a non-MEMS device is capable of producing approximately 10 μW of electric energy that can easily satisfy the power need of a miniature device such as a RFID tag.

This field is still emerging, and its beginning was anchored in the interest of Defense Advanced Research Project Agency (DARPA, a major research arm of the US military) to develop a source of energy that is completely independent of fossil fuel and could be readily available under field conditions. The piezoelectric energy harvester is the result of this interest. It is 100% nonpolluting and does not depend upon any other input source to produce electricity. This technology is ideally suited to meet many needs of a soldier in the battlefield, where there is no electrical outlet to be found far and away. We can now easily imagine conditions under which a vast population of the world could use such a source of power.

Piezoelectric effects are used for producing electrical energy primarily from vibrations. There can be many origins of vibrations. Each of them can be used for energy harvesting. In Table 9.1, we list some of these

Table 9.1 Types of vibrations that can be used for energy harvesting using piezoelectric effects.

Sources of vibration	Potential to produce power (μW cm^{-3})
Humans	≈350
Airflow	≈400
Mechanical	≈350
Pressure	≈200

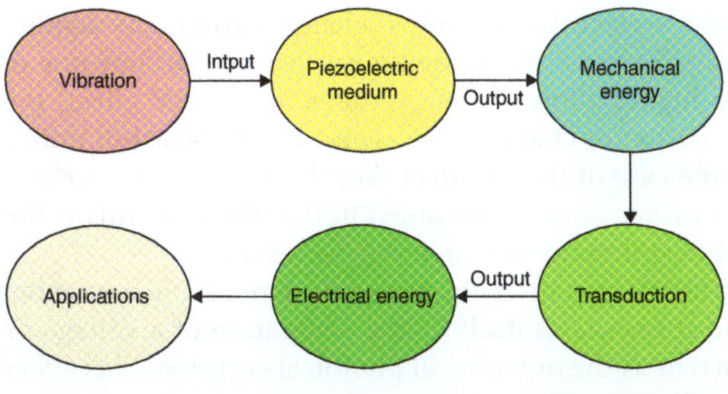

Figure 9.7 Steps involved in piezoelectric energy harvesting technology.

sources with the approximate amount of power they can generate.

How vibrations can be utilized by piezoelectric materials to generate power that can find its acceptance in technology is outlined in Figure 9.7. First, a mechanical energy is produced by vibration as a consequence of the converse piezoelectric effect that is subsequently converted into electrical energy by the transducing property of a piezoelectric material.

The question remaining to be answered is how efficient is such a conversion? For this, we need to know the electromechanical coupling coefficient of the piezoelectric material we are using. It is a material parameter that is represented by the symbol k, which is always <1. It is a measure of the ability of a piezoelectric material for converting mechanical energy into electrical energy or electrical energy into mechanical energy. This parameter has been defined in Chapter 5 by Eq. (5.40). We can rewrite the equation here in slightly modified form.

$$k^{-2} \approx \frac{M_{\text{input}}}{E_{\text{output}}} \approx \frac{\text{mechanical energy}}{\text{electrical energy}} > 1 \quad (9.11)$$

In Table 9.2, we tabulate the values for the electromechanical coupling coefficient, k, for some leading piezoelectric materials.

We find from this table that PZT has the highest value for the coupling coefficient and therefore, it is the most suitable material for energy harvesting. The other materials that are also considered for energy harvesting are lead-titanate ($PbTiO_3$), lead zirconate ($PbZrO_3$), and barium titanate ($BaTiO_3$). It is interesting that all of them are the members of the ferroelectric-piezoelectric perovskite family.

Piezoelectric generated energy is good for low-frequency vibrations (<250 Hz). The relationship between the power generated and the frequency is plotted in Figure 9.8 for energy produced by subjecting a piezoelectric substrate to vibrations. Here we find that the power increases with increasing frequency reaching its peak value at a frequency corresponding to the natural resonance of the piezoelectric capacitor. Above this frequency, it begins to decrease. The narrow bandwidth (shaded part of the curve) is chosen as the operating range for the device because in its vicinity the power output changes very little. Therefore, for the best results one would need to choose an application that corresponds

Table 9.2 Values of coupling coefficients for selected materials.

Piezoelectric material	Electromechanical coupling coefficient, *k* (it is a unitless quantity)	Comments
Silicon dioxide, SiO_2	0.099	It is perhaps the oldest piezoelectric material discovered. It is found naturally. Synthetically grown single crystals of this material are widely used in many applications including electronic wrist watches. Because of its low value of coupling coefficient it is not suitable for energy harvesting
Lithium niobate, $LiNbO_3$	0.035	Used in SAW and electro-optic devices
Lithium tantalate, $LiTaO_3$	0.10	Acoustical amplifiers and acousto-optical devices
Lead zirconate titanate (PZT)	0.66	Energy harvesting and the most used piezoelectric material for hosts of applications
Lead titanate, $PbTiO_3$	—	Transducers and acoustic amplifiers

Source: Kao 2004 [4].

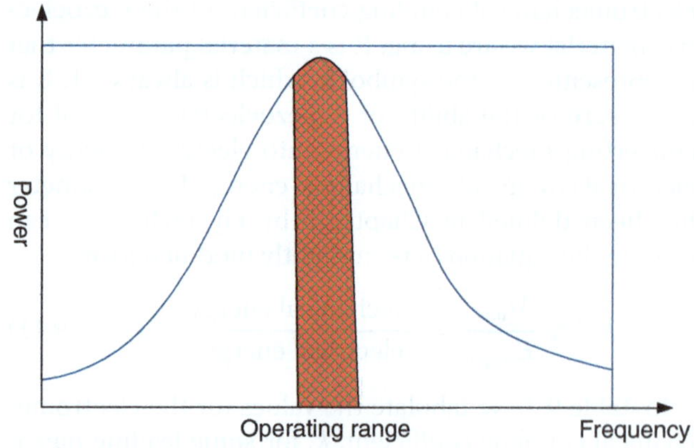

Figure 9.8 Frequency dependence of power generated by a piezoelectric material in response to vibrations.

to the operating bandwidth of the piezoelectric power generator.

9.3.4 MEMS Power Generator

A micromachined (MEMS) cantilever structure of a piezoelectric film with both ends clamped is widely used for power generation. We have discussed a MEMS actuator in some detail in Chapter 5. There a cantilever structure with one end clamped was used for actuation, which is achieved by the displacement of the unclamped other end in response to a voltage applied to the piezoelectric cell. But for power generation, we need a cantilever with both ends clamped as shown in Figure 9.9.

The principle of operation is simple and based on the direct piezoelectric effect. First, the top electrode is exposed to a mechanical stress by applying a mechanical pressure or an impact or even an acoustic wave. As a result an electric charge (Q) is generated that can be collected and stored in a capacitor for future use? Energy density as high as 35 mJ cm^{-3} has been produced by such a structure that is sufficient to meet the requirements of many devices that do not require a large amount of power to function.

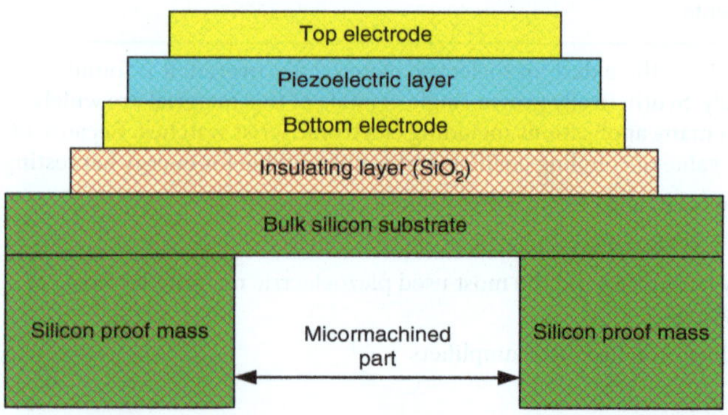

Figure 9.9 Configuration of micromachined device for energy harvesting via the piezoelectric effect.

9.3.5 Ferroelectric Photovoltaic Devices

The PV effect is analogous to the photoelectric (PE) effect. From the discussions in Chapter 1, we know that in the PE effect the electrons are ejected from the surface of a metal under high vacuum when bombarded with radiation. The absorbed photon energy causes the surface electrons or other charge carriers to acquire enough energy to escape the surface and occupy one of the higher allowed energy states. In the PV effect, the excitation of charge carriers also takes place, but unlike in the case of the PE effect they do not leave the surface, but rather remain contained in the material. This is the subtle difference between the two effects.

The PV effect is a physical and chemical phenomenon, and it manifests itself by the generation of a voltage or current as the outcome of photon absorption. This effect was discovered by the famous French physicist, A.E. Becquerel, in 1839 while experimenting with electrochemical cells. But in solids, it was discovered much later; about 50–60 years ago. By now it has made a profound impact on the advancement of renewable energy. Silicon solar cells are the prime example of this effect being used for the good of mankind and our planet. Large solar panels are commercially available and are being installed all over the world at a rapid rate. Energy-hungry countries like China and India are aggressively pursuing this technology, and the western world is not far behind.

Compared to the conventional solar cell technology, which uses the p–n junction effect of a semiconductor for conversion of solar energy into electrical power, the ferroelectric-photovoltaic (Fe-PV) technology is in its infancy. Because it is a late comer in this arena, it will take many more years before it becomes a viable option for producing green energy using solar radiation. Some ferroelectric oxides exhibit the development of voltage when exposed to solar radiation. But the photo-current is very small, and the conversion rate very low. As a result this physical phenomenon has remained just an academic curiosity and uninteresting for practical applications.

The field of Fe-PV effect made a quantum leap when *giant PV* was discovered in $LiNbO_3$ in 1960. It is also known as *anomalous PV effect* and *bulk PV* effect. Experimenting with the single crystals of $LiNbO_3$, it was found that by exposing it to solar radiation a photo-voltage as large as 105 V could be produced. Naturally, this discovery led to the race of making the Fe-PV affect a viable alternative to silicon solar cells. The other ferroelectrics in which anomalous PV has been observed are

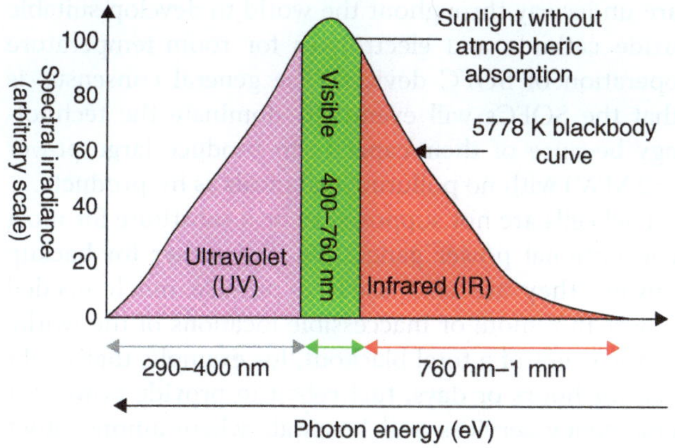

Figure 9.10 Wavelength vs. spectral irradiance of solar spectrum.

barium titanate ($BaTiO_3$), lead-zirconate-titanate (PZT), and bismuth ferrite ($BiFeO_3$). The physical mechanism behind the PV effect is the response of the switchable polarization of ferroelectrics to radiation. Rapid changes are induced macroscopically in the polarization because of the photon energy absorbed by the sample. A change in the polarization states gives rise to a voltage. Compared to this the p–n junction in a semiconductor material like Si responds to solar radiation giving rise to a photocurrent. The physical mechanism is here quantum mechanical in nature and not macroscopic as is the case with the Fe-PV effect. This subtle distinction between these two classes of materials must be taken into consideration while developing a workable technology based on the Fe-PV effect. One important advantage of the Fe-PV effect is that the polarity of the photo-voltage can be easily switched simply by changing the sign of the polarization.

In Figure 9.10, we reproduce the solar spectrum that consists of the entire range of electromagnetic waves beginning with ultraviolet (UV) and ending in infrared (IR). The output of the solar irradiance is measured as *spectral irradiation* that is defined as the power per unit area for all wavelengths that are incident on the Earth's upper atmosphere. Its unit is $W\,m^{-3}$.

It is fortunate that of all the solar radiation received by the upper atmosphere only 47% of it reaches the Earth; the rest is absorbed in the upper atmosphere. If it were not the case, our planet would have been inhospitable and uninhabitable.

The configuration of Figure 9.11 can be used to monitor the photo-voltage generated by a ferroelectric sample when it absorbs the appropriate level of photon energy. It is measured between the top and bottom electrodes.

The relationship between photovoltage and photon wavelength is shown in Figure 9.12. Its magnitude changes with the change in wavelength of radiation. Here we assume that the three wavelengths (λ) are such

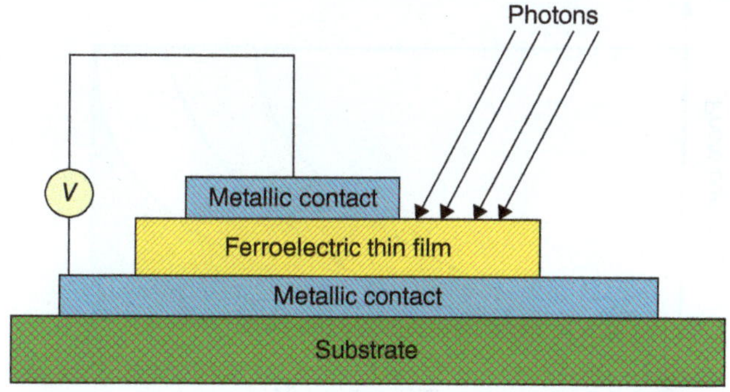

Figure 9.11 Experimental setup for monitoring photovoltaic effect.

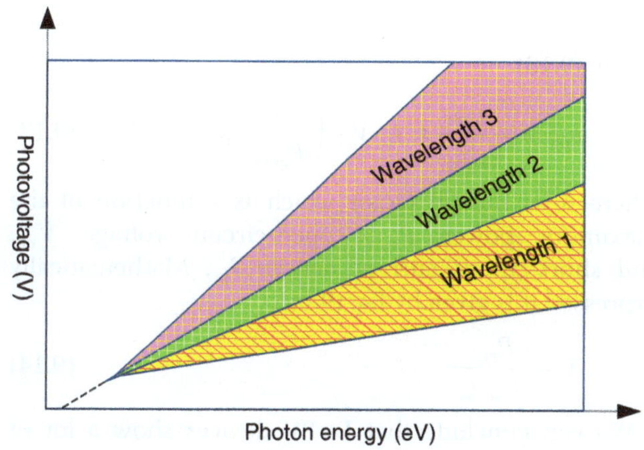

Figure 9.12 Photovoltage as a function of photon energy for Fe–PV device. $\lambda_1 < \lambda_2 < \lambda_3$.

that $\lambda_1 > \lambda_2 > \lambda_3$. If photon energy of λ_1 is $E_{\lambda 1}$, $E_{\lambda 2}$ for the second wavelength and $E_{\lambda 3}$ for the third wavelength, then $E_{\lambda 1} < E_{\lambda 2} < E_{\lambda 3}$. That tells us that the magnitude of the photo voltage increases with the increase in the input energy.

If the sample is biased, then we can monitor the photocurrent. Photocurrent is strongly dependent on the intensity of light as seen from Figure 9.13.

The photo-voltage, V_{ph}, is defined in Eq. (9.12), where I_{ph} is the photo current, σ_d is the photo conductivity of dark current, σ_{ph} is the photo conductivity, and d is the separation between the top and bottom electrodes. Normally, $\sigma_d \ll \sigma_{ph}$ and can be neglected from further considerations ion.

$$V_{ph} \approx \left(\frac{I_{ph}}{\sigma_d + \sigma_{ph}}\right) \cdot d \approx \left(\frac{I_{ph}}{\sigma_{ph}}\right) \cdot d \qquad (9.12)$$

The other parameter of importance for PV devices is the power conversion efficiency (or coefficient) η_{ph}. It is the ratio between the output power, P_{output}, and input power, P_{input}. Therefore, it is also called the figure-of-merit of a PV device. This parameter is defined

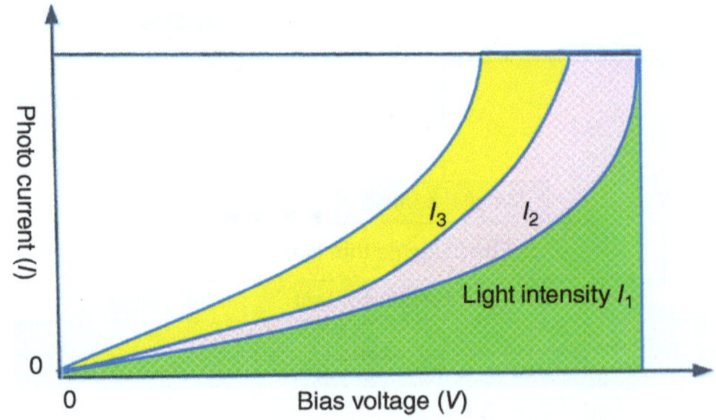

Figure 9.13 Photocurrent vs. bias voltage with varying light intensity. The intensity of radiation bear the relationship $I_1 < I_2 < I_3$.

by Eq. (9.13).

$$\eta_{ph} \approx \frac{P_{output}}{P_{input}} \approx J_{sc} V_{oc} \left(\frac{X}{P_{input}} \right) \tag{9.13}$$

where X is the fill factor which is a function of the maximum power P_{max}, open-circuit voltage V_{oc}, and short-circuit current density J_{sc}. Mathematically expressed it is given by Eq. (9.14).

$$X \approx \frac{P_{max}}{V_{oc} J_{sc}} \tag{9.14}$$

We can conclude that Fe-PV devices show a lot of potential for developing practical solar cells if the output photocurrent (I_{ph}) and conversion coefficient (η_{ph}) could be improved, which, of course, entails making vast improvements in materials processing or even discovering new Fe-PV materials. Nevertheless, this device is capable of generating large voltages when radiated with solar radiation or a UV laser.

9.3.6 Solid Oxide Fuel Cells (SOFC)

A fuel cell is a very interesting electricity generating device that was discovered in 1838. This technology remained in oblivion for almost 100 years, but gained in importance only after space exploration began. NASA uses t for providing power to satellites and space capsules. In the meantime, the range of its applications has expanded, and it is routinely used to power submarines, automobiles, motorcycles, buses, boats, and many other devices. There are different types of fuel cells, some of which are proton exchange membrane fuel cell (PEMFC), alkaline fuel cell (AFC), (solid oxide fuel cell (SOFC), and microbial fuel cell. The PEMFC technology is already highly advanced, and these cells are now routinely used for powering automobiles and other motorized vehicles. SOFC cells for the present are suitable for high-temperature operations, but not for room-temperature operation. But intensive R&D efforts are underway throughout the world to develop suitable oxide cathode and electrolytes for room-temperature operation of SOFC devices. The general consensus is that the SOFCs will eventually dominate the technology because of their capacity to produce large power (>2 MW) with no polluting chemicals as by-products.

Fuel cells are not supposed to be a substitute for most conventional power generation, but rather for backup power. They are also used to provide much needed power in remote or inaccessible locations of the world. In the event of a total blackout, for example, that might last for hours or days, fuel cells can provide power for emergency services and hospitals where among other critical needs a lifesaving surgery might be in progress.

A fuel cell is an electrochemical device just like a battery. Both can convert chemical energy spontaneously without any external assistance. Batteries and fuel cells are different from each other in their operating mechanism. In a battery, the chemical energy is the result of chemicals present in its housing, whereas in a fuel cell the chemical energy is generated by the dissociation at the electrolyte as positively charged hydrogen ions (protons) and negatively charged oxygen ions (electrons).

Figure 9.14 is the configuration of a typical PEMFC device. All fuel cells have three common components that are an anode (+), a cathode (−) and an electrolyte. Electrolyte is sandwiched between the two electrodes. An electrolyte is a chemical medium to produce negatively charged (electrons) and positively charged (protons) ions on dissociation of a molecule such as water, H_20. Common examples of electrolytes are acids, bases, and salts that ionize when dissolved in a solvent. Salts such as sodium chloride (NaCl) and silver iodide (AgI_2) become electrolytes in their molten phase. Electrolytes can also

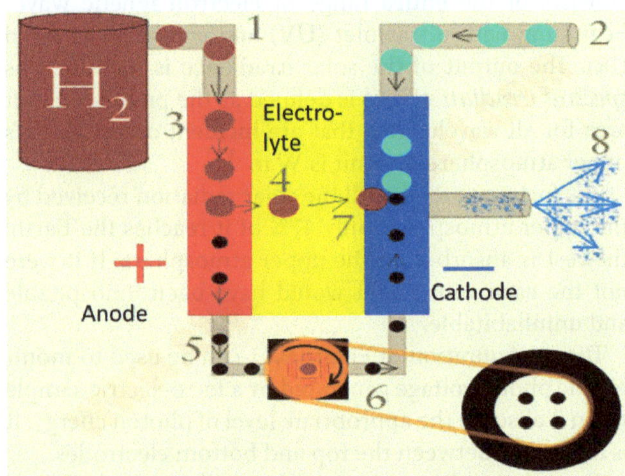

Figure 9.14 Basic configuration of a PEM fuel cell. 1. H_2 source; 2. O_2 from air; 3. Anode (Pt); 4. Electrolyte; 5. Electric circuit with a load; 6. Motor; 7. Cathode and 8. Water (H_2O). Source: http://cdn4.explainthatstuff.com/how-fuel-cell-works.png.

be solids such an oxide ceramic, which is the basis for *SOFC*.

The operating mechanism of this device is simple and straightforward. It is all chemistry with minimal involvement of mathematics. First, we describe here the operational principles behind the PEMFC devices and then we extend our description to understand how a SOFC works.

In order for the fuel cell to generate electricity, hydrogen gas is first introduced in the reaction chamber through a pipe (1). From another pipe in the opposite side (2) air is pumped through which reaches the cathode. There it becomes the source for oxygen. The anode is made of platinum, which also acts as a catalyst to speed up the chemical reaction occurring in the cell compartment. On reaching the anode, the hydrogen molecule splits up as positive ions (proton, p^+) and negative ions (electron, e^-). The positively charge p^+ ions are drawn to the negative terminal and travel through the electrolyte to reach the cathode. At the same time, electrons travel from the anode to cathode through an external electric circuit. In the process, a current is generated which then can be used to power a motor. Finally, electrons and protons recombine with O_2 at the cathode forming as water.

The reaction chambers of the PEMFC and SOFC devices are identical in their configuration; however, they differ from each other in the materials used for cathode and electrolyte. Pt remains as the anode terminal for both these devices. For the PEMFC devices, the electrolyte is a polymer membrane through which only protons can pass. It is replaced by doped yttria stabilized zirconia with the trade name of YSZ as the electrolyte for SOFC devices. YSZ is a prominent member of the oxide ceramic family. Chemically, it is a complex mixture of zirconium oxide, ZrO_2, and yttrium oxide, Y_2O_3. The cathode for a SOFC is an oxide of the perovskite group with the formulation of ABO_3. So far lanthanum manganites ($LaMnO_3$: LSM) have proven to be the best material for SOFC cathodes.

A SOFC is a versatile fuel cell, and it is projected to be the most efficient device among all other fuel cells. It is capable of producing power from 100 W to approximately 2 MW. Moreover, being based on oxide ceramic, it can sustain high temperatures and therefore, can operate even at elevated temperatures. Like the Fe-PV solar cells and piezoelectric power generator, it is environmentally friendly and 100% nonpolluting. The energy efficiency of a SOFC is generally in the range of 40–60%. It is hoped that by proper management of heat in the reaction chamber an efficiency of about 85% can be achieved.

9.3.7 Antiferroelectric Energy Storage

We have already discussed that a linear capacitor is a very convenient component of an electrical circuit to store energy that can be released at will. We recall that the energy density of a capacitor, see Eq. (9.9), is directly proportional to the dielectric constant of the capacitor material and to the square of the electric field applied. We cannot keep on increasing the field arbitrarily, otherwise, a point will come when the dielectric breakdown will set in. Therefore, we need to find a class of dielectric with very high dielectric constant that can tolerate moderately high fields before breaking down for fabricating capacitors for large energy storage. The supercapacitor CCTO, as we discussed, is such a material. However, it has the drawback of discharging fast. Nonlinear dielectrics with spontaneous polarization are very attractive alternatives to a liner dielectric material. These are the ferroics that we discussed in Chapter 6 in some detail.

Many of such materials have very large dielectric constant that is comparable to the CCTO supercapacitor and exhibit tolerance to very high electric fields. We can then safely assume that capacitors made out of such materials would be capable of storing large energy in a small volume. This has been shown to be true by numerous experimental data.

A ferroelectric material is a prominent member of the ferroic group. It is the classical example of a nonlinear dielectric with the unique property of being able to exhibit hysteresis loops between its polarization and electric field applied. This unique property is the basis for many technically important applications which we also have discussed in Chapter 6. Now let us examine how we can make use of the hysteresis loop of a ferroelectric capacitor for storing large amount of energy. Because we will be dealing with a material with large dielectric constant our capacitors can be very small in size that can pack large energy with high density.

There are two subsets of ferroelectrics which are called relaxors and antiferroelectrics. We have dealt with these topics also in Chapter 6. Just like ferroelectrics, relaxors too show hysteresis loops. But antiferroelectrics do not show any hysteresis loops at a low field but can display double hysteresis loops at fields greater than a critical field. These three types of loops are shown in Figures 9.15a,b, and 9.16, respectively, with energy lost and energy stored identified. Let us first discuss ferroelectric and relaxor loops and then we will come back to the topic of antiferroelectric loops.

The main difference between these two loops is in their respective shapes. The loop is wider for ferroelectrics and narrower for relaxors with a much smaller value of the coercivity. This difference is the origin of many of their applications; some of which we have already covered in Chapter 6.

Let us now consider how the electrical energy can be stored and lost in a loop of ferroic materials. The energy lost, E_l in a hysteresis loop is equal to the dielectric loss

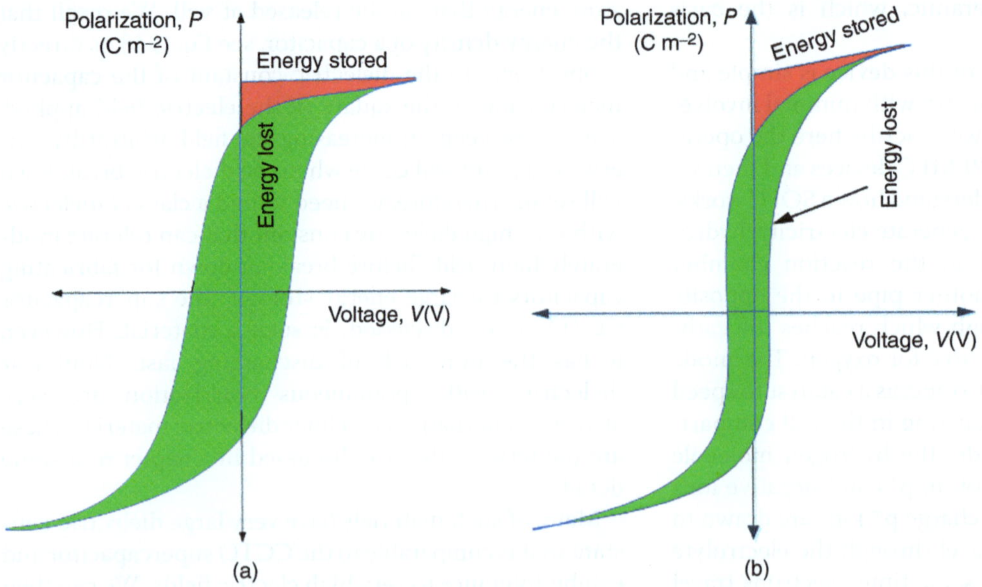

Figure 9.15 (a) Energy stored and lost in the hysteresis loop of a ferroelectric material. (b) Energy stored and lost in hysteresis loop of a relaxor material.

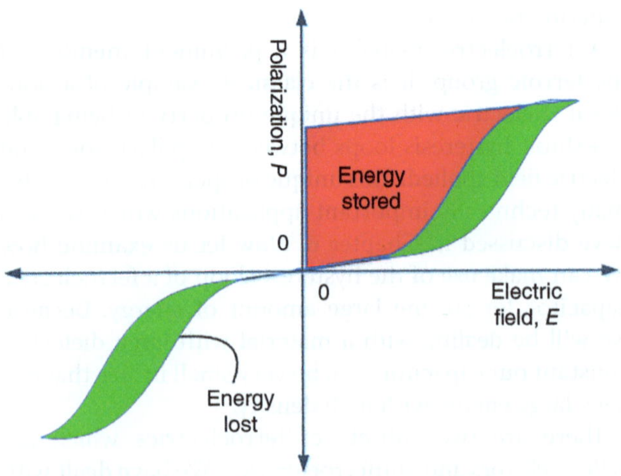

Figure 9.16 Energy stored and lost in the double hysteresis loop of an antiferroelectric material.

which is equivalent to its area as given by Eq. (9.15).

$$E_l \approx \oint E \cdot dP \approx \text{loop area} \tag{9.15}$$

here P is the dielectric polarization.

Similarly the energy stored, E_s, is given by Eq. (9.16).

$$E_s \approx \int_0^{P_m} E\, dP \approx \text{area enclosed} \tag{9.16}$$

here P_m is the maximum polarization.

The energy stored for these two ferroic materials is identified by the red areas of (a) and (b) in Figure 9.15. We find that it lies above the main hysteresis curve once the maximum value of polarization is achieved. A qualitative analysis shows that the energy stored by the two ferroics appears to be small and may not be enough to meet any practical requirements. Any linear capacitor with high dielectric constant can give us a much larger value of energy storage than these nonlinear dielectrics.

The picture, however, changes drastically once we consider an antiferroelectric material. We would recall from Chapter 6 that its signature property is a double hysteresis loop as shown in Figure 9.16. The two loops appear only when an electric field greater that a critical field is applied. The critical field is a material constant and varies from one material to another. Below the critical field there is no hysteresis loop present in an antiferroelectric material and therefore, there is no polarization. Once the double loops are displayed above the critical field on the positive and negative sides of electric field, they resemble the typical loop of a ferroic material.

Above the critical field, the spontaneous polarization sets in. From this point, the material becomes an interesting energy storage device. Because of the shape of the loop, we find that an antiferroelectric material can store a large amount of energy; in fact much larger than the combined storage achieved by a ferroelectric and a relaxor. It is believed that the antiferroelectric materials have the potential of becoming strong contenders in the field of capacitor technology geared toward green energy. Experimental evidence exists to support the fact that nonlinear capacitors fabricated from an antiferroelectric material are attractive viable candidates for energy storage. The important attributes of such an energy storage capacitor would be (i) potential for storing a large amount of energy with high density, (ii) low values of remnant polarization and coercive field, and (iii) low dielectric loss.

The energy stored, E_s, in an antiferroelectric material can be evaluated using the modified relationship of Eq. (9.8),

$$E_s \approx \frac{1}{2}\varepsilon_r E_b^2 \tag{9.17}$$

here E_b is the breakdown field for the material and ε is its permittivity. The efficiency, T_d, of the storage device is given by Eq. (9.18).

$$T_d \approx \frac{\text{Energy stored}}{\text{Total energy}} \approx \frac{E_S}{E_s + E_l} \tag{9.18}$$

This equation tells us that by minimizing the energy loss, we can maximize the efficiency of the device. This, of course, can be done by implementing improved conditions for ceramic processing. Capacitors based on lead-zirconate (PZ) films have shown the potential of storing energy with density of 50 J cm^{-3} which is comparable to the values for electrochemical supercapacitors. Doped lead-zirconate-titanate (PZT) and perovskite $Bi_{1-x}Nd_xFeO_3$ (BNdFeO) are also being actively studied for high-energy density devices. Modeling and simulation show that Bi-based perovskite can have energy density as high as 150 J cm^{-3} with about 85% efficiency for an electric field of about 3 MV cm^{-1}.

Glossary

Anode A terminal in an electrochemical cell with positive electrical polarity such that negative ions generated in an electrolyte are attracted to it. This is equivalent to the positive terminal of a power supply.

Cathode A terminal with negative polarity to which positive ions are attracted. It is the counterpart of an anode in an electrochemical cell. A cathode is equivalent to a negative terminal of a power supply.

Electrolyte A medium in which chemical energy is generated by the dissociation of a chemical into positively charged ions (protons) and negatively charged ions (electrons).

Energy density It is a measure of energy per unit volume. Its unit is J m^{-3}.

Irradiation A measure of electrical power per unit area. Its unit of W m^2 is associated with a radiation. Solar irradiation is normally expressed as power per unit area per wavelength and its unit is W m^{-2}.

PEMFC An acronym for "proton exchange membrane fuel cell." It is a very successful energy storage device and is the favorite of the automobile industry.

Photovoltaic A physical effect, similar to the photoelectric effect, in which a voltage or an electrical current is created when a material surface is exposed to a radiation of suitable photonic energy ($h\nu$).

Power density The amount of electrical power ($V \times I$) per unit volume. Its unit is W m^{-3}. Also W kg^{-1} is sometimes used.

SOFC An acronym for "solid oxide fuel cell."

Solar spectrum The range of photonic energy or wavelengths covering the radiation beginning with ultraviolet (UV) and ending in far infrared (IR).

Problems

9.1 What is meant by the terms *energy density* and *power density*? Why both of them must be considered in determining the practical applications of an energy storage device. Justify your answer by giving some examples.

9.2 Consider an energy storage cell in the shape of a cube with length of 5 mm. Its energy density is 1000 Wh m^{-3}. What the value of the current when the device is subjected to a potential of 10 V? Find the ratio between the energy density and the power density for the device.

9.3 Consider a capacitor in the shape of a 1 mm^3 cube. The area of the electrode on the top and bottom faces is 10 μm^2. Calculate the energy density of the device assuming the thickness to be 5 mm, its dielectric constant to be 5000 and potential applied equal to 10 V.

9.4 What is a piezoelectric energy harvester? Show that its transducer property allows us to extract electrical energy by exposing it to storms. Can it ever produce energy on a large scale but for short duration?

9.5 Consider a PZT energy harvester with the coupling constant of 0.66. It is subjected to a mechanical energy of 5 mJ. Calculate the value of the potential that must be applied to generate a current 1 μA in the device.

9.6 Describe the configuration and operating principles of a ferroelectric photovoltaic device. Define the terms *photo-voltage, photo current, photoconductivity,* dark current.

9.7 Consider a thin-film photovoltaic device that responds to solar radiations. Its thickness is 1 μm, and the electric field that develops is $1\,\mathrm{mV\,m^{-1}}$. Find the value of photoconductivity assuming the photocurrent to be 1 μA.

9.8 What is a fuel cell? Describe the operation of a solid oxide fuel cell and compare it to other energy storage devices.

9.9 How can we configure ferroelectrics, antiferroelectrics, and relaxors for energy storage? How practical are these devices and can they complete with other more established energy storage media?

9.10 Consider an antiferroelectric thin-film capacitor with thickness of 5 μm for storing energy. The dielectric constant is 10^3, the breakdown voltage is $50\,\mathrm{kV\,cm^{-1}}$, and the device efficiency is 50%. Calculate the total energy stored.

References

1 Zepf, V., Simmons, J., Reller, A. et al. (2014). *Materials Critical to the Energy Industry: An Introduction*, 2e. London: BP p.l.c. ISBN: 978-0-9928387-0-6.
2 Ragone, D.V. (1968). Review of Battery Systems for Electrically Powered Vehicles. *SAE Technical Paper*. 10.4271/680453. 680453
3 Pandey, R.K., Stapleton, W.A., Tate, J. et al. (2013). *AIP Adv.* 062126: 3.
4 Kao, K.C. (2004). *Dielectric Phenomena in Solids*. Elsevier Academic Press.

10

Electroceramic Magnetics

CHAPTER MENU

Equipped with his five senses, man explores the universe around him and calls the adventure Science.
Edwin Powell Hubble

10.1 Introduction

Magnetism is an ancient field and has played a vital role in the advancement of human cultures over the many past centuries. Already in 206 BCE the magnetic power of loadstone (magnetite, Fe_3O_4) was recognized in China. Magnetite is a mixed iron ore consisting of ferrous oxide (FeO) and hematite (Fe_2O_3).

The first magnetic compass appeared during the Han Dynasty and this ancient device was subsequently adopted for navigation by the Chinese during the rule of the Song Dynasty during the eleventh century. A legend tells us that the sea pirates in ancient China used big loadstone rocks to pull nails from ships and boats to incapacitate the vessels and then rob them. This compass remained unknown to the western world until about the thirteenth century. Our own GPS system of today uses the concept of magnetic compass.

Until the appearance of atomic physics and quantum mechanics in early twentieth century, the real origin of magnetism was not known to science though it was in wide spread use in motors and generators. It is now well established that the origin of magnetism is embedded in the quantum mechanical properties of an electron especially in its spin quantum number.

Nature produces magnetic elements. Prominent among them in our daily lives are iron, cobalt, and nickel. The impact of iron and its alloys for the development of modern society can hardly be understated. It permeates every aspect of our daily life. The well-known magnetic ferrites are all oxides and therefore ceramics. They can be processed in large volumes with excellent reproducibility using the standard ceramic processing methods. Ferrites are firmly established since over 50 years as indispensable members of electronics and electrical machinery technologies

10.2 Magnetic Parameters

All materials get magnetized when subjected to a magnetic field (*H*). This gives rise to a parameter known as magnetization (*M*). The field-induced magnetization (*M*) is either parallel or antiparallel to the magnetic field (*H*). The ratio between *M* and *H* ($\approx \Delta M/\Delta H$) is called the *magnetic susceptibility* (χ_m). The magnetic susceptibility plays a vital role in classifying magnetic materials.

All materials are essentially magnetic. There is nothing in nature that is not magnetic be it a solid or a gas or chemically organic or inorganic. Magnetic susceptibility χ_m, is their inherent property as also is the magnetic permissibility. Both these parameters are used for classifying

Fundamentals of Electroceramics: Materials, Devices, and Applications, First Edition. R. K. Pandey.

Companion website: www.wiley.com/go/Pandey/Fundamentals_Electroceramics

different types of magnetic materials. All materials can be divided into the following magnetic groups.

Diamagnetic: $\chi < -1$. For these materials, M is antiparallel to H and therefore, the magnetic field cannot penetrate them. Instead, it is repelled as is the case with a superconducting material that is capable of floating a magnet above its surface. Examples of diamagnetic materials are Cu, Au, and Ag, as well as most of the organic materials, and the noble gases He, Ne, and Ar.

Paramagnetic: χ is small and positive in sign. Here M is parallel to H. Examples are Al, Cr, and Ti. The relationship between M and H is linear.

Ferromagnetic: $\chi \gg 1$ and positive in sign. These are spontaneously magnetized materials because of the parallel arrangement of the magnetic spins in their crystal lattices. Examples of these materials are Fe, Co, and Ni; and rare-earths such as Gd, Er, and Eu. Their signature property is the hysteresis loop between M (and B) and H.

Ferrimagnetic: $\chi \gg 1$ but less than of a ferromagnet. It is a subset of ferromagnet. Here the spins have antiparallel arrangement, and they occupy two different sites in the crystal lattice. Its spontaneous magnetization (M_s) is smaller than that of ferromagnets. The classical examples of these materials are magnetite (Fe_3O_4), nickel ferrite ($NiFe_2O_3$), and yttrium iron garnet ($Y_3Fe_5O_{12}$). A large number of magnetic oxides exhibit ferrimagnetism. Collectively, they are called ferrites. They are based on magnetite and are synthetically processed using ceramic processing methods. There are large numbers of applications in which these electroceramics are used exclusively. They also exhibit a hysteresis loop just like a ferromagnetic material but with a smaller area enclosed in its loop than is the case for a ferromagnetic material.

Antiferromagnetism: χ = small and positive. In the absence of an external magnetic field, antiferromagnetic materials exhibit zero magnetization at $T = 0\,K$ because its internal structure consists of perfectly aligned anti-parallel domains. The Néel point is its inherent property below which antiferromagnetism prevails and above which it disappears. Below this point small net magnetization appears because of the spin canting effect. Its magnetic susceptibility shows a maximum at the Néel temperature. Antiferromagnetism occurs among oxides and transitional metal compounds. Some examples are hematite (Fe_2O_3), nickel oxide (NiO), manganese-iron alloy (MnFe), and organic molecules.

Superparamagnetism: This type of magnetism is observed only in nanoparticles of a ferromagnetic or ferrimagnetic material when they consist of single domain magnetic particles. It is more related to a ferromagnetic material than a paramagnetic material. When subjected to an external magnetic field, it assumes a net magnetization with an M–H curve similar to that of a ferromagnetic material or a ferrimagnetic material, but without any memory effect. The M (or B)–H curve is represented by a single line that saturates at a sufficiently high field as shown in Figure 10.1.

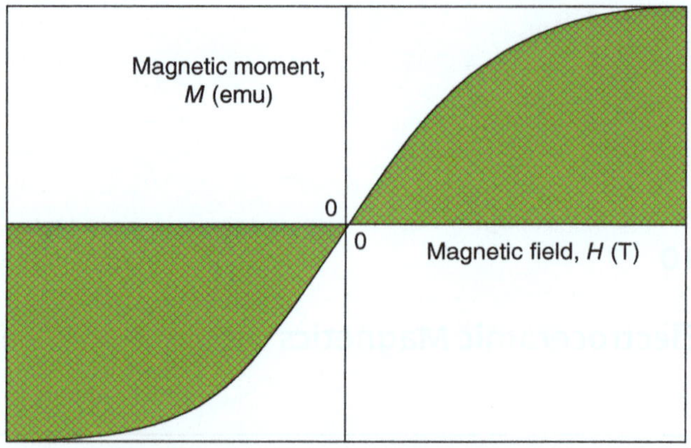

Figure 10.1 Magnetization vs. magnetic field plot of superparamagnetic material.

In Table 10.1, we provide a list of selected elements and oxides with values of their respective magnetic susceptibility.

Notice how oxygen alters the fundamental magnetic nature of some metals.

10.3 Relationship Between Magnetic Flux, Susceptibility, and Permeability

Besides susceptibility (χ_m), permeability (μ) is the other intrinsic magnetic property of a material. They are interdependent parameters. Equation (10.1) shows how magnetization, M, of a ferromagnetic material is related to the magnetic flux, B, in the presence of an external magnetic field, H.

$$B = \mu_0(H + M) \approx M \tag{10.1}$$

Because the product $\mu_0 H \ll M$ for a ferromagnetic material, $B \approx M$. The reason for this will be apparent in the section in which we will discuss the theoretical basis of this phenomenon. Using the concepts of B and M, we can arrive at a very useful formula (Eq. (10.2)).

$$B = \mu_0(1 + \chi_m)H \approx \mu_0\mu_r H \approx \mu H \tag{10.2}$$

where $\mu_r = {}^{\mu}/_{\mu_0}$. μ is the actual permeability of a material and μ_0 that of the free space. μ_r is defined as the relative permeability for a magnetic material, and it is a unitless

Table 10.1 Magnetic susceptibility of some metals and oxides.[a)]

Name	Formula	Susceptibility, χ_m (cm^3 mol^{-1})	Comment
Aluminum	Al	+16.5	Paramagnetic
Aluminum oxide	Al_2O_3	−37	Diamagnetic
Barium	Ba	+20.6	Paramagnetic
Barium oxide	BaO	−29.1	Diamagnetic
Cobalt	Co	+very large	Ferromagnetic
Cobalt(III) oxide	Co_3O_4	+7380	Paramagnetic
Copper	Cu	−5.46	Diamagnetic
Copper(I) oxide	CuO	−20	Diamagnetic
Germanium	Ge	−11.6	Diamagnetic
Germanium(II) oxide	GeO	−28.8	Diamagnetic
Iron	Fe	+very large	Ferromagnetic
Iron(II) oxide	FeO	+7200	Paramagnetic
Nickel	Ni	+very large	Ferromagnetic
Nickel(II) oxide	NiO	+660	Paramagnetic
Oxygen	O_2	+3449	Paramagnetic

Notice how oxygen alters the fundamental magnetic nature of some metals. a) Extracted from the table published in https://www.google.com/url?sa=t&rct=j&q=&esrc=s&source=web&cd=3&cad=rja&uact=8&ved=0ahUKEwjvuPTg9_XVAhVD7WMKHXppACgQFgg1MAI&url=https%3A%2F%2Fen.wikipedia.org%2Fwiki%2FMagnetic_susceptibility&usg=AFQjCNG2-XnkAXQMxMAP0vjmpxpCrw5i-w.

quantity. The significance of the relationship $B \approx \mu H$ is that the magnetic field acting on a material is actually B and not H as is sometimes erroneously assumed. Both B and H have the same unit of Tesla, T. From Eq. (10.2), another very useful relationship follows that is given by Eq. (10.3).

$$\chi_m \approx (\mu_r - 1) \tag{10.3}$$

The units for many of these parameters can be quite confusing because it is not uncommon that the Gaussian units (CGS system) are still used instead of the SI unit (MKS system), which is the internationally adopted unit system. To remedy this situation, we summarize the units for magnetic parameters in Table 10.2, in both Gaussian and SI systems along with some very fundamental equations. Also the electrical counterparts to magnetic parameters are included for comparison.

The values for relative permeability for a few materials are given in Table 10.3.

We find from this table that the relative permeability for ferrites and ferromagnetic materials (such as Ni) are orders of magnitude greater than for nonferromagnetic metals and air. We will explore a bit later what causes these values to be anomalously large for these materials. Such a study will give us an insight in the mechanism

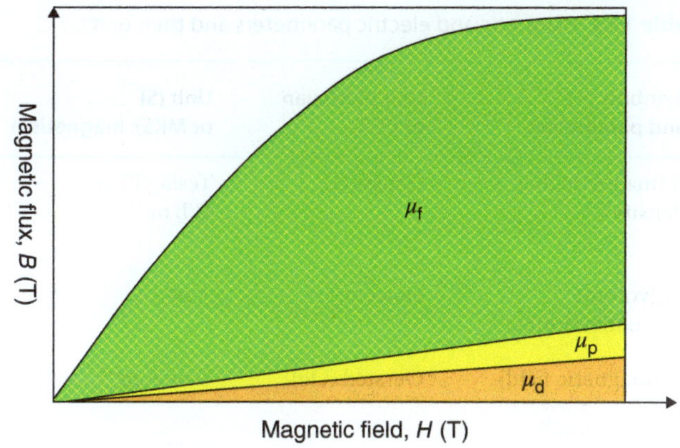

Figure 10.2 Qualitative comparison between the relative permeability of a diamagnetic (μ_d), paramagnetic (μ_p), and ferromagnetic/ferrimagnetic (μ_f) materials.

behind the unique phenomena of ferromagnetism and ferrimagnetism. In Figure 10.2, we illustrate a qualitative relationship between the type of a magnetic material and its permeability.

10.4 Signature Properties of Ferrites

There are four properties that are unique to ferrites. First and foremost is the hysteresis loop between magnetization (of flux) and the applied magnetic field. The second signature property is the Curie point that is defined as that critical temperature at which the hysteresis loop collapses and is reduced to a single line indicating that the material has transformed from the ferrimagnetic state (nonlinear state) to a paramagnetic state (linear state). The third property is the existence of *magnetic domains* at the atomic level. The fourth unique property is the very high value of electrical resistivity making ferrites good examples of magnetic insulators. Being oxides, ferrites are essentially ceramic in nature.

We would recall that we have encountered hysteresis loops previously many times especially in Chapter 5 while discussing ferroic materials (Figure 5.6) and the hysteresis loop of a ferroelectric material (Figure 6.3). Since ferrites are a subset of ferromagnetic materials, they share the first three properties with ferromagnetism. We know that both ferromagnetics and ferrites are prominent members of the ferroic groups with spontaneous magnetization as their order parameter. We also became familiar with the word *domains* in Chapter 6, while learning about ferroelectricity.

The typical hysteresis loop of a ferrite is displayed in Figure 10.3. These loops are defined with the help of three parameters that are their intrinsic attributes. They are the maximum flux (B_{max}), remnant flux (B_{rem}), and coercive force (H_c).

Table 10.2 Magnetic and electric parameters and their units.

Symbols and parameters	Unit (Gaussian or CGS)	Unit (SI or MKS): magnetism	Basic magnetic formula	Electrical counterpart	Units (SI): electricity
B (magnetic flux density)	Gauss (G)	Tesla (T) or $\mathrm{Wb\,m^{-2}}$	$B = H + 4\pi M$ (Gaussian) $B = \mu_0(H + M)$ (SI)	D (displacement)	$\mathrm{C\,m^{-2}}$
M (volume magnetization)	$\mathrm{emu\,cm^{-3}}$	$\mathrm{A\,m^{-1}}$	$M = \chi H$ (Gaussian) $M = \chi H$ (SI)	P (polarization)	$\mathrm{C\,m^{-2}}$
H (magnetic field)	Oersted (Oe)	T or $\mathrm{A\,m^{-1}}$	$B = \mu H$ (Gaussian) $B = \mu H$ (SI)	E (electric filed)	$\mathrm{V\,m^{-1}}$
χ (mass magnetic susceptibility)	$\mathrm{emu\,g^{-1}}$	$\mathrm{H\,m^2\,kg}$	$\mu = (1 + 4\pi\chi)$ (Gaussian) $\frac{\mu}{\mu_0} = \mu_r = (1 + \chi)$ (SI)	χ_e (electrical susceptibility)	Dimensionless
μ (magnetic permeability)	Dimensionless	$\mathrm{H\,m^{-1}}$	—	ϵ (permittivity)	$\mathrm{F\,m^{-1}}$

Table 10.3 Relative permeability at room temperature for few selected materials.[a)]

Materials	Relative permeability, μ_r	Comments
Mn–Zn ferrite	>650	100 kHz–1 MHz
Ni–Zn ferrite	16–650	100 kHz–1 MHz
Nickel, Ni	100–600	$H = 2 \times 10^{-3}$ T
Platinum, Pt	≈1	
Aluminum, Al	≈1	
Air	≈1	

a) Extracted from the table published in https://en.wikipedia.org/wiki/Permeability_(electromagnetism).

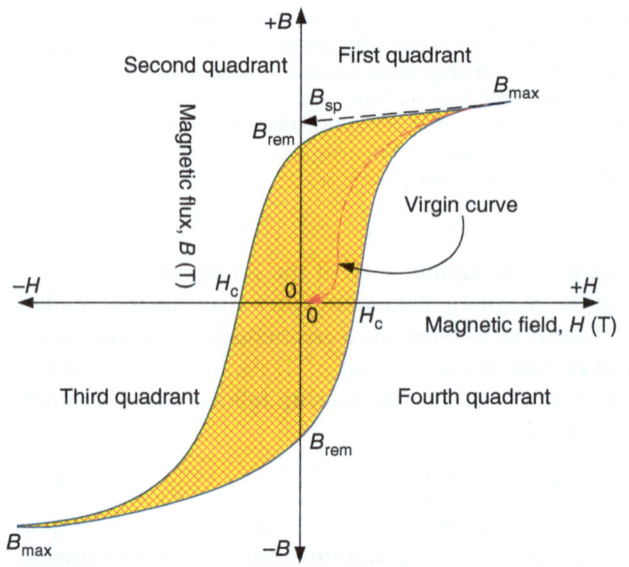

Figure 10.3 Typical hysteresis loop of a ferrimagnetic or a ferromagnetic material.

The parameters B_{rem} and H_c are called the retentively and the coercivity, respectively. Each of these three parameters comes as twins with the same magnitude but opposite in sign. They are of great significance to the magnetic technology and must be taken into consideration while designing a magnetic device.

When a virgin ferrimagnetic (or ferromagnetic) material is subjected to an external magnetic field first of all the initial magnetization curve follows the path as indicated by the dashed line in the two figures. This is called the virgin curve. It reaches the level of maximum flux (B_{max}). It remains there constant in magnitude irrespective of the increase in the value of the magnetic field applied. When the field is reduced slowly the initial plot is not repeated; instead it follows the path indicated by the solid line in first quadrant. When it reaches the (0, 0) point it does not go to zero, rather stays at the remnant point B_{rem}. A further increase of field in the negative direction gives the solid curve in the second quadrant. It intersects there with the field-axis at the coercive point H_c. Notice that at this point, the magnetization disappears completely, and the material is technically in the demagnetized state. The concept of demagnetization is very important for technical applications particularly for designing a magnet. Further manipulation of the field generates the curves resulting in the shape that is identical to the shape of the first two quadrants but of opposite polarity. Seen as a whole, a complete hysteresis loop is generated with two equal in value but opposite in sign for remanence, coercivity, and maximum magnetization. Such a loop is the characteristic of a ferrite and a ferromagnetic material. The same loop will be generated without any change in the magnitude of the three parameters no matter how many times the material is magnetized and demagnetized.

One should be careful in calling the B_{max} point the spontaneous magnetization (M_{sp}). That would be wrong because the spontaneous magnetization is a concept with its origin anchored in the quantum mechanical nature of the atomic magnetization. The unique aspect of it is that it does not need an external magnetic field for its existence. Theoretically, the exact magnitude of the spontaneous magnetization can be obtained only at 0 K with zero magnetic fields present. Therefore, in practice a compromise is made and its value is determined by extrapolating the upper solid curve in the first quadrant to intersect the B-axis at B_{sp}. This point is then defined as the technical spontaneous flux or magnetization.

The bipolar states identified as the two B_{rem} points in a hysteresis loop make ferromagnetic materials and ferrites well suited for magnetic memory and data storage. Historically, one would remember that the first computers were based on ferrite core memories. Even today, ferrites are used for the development of powerful and high-density nonvolatile memory and data storage. One uses for example the $+B_{ret}$ point for "write in" and $-B_{ret}$ point for "read out." Therefore, $+B_{ret}$ and $-B_{ret}$ states are also called the "on" and "off" states of the computer memory.

In generating a hysteresis loop, a work (W) is done that is given by Eq. (10.4). We would recall using a similar equation in Chapter 9, while considering energy stored in a hysteresis loops of nonlinear dielectrics (see Eq. (9.15)).

$$W = \oint B\,\mathrm{d}H \approx \text{loop area} \tag{10.4}$$

The area of the loop is also equivalent to the energy lost in generating a full complete cycle of the hysteresis loop.

Based on the shape of the hysteresis curves, ferrites are classified either as a soft ferrite or a hard ferrite. This classification is commonly used for technical applications. Soft ferrites have low coercivity, which means that they can be demagnetized without dissipating much energy. Typically, they have small hysteresis loss, low coercivity, and high resistivity that prevent the onset of eddy currents to develop. Typical examples of these ferrites are Mn–Zn-ferrites, Ni–Zn-ferrites. Hard ferrites have high coercivity, which makes it difficult to demagnetize them. Their hysteresis loss is also high as well as they exhibit high magnetic permeability. The examples of hard ferrites are Sr-ferrite ($SrFe_{12}O_{19}$), Ba-ferrite ($BaFe_{12}O_{19}$), and Co-ferrite ($CoFe_2O_4$). They are important engineering materials suitable for producing a wide variety of magnetic devices such as a permanent magnet and a data storage device.

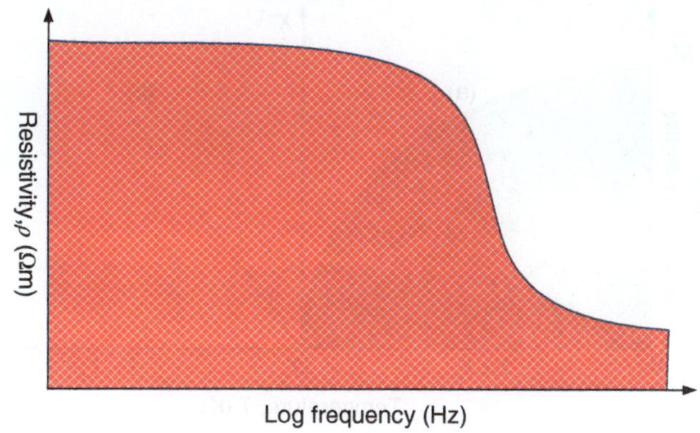

Figure 10.4 Frequency dependence of the electrical resistivity of a ferrite.

For ferromagnetic materials that are metallic, the heat loss ($=I^2R$) can be substantial. To minimize the heat generated in the process of cycling through the loop millions of times such as in a transformer, oil is used in the chamber to cool the magnetic cores. Because ferrites are ceramic materials with high resistance, the heat loss associated with a loop is minimal and does not pose a problem for efficient operation of ferrite devices.

The resistivity of ferrites at room temperature varies between $10^{-1} < \rho < 10^6$ Ωm which is orders of magnitude higher than of ferromagnetic materials. Therefore, they are well suited for many applications, some of which we will discuss in another section. The resistivity of ferrites is highly frequency dependent as shown in Figure 10.4.

The relative permeability and loss tangent also show strong dependence on frequency as shown in Figure 10.5. These three frequency-dependent parameters are

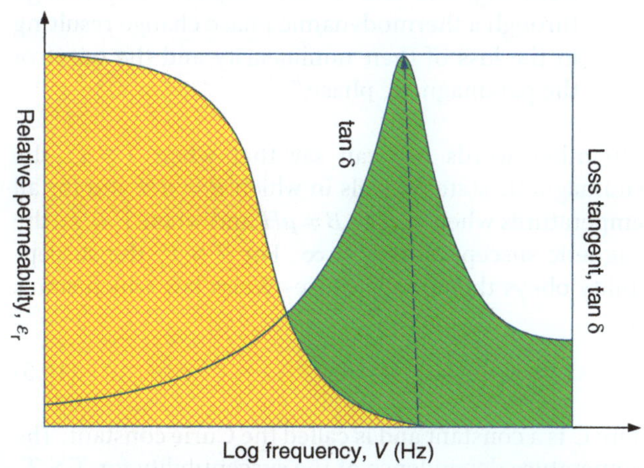

Figure 10.5 Frequency dependence of permeability and loss tangent of a ferrite material.

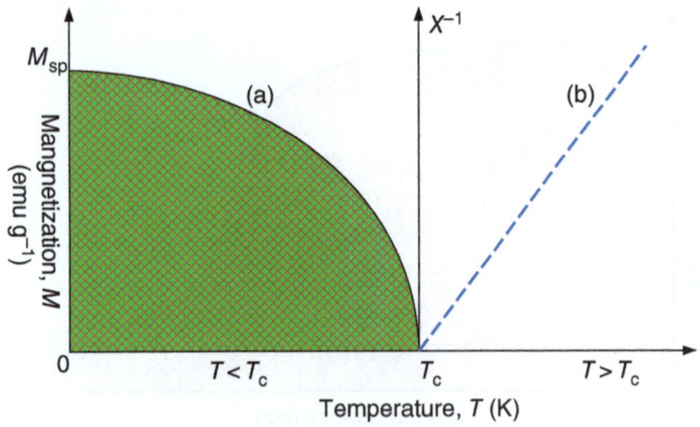

Figure 10.6 Temperature dependence of (a) magnetization and (b) magnetic susceptibility of a ferrimagnetic (also ferromagnetic) material.

exploited for many applications some of which we will discuss later in this chapter.

10.4.1 Temperature Dependence of Magnetic Parameters

Like many other materials properties, also magnetic properties show strong dependence on temperature. In Figure 10.6, the thermal behavior of magnetization and susceptibility for ferrites are plotted. As the temperature increases from its initial values at 0 K the spontaneous magnetization decreases following the solid curve (a) and disappears completely at the Curie point, T_c. As already discussed, the Curie temperature is one of the defining properties of ferromagnetic and ferrimagnetic materials. It can be defined in many different ways some of which we have already discussed. Its most precise definition can be offered by the following statement.

> "The Curie point is that temperature at which a ferromagnetic and a ferrimagnetic material go through a thermodynamic phase change resulting in the loss of their nonlinearity and the onset of the paramagnetic phase."

In other words, we can say that when $T > T_c$ the paramagnetic state prevails in which $B = \mu H$ and for all temperatures when $T < T_c$ $B \neq \mu H$ and when $T = T_c$ the magnetic susceptibility $\chi \approx \infty$. For $T > T_c$ the susceptibility obeys the famous Curie–Weiss law which states that:

$$\chi \approx \frac{C}{T - T_c} \quad (10.5)$$

Here C is a constant and is called the Curie constant. The temperature dependence of the susceptibility for $T > T_c$ is shown as the dashed line (b) in Figure 10.6. An expression similar to Eq. (10.5) describes the temperature dependence of a ferroelectric material for $T > T_c$ (Eq. (6.1)). In Table 10.4, we tabulate the values of the Curie temperature for some prominent ferrites. It also includes the Curie temperatures for leading ferromagnetic materials for compassion.

Table 10.4 Values of Curie temperature of selected magnetic materials.

Material	Curie temperature (K)	Comment
Fe (iron)	1043	Ferromagnetic and metallic
Co (cobalt)	1400	Ferromagnetic and metallic
Ni (nickel)	627	Ferromagnetic and metallic
Fe_2O_3 (hematite)	948	Ferrite and insulating
Fe_3O_4 (magnetite)	858	Ferrite and insulating
$MnO–Fe_2O_3$	573	Mn-ferrite and insulating
$NiO–Fe_2O_3$	858	Ni-ferrite and insulating
$Y_3Fe_5O_{12}$ (yttrium iron garnet)	560	Ferrite and insulating

Source: https://en.m.wikipedia.org/wiki/Curie_temperature (selected data extracted from a table).

This is worth remembering that the law described by Eq. (10.5) is named after the same Pierre Curie who discovered piezoelectricity and shared Nobel Prize in Physics with his physicist wife, Madame Marie Curie in 1903 for explaining radioactivity. Pierre-Ernest Weiss was also a French physicist who developed the domain theory of ferromagnetism in 1907. The theory is also now known as the Weiss-mean-field theory or the molecular field theory.

10.5 Typical Structures Associated with Ferrites

Ferrites crystallize in four structures that are spinel, garnet, ortho ferrites, and hexagonal magneto-plumbite.

Spinels: They are represented by the chemical formula of AB_2O_4 where A is a divalent metal, whereas B is a trivalent metal. Examples of A^{+2} metals include Fe, Co, Cu, Cu, Zn, Cd, etc.; and that of B^{+3} metals are Al, Ga, Cr, etc. It is a complex structure with the basic unit cell is fcc (face-centered-cube). One spinel unit cell consists of 8 fcc cells made of O_2-ions in the configuration of $2\times 2\times 2$. It is a big structure consisting of 32 O_2 atoms, 8 A atoms, and 16 B atoms. A spinel structure

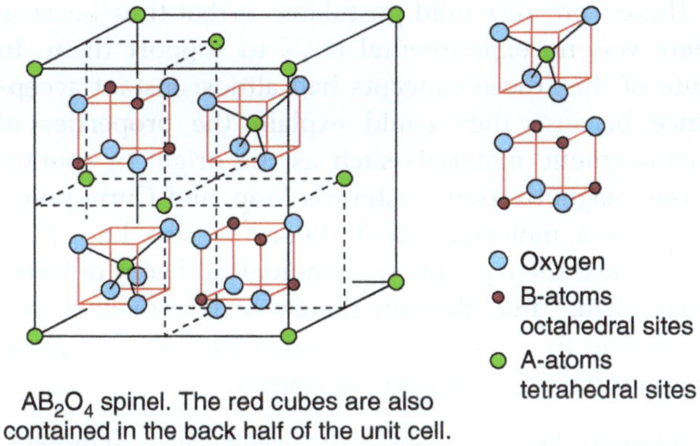

Figure 10.7 Unit cell of a normal spinel crystal. Source: http://nptel.ac.in/courses/113104005/lecture3/figure/figure10.jpg.

can be *normal* or *inverse* depending on whether the positive ions (cations) occupy tetrahedral or octahedral sites. The unit cell of a normal spinel crystal is shown is Figure 10.7. The A^{+2} ions occupy the tetrahedral sites and B^{+3} ions the octahedral sites. Many aluminates crystallize in this structure. Examples of normal spinel ferrites are Zn-ferrite ($ZnFe_2O_4$) and Cd-ferrite ($CdFe_2O_4$). The inverse spinels are represented by the formula of $B(AB)O_4$. Notice the subtle distinction between the way chemical formula for normal and inverse spinel are written. In this structure, all A^{+2} ions occupy the octahedral site; ½ of the B^{+3} ions settle on tetrahedral sites and the other half on the octahedral sites. This is the prime structure for ferrites as most of them crystallize as inverse spinels. The classic example is the magnetite with the formula of Fe_3O_4 that can also be written as the mixture of $Feo{\cdot}Fe_2O_3$. Ni-ferrite ($NIFe_2O_4$) and Co-ferrite ($CoFe_2O_4$) also have the inverse spinel structure.

Garnets: Their general formula is $R_3Fe_3O_{12}$. R stands for rare earth elements that are yttrium (Y), neodymium (Nd), samarium (Sm), gadolinium (Gd), holmium (Ho), erbium (Er), etc. The garnets are excellent magnetic insulators and are therefore used for high-frequency applications in spite of the fact they are more expensive than ceramics.

Hexagonal magneto-plumbite: For these ferrites, the chemical formula is given by $MFe_{12}O_{19}$, where M is a divalent metal of the alkali earth group (Group II A of the periodic table). Ba and Sr are commonly used for producing hexagonal magneto-plumbite ferrites. These are excellent materials for permanent magnets because high magnetic fields (≈1.7 T) are required to rotate their spontaneous magnetization parallel to the basal plane. Their hysteresis loops are broad encompassing a large area.

Ortho ferrites: It is given by the formula of $RFeO_3$. Notice the similarity between this and perovskite formula of ABO_3. Once they were the lead materials for "bubble memories" but were later replaced by garnets.

10.6 Essential Theoretical Concepts

In order to gain some insight into the properties exhibited by ferromagnetic, anti-ferromagnetic, and ferrites, we need to understand the origin of these magnetic phenomena. There are three models that can explain ferromagnetism. These are Weiss model, Heisenberg model, and Ising model. For a full treatment of any of these theories, one needs a solid background in physics and advanced mathematics. Interested readers can consult any of the excellent books written on magnetism. Our objective would be met by mastering the principal concepts behind these theories.

Ferrites and ferromagnetism are analogous phenomena. Ferrites behave like ferromagnetic materials. They differ vastly in their origin. Ferrites are magnetic insulators, whereas ferromagnetism is found in metallic systems. Antiferromagnetism is a different story; however, antiferromagnets and ferrites share similar ordering structures. We would recall that we have classified them as members of the ferroic groups in Chapter 6. As such, they are important in the emerging magnetoelectric technology.

For understanding different types of magnetic materials, let us first discuss the electronic magnetization that would make it easier for discussing theory of magnetism.

10.7 Magnetic Nature of Electron

Even though magnetism has been known to science since ancient times, its real origin remained unknown until the quantum mechanics appeared on the scene. Until then classical physics was not in a position to offer an acceptable explanation of this physical phenomenon. Only after the true nature of electron was understood, four quantum numbers identified and Pauli's selection rules became known physics began to pinpoint the origin of magnetism. Today, it is universally accepted that magnetism originates because of the spinning electron. We recall that the spin quantum number, $s = \pm 1/2$. Since the electron is a charged particle with the elementary charges of e^-, its angular momentum is the combination of its spin rotation and orbital motion. We know from classical electrodynamics that a rotating electrical charge creates magnetic dipoles with equal but opposite polarity. Applying this analogy, it is easier to comprehend how an electron would behave like an elementary bar magnet.

The magnetic diploe moment, μ_e, is given by Eq. (10.6).

$$\mu_e \approx -\frac{e}{2m_e} gS \tag{10.6}$$

Here g is the gyromagnetic ratio or g-ratio that has the value of 2 and S the quantized angular momentum of the electron. Both S and μ_e are vectors. Since electron is a $s - 1/2$ particle, it turns out that $S = \hbar/2$.

It is customary in quantum mechanics to express magnetic moments in terms of reduced Plank constant $\hbar$ and Bohr magneton, μ_B, which is the most fundamental magnetic moment because it is the magnetic moment of the hydrogen atom (which is a one electron system) in its ground state when the orbital angular momentum, $l = 0$ and magnetic quantum number, m_l is also 0. It is defined by Eq. (10.7).

$$\mu_B \approx -\frac{e\hbar}{2m_e} \tag{10.7}$$

μ_B is called Bohr magneton after Niels Bohr who gave us the theory of H-atom.

Combining Eqs. (10.6) and (10.7) and substituting for S and μ_B, we arrive at a very useful result of $\mu_e = \mu_B$.

$$\frac{\mu_e}{\mu_B} \approx \frac{1}{\hbar} gS \approx \frac{2}{\hbar} x \left(\frac{\hbar}{2}\right) \approx 1 \tag{10.8}$$

Exercise 10.1
Compute the value of Bohr magneton.

Solution: According to Eq. (10.7), $\mu_B \approx \frac{e\hbar}{2m_e}$. The electronic charge, $e = 1.60 \times 10^{-19}$ C and electron mass, $m_e = 9.11 \times 10^{-31}$ kg. Substituting these numbers in the equation, we get $\mu_B = 9.27 \times 10^{-24}\ \mathrm{J\,T^{-1}} = 5.78 \times 10^{-5}\mathrm{eV\,T^{-1}}$.

Exercise 10.2
Calculate the saturation magnetization, M_s of nickel considering that its atomic weight (A) is 58.71, density, d, is 8.90 g cm^{-3}, atomic magnetic moment = 0.60 μ_B and Avogadro's number, $N_A = 6.022 \times 10^{23}$ mol^{-1}.

Solution: Assume that there are n number of atoms per unit volume. Then $n = \left(\frac{dN_A}{A}\right) \approx 9.12 \times 10^{28}$. Therefore, the saturation moment, $M_s = n \times 0.60 \times \mu_B \approx 5.1 \times 10^5\ \mathrm{A\,m^{-1}}$.

10.7.1 Molecular Field Theory

In 1907, Pierre Weiss proposed the concepts of *magnetic domains* and *internal magnetic field* to explain the phenomenon of ferromagnetism. He postulated that:

1) The domains are formed because the atoms interact with each other with the mechanism of *exchange interaction* that is present in a ferromagnetic material;
2) The spontaneous magnetization found in ferromagnetic materials has its origin in the in-built internal magnetic field of the material which is large and very strong.

These were very bold postulates at that time because there was no experimental basis to support them. In spite of this, these concepts had almost instant acceptance because they could explain the properties of ferromagnetic materials such as the origin of spontaneous magnetization, hysteresis loop, and Curie point. The Weiss molecular field theory is also known as "mean field theory," "phenomenological theory of ferromagnetism," and "domain theory of ferromagnetism." According to the molecular field theory, the magnetic domains have the following attributes.

1) Domain structure consists of large number of individual domains that are isolated from each other because of the *domain walls* that exist between the two neighboring domains.
2) The domains tend to align and grow in size in the direction of the applied magnetic field. Conversely, they shrink in size opposite to the magnetic field. Alternatively, we can say, that domain boundaries move to expand the favorable domains.
3) They rotate to orient themselves in the direction of the magnetic field.
4) When the field is weak, the magnetization occurs mostly by the process of domain growth.

The validation of the domain concept is provided by experiments. An example of experimentally observed domain structure for a (111) oriented YIG single crystal is depicted in Figure 10.8.

Notice the complex structure of domains when $H = 0$ (Figure 10.8a) and how both their shapes and orientations change when $H > 0$ Oe (Figure 10.8b). In the presence of the field, all domains are parallel to each other and separated by the domain walls (dark areas).

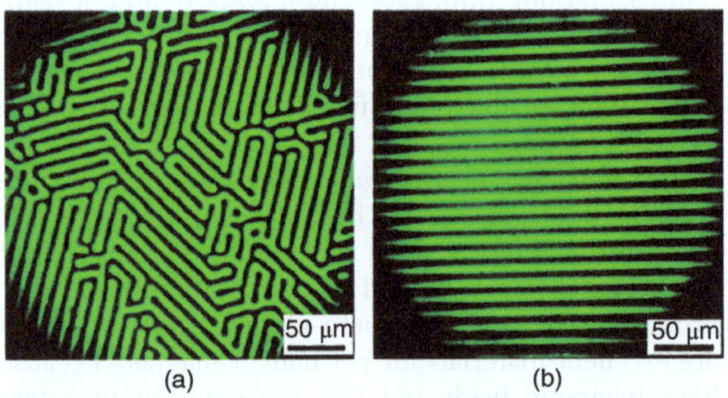

Figure 10.8 Magnetic domain structures of (111) oriented YIG film observed by magneto-optic Faraday effects. (a) $H_0 = 0$ Oe. (b) $H_n = 4$ Oe. Source: https://www.researchgate.net/profile/Maksym_Popov/publication/238616732/figure/fig2/AS:298739236392973@1448236392819/FIG-2-Domain-structure-observed-by-magneto-optic-Faraday-effects-in-the-111-YIG-film.png.

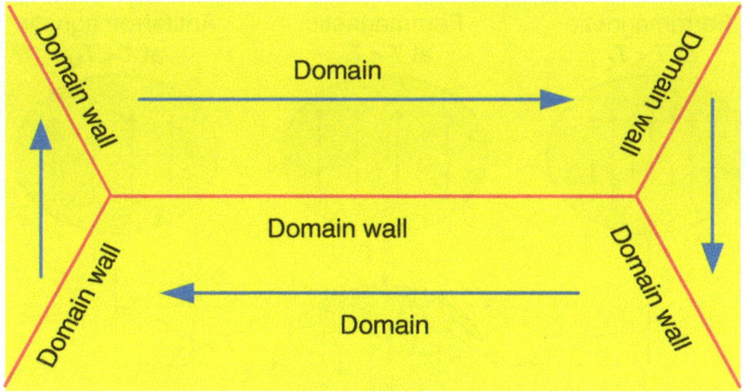

Figure 10.9 Idealized domain configuration of a ferrimagnetic material in its demagnetized state at room temperature.

At room temperature, a magnetic material is in demagnetized state because of the thermal effect that destroys the magnetic configurations at 0 K. As a consequence, the neighboring domains assume an antiparallel configuration as can be seen from Figure 10.9. Antiparallel configuration of domains cancel the magnetic moments resulting in zero net magnetization.

They are isolated from each other by the domain walls (red). In the demagnetized state, the net $B = 0$ at room temperature. The material must be magnetized using an external magnetic field before it can display a hysteresis loop.

According to the Weiss theory, the net magnetic field (B_n) acting on a ferromagnetic material is the sum of the external field applied (B_a) and the Weiss molecular field (B_M). We can write it as Eq. (10.9),

$$B_n \approx B_a + B_M \tag{10.9}$$

Now we need to examine the nature of the molecular field and how can it contribute to the net magnetization. We know from experiments that the spontaneous magnetization goes to zero at the Curie point (T_c) and paramagnetism prevails for $T > T_c$. This can happen only when the thermal energy is sufficient to destroy the magnetic ordering. Let us assume that the atomic magnetic moment is 1 μ_B, then according to Eq. (10.10) we get.

$$\mu_B B_M \approx k_B T_c \tag{10.10}$$

Substituting for $T_c \approx 10^3$ K, $\mu_B = 9.27 \times 10^{-24}\,\mathrm{J\,T^{-1}}$ and $k_B = 1.38 \times 10^{-23}\,\mathrm{J\,K^{-1}}$, we get for the molecular field, $B_M \approx 10^3$ T. This is a very large magnetic field which is at least 1 order of magnitude greater than the highest field of 10^2 T produced in a laboratory.[1] This field is also called the *molecular exchange field* because it is shared by all magnetic ions in the crystal lattice. How does this magnetic field become so large? To explain this, Weiss assumed further that the molecular field, B_M, acting on each magnetic atom is proportional to the magnetization, M. Then we can write it as Eq. (10.11).

$$B_M \approx \lambda M \tag{10.11}$$

Here λ is temperature-independent constant called *mean field constant*.

Now the questions to ask ourselves are (i) What is the magnitude of λ?; (ii) Can we completely ignore the dipole–dipole interaction from any further consideration. After all, tiny atomic moments are equivalent to an infinitely small dipole, and it is conceivable to imagine that they might be contributing to ferromagnetism?

To find the answers, let us consider the case of iron, which is the strongest ferromagnetic material. Its crystal structure is body-centered cubic (bcc) with the lattice constant of 0.287 nm. Each magnetic Fe-atom will produce a filed $M \approx \frac{\mu_B}{(0.287)^3} \approx 10^{-1}$T. This is equivalent to 10^3 G, which is a very large field to be produced by a single Fe-atom. This clearly demonstrates that the ferromagnetic phenomenon is certainly not because of the dipole–dipole interaction. Instead, its origin is anchored in the molecular exchange field, B_M.

Substituting for B_M equal to 10^{-1} T, we get from Eq. (10.11), $\lambda \approx 10^4$. In summary, we can say that the Weiss molecular field theory provides us with a solid theory for ferromagnetism. We will subsequently deal with the phenomena of antiferromagnetism and ferrimagnetism in the next section while making use of some of the concepts of the Weiss molecular filed theory.

10.7.2 Antiferromagnetism and Ferrimagnetism

Antiferromagnetism is an interesting phenomenon exhibited by transition metal alloys and oxides. The classic examples are manganese oxide (MnO), iron (II) oxide (FeO), nickel oxide (NiO), and iron manganese (Fe–Mn). They are unique materials in the sense that though they are counted among the magnetic materials with high degree of magnetic ordering yet they display zero magnetization in the absence of an applied magnetic field. This uniqueness was explained by Louis Néel of France in 1948 in his celebrated theory for which he earned the Nobel Prize in Physics in 1970. Like Weiss before him, Néel also introduced the concept of molecular field to explain antiferromagnetism. He proposed that at $T = 0$ K, the magnetic moments in these materials align themselves in an antiparallel configuration resulting in zero magnetic moment. He furthermore assumed that there are two species of magnetic spins, one being type A and the other type B. The effective magnetic fields for each of these two species can be expressed with the help of Eqs. (10.12) and (10.3).

$$B_{\mathrm{eff}}^{A} \approx B_a - \mu_0 \lambda M_B \tag{10.12}$$

1 National High Magnetic Field Laboratory at the DOE's (US Department of Energy) Los Alamos National Labs in New Mexico, USA.

$$B_{eff}^{B} \approx B_a - \mu_0 \lambda M_A \tag{10.13}$$

In these equations, B_a is the applied magnetic field, M_A and M_B their magnetic moments of A and B species, respectively, and B_{eff}^{A} and B_{eff}^{B} their effective fields. According to the Néel theory, the temperature dependence of the susceptibility for antiferromagnetic materials is given by Eq. (10.14)

$$\chi \approx \frac{C}{T + T_N} \tag{10.14}$$

where C is a constant and T_N is the Néel temperature which is analogous to the Curie temperature of ferromagnetic materials. The Néel temperature usually occurs at low temperatures. At $T = -T_N$, the magnetic susceptibility reaches its maximum value and for $T > T_N$ it goes through a phase transition from an antiferromagnetic state to a paramagnetic state.

The Néel theory of antiferromagnetism is also applicable for ferrimagnetic materials. Here too the spontaneous magnetism has its origin in the two types of magnetic ions, A and B. They reside in two sublattices of a crystal. However, unlike the magnetic ions of antiferromagnetic materials, these are of two different magnitudes resulting in unequal magnetic moments. At $T = 0$ K, the atomic moments align themselves parallel to each other and because of the unequal atomic moments they manifest themselves in the onset of the spontaneous magnetization. For ferrites also the spontaneous magnetization disappears at the Curie point and above which paramagnetism prevails. A B–H hysteresis loop as we know is also the signature attribute of a ferrimagnetic material. The ordering of the atomic magnetic moments for ferromagnets, ferrimagnets, and antiferromagnets are shown in Figure 10.10. We notice here that each of these three materials go into the paramagnetic phase above their critical temperatures. In that phase, the magnetic ordering disappears and materials become an ordinary linear magnetic material.

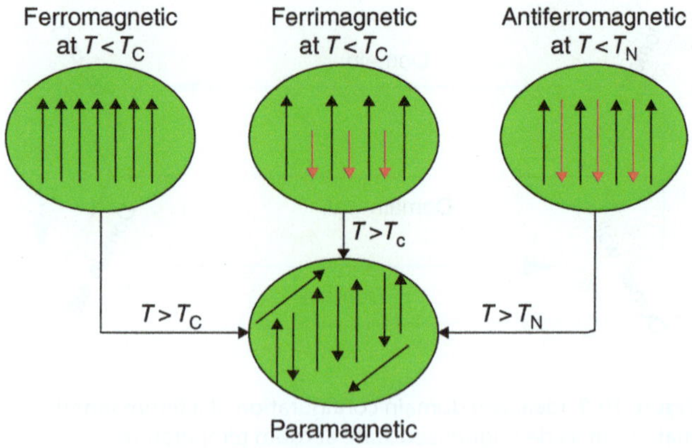

Figure 10.10 Ordered configurations of ferromagnetic, ferromagnetic, and antiferromagnetic materials below transition temperatures, and their disordered orientation in paramagnetic state above transition temperatures.

10.7.3 Quantum Mechanics and Magnetism

Earlier we discussed the nature of the magnetic properties of an electron and found that its magnetic moment is equal to the Bohr magneton. Magnetism at its atomic level is based on the three fundamental principles of quantum mechanics which are Pauli exclusion principle, Aufbau principle, and Hund's rule.

We know from our discussions in Chapter 1 that according to the Pauli exclusion principle no two electrons can occupy the same orbital; n, l, m_l can be the same but the spins must be antiparallel (↑↓, or $\pm\frac{1}{2}$). The Aufbau (it is a German word meaning *to build up*). According to this principle, the filling of states first begins at the bottom and once the available states are filled only then the occupation of the next higher state can begin. However, according to the Hund's rule if multiple orbitals have the same energy, then first of all one electron goes into each of them before they start to double up.

These three rules are at the heart of the quantum mechanical nature of magnetic moment, but the Hund's rule is the most consequential of them all for understanding magnetism. Hund's rule was formulated in 1927 by Friedrich Hund of Germany. For a better understanding of his rule, let us split it into three parts:

Part 1: For a given electron configuration, the term with the maximum multiplicity has the lowest energy. The multiplicity is defined as equal to $2S + 1$, where S is the total spin angular momentum for all electrons. Therefore, the term with maximum S corresponds to the lowest energy state.

Part 2: For a given multiplicity, the term with the largest value of the total orbital angular momentum, L, corresponds to the lowest energy state.

Part 3: If the outermost shell is more than half filled, the lowest energy corresponds to the level with the higher value of J ($J = L + S$).

Simply stated we can sum up the three parts as follows: (i) atoms with filled shells have zero magnetic moment because the various electronic contributions cancel each other; (ii) the spins arrange themselves to produce the maximum possible value consistent with the Pauli exclusion principle, and (iii) the spins favor the half-filled states.

The Hund's rule is applied to configure the magnetic moments of elements in the entire periodic table; certainly not a trivial success of a theory. According to Hund's rule, He with two electrons in its s-shell has no magnetic moment because the two s-states are oriented antiparallel to each other. One by one it is possible to find out the magnetic moments of all elements by applying the three parts of Hund's rule. It becomes really interesting when we come to the 3d-states. We illustrate this point in Figure 10.11, where for convenience, we have divided the 3d-states in two parts as 3d(+) and 3d(−) states. In Fe only, the 6 states out of total of 10 states available in the 3-d shell are occupied. That leaves us with four states unoccupied. According to the Hund's rule first the five sites in 3d(+)-states will be occupied and then the last left over electron will go to occupy the first site of the 3d(−)-states having the antiparallel orientation with respect to other electrons. Consequently, the atomic magnetic moment of Fe amounts to $4\mu_{B.}$ Obviously, the four empty sites in the 3d(−)-states will contribute nothing to the total magnetic moment. Let us now apply the Hund's rule to Co and Ni. We find that the atomic magnetic moment for Co turns out to be $3\mu_B$ and for Ni it is $2\mu_B$ (as shown in Figure 10.11). Fe, Co, and Ni are industrially the most important ferromagnetic metals and host of applications are based on them.

The values of the atomic magnetic movements for Fe, Co, and Ni based on Hund's rule are found to be twice as large as the experimentally determined values which are 2.22 μ_B for Fe, 1.7 μ_B for Co, and 0.61 μ_B for Ni. The discrepancy could be attributed to the fact that contributions made by the spin–orbit coupling and electronic conduction have been ignored and not taken into account in the formulation of the Hund's rule. But this shortcoming does not diminish in the least the importance of Hund's rule and its contributions to magnetism.

10.8 Classical Applications of Ferrites

Figure 10.12 shows the typical applications of ferrite materials. Large and small transformers, permanent magnets, and inductors can be lumped into one group, whereas the electronic devices such as microwave devices, devices for information storage and optical signal processing, and antennas form the other group.

The operating frequency of a ferrites device can be used classifying applications of ferrites. For example, signal transformers operate at high frequencies and are of small sizes, whereas a power transformer is big in size and operates at low frequencies. The ferrite cores that are designed for operation in the frequency range of 1–200 kHz are large in size compared to the cores for operation in the range of 1 kHz to many MHz.

In designing a ferrite core, one needs to account also for the *eddy currents* loss. When a ferrite core is subjected to a time-dependent magnetic field, eddy currents are generated according to the Lenz law. These currents appear in loops because of the Faraday's law of induction. This leads to the heating of the cores (I^2R-loss), which can contribute eventually to the failure of the device. Therefore, it is important that ferrite core materials must be selected with care in which the power dissipation due to eddy currents is small. The power dissipation (P) due to the eddy currents for a wire or a thin sheet can be estimated using Eq. (10.15). This empirical equation is valid only when there is no skin-effect taking place (that is, when the frequency of the magnetic field does not penetrate the material) and when electromagnetic waves can fully penetrate the material.

$$P = \frac{\pi^2 B_p^2 d^2 f^2}{6k\rho D} \tag{10.15}$$

Here P is the power lost per unit mass (W kg^{-1}), B_p is the peak magnetic field (T), d is the thickness of the sheet or diameter of the wire (m), f the frequency (Hz), k is an arbitrary constant set as 1 for a thin sheet and 2 for a thin wire, ρ the resistivity (Ω m) and D is the density of the material (kg m^{-3}).

Both soft and hard ferrites are used for applications. They are listed in Table 10.5.

Table 10.5 Ferrites and their applications.

Material	Type of ferrite	Applications
Mn–Zn-ferrites	Soft	Used as cores for RF transformers and inductors
Ni–Zn-ferrite	Soft	Same as above
Sr-ferrite	Hard	Small electric motors, microwave devices, recording media, magneto-optic media, telecommunication and electronic components
Ba-ferrite	Hard	Ceramic magnets for loud speakers, and magnetic recording on magnetic stripe cards
Co-ferrite	Hard	Magnetic recording

10.9 Novel Magnetic Technologies

In the last 20 years, three new discoveries in the field of magnetism have appeared which can be exploited for a host of novel magnetic applications. They are the *giant magnetoresistance effect* (GMR) effect; the *colossal magnetoresistance effect* (CMR); and *spintronics*. The GMR and CMR effects have their origins in the magneto-resistive property of materials. All materials

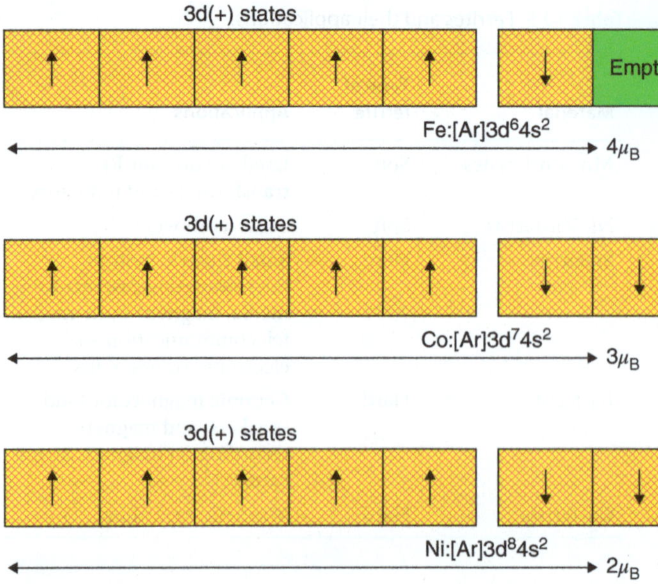

Figure 10.11 Electronic configurations of 3-d states for Fe, Co, and Ni.

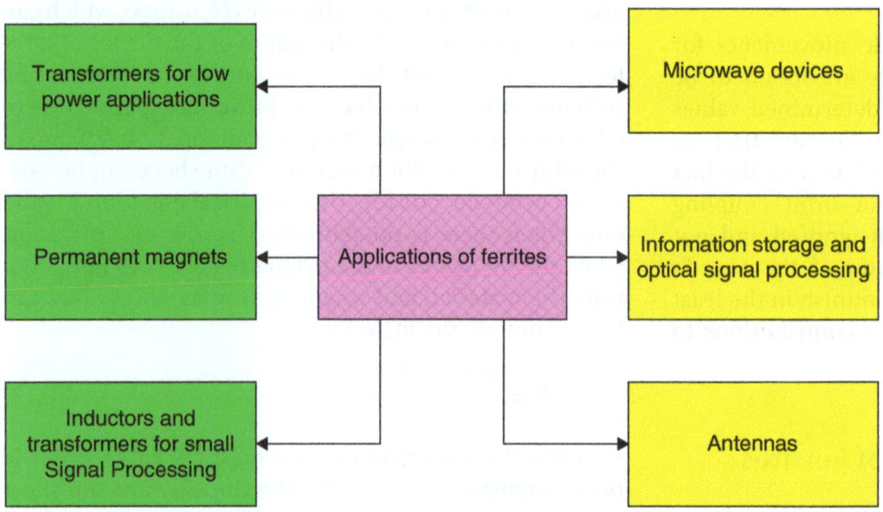

Figure 10.12 Classical applications of ferrite materials.

exhibit some change in their resistivity under the influence of a magnetic field. The effect is usually small and of no special significance for technology. We have briefly discussed these materials' properties in Chapter 8 where we have defined it with the help of Eq. (8.44). We can once again rewrite it here as Eq. (10.16).

$$\mathrm{MR} \approx \frac{R_0 \pm R_\mathrm{H}}{R_0} \times 100\% \tag{10.16}$$

where MR is the magneto-resistance, R_H is the resistance with magnetic field applied, and R_0 the original resistance with zero magnetic field. MR is usually expressed as percentage change. MR can be either positive or negative depending on how a material responds to a magnetic field. For some materials, $R_\mathrm{H} > R_0$ and for others $R_\mathrm{H} < R_0$. Normally, MR for most materials is less than 5%. But the change is very large for GMR and CMR materials.

10.9.1 GMR Effect

It was discovered in 1988 independently by Albert Fert of France and Peter Grünberg of Germany. It is based on the quantum mechanical principle of electron tunneling through a barrier. Fert and Grünberg received the Nobel Prize in Physics in 2007 for the discovery of the GMR effect.

The effect was first identified at 4.22 K, when it was observed that the magneto-resistance of a superlattice changed by approximately 80%. The structure consisted of layers of 3 nm thick Fe and 0.9 nm thick Cr. Large

MR was also reported for two other compositions of Fe/Cr superlattices. The observed MR effect is highly anisotropic. It is large when the magnetizations of the two layers are antiparallel to each other, and it is small when they are parallel to each other. Its importance in sensor technology and computer memory was almost immediately realized. A paradigm shift in sensor technology occurred. The GMR superlattice device is nicknamed *spin valves*.

A spin valve is the classic example of a spintronic device in which the magnetic state of materials can be altered at the atomic level. Its discovery at IBM paved the landscape of magnetic data storage with enormous capacity. It is now present in every hard disk drive. Spin valves can detect minute magnetic impulses. It has enabled the technology of *cloud computing*.

Another important application of the GMR effect is in the development of high speed and high density MRAM (magnetoresistive random access memory). A competing technology has now appeared on the horizon. It is called MeRAM which stands for magnetoelectric random access memory. We have described this emerging device in Chapter 5 (Section 5.15). A MeRAM consists of a ferroelectric and antiferromagnetic thin film structure (see Figure 5.9). Biosensors and MEMS sensors also make use of spin valves.

10.9.2 CMR Effect

Soon after the GMR effect was discovered, in 1994, it was reported that some oxides with the crystal structure of perovskite showed colossal magneto-resistance of the order of 10^{3-4}. The CMR effect is found predominantly in oxides, whereas the GMR effect occurs in metallic nm thin metallic films.

The CMR effect was first reported in epitaxial films of $LaCa_{0.77}Mn_{0.33}O_x$ grown by the laser ablation method on $LaAlO_3$ substrate [1]. The highest value of $MR > 10^8\%$ was reported for perovskite La–Y–Ca–Mn–O films in 1996 [2]. It was found that its MR value dropped from $10^8\%$ at 60 K to $10^3\%$ at room temperature, which is still a very good value. By now, many more oxides have been found to exhibit CMR effect and perovskite structure seems to be the leading structure for CMR. So far the CMR materials have not been harnessed for practical applications though they appear to be extremely attractive for sensors and disk read-write heads. But it is just the matter of time that engineering will find a way to make use of such a good effect.

10.9.3 Spintronics

The word spintronics was coined in 1990 to describe electronic devices that take advantage of spins of an electron. This is an emerging technology that is poised to make quantum computing a possibility and may lead to the invention of number of novel devices that will revolutionize the field of data processing and high-speed transistors.

We have briefly discussed the topic of spintronics in Chapter 8 in conjunction with magnetic transistors. Spintronics is the technology that is based on two fundamental properties of electrons that are its charge and its spin. In our conventional electronics, we make use of its charge property for transport of current. In spintronics, not only the charge of an electron is manipulated for transport of current but also the spins are used as an additional degree of freedom to perform logic operations and for storage of information. In spintronic technology, the spin-dependent electronic properties of magnetic and semiconductor materials are manipulated for smooth operation of spintronic devices. We have mentioned the landmark discovery of spintronic transistor in Chapter 8 by Datta et al. [3]. For their work, they used a configuration similar to that of a standard MOSFET device with the difference that the metallic iron contacts were used and the Schottky gate was of InAlAs. InGaAs semiconductor served as the substrate. Ferromagnetic iron contacts were used for injecting polarized spins in the semiconductor substrate.

Both GMR and CMR effects have special roles to play in this emerging technology. Another effect that is also of great importance to spintronics is the *tunnel magnetoresistance effect* (TMR). The TMR effect is observed in *magnetic tunnel junction* (MTJ). This device, as also the spin valve, is based on the quantum mechanical phenomenon of electron tunneling through a very thin insulating layer which is sandwiched between two ferromagnetic metallic electrodes. The electrons from one electrode have to tunnel through the insulating barrier to reach the other electrode. The TMR is defined as the variation of resistance between the parallel and antiparallel states of magnetization of the two electrodes. This we can express as in Eq. (10.17).

$$\mathrm{TMR} = \left(\frac{R_{\mathrm{AP}}}{R_{\mathrm{P}}}\right) - 1 \tag{10.17}$$

Here R_{AP} is the resistance of the antiparallel state and R_{P} that of the parallel state. If the spin polarization of one electrode is P_1 and that of the other P_2, then TMR is given by Eq. (10.18).

$$\mathrm{TMR} = \frac{2P_1P_2}{1 - P_1P_2} \tag{10.18}$$

Equation (10.18) is known as Julliére's formula [4]. The polarization represents the effective spin polarization of the tunneling probability. It is not an intrinsic property of the ferromagnetic electrodes but depends on the nature of the insulating barrier.

We would recall that while discussing the Josephson junction effect in superconductivity in Chapter 1, we briefly discussed the electron tunneling. The famous Josephson junction device works on the principle of electron tunneling through a very thin insulating barrier sandwiched between a metal and a superconductor. The magnetic tunneling junction is slightly different in its configuration from the superconducting tunneling junction. The first TMR tunneling effect was demonstrated experimentally in 1996 by a group of scientists from IBM [5]. TMR as high as 83% was found for a configuration consisting of a thin insulating layer of $SrTiO_3$ (STO) and two electrodes of $La_{0.67}Sr_{0.33}MnO_3$ (LSMO). The LSMO electrodes were separated by the STO layer of about 5 nm thick. This landmark experiment set the stage for magnetic oxide-based spintronic devices. Oxides of perovskite structure are important materials for this emerging technology.

Spintronics is poised to dominate the area of transistors. It is supposed to be of very high speed, high density, and nonvolatile. It is presumed to be far superior to any MOSFET transistor which is currently the work horse of electronic technology. Spin-polarized injection should make it possible to produce spintronic transistors with steeper subthreshold slope than found for MOSFET transistors. Another advantage of such transistors would be that it could produce controllable circularly polarized coherent light. Many novel applications for spintronics have been proposed; some possible and others speculative. An excellent review of oxide spintronics is given by Bibes and Barthélémy [6]. I highly recommend it for those interested in research in this area.

A list of oxides for applications in spintronics is given in Table 10.6.

Table 10.6 Examples of some spin polarized oxides.

Materials	Comments
$La_{1-x}Sr_xMnO_3$	It is a member of the manganite family; it is a half-metallic antiferromagnetic material. Its structure is perovskite. $LaMnO_3$ is its typical example
Ca_2FeReO_6, Sr_2CrRO_6	They are classified as double-perovskites; they are possibly half-metallic
CrO_2	Chromium double oxide is metallic binary oxide. Its crystal structure is tetragonal. It is a prototype of double-exchange system. It is ferromagnetic with $T_c = 395$ K
Fe_3O_4	Magnetite is the oldest magnetic material with $T_c = 858$ K. It crystallizes in spinel structure. Being a mixture of Fe^{2+} and Fe^{3+} ions it is ideally suited for studies of magnetic double exchange interactions
$NiFe_2O_4$	Nickel ferrite has inverse spinel structure, and it is an insulator. But it becomes conductive when its films are grown on $SrTiO_3$. In film form, it has been used for magnetic tunnel junction experiments
$SrRuO_3$	It is metallic ferromagnet with $T_c = 160$ K. Its structure is perovskite. It has good electrical conductivity and is used as an electrode material
ZnO Co-TiO_2	When doped with transition metals, they become ferromagnetic. ZnO exhibits long spin lifetime and is expected to play a key role in spintronics. Co-TiO_2 is also a promising spintronic material.
Co-$La_{1-x}Sr_xTiO_3$	Cobalt doped LSTO shows promise for spin-injected applications
$BiFeO_3$	Bismuth ferrite is a prominent member of the multiferroic family. $BiFeO_3$/CoFeB bilayer has been identified as a promising spintronic material

Glossary

Antiferromagnetism It is represented by antiparallel spins that amounts to zero magnetic moment in the absence of an applied magnetic field. Below Néel point, it assumes a high degree of magnetic order making it a member of the ferroic family.

Bohr magneton It is a measure of the elementary atomic magnetic moment. Magnetic moment of an electron is equal to one Bohr magneton.

CMR effect It is the abbreviation for "colossal magneto-resistive" effect. It is observed in some oxides in which the value of magneto-resistance (MR) can reach 1000% or more. These materials are promising materials for many applications including novel sensors.

Curie temperature All ferromagnetic and ferrimagnetic materials lose their spontaneous magnetization at a critical temperature called the Curie point. Below this point ferro- and ferrimagnetism dominates and above it paramagnetism prevails.

Diamagnetism It is a magnetic phenomenon associated with diamagnetic materials. For these materials, the induced magnetization is antiparallel to the field giving rise to a negative susceptibility ($\chi \approx -\Delta M/\Delta H$). No magnetic flux is generated because the magnetic field cannot enter these materials; instead they are repelled.

Double exchange interaction A magnetic exchange that might arise between ions of different oxidation states. The mixed valent Fe-ions in magnetite can induce double exchange interaction. Many ferrites are good candidates for such a study.

Ferrimagnetism These are magnetic oxides with high resistivity. It is represented by antiparallel spin ordering with unequal spins. Magnetite, Fe_3O_4 is the classic example of ferrites. Practically, all ferrites contain magnetite in some proportion with other oxides.

Ferromagnetism It is a physical phenomenon associated with ferromagnetic materials. These materials are represented by parallel spins at the atomic level. They exhibit a high degree of spin ordering. It is the most prominent member of the multiferroic family. Ferromagnetic materials have typically very high level of spontaneous magnetization and high Curie points.

Gauss Magnetic fields are often expressed in units of Gauss in the CGS system of units.

GMR effect The "giant magnetoresistance effect" is abbreviated as GMR. This effect is observed at low temperatures in nm thin films of metallic ferromagnets This is a quantum mechanical effect. The magneto-resistive change is very large compared to very small changes observed for normal metals. These materials are very attractive for sensor and many other applications. Spin valves are GMR based devices.

Magnetic moment All materials get magnetized when subjected to a magnetic field. The resulting magnetization is expressed as magnetic moment per unit volume or per unit mass.

Magnetic susceptibility It is magnetic parameter used for the classification of magnetic materials. It is expressed as the ratio between the magnetic moment and magnetic field.

Néel temperature It is analogous to the Curie temperature of a ferromagnetic material. The critical temperature associated with an antiferromagnetic material is called Néel point or Néel temperature. At this point, an antiferromagnetic material goes through a phase change from the antiferromagnetism to the state of paramagnetism.

Paramagnetism Paramagnetic materials are represented by a chaotic distribution of atomic magnetic moments. They acquire magnetic moment when a magnetic field is applied to it, and once the field is withdrawn the magnetization disappears. Its induced magnetization is a linear function of the applied magnetic field.

Spintronics It is an electronic technology based on polarized magnetic spins. It was discovered in 1996. It is poised to revolutionize the sensor and data processing technology.

Superparamagnetism This is a unique magnetic phenomenon observed at nanometre scale. Nanoparticles of a ferromagnetic or ferrimagnetic material assume this property when they consist of single domain magnetic particles. These materials also exhibit nonlinear *B–H* curve, but without the memory effect.

Tesla It is the unit used for magnetic field and magnetic flux (*B*) in the SI unit system.

Problems

10.1 How many types of magnetic materials are there? Briefly describe each of them. What distinguishes them from each other? Give a few examples of each type.

10.2 Identify the signature properties of ferrites and discuss the meaning of various parameters that define the hysteresis loop of a ferrite material.

10.3 What do you understand by the term *magneto-electric effect*? Describe on the basis of *E–X–H* diagram, how you would alter the spontaneous polarization of a ferroelectric material with the help of an external magnetic field?

10.4 What is meant by the term *demagnetized* state of a ferrite material? How can we technically demagnetize a ferrite and what advantages we will get by doing so?

10.5 What is the leading crystal structure in which most of the ferrites crystallize? Briefly describe its significant features and show how does it differ from a typical bcc structure?

10.6 Calculate the saturation magnetization of a ferrite sample assuming its atomic weight to be 30, density equal to $5\,g\,cm^{-3}$, and atomic magnetic moment equal to $0.75\mu_B$.

10.7 The temperature dependence of saturation magnetization of Fe–Si transformer material is given in the following table. Graphically determine the Curie point. Is the transition from ferromagnetic to paramagnetic state of the first or second order? Explain.

Temperature (°C)	Saturation magnetization, M_s (G cm^3 g^{-1})
50	203.1
100	200.0
300	181.9
500	162.0
600	138.1
700	86.9
730	39.0
750	16.8
800	6.0
900	1.9

10.8 Briefly describe the Weiss molecular field theory of magnetism. Discuss the validity of the assumption that ferromagnets are spontaneously magnetized because there is a large internal magnetic field in ferromagnetic materials. Can it be experimentally determined?

10.9 The temperature dependence of magnetic susceptibility for a Fe–Si sample is given in the table below. Show that it obeys the Curie–Weiss law and determine the value of the Curie constant.

Temperature, T (°C)	Inverse susceptibility, χ^{-1} (× 10^{-4} g cm^{-3})
900	0.49
1000	1.08
1100	1.56
1200	2.01
1300	2.49
1400	2.79
1500	3.49

10.10 What are magnetic domains? Distinguish between the magnetic materials on the basis of domain ordering. Based on the domain ordering, give a general formula for the magnetic moment of a ferrite.

10.11 Define magneto-resistance. Discuss the concept of a spin valve and explain its operating mechanism.

10.12 The origin of ferromagnetism and ferrimagnetism are based on the spin quantum number. Compute the spin magnetic moment.

10.13 What is Hund's rule? Discuss that it is complimentary to the Pauli Exclusion Principle and not in conflict.

10.14 What are the orbital quantum numbers (l) and magnetic quantum numbers (m_l) corresponding to the principal quantum number $n = 4$? Write all their values and their corresponding spectrographic designations.

10.15 Write the electron configuration of Cr, Mn, and Cu that are also members of the 3d-transition metals (period 4) of the periodic table as are Fe, Co, and Ni. Discuss why Fe, Co, and Ni are ferromagnetic, whereas Cr, Mn, and Cu are not.

10.16 (a) Define eddy currents; and (b) the resistivity of Fe increases when Si atoms are introduced in its crystal structure. Find the eddy current loss if the Si content is increased from 2% to 5% in the Fe–Si transformer sheet. Consider that the resistivity of a material changes when impurities are added to it. For small impurity concentration the following relationship is valid: $\sigma = \sigma_i/(1 + \alpha c)$ where σ_i is the intrinsic conductivity of the material, α is material specific constant, and c is the impurity concentration. Assume that for Fe–Si sample $\alpha = 117$.

References

1 Jin, S., Tiefel, T.H., McCormac, M. et al. (1994). Thousand fold change in resistivity in magnetoresistance La-Ca-Mn-O films. *Science* 264: 413.

2 Chen, L.H., Tiefel, T.H., Jin, S. et al. (1996). Colossal magnetoresistance in La-Y-Ca-Mn-O films. *IEEE Trans. Magn.* 32 (5): 4692–4694.

3 Datta, S. and Das, B. (1990). Electronic analog of the electro-optic modulator. *Appl. Phys. Lett.* 56: 665–667. http://dx.doi.org/10.1063/1.102730.

4 Jullére, M. (1975). *Phys. Lett. A* 54A: 225.

5 Lu, Y., Li, W., Gong, G. et al. (1996). *Phys. Rev. B* 54: R8357.

6 Bibes, M. and Barthélémy, A. (2006). Oxide Spintronics. In: *IEEE Trans. Electron Devices, X(X1)*: 1–18.

Further Reading

Kasap, S.O. (2006). *Principles of Electronic Materials and Devices*, 3e. McGraw Hill.

Solymar, L. and Walsh, D. (2010). *Electrical Properties of Materials*, 8e. Oxford University Press.

11

Electro-optics and Acousto-optics

CHAPTER MENU

The light of knowledge is fully capable of destroying the darkness of ignorance. This also helps us in overcoming all the difficulties and in achieving success in all our endeavors.

Sam Veda

11.1 Introduction

Electro-optics and acousto-optics are specialized photonics phenomena in which electroceramics play a prominent role. They facilitate high-speed optical communication and integrated optics. Advancement in materials processing have resulted in producing high-quality single crystals of nonlinear materials that have paved the way for finding new and novel applications of electro-optics and acousto-optics effects. So far as the field of electro-optics is concerned lithium niobate, $LiNbO_3$, has emerged, as we will see later, as its workhorse.

In this chapter, we will first discuss the nature of light, some of the important properties that arise when light interacts with nonlinear crystals, define what is meant by the terms *refractive index* and *birefringence*. We will then discuss briefly the two famous electro-optic effects which are Pockels effect and Kerr effect. The first one is a linear effect, whereas the Kerr effect is the quadratic effect. Pockels effect, in particular, becomes the basis for the development of large number of electro-optic devices and their applications.

11.2 Nature of Light

Light is only a small part of the electromagnetic spectrum that includes γ ray on the one end and the long radio waves on the other end. The wavelength of γ rays is approximately 10^{-12} m, whereas it is approximately10^5 m for long radio waves. It is a huge spectrum covering a variation of the order of 10^{17} m in wavelength. In other words, the electromagnetic spectrum encompasses a range of frequency varying from 10^3 to 10^{10} Hz. Our light covers only a small part of the electromagnetic spectrum. It begins with ultraviolet (UV) and ends with infrared (IR) with the wavelength varying between 10^{-8} for UV to 10^{-4} m for IR. The visible part that lies between these two extremes covers only the wavelength of 10^{-7} to 10^{-5} m. In Table 8.3, we have given more precise values of wavelengths and their corresponding frequencies for the visible radiation.

Why the light waves are classified as electromagnetic? As the name suggests the light waves consist of electric and magnetic components, and they are inseparable. They are always present and must be dealt with while dealing with radiation. Figure 11.1 describes the electric and magnetic components, and how they propagate in tandem through the space.

Natural light is unpolarized such as the solar radiation and light from incandescent lamps. Natural light can be converted to a linearly polarized light after allowing it to pass through an optical filter. These waves travel through space as plane waves. Plane waves can further be transformed into circularly or elliptically polarized waves by passing through optical filters. In circularly polarized

Fundamentals of Electroceramics: Materials, Devices, and Applications, First Edition. R. K. Pandey.

Companion website: www.wiley.com/go/Pandey/Fundamentals_Electroceramics

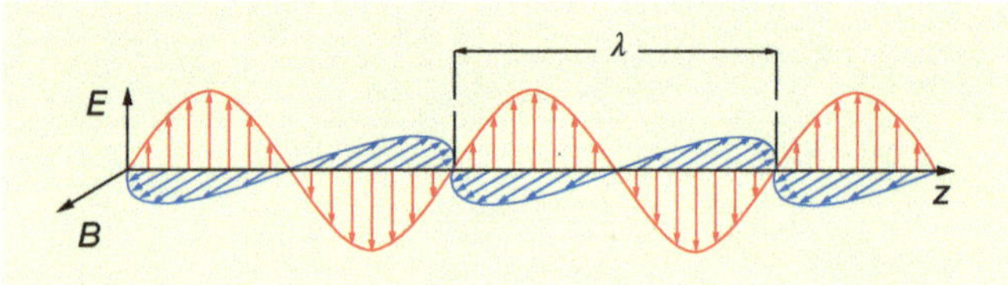

Figure 11.1 Linearly polarized electromagnetic wave propagating through the space. The electric field, *E*, oscillates in the vertical direction whereas the magnetic field, *B*, oscillates perpendicular to the *E*-field oscillation. Both *E* and *B*-fields are perpendicular to the *z*-direction of propagation. λ defines the wavelength of the radiation. Source: https://upload.wikimedia.org/wikipedia/commons/0/0a/Electromagnetic_wave.png.

light, the phase shift is ±90°. If two plane waves with different amplitudes are related in phase by 90°, then it is called elliptically polarized light. Also when the relative phase in other than 90° the light wave is considered to be elliptically polarized. Circularly polarized light may also be produced by passing a linearly polarized light through quarter wave plates.

11.2.1 Fundamental Optical Properties of a Crystal

Whenever a light beam encounters a solid surface, one part is reflected back and other part exits the sample as a refracted component of the original beam. The process involves two media: one is the solid itself and the other is normally air. Reflection and refraction are the two natural outcomes, whenever a light beam strikes a solid surface. The reflected and refracted components bear certain mathematical relationship to each other giving rise to a fundamental optical property called the *refractive index*. It is a material property and unique to each material. No two materials have the same value of the refractive index. The relationship between the angle of incidence (i) and angle of refraction (r) is given by the Snell's law which states that:

$$n = \frac{\sin i}{\sin r} = \frac{\text{wave velocity in solid}}{\text{wave velocity in air}} \tag{11.1}$$

Here n is the refractive index of the solid sample with respect to air. The refractive index of air is 1.000 28 which is for all practical purposes considered to be 1.

One of the most important significance of Eq. (11.1) is that when r equals 90°, i attains its critical value expressed as i_c. Then Eq. (11.1) takes the form of Eq. (11.2).

$$\sin i_c \approx n \tag{11.2}$$

When $i = i_c$ the light beam cannot exit the solid and enter air. It remains confined within the solid and goes through the process of *internal reflection*. Internal reflection is a unique optical phenomenon and is the basis for many optical applications. Values of refractive index of some selected materials are given in Table 11.1.

The refractive index can also be defined as Eq. (11.3).

$$n = \frac{\text{Velocity of light in vacuum}}{\text{Velocity of light in material}} \approx \frac{c}{v} \tag{11.3}$$

Table 11.1 Refractive index of selected materials.

Material	Wavelength of light	Refractive index, n
Vacuum		1 (by definition)
Air		1.000 28
Water	589.29	1.330
Rutile, TiO_2	589.29	2.614
Diamond	589.29	2.417
Strontium titanate, $SrTiO_3$	589.29	2.41
Fused quartz, SiO_2	589.29	1.458

Source: https://en.wikipedia.org/wiki/List_of_refractive_indices.

At very high frequencies, such as in optical regime, refraction index is related to the dielectric constant as stated in the following equation.

$$n \approx \sqrt{\epsilon_r \mu_r} \approx \epsilon' \tag{11.4}$$

We can consider $n^2 \approx \epsilon$ because for most of the dielectric materials $\mu_r \approx 1$. The three equations, Eqs. (11.1)–(11.3), may look very simple but have profound implications in understanding the interactions of light with a material.

Exercise 11.1
From Table 11.1 use the values of n given for $SrTiO_3$ and quartz. Using Eq. (11.3) show that the velocity of light in quartz is higher than in $SrTiO_3$.

Solution:
We can rewrite Eq. (11.3) assuming that n_1 is the refractive index of $SrTiO_3$ and n_2 is the refractive index of quartz, and V_1 is the velocity in $SrTiO_3$ and V_2 that is quartz. This leads us to the following equation:

$$\frac{n_1}{n_2} \approx \left(\frac{c}{V_1}\right) x \left(\frac{V_2}{c}\right) \approx \left(\frac{V_2}{V_1}\right)$$

and,

$$\frac{V_2}{V1} \approx \frac{n_1}{n_2} \approx \frac{2.41}{1.458} \approx 1.653$$

That tells us that the velocity in quartz is greater than in $SrTiO_3$ by a factor of 1.653 (=65.3%). *Comment:* The light velocity will slow down in a material with high value of n.

In an isotropic dielectric medium such as a slab of glass, the induced electric polarization is always parallel to the electric field. However, this is not the case when we consider polarization in an anisotropic medium such as a noncubic crystal. In such cases, the polarization is a function of both the direction and magnitude of the applied field. The three components of polarization may be written as in Eq. (11.5):

$$\begin{aligned} D_x &= \epsilon_{11}E_x + \epsilon_{12}E_y + \epsilon_{13}E_z \\ D_y &= \epsilon_{21}E_x + \epsilon_{22}E_y + \epsilon_{23}E_z \\ D_z &= \epsilon_{31}E_x + \epsilon_{32}E_y + \epsilon_{33}E_z \end{aligned} \tag{11.5}$$

This obviously can be abbreviated into a tensor format with ϵ_{ij} being a second rank tensor.

$$D_i = \epsilon_{ij}E_j \tag{11.6}$$

We would recall using such tensor formulations while studying the piezoelectric effect in Chapter 5. Equation (11.5) is the same thing as Eq. (5.25) and Eq. (11.6) is the same as Eq. (5.24).

Another optical property of solids of importance to our studies is called *birefringence* (also double refraction). Usually isotropic crystals have only one unique value of the refractive index, *n*. However, there are large numbers of anisotropic crystals in which the refractive index can assume two or more values. They are important to electro-optic technology. This effect was discovered by Rasmus Bartholm of Denmark in 1669 while experimenting with a calcite, $CaTiO_3$ crystal. However, its relationship with polarization and direction of propagation of light was not known until the nineteenth century. It was Augustin-Jean Fresnel of France who showed how it is a function of both polarization and the direction of light propagation. The birefringence effect in a calcite crystal is reproduced in Figure 11.2. When a beam of unpolarized light strikes a surface of a calcite prism it exits on the opposite side as an ordinary ray, which passes through the crystal unchanged. Its refractive index is n_0. However, another ray develops and emerges out of the prism after suffering refraction. This is called the extraordinary ray having the refractive index of n_e.

The difference between n_e and n_0 is the measure of the birefringence or double refraction as given by Eq. (11.7).

$$\Delta n = n_e - n_0 \tag{11.7}$$

Based on the birefringence effect crystals are optically classified as *uniaxial or biaxial* as shown in Table 11.2. The seven types of crystals are divided into two groups of *uniaxial and biaxial*. The uniaxial group consists of tetragonal, hexagonal, and trigonal crystals. The biaxial group includes orthorhombic, monoclinic, and triclinic crystals. The table also includes the characteristic symmetry elements for each crystal type. That leads to the order of symmetry that is highest for cubic and lowest for triclinic.

A list of selected uniaxial crystals is given in Table 11.3. All these materials are oxides and potentially suitable for electro-optic applications. Calcite and rutile exhibit high values for the birefringence, Δn. It is −0.172 for calcite and +0.387 for rutile. Also notice that all these oxides belong to either trigonal or tetragonal crystal class. None of them are cubic which by definition excludes point of inversion (non-centro-symmetry).

The uniaxial crystals have two values of refractive indices, n_0 and n_e leading to only one value of Δn which can be either positive or negative. When $n_e < n_0$ the crystal is said to have negative birefringence; similarly when $n_e > n_0$ the birefringence is positive. Whether it is positive or negative refers to the ellipsoids with respect to the optic axis. We need not go into the details of this because it does not serve our purpose. The sign of Δn is important to geologists and mineralogists in classifying gemstones and minerals.

Biaxial crystals are defined as having three different values of refractive indices that are labeled as n_x, n_y, and n_z which are parallel to *x*, *y*, *z* directions of a crystal. Only calcium titanate, which is perovskite, comes as a biaxial crystal. Others are primarily minerals and not much of our interest.

11.2.2 Electro-optic Effects

Two electro effects are of interest to us. We referred to them previously as linear and quadratic effect. The linear effect is also called the Pockels effect, and it states that the birefringence Δn, is directly proportional to the applied electric field, *E*. A German physicist named Friedrich Carl Alvin Pockels discovered this effect in 1893. Pockels cells operate on the basis of this effect.

The another effect is the Kerr effect that states that changes in the birefringence of a material is proportional

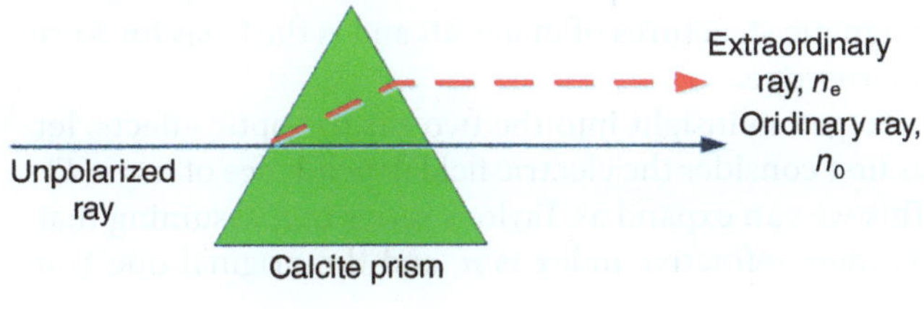

Figure 11.2 Production of birefringence in a calcite crystal.

Table 11.2 Optical classification of crystals based on birefringence effect.

Crystal type	Optical classification	Characteristic symmetry	Order of symmetry
Cubic	Isotropic	4 threefold axes	Highest
	Uniaxial group		
Tetragonal	Uniaxial	1 fourfold axis	
Hexagonal	Uniaxial	1 sixfold axis	
Trigonal (rhombohedral)	Uniaxial	1 threefold axis	
	Biaxial group		
Orthorhombic	Biaxial	Three mutually perpendicular twofold axes; no axis of higher order	
Monoclinic	Biaxial	1 twofold axis	
Triclinic	Biaxial	A center of symmetry or no symmetry at all	Lowest

Source: From Pandey 2005 [1].

Table 11.3 Selected uniaxial crystals with their optical parameters.

Crystal	Formula	Type	n_o	n_e	Δn	Positive/ Negative
Calcite	$CaCO_3$	Trigonal	1.658	1.486	−0.172	Negative
Lithium niobate	$LiNbO_3$	Trigonal	2.272	2.187	−0.085	Negative
Quartz	SiO_2	Trigonal	1.544	1.553	+0.009	Positive
Ruby	Al_2O_3	Trigonal	1.770	1.762	−0.008	Negative
Rutile	TiO_2	Tetragonal	2.616	2.903	+0.287	Positive
Sapphire	Al_2O_3	Trigonal	1.768	1.760	−0.008	Negative

Source: https://en.wikipedia.org/wiki/Birefringence#Fast_and_slow_rays.

to the square of the electric field, E. This effect was discovered in 1875 by John Kerr of Scotland. The Kerr effect is also known as Kerr nonlinear optical effect or quadratic electro-optic effect. The Kerr effect is exhibited by all materials but the Pockels effect can be observed only in crystals that lack center of symmetry. Potentially all 20 piezoelectric crystals are candidates for observing this effect (see Table 4.6).

A related effect but not classified as an electro-optic effect is the magneto-optic Kerr effect. It deals with the changes induced in light that reflects from a magnetized surface. This is a powerful tool for investigating the magnetic structures of materials and is the basis for Kerr microscope.

To get an insight into the two electro-optic effects, let us first consider the electric field dependence of $n = n(E)$. This we can expand as Taylor's series in E assuming that the new refractive index is n' and the original one is n such that the birefringence is $\Delta n = n' - n$. Then we have,

$$\Delta n = a_1 E + a_2 E^2 + a_3 E^3 + \cdots \quad (11.8)$$

From this, we see that the Pockels effect is $\Delta n \approx a_1 E$ with a_1 being a coefficient. The Kerr effect is given by $\Delta n \approx a_2 E^2$ where a_2 is another coefficient. It is customary to substitute the coefficient a_2 with λK where K is called the Kerr coefficient and λ is the wavelength of light. That leads us to the standard formulation of the Kerr effect as given in Eq. (11.9).

$$\Delta n \approx \lambda K E^2 \quad (11.9)$$

Exercise 11.2
Find the change in the birefringence of a material when the electric field increases by 10%. Assume the wavelength and the Kerr coefficient to remain constant.

Solution:
Let Δn_1 correspond to E_1 and Δn_2 correspond to $1.1E_1$. Substituting these values in Eq. (11.9), we have

$$\frac{\Delta n_2}{\Delta n_1} \approx \frac{(1.1E_1)^2}{E_1^2} \approx 1.21$$

The change in the birefringence is 21%.

11.2.3 Selected Electro-optic Applications

We have discussed previously photoelectric effect, photovoltaic effect, light-emitting diode, and transparent transistors. They all deal with interaction of light with solids leading to many unique and novel applications. In electro-optics it is not only the refractive index that plays a role rather also the birefringence. One of the most exciting applications of electro-optics is the emergence of the field of *integrated optics*. The space available to us does not allow us to go into details of this vast field, but we will do some justice to the field by getting acquainted with a select few devices. Advanced readers interested in gaining deeper insight into the field of electro-optic devices and applications should refer to specialized books of which there are many. In this chapter, we propose to become familiar with commonly used devices and their applications. Many of these devices are based on the Pockels effect.

Remembering that $n^2 = \epsilon$, we can also express the Pockels effect with the equation as in Eq. (11.10).

$$\Delta\left(\frac{1}{\epsilon_r}\right) \approx a_1 E \tag{11.10}$$

Alternatively, it can also take the following form.

$$\Delta n \approx -\frac{1}{2} n^3 a_1 E \tag{11.11}$$

Considering that n for $LiNbO_3$ is 2.272, $\Delta n \approx -0.085$ and for certain orientations the coefficient a_1 is 3.08×10^{-11} m V^{-1}, we get $E \approx 4 \times 10^8$ V m^{-1} from Eq. (11.10). If the field is reduced to 4×10^6 V m^{-1}, which is a reasonable field that can be applied in practice; then the resulting values of $\Delta n \approx -0.001\,24$. This is a very small change and does not look like to be of any practical value. As it turns out that large number of very practical and useful devices can be produced using such small values for birefringence.

In Table 11.4, we give a list of important electro-optic crystals with some of their properties. Notice the large values for the electro-optic coefficient and dielectric constant for ferroelectric-piezoelectric barium titanate, $BaTiO_3$ and potassium niobate, $KNbO_3$. Both belong to the class of perovskite crystals and used for electro-optic applications.

11.2.3.1 Optical Waveguides

These devices, as their name suggests, guide electromagnetic waves in the optical spectrum. Optical fibers and rectangular dielectric slab waveguides are the two common configurations. The principle of operation is rather straightforward and simple. If a material of higher refractive index is surrounded by a material of lower refractive index then by using total internal reflections, a wave may be guided in the material having higher refractive index. These devices are used in optical integrated circuits and in optical communication systems. A simple representation of $LiNbO_3$-based waveguide is shown in Figure 11.3. $LiNbO_3$ single crystal is the leading material for fabrication of slab waveguides. The titanium film of $n = 2.612$ is interdiffused on the surface of the $LiNbO_3$ surface with the refractive index of 2.303 (note this value is lower than the value quoted in Table 11.4. Here the value has been taken at 587.6 nm of wavelength. We know n is a strong function of wavelength). The titanium film provides the channel with $n > n$ of the substrate. When a light wave is launched, it suffers multiple internal reflections in the channel satisfying the condition for the waveguide.

Table 11.4 Selected elector-optical crystals with their properties.

Crystal	Formula	Wavelength of light (nm)	Electro-optic coefficient (10^{-12} m V^{-1})	Refractive index, n	Dielectric constant, ϵ
Bismuth silicate	$Bi_{12}SiO_{20}$	514 (green)	2.3	2.22	56
Barium titanate	$BaTiO_3$	514 (green)	820	2.49	4300
Cadmium telluride	CdTe	1.0 (infrared)	4.5	2.84	9.4
Gallium arsenide	GaAs	1.15 (infrared)	1.43	3.43	12.3
Potassium niobate	$KNbO_3$	633 (orange)	380	2.33	50
Lithium niobate	$LiNbO_3$	633 (orange)	32.6	2.29	78
Zinc oxide	ZnO	633 (orange)	2.6	2.01	8.15

Source: From Solymar and Walsh 2010 [2].

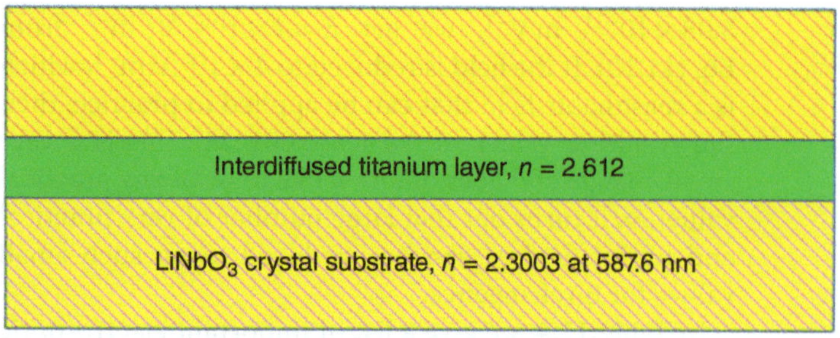

Figure 11.3 Configuration of a slab optical waveguide.

11.2.3.2 Phase Shifters

Light waves propagate in space as sinusoidal waves. If for any reason, the original wave shifts horizontally with respect to its original position, then a phase shift has occurred. A full cycle of a wave is 360°. A phase shift is measured in degrees. Also because the horizontal axis of a sinusoidal wave represents time, a phase shift can also be measured in time. In an electrical circuit, for example, the phase shift between input and output is measured in time. Optical phase shifter is based on the Pockels effect. Let us consider the waveguide we just discussed and show that it can be easily transformed into a phase shifter. All we need to do is to add two metallic electrodes on the surface of the $LiNbO_3$ substrate as shown in Figure 11.4. The waveguide that we produced by diffusing Ti in the substrate is now sandwiched between the two electrodes. This now gives us the option of inducing changes in the refractive index of Ti waveguide by applying a potential between the two electrodes.

By doing so, we will succeed in introducing a phase shift $\Delta\phi$ between the original wave (zero potential applied) and the wave with potential applied. The phase shift can be calculated using the following equation:

$$\Delta\phi \approx \frac{2\pi L \Delta n}{\lambda} \tag{11.12}$$

Here L is the length of the electrode.

11.2.3.3 Electro-optic Modulators

We know that by applying a voltage to an electro-optic crystal, we can easily induce changes in the refractive index because of the Pockels effect. This is turn can facilitate a phase modulation of the transmitted beam. Amplitude modulation can also be achieved by putting the sample between two linear polarizers. The same effect can be achieved also by putting a Mach–Zehnder interferometer in the path of the voltage biased sample.

Electro-optic deflectors: These devices are built by using a prism of an electro-optic crystal. By using the Pockels effect, one can change the refractive index that will cause a change in the direction of the propagation of the light beam inside the crystal. These devices have fast response time.

Second harmonic generator (SHG): It is a very interesting application of nonlinear electro-optical crystals. Photons interacting with a nonlinear crystal are effectively allowed to combine to form new photons with twice the energy and therefore twice the frequency. That is why these devices are also called *frequency doubles*. The wavelength of the new photon has to be obviously half of the original photon. SHG has found applications in biological systems. It is commonly used to convert green laser with 532 nm wavelength to a wavelength of 1064 nm. Nd-doped yttrium aluminum garnet (Nd:YAG) and KDP (potassium dihydrogen phosphate), are the two nonlinear materials that are mostly used to produce SHG.

11.2.3.4 Night Vision Devices (NVD)

These devices allow images to be produced in level of light from 100% bright to almost total dark. The conversion of images can be from visible light to near-IR and vice versa. The images produced are signature of thermal imaging of the object and therefore can be easily detected by thermal IR. The images produced are monochromatic. These devices are routinely used by military and law enforcement for night vision. They also

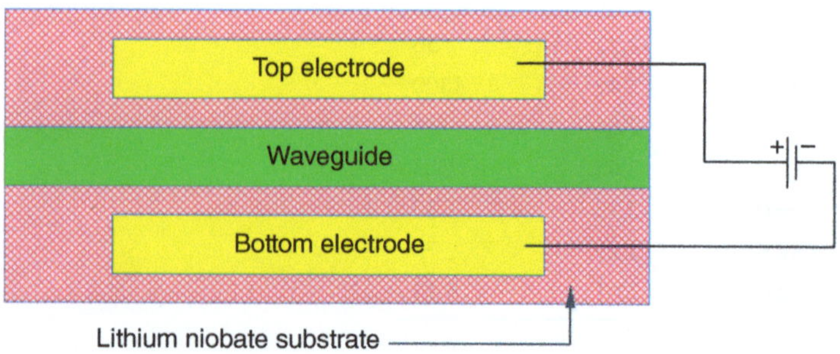

Figure 11.4 Configuration of an optical phase shifter.

have found their applications in civilian sectors. Night vision goggles, cameras, and binoculars are commercially available.

Lead lanthanum zirconate titanate (PLZT) hot pressed ceramic are widely used to produce night vision devices (NVD) devices. PLZT hot pressed ceramic besides being active electros-optic medium also exhibits another photo-related effect called *photochromism*. It refers to a reversible change in the color of a material when exposed to visible or UV light. This effect in PLZT is strong enough for applications.

11.2.4 Acousto-optic Effect and Applications

Like light sound waves can also interact with materials giving rise to the field of acousto-optics. Already in 1922, the famous French physicist Brillouin (Léon Nicolas) predicted the diffraction of light produced by acoustic waves. Ten years later in 1933, it was confirmed by Debye and Sears. In general, the field of acousto-optics deals with the change in the refractive index of a material in the presence of sound waves. The classical work of C.V. Raman and Nagendra Nath of India demonstrated experimentally that the refractive index can be changed by several orders by subjecting the sample to sound waves. Raman received the Nobel Prize in Physics in 1930 for his work.

A unique property of an acousto-optic effect is that while propagating through a crystal it produces a refractive index grating. This is equivalent to producing variations in refractive index due to the pressure fluctuations in the crystal. For grating to be formed, ultrasound waves having the frequency >20 kHz are needed. Ultrasound waves can be generated by piezoelectric effect or even by magnetostriction.

The acousto-optic gratings are similar to the conventional gratings that are extensively used in the field of optics. These are simple devices with periodic structures that are usually ridges or rulings on the surface of a glass slab. This is the simplest configuration of a grating. There are many different types of grating; one more sophisticated than the other.

The periodic structures are called grating elements. An element is equal to the wavelength of the ultrasound waves. The element by definition is the width of the grating. We can easily calculate the width by using Bragg's equation. Let us assume that the grating element is d, wavelength of the optical wave is λ_{op}, and θ is the angle of diffraction when the ultrasound waves pass through the grating. Then the grating spacing, d, is given by Eq. (11.13).

$$d \approx \frac{\lambda_{op}}{2n \sin \theta} \tag{11.13}$$

Here n is the order of diffraction (and not refractive index) having the values of 1, 2, 3, etc.

The relationship between the optical wave and the acoustic wave is given by Eq. (11.14).

$$\lambda_{ac} \approx \frac{\lambda_{op}}{2n \sin \theta} \approx d \tag{11.14}$$

If the sound wavelength changes from λ_{ac} to $\Delta\lambda_{ac}$ there will be a corresponding change in the Bragg angle, θ. It will increase from θ to $\Delta\theta$. Then Eq. (11.14) will take the form of Eq. (11.15).

$$\lambda_{ac} + \Delta\lambda_{ac} \approx \frac{\lambda_{op}}{2n}[\sin(\theta + \Delta\theta)]^{-1} \tag{11.15}$$

This is an important equation. This is used in inducing a change in the deflection of optical waves. By launching acoustic waves with the help of appropriate launching arrays it is possible to induce a deflection in the optical wave. As its consequence the working equation is given by Eq. (11.16).

$$\lambda_{ac} \approx \frac{1}{2n}\lambda_{op}\left[\sin\left(\frac{\Delta\theta}{2}\right)\right]^{-1} \tag{11.16}$$

The total deflection $\Delta\theta$ must be substituted by $\left(\frac{\Delta\theta}{2}\right)$ to meet the conditions of Bragg's diffraction. This becomes necessary because of the geometry of the launching arrays.

The velocity of the sound wave with which the deflection of the optical wave is accomplished can also be evaluated. Let us say that its velocity is V_{ac} and its frequency is ν_{ac} and wavelength λ_{ac}. Then $V_{ac} = \lambda_{ac}\nu_{ac}$. Then we can rewrite Eq. (11.14) as Eq. (11.17).

$$V_{ac} \approx \frac{\lambda_{op}\nu_{ac}}{2n}(\sin \theta)^{-1}. \tag{11.17}$$

Lists of materials that exhibit acousto-optic effect are tabulated in Table 11.5.

When an acoustic wave enters a medium and propagates through it a strain (S) is induced which can contribute to the change in the dielectric constant (ϵ_r) and then ultimately its refractive index (n). As we know the relationship between the dielectric constant and strain is best described with the help of tensors but instead of going into the elaborate mathematical steps involving multiple tensor elements we will take a short cut that will serve our purpose.

In practice, at any one time, we need to consider only one coefficient. Then the equation relating to the strain and refractive index reduces to the simple form as given by Eq. (11.18).

$$\Delta\left(\frac{1}{\epsilon_r}\right) \approx pS \tag{11.18}$$

Here p is the photoelastic coefficient.

Acousto-optic effect is used extensively for the measurement of ultrasound waves. It is also used for

Table 11.5 Acousto-optic materials and their properties.

Materials	Formula	Light wavelength (nm)	Density (10^3 kg m^{-3})	Refractive index	Sound velocity (10^3 m s^{-1})
Water	H_2O	633	1	1.33	1.5
Fused quartz	SiO_2	633	2.2	1.46	5.95
Gallium arsenide	GaAs	1.15	5.34	3.43	5.15
Lithium niobate	$LiNbO_3$	633	4.7	2.29	6.57
Lithium Tantalate	$LiTaO_3$	633	7.45	2.18	6.19
Lead molybdate	$PbMoO_4$	633	6.95	2.4	3.75
Zinc sulfide	ZnS	633	4.10	2.35	5.51

Source: From Solymar and Walsh 2010 [3].

deflection, modulation, signal processing, and frequency shifting of light beams.

Another complimentary optical effect is the *photorefractive effect* found in some select group of crystals. The photorefractive materials respond to light waves by undergoing change in refractive index. Barium titanate, lithium niobate, vanadium doped zinc telluride are leading photorefractive materials. Some polymers and organic materials show this effect. Photorefractive effect is used to store temporary holograms that can be erased and recreated. Holograms are used for data storage. There are many other applications of photorefractive effect. They include signal processing, integrated optics, optical spatial soliton, and phase-conjugate mirrors.

Glossary

Acousto-optics It is that branch of optics in which the refractive index of a material is changed by subjecting the sample to high frequency sound waves such as ultrasound with frequency greater than 20 kHz.

Birefringence Crystals lacking in inversion point (center of symmetry) can exhibit two or three different values of refractive index. For example, calcite has two values of refractive indices called ordinary (n_0) and extraordinary (n_e). The difference between ($n_e - n_0$) is called birefringence and is represented by the symbol Δn. The birefringence is also referred to as *double refraction*.

Diffraction grating It is an optical component with a periodic structure that is used for splitting and diffracting a light beam into multiple beams propagating in different directions. The periodic structures are usually ridges or rulings on the surface of a glass slab. This is the simplest type of a diffraction grating.

Electro-optics It is that branch of optics in which changes in refractive index are induced by the application of an electric field. It is a vast field by itself having many important and novel applications such as integrated optics, wave guides, deflectors, and sensors just to name a few.

Electromagnetic spectrum This includes electromagnetic waves varying in very small wavelengths (such as γ rays) to very large wavelengths (such as radio waves). These waves consist of electric and magnetic components and hence the name *electromagnetic waves*. Our visible light is part of this vast spectrum.

Kerr effect Kerr effect is the quadratic electro-optic effect in which the birefringence changes as the function of the square of the applied electric field.

Pockels effect This is the linear electro-optic effect. Here the change in the birefringence of a material is directly proportional to the electric field applied.

Polarized light Normally light waves propagate through space with both electric and magnetic fields oscillating. The two fields are perpendicular to each other, and both are perpendicular to the direction of the propagation. In contrast, the unpolarized light has electric field oscillating in all directions. An unpolarized light can be made polarized by passing it through special filters. Examples of unpolarized light are sunlight and light from filament lamps.

Refractive index It is the optical property of all materials and is usually assigned the symbol n. Each material has its own unique value of refractive index. By definition, the refractive index of vacuum is assigned the value of 1. The refractive index of air is very slightly larger than 1 but can be considered 1 for all practical purposes.

Total internal reflection When the angle of incidence is equal to a critical value, total internal reflection occurs. When it happens, the light beam cannot exit the medium and therefore, does not suffer any refraction. It remains confined in the medium and goes through multiple internal reflections.

Ultrasound These are sound waves with frequencies greater than 20 kHz. Human ears cannot detect it. It is a very important sound wave with a large number of practical applications. Ultrasound imaging is routinely used in medicine for diagnostic purposes. The field of acousto-optics is based on the changes in the refractive index of a crystal when ultrasound waves propagate through them.

Problems

11.1 Distinguish between electro-optics and acousto-optics emphasizing their distinguishing features.

11.2 Describe some typical applications based on the two effects discussed in the previous problem.

11.3 Find the grating spacing in lithium niobate that has the refractive index of 2.18 when a laser beam of 900 nm is deflected by 5°.

11.4 What is the frequency of the acoustic wave interacting with a $LiNbO_3$ crystal in the above problem? Assume the sound velocity to be 6.57×10^3 m s^{-1} for the $LiNbO_3$ crystal.

11.5 Consider a ZnS crystal with $n = 2.35$, and wavelength of light to be 633 nm. Find: (a) the grating spacing and (b) the wavelength of the acoustic wave that can deflect the optical wave by 6°. Consider also that the Bragg angle is 2° for the optical wave.

11.6 Calculate the amount of phase shift that can be produced in a TiO_2 crystal with $\Delta n \approx 0.287$. Consider that the TiO_2 waveguide is 5 μm long and the sample is exposed to an optical wavelength of 900 nm.

References

1 Pandey, R.K. (2005). Ferroelectric materials. In: *Encyclopedia of RF and Microwave Engineering*, vol. 2 (ed. K. Chang), 1505–1519. Wiley.

2 Solymar, L. and Walsh, D. (2010). *Electrical Properties of Materials*, 8e. Oxford University Press Source: Table # 13.20.

3 Solymar, L. and Walsh, D. (2010). *Electrical Properties of Materials*, 8e. Oxford University Press Source: Table 13.3.

Further Reading

Kasap, S.O. (2006). *Principles of Electronic Materials and Devices*, 3e. McGraw Hill.

Appendix A

Periodic Table of the Elements

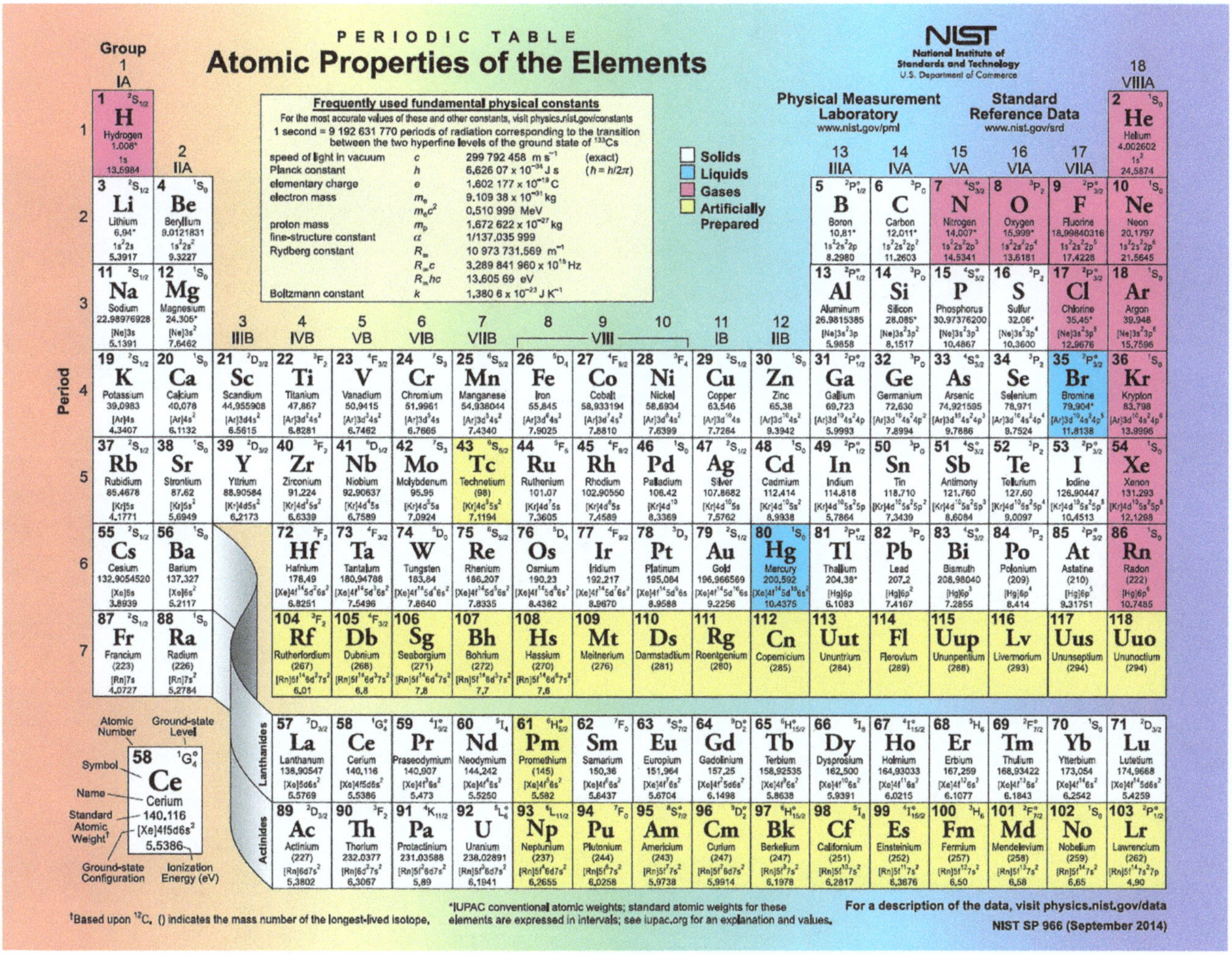

Source: https://www.nist.gov/sites/default/files/illo_for_2014_pt_1.png.

Companion website: www.wiley.com/go/Pandey/Fundamentals_Electroceramics

Appendix B

Fundamental Physical Constants and Frequently Used Symbols and Units (Rounded to Three Decimal Points)

Name	Symbol	Value
Velocity of light	c	3×10^8 m s^{-1}
Planck constant	h	6.626×10^{-34} J s
Planck constant	h	4.136×10^{-15} eV s
Reduced Planck constant	$\hbar$	6.582×10^{-16} eV s
Boltzmann constant	k_B	1.381×10^{-23} J K^{-1}
Boltzmann constant	k_B	8.617×10^{-5} eV K^{-1}
Avogadro's number	N_A	6.022×10^{23} mol^{-1}
		6.022×10^{26} molecules kg^{-1} mol^{-1}
Charge of electron	e	1.602×10^{-19} C
Permeability of vacuum	μ_0	$4\pi \times 10^{-7}$ H m^{-1}
Permittivity of vacuum	ϵ_0	8.854×10^{-12} F m^{-1}
Mass of electron	m_e	9.109×10^{-31} kg
Mass of proton	m_p	1.673×10^{-27} kg
Mass of neutron	m_n	1.675×10^{-27} kg
Bohr magneton	μ_B	9.27×10^{-24} J T^{-1}
Bohr magneton	μ_B	5.789×10^{-5} eV T^{-1}
kT at 290 K*	—	0.025 eV or 4.05×10^{-21} J
Wavelength of photon having 1 eV energy*	—	1 240 nm
Standard atmosphere	atm	101 325 Pa

Source: http://hyperphysics.phy-astr.gsu.edu/hbase/hph.html; except with*.

Fundamentals of Electroceramics: Materials, Devices, and Applications, First Edition. R. K. Pandey.

Companion website: www.wiley.com/go/Pandey/Fundamentals_Electroceramics

Appendix C

List of Prefixes Commonly Used

Symbol	Abbreviation	Unit
ato	a	10^{-18}
femto	f	10^{-15}
pico	p	10^{-12}
nano	n	10^{-9}
micro	μ	10^{-6}
milli	m	10^{-3}
centi	c	10^{-2}
kilo	k	10^{3}
mega	M	10^{6}
Giga	G	10^{9}
Tera	T	10^{12}
Peta	P	10^{15}

Source: Solymar and Walsh 2010 [1].

Reference

1 Solymar, L. and Walsh, D. (2010). *Electrical Properties of Materials*, 8e. Oxford University Press.

Fundamentals of Electroceramics: Materials, Devices, and Applications, First Edition. R. K. Pandey.

Companion website: www.wiley.com/go/Pandey/Fundamentals_Electroceramics

Appendix D

Frequently Used Symbols and Units

Quantity	Symbol	Unit	Abbreviation
Current	I	Ampere	A
Voltage	V	Volt	V
Charge	q	Coulomb	C
Capacitance	C	Farad	F
Inductance	L	Henry	H
Energy	E	Joule	J
Energy (eV)	1 eV ($=1.60 \times 10^{-19}$ J)	Electron volt	eV
Resistance	R	Ohm	Ω
Power	P	Watt	W
Electric field	E	Volt/meter	$\mathrm{V\,m^{-1}}$
Magnetic field	H	Ampere/meter	$\mathrm{A\,m^{-1}}$
Electric flux density	D	Coulomb/meter^2	$\mathrm{C\,m^{-2}}$
Magnetic flux density	B	Tesla	T
Frequency	f	Hertz	Hz
Wavelength	λ	Meter	m
Temperature	T	Kelvin	K

Source: Solymar and Walsh 2010 [1].

Reference

1 Solymar, L. and Walsh, D. (2010). *Electrical Properties of Materials*, 8e. Oxford University Press.

Fundamentals of Electroceramics: Materials, Devices, and Applications, First Edition. R. K. Pandey.

Companion website: www.wiley.com/go/Pandey/Fundamentals_Electroceramics

Index

Fundamentals of Electroceramics: Materials, Devices, and Applications, First Edition. R. K. Pandey.

Companion website: www.wiley.com/go/Pandey/Fundamentals_Electroceramics

d

g

h

q

r